Water Quality Engineering and Wastewater Treatment III

Water Quality Engineering and Wastewater Treatment III

Guest Editors

Yung-Tse Hung
Hamidi Abdul Aziz
Issam A. Al-Khatib
Rehab O. Abdel Rahman
Tsuyoshi Imai

Basel • Beijing • Wuhan • Barcelona • Belgrade • Novi Sad • Cluj • Manchester

Guest Editors

Yung-Tse Hung
Civil and Environmental
Engineering
Cleveland State University
Cleveland
United States

Hamidi Abdul Aziz
Civil Engineering
Universiti Sains Malaysia
Nibong Tebal
Malaysia

Issam A. Al-Khatib
Institute of Environmental
and Water Studies
Birzeit University
Birzeit
West Bank

Rehab O. Abdel Rahman
Hot Lab. & Waste
Manag. Center
Atomic Energy Authority
of Egypt
Cairo
Egypt

Tsuyoshi Imai
Division of Construction and
Environmental Engineering
Yamaguchi University
Ube
Japan

Editorial Office
MDPI AG
Grosspeteranlage 5
4052 Basel, Switzerland

This is a reprint of the Special Issue, published open access by the journal *Water* (ISSN 2073-4441), freely accessible at: www.mdpi.com/journal/water/special_issues/P7KM0KPU10.

For citation purposes, cite each article independently as indicated on the article page online and using the guide below:

Lastname, A.A.; Lastname, B.B. Article Title. *Journal Name* **Year**, *Volume Number*, Page Range.

ISBN 978-3-7258-3488-4 (Hbk)
ISBN 978-3-7258-3487-7 (PDF)
https://doi.org/10.3390/books978-3-7258-3487-7

© 2025 by the authors. Articles in this book are Open Access and distributed under the Creative Commons Attribution (CC BY) license. The book as a whole is distributed by MDPI under the terms and conditions of the Creative Commons Attribution-NonCommercial-NoDerivs (CC BY-NC-ND) license (https://creativecommons.org/licenses/by-nc-nd/4.0/).

Contents

About the Editors . . . . **vii**

Yung-Tse Hung, Rehab O. Abdel Rahman, Hamidi Abdul Aziz, Issam A. Al-Khatib and Tsuyoshi Imai
Water Quality Engineering and Wastewater Treatment III
Reprinted from: *Water* **2025**, *17*, 580, https://doi.org/10.3390/w17040580 . . . . **1**

Paula Núñez-Tafalla, Irene Salmerón, Silvia Venditti and Joachim Hansen
Assessing the Synergies of Photo-Fenton at Natural pH and Granular Activated Carbon as a Quaternary Treatment
Reprinted from: *Water* **2024**, *16*, 2824, https://doi.org/10.3390/w16192824 . . . . **5**

Meseret Dawit Teweldebrihan and Megersa Olumana Dinka
Methyl Red Adsorption from Aqueous Solution Using Rumex Abyssinicus-Derived Biochar: Studies of Kinetics and Isotherm
Reprinted from: *Water* **2024**, *16*, 2237, https://doi.org/10.3390/w16162237 . . . . **23**

Thomas Neuner, Michael Meister, Martin Pillei, Thomas Senfter, Simon Draxl-Weiskopf and Christian Ebner et al.
Impact of Design and Mixing Strategies on Biogas Production in Anaerobic Digesters
Reprinted from: *Water* **2024**, *16*, 2205, https://doi.org/10.3390/w16152205 . . . . **42**

Yang Yang, Cancan Jiang, Xu Wang, Lijing Fan, Yawen Xie and Danhua Wang et al.
Unraveling the Potential of Microbial Flocculants: Preparation, Performance, and Applications in Wastewater Treatment
Reprinted from: *Water* **2024**, *16*, 1995, https://doi.org/10.3390/w16141995 . . . . **65**

Carmen Ka-Man Chan, Chris Kwan-Yu Lo and Chi-Wai Kan
A Systematic Literature Review for Addressing Microplastic Fibre Pollution: Urgency and Opportunities
Reprinted from: *Water* **2024**, *16*, 1988, https://doi.org/10.3390/w16141988 . . . . **83**

Claudia Hledik, Yilan Zeng, Tobias Plattner and Maria Fuerhacker
Mercury Concentrations in Dust from Dry Gas Cleaning of Sinter Plant and Technical Removal Options
Reprinted from: *Water* **2024**, *16*, 1948, https://doi.org/10.3390/w16141948 . . . . **112**

Rehab O. Abdel Rahman, Ahmed M. El-Kamash and Yung-Tse Hung
Permeable Concrete Barriers to Control Water Pollution: A Review
Reprinted from: *Water* **2023**, *15*, 3867, https://doi.org/10.3390/w15213867 . . . . **125**

Njabulo Thela, David Ikumi, Theo Harding and Moses Basitere
Growing an Enhanced Culture of Polyphosphate-Accumulating Organisms to Optimize the Recovery of Phosphate from Wastewater
Reprinted from: *Water* **2023**, *15*, 2014, https://doi.org/10.3390/w15112014 . . . . **160**

Huy Thanh Vo, Tsuyoshi Imai, Masato Fukushima, Tasuma Suzuki, Hiraku Sakuma and Takashi Hitomi et al.
Utilizing Electricity-Producing Bacteria Flora to Mitigate Hydrogen Sulfide Generation in Sewers through an Electron-Pathway Enabled Conductive Concrete
Reprinted from: *Water* **2023**, *15*, 1749, https://doi.org/10.3390/w15091749 . . . . **172**

Tonni Agustiono Kurniawan, Mohd Hafiz Dzarfan Othman, Mohd Ridhwan Adam, Xue Liang, Huihwang Goh and Abdelkader Anouzla et al.
Chromium Removal from Aqueous Solution Using Natural Clinoptilolite
Reprinted from: Water 2023, 15, 1667, https://doi.org/10.3390/w15091667 187

Jasmina Ćetković, Miloš Knežević, Radoje Vujadinović, Esad Tombarević and Marija Grujić
Selection of Wastewater Treatment Technology: AHP Method in Multi-Criteria Decision Making
Reprinted from: Water 2023, 15, 1645, https://doi.org/10.3390/w15091645 202

Cristina E. Almeida-Naranjo, Víctor H. Guerrero and Cristina Alejandra Villamar-Ayala
Emerging Contaminants and Their Removal from Aqueous Media Using Conventional/Non-Conventional Adsorbents: A Glance at the Relationship between Materials, Processes, and Technologies
Reprinted from: Water 2023, 15, 1626, https://doi.org/10.3390/w15081626 227

Hamidi Abdul Aziz, Siti Fatihah Ramli and Yung-Tse Hung
Physicochemical Technique in Municipal Solid Waste (MSW) Landfill Leachate Remediation: A Review
Reprinted from: Water 2023, 15, 1249, https://doi.org/10.3390/w15061249 274

About the Editors

Yung-Tse Hung

Prof. Dr. Yung Tse Hung, Ph.D., P.E., DEE, Fellow-ASCE, served as a Professor of Civil Engineering at Cleveland State University from 1981 to 2024. He earned his B.S. and M.S. in Civil Engineering from Cheng Kung University, Taiwan, and his Ph.D. from the University of Texas at Austin. Prof. Hung has taught at 16 universities across 8 countries and started the public health engineering program at the University of Canterbury, New Zealand, in 1972. He has served on the faculties of numerous universities globally, including in New Zealand, the USA, Hong Kong, the UAE, Singapore, Australia, Russia, and Kyrgyzstan. Prof. Hung's research focuses on biological wastewater treatment, industrial water pollution control, and municipal wastewater treatment. He has published approximately 40 books, 242 book chapters, 198 refereed publications, and 352 other scholarly works, totaling around 811 publications and presentations. He is a Fellow of ASCE, a Diplomate of AAEE, a Fellow of the Ohio Academy of Science, a member of AEESP, and a Life Member of WEF. He serves as Editor-in-Chief for several international journals and books and is a registered professional engineer in Ohio and North Dakota.

Hamidi Abdul Aziz

Prof. Dr. Hamidi Abdul Aziz is a distinguished professor in Environmental Engineering at the School of Civil Engineering, Universiti Sains Malaysia. He earned his Ph.D. in Civil Engineering (Environmental) from the University of Strathclyde in 1992. Currently, he heads the Solid Waste Management Cluster (SWAM) at Universiti Sains Malaysia. With 29 years of teaching and research experience in environmental engineering, Professor Aziz has made significant contributions in areas such as solid waste management, landfill technology, water and wastewater treatment, leachate treatment, bioremediation, pollution control, and environmental impact assessment. To date, he has mentored over 100 Ph.D. and MSc students and has published over 200 ISI papers and several books. Additionally, he serves as an editor and editorial board member for several international journals. In recognition of his outstanding research contributions, the Malaysia Academy of Sciences named him a Top Research Scientist of Malaysia in 2012. In 2020, he was listed among the top 2% of scientists in his field globally, as compiled by the prestigious Stanford University.

Issam A. Al-Khatib

Prof. Dr. Issam A. Al-Khatib is a faculty member at the Institute of Environmental and Water Studies, Birzeit University, Palestine. His expertise spans across water resource management, environmental assessment, wastewater management, and climate change, with a focus on environmental health and sustainable development. Dr. Al-Khatib has led numerous research and evaluation projects on topics such as medical waste management, solid waste, water and sanitation, and urban environments. He is dedicated to promoting public environmental awareness through training, workshops, and community campaigns. He has contributed to national committees on environmental standards, including swimming pool water quality regulations. Dr. Al-Khatib has supervised over 40 MSc theses, focusing on these critical environmental issues, and is an expert in academic quality evaluation. Currently, he is working on a project to evaluate the attitudes, perceptions, and behaviors of Birzeit University students toward the management of single-use plastic waste. Prof. Al-Khatib is ranked among the top 0.5% of researchers globally according to Scholar GPS in 2025, recognized for his exceptional contributions to environmental science, monitoring, and waste management.

Rehab O. Abdel Rahman

Prof. Dr. Rehab O. Abdel Rahman is Executive project manager, NLSD; Prof.; and Head of the Radioactive Waste Management Department, EAEA, Egypt. She received her Ph.D. degree in Nuclear Engineering and participated in international projects and meetings on safety cases and safety assessments especially for radioactive waste disposal facilitates. Her published research focuses on different aspects of safe radioactive and hazardous waste management, including liquid wastewater treatment and the immobilization and stabilization of hazardous and radioactive wastes, assessing the performance of disposal barriers. She supports national capacity building, where she supervises postgraduate students; taught undergraduate courses; and mentored the safety assessment team. She serves as a member in international scientific committees, editor in international journals, and (co)-editor for several books; contributed to the publication several book chapters; and is a reviewer for scientific papers and grant applications. She was awarded the State Encouragement Award in Engineering sciences, Egypt, in 2011.

Tsuyoshi Imai

Prof. Dr. Tsuyoshi Imai is a distinguished academic at Yamaguchi University in Japan. He is a Professor in the Division of Environmental Engineering, Graduate School of Sciences and Technology for Innovation. Prof. Imai has an impressive educational background, having earned his Bachelor of Engineering (Sanitary Engineering) in March 1989, Master of Engineering (Sanitary Engineering) in March 1991, and Doctor of Engineering (Environmental Engineering) in September 1995, all from Kyushu University, Japan. He began his career as a Research Associate at the Faculty of Engineering, Yamaguchi University, in 1994. After thirteen years in various roles, he became a Professor in 2007. Currently, he serves as the Chairman of the Department of Sustainable Environmental Engineering, Faculty of Engineering, Yamaguchi University, a position he has held since 2014. Prof. Imai's research interests include environmental engineering, with a focus on topics such as eutrophication, sedimentation in reservoirs, and zero solid waste management in wastewater treatment systems.

Editorial

Water Quality Engineering and Wastewater Treatment III

Yung-Tse Hung [1], Rehab O. Abdel Rahman [2,*], Hamidi Abdul Aziz [3], Issam A. Al-Khatib [4] and Tsuyoshi Imai [5]

[1] Department of Civil and Environmental Engineering, Cleveland State University, Cleveland, OH 44115, USA; y.hung@csuohio.edu
[2] Hot Laboratory Center, Atomic Energy Authority of Egypt, Inshas, Cairo P.C. Box 13759, Egypt
[3] School of Civil Engineering, Engineering Campus, Universiti Sains Malaysia, Nibong Tebal 14300, Malaysia; cehamidi@usm.my
[4] Institute of Environmental and Water Studies, Birzeit University, Birzeit P.O. Box 14, Palestine; ikhatib@birzeit.edu
[5] Graduate School of Sciences and Technology for Innovation, Yamaguchi University, 2-16-1, Tokiwadai, Ube City 755-8611, Japan; imai@yamaguchi-u.ac.jp
* Correspondence: alaarehab@yahoo.com

1. Introduction

The provision of clean water is a vital element to ensure life sustainability; this can be achieved by designing and implementing effective prevention and control measures to protect water resources. Designing efficient wastewater technologies is a crucial tool to prevent potential pollution releases into water resources, whereas setting controls on pollution can be achieved by designing efficient remediation technologies that mitigate the spread of the contaminants of concern. Despite the huge accumulated knowledge in the field of water quality engineering and wastewater treatment, this field is still challenged by difficulties in addressing emerging contaminants [1], the need to improve the efficiency for conventional technologies to meet the strengthen regulatory standard on effluent releases [2–4], and need to advance innovative technologies for large-scale applications [5].

2. Overview on This Special Issue

This Special Issue includes this editorial and 13 papers (eight research papers and five review papers) {C1–C13}. The focus of this Special Issue is directed towards addressing recent advances in wastewater treatment technologies, including chemical precipitation {C4}, flocculation {9}, biological treatment {C3,C5,C8}, sorption {C2,C7,C12}, and a combination of advanced oxidation process (AOP) with sorption {C1}. Moreover, the control of pollution sources toward achieving better water quality is addressed in this Special Issue in four papers that focus on the mitigation of sewage system corrosion {C6}, analyzing the knowledge gap to mitigate the micro-plastic pollution {C10}, the use of permeable barriers to control water pollution {C11}, and leachate treatment {C13}.

Advances in primary wastewater treatment methods, i.e., chemical precipitation and flocculation, are addressed in two papers {C4,C9}. The first studies Hg removal from dry gas cleaning by-products from sinter plants in the steel industry {C4}. Hg concentration values in the leachate of the by-products were measured, and its removal efficiency by chemical precipitation with organic or sodium sulfides followed by filtration was determined. The study reveals the efficiency of the process in the high salt concentration conditions, and referrers to the importance of the precipitator dosage optimization to mitigate the Hg re-dissolution {C4}. A review paper addressed the potential role of the microbial flocculants (MBF), as eco-friendly materials, in the treatment of wastewater {C9}. An overview of the state of the art of the MBF sources, categories, production, and their performance and mechanisms is presented. MBF applications in the removal of suspended solids, dyes, and heavy metals from municipal and industrial effluents were presented. The challenges that face the wide applications of these materials were summarized, and the way forward, towards the advancement of the technology's readiness, is outlined {C9}.

Three papers present the advances in the secondary wastewater treatment, where research on the optimization of biogas production in an anaerobic digester (AD), enhanced biological removal of phosphorus, and optimizing the implementation of the innovative moving bed bio-film reactor (MBBR) are included {C3,C5,C8}. In this respect, an analysis was conducted to compare and assess the performance of 34 AD implemented in the domestic wastewater treatment plants in Austrian alpine region {C3}. The study addressed the reactor shapes, mixing methods, energy efficiency, and biogas production. That work highlights that the process optimization can enhance the role of the AD technology as green energy source by increasing the gas production and reducing the operating costs and gas losses {C3}. In another work, a design for an enhanced biological phosphorus removal (EBPR) process was studied, where a culture of phosphorous accumulating organisms (PAO) were grown in a membrane–bioreactor-activated sludge system. The process was optimized by assessing the effect of the substrate on the PAO growth {C5}. The results indicate that increasing the dosage of the propionate substrate has linear correlation with the PAO concentration {C5}. In the last paper in under this topic, the optimization of the innovative MBBR technology to improve the capacity of the wastewater treatment plant in Dojran, North Macedonia, was presented {C8}. The analytical hierarchy process (AHP) was used to optimize the design of the plant; the analysis considered, among other factors, the condition assessment and efficiency of the existing plant, the available space at the site, and financial parameters. The results reveal that the optimal solution to improve the capacity of the existing plant is the construction of a new MBBR with 6000 equivalent inhabitants capacity {C8}.

The use in the tertiary and quaternary treatment technologies in the removal of methyl red (MR) dye, chromium, and emerging contaminant (EC) is presented in four papers {C1,C2,C7,C12}. The use of biochar derived from Rumex-abyssinicus for the removal of MR was investigated {C2}. In that work, the biochar was prepared, characterized, and tested to identify the effect of the operating conditions on the removal performance. The study concluded on the feasibility of using this biochar for the treatment of industrial wastes containing MR {C2}. In another study, Cr removal using pre-treated clinoptilolite was studied, where the compliance with the discharge limits and regeneration feasibility were investigated {C7}. The study reveals the ability of the pre-treated clinoptilolite to treat the wastewater in compliance with relevant regulations {C7}. The third paper is a review paper that addresses the EC and their removal from aqueous media using sorbents {C12}. It overviewed the performance of different conventional (e.g., activated carbon, clay) and non-conventional (e.g., industrial and agro residues) materials in EC removal. The study highlights the promising role of agro-industrial residues in EC removal, and linked that to their availability, low toxicity, and sorption capacities {C12}. The combined implementation of Photo-Fenton and use of Granular Activated Carbon (GAC) was assessed for the removal of micro-contaminants from wastewater effluents {C1}. The study encompassed the investigation of four advanced oxidation process scenarios coupled with GAC filtration. The analysis of the toxicity and phyto-toxicity of the treated wastewater proves the efficiency of the tested processes {C1}.

A research paper addressed the mitigation of the sewage system corrosion by controlling the hydrogen sulfide (H_2S) via biological oxidation {C6}. The experimental data showed that the conductive concrete is providing electron pathway for biological oxidation of H$_2$S {C6}. In another work, the knowledge gap to mitigate micro-plastic pollution was reviewed {C10}. That work analyzed related literature to support the prioritization of the efforts to reduce this pollution type. It concluded that the knowledge gaps include, among others, a lack of information about non-standardized test methodologies addressing life-cycle impacts and real hotspots {C10}. Another review paper presented the role of permeable concrete barriers in controlling the water pollution {C11}. It overviewed the advances in the studying the mix-design of conventional and innovative permeable concrete, their characterization techniques, and their performance in the pollution control. The work recommended the optimization of the mix-design of the permeable reactive barriers in a way that balances the performance measures and the durability of the barrier over its service life {C11}. In the last paper in this Special Issue, techniques for municipal solid landfill leachate remediation were reviewed {C13}. This work focused on the use of contemporary techniques to remediate leachate of high COD content, colored, and NH$_3$-N with low biodegradability. Finally, the challenges and future prospects for the semi-aerobic landfill design were presented {C13}.

Author Contributions: Conceptualization and writing—original draft preparation, Y.-T.H. and R.O.A.R.; writing—review and editing, all authors. All authors have read and agreed to the published version of the manuscript.

Conflicts of Interest: The authors declare no conflicts of interest.

List of Contributions

1. Núñez-Tafalla, P.; Salmerón, I.; Venditti, S.; Hansen, J. Assessing the Synergies of Photo-Fenton at Natural pH and Granular Activated Carbon as a Quaternary Treatment. *Water* **2024**, *16*, 2824.
2. Teweldebrihan, M.D.; Dinka, M.O. Methyl Red Adsorption from Aqueous Solution Using Rumex Abyssinicus-Derived Biochar: Studies of Kinetics and Isotherm. *Water* **2024**, *16*, 2237.
3. Neuner, T.; Meister, M.; Pillei, M.; Senfter, T.; Draxl-Weiskopf, S.; Ebner, C.; Winkler, J.; Rauch, W. Impact of Design and Mixing Strategies on Biogas Production in Anaerobic Digesters. *Water* **2024**, *16*, 2205.
4. Hledik, C.; Zeng, Y.; Plattner, T.; Fuerhacker, M. Mercury Concentrations in Dust from Dry Gas Cleaning of Sinter Plant and Technical Removal Options. *Water* **2024**, *16*, 1948.
5. Thela, N.; Ikumi, D.; Harding, T.; Basitere, M. Growing an Enhanced Culture of Polyphosphate-Accumulating Organisms to Optimize the Recovery of Phosphate from Wastewater. *Water* **2023**, *15*, 2014.
6. Vo, H.T.; Imai, T.; Fukushima, M.; Suzuki, T.; Sakuma, H.; Hitomi. T.; Hung, Y.-T. Utilizing Electricity-Producing Bacteria Flora to Mitigate Hydrogen Sulfide Generation in Sewers through an Electron-Pathway Enabled Conductive Concrete. *Water* **2023**, *15*, 1749.
7. Kurniawan, T.A.; Othman, M.H.D.; Adam, M.R.; Liang, X.; Goh, H.; Anouzla, A.; Sillanpää, M.; Mohyuddin, A.; Chew, K.W. Chromium Removal from Aqueous Solution Using Natural Clinoptilolite. *Water* **2023**, *15*, 1667.
8. Ćetković, J.; Knežević, M.; Vujadinović, R.; Tombarević, E.; Grujić, M. Selection of Wastewater Treatment Technology: AHP Method in Multi-Criteria Decision Making. *Water* **2023**, *15*, 1645.
9. Yang, Y.; Jiang, C.; Wang, X.; Fan, L.; Xie, Y.; Wang, D.; Yang, T.; Peng, J.; Zhang, X.; Zhuang, X. Unraveling the Potential of Microbial Flocculants: Preparation, Performance, and Applications in Wastewater Treatment. *Water* **2024**, *16*, 1995.
10. Chan, C.K.-M.; Lo, C.K.-Y.; Kan, C.-W. A Systematic Literature Review for Addressing Microplastic Fibre Pollution: Urgency and Opportunities. *Water* **2024**, *16*, 1988.
11. Abdel Rahman, R.O.; El-Kamash, A.M.; Hung, Y.-T. Permeable Concrete Barriers to Control Water Pollution: A Review. *Water* **2023**, *15*, 3867.
12. Almeida-Naranjo, C.E.; Guerrero, V.H.; Villamar-Ayala, C.A. Emerging Contaminants and Their Removal from Aqueous Media Using Conventional/Non-Conventional Adsorbents: A Glance at the Relationship between Materials, Processes, and Technologies. *Water* **2023**, *15*, 1626.

13. Aziz, H.A.; Ramli, S.F.; Hung, Y.-T. Physicochemical Technique in Municipal Solid Waste (MSW) Landfill Leachate Remediation: A Review. *Water* **2023**, *15*, 1249.

References

1. Jiang, T.; Wu, W.; Ma, M.; Hu, Y.; Li, R. Occurrence and distribution of emerging contaminants in wastewater treatment plants: A globally review over the past two decades. *Sci. Total Environ.* **2024**, *951*, 175664. [CrossRef] [PubMed]
2. Austria, H.F.M.; Young, J.O.; Setiawan, O.; Huang, T.H.; Caparanga, A.R.; Pamintuan, K.R.S.; Hung, W.S. Thin film nanocomposite loose nanofiltration membranes with graphene oxide interlayer for textile wastewater treatment. *Sep. Purif. Technol.* **2025**, *354*, 129053. [CrossRef]
3. Debroy, P.; Majumder, P.; Seban, L. A simulation based water quality parameter control of aquaponic system employing model predictive control strategy incorporation with optimization technique. *Environ. Prog. Sustain. Energy* **2025**, *44*, e14530. [CrossRef]
4. Ke, X.; Qin, Z.; Chen, A.; Tian, Y.; Yang, Y.; Zhang, H.; Qiu, G.; Wu, H.; Wei, C. Triple strategies for process salt reduction in industrial wastewater treatment: The case of coking wastewater. *Sep. Purif. Technol.* **2025**, *355*, 129614. [CrossRef]
5. Shamshad, J.; Rehman, R.U. Innovative approaches to sustainable wastewater treatment: A comprehensive exploration of conventional and emerging technologies. *Environ. Sci. Adv.* **2025**, *4*, 189–222. [CrossRef]

Disclaimer/Publisher's Note: The statements, opinions and data contained in all publications are solely those of the individual author(s) and contributor(s) and not of MDPI and/or the editor(s). MDPI and/or the editor(s) disclaim responsibility for any injury to people or property resulting from any ideas, methods, instructions or products referred to in the content.

Article

Assessing the Synergies of Photo-Fenton at Natural pH and Granular Activated Carbon as a Quaternary Treatment

Paula Núñez-Tafalla *[ID], Irene Salmerón [ID], Silvia Venditti [ID] and Joachim Hansen [ID]

Faculty of Science, Technology and Medicine, University of Luxembourg, 6, Rue Richard Coudenhove-Kalergi, L-1359 Luxembourg, Luxembourg; irene.salmeron@ext.uni.lu (I.S.); silvia.venditti@uni.lu (S.V.); joachim.hansen@uni.lu (J.H.)

* Correspondence: paula.nunez@uni.lu; Tel.: +352-46-66-5560

Abstract: The challenge of microcontaminants (MCs) in wastewater effluent has been addressed by using different technologies, including advanced oxidation processes (AOPs) and adsorption. This work evaluates the benefits and synergies of combining these two processes. The AOPs were photo-Fenton and UV/H_2O_2 operated under natural pH but with different reagents dosages, lamps, and chelating agents. Chelating agents were used at analytical (ethylenediamine-N,N-disuccinic acid and citric acid) and technical grade (citric acid) to simulate scaling-up conditions. The adsorption process was studied via granular activated carbon (GAC) filtration using fresh and regenerated GAC. Four AOP scenarios were selected and coupled with GAC filtration, showing benefits for both processes. AOP treatment time decreased from 10–15 min to 5 min, resulting in a reduction in energy consumption of between 50 and 66%. In the photo-Fenton process, it was possible to work with low reagent dosages (1.5 mg L^{-1} iron and 20 mg L^{-1} of H_2O_2). However, the use of UV/H_2O_2 showed close removal, highlighting it as a real alternative. An extension of the GAC lifetime by up to 11 times was obtained in all the scenarios, being higher for regenerated than for fresh GAC. Furthermore, the toxicity and phytotoxicity of the treated wastewater were evaluated, and no acute toxicity or slight variation in the phytotoxicity was observed in the combination of these processes.

Keywords: contaminants of emerging concern; micropollutant; process combination; real wastewater; water remediation

Citation: Núñez-Tafalla, P.; Salmerón, I.; Venditti, S.; Hansen, J. Assessing the Synergies of Photo-Fenton at Natural pH and Granular Activated Carbon as a Quaternary Treatment. *Water* 2024, 16, 2824. https://doi.org/10.3390/w16192824

Academic Editor: Alejandro Gonzalez-Martinez

Received: 3 September 2024
Revised: 18 September 2024
Accepted: 19 September 2024
Published: 4 October 2024

Copyright: © 2024 by the authors. Licensee MDPI, Basel, Switzerland. This article is an open access article distributed under the terms and conditions of the Creative Commons Attribution (CC BY) license (https://creativecommons.org/licenses/by/4.0/).

1. Introduction

The growing presence of microcontaminants (MCs) in surface water poses a challenge to their mitigation, identifying the point source, and implementing technical solutions as for the wastewater treatment plants (WWTPs) [1]. MCs are organic molecules, such as personal care products, pharmaceuticals, pesticides, and herbicides, that are highly recalcitrant and are only partially removed in conventional WWTPs [2]. Despite their low concentration (μg or ng L^{-1}), they can potentially affect water bodies due to their risk to aquatic life and human health [3,4].

In response to this concern, the European Union (EU) has developed a new proposal for a Directive of the European Parliament and of the Council concerning urban wastewater treatment (UWWTD) [5]. The new recast of the exiting UWWTD tackles the MCs problem for the first time, alongside more restrictive limits for nitrogen and phosphorous emissions, including the necessity to achieve energy neutrality in WWTPs. Regarding MCs, quaternary treatments should be installed in WWTPs treating a load of 150,000 p.e. and above to reduce the MC emissions and preserve the quality of receiving water bodies. However, the necessity of installing quaternary treatments in rural areas (up to 10,000 p.e.) is also proposed in urban WWTPs that represent a risk to human health or the environment. The required removal was defined as a minimum of 80% of 12 MCs. which were selected as the most representative, that should be measured in the dry season when a higher MC concentration is expected. The new proposal defines WWTPs as relevant actors in the

European Green Deal; consequently, it marks the necessity of achieving energy neutrality by 2045. Therefore, quaternary treatment should not increase the consumption of energy significantly.

Previously to the recast of the existing UWWTD, several countries, such as Luxembourg and Switzerland, adopted precautionary measures and regulated the removal of MCs from wastewater [6,7]. Luxembourg proposed the 80% removal of four mandatory MCs—diclofenac, carbamazepine, clarithromycin, and benzotriazole—from the inlet to the outlet of WWTPs, selected due to their relevance for local water bodies. This choice anticipated the EU strategy, as the four MCs are included in the 12 selected MCs from the new directive. The Luxembourgish regulation is expected to be adapted to new EU requirements in the coming years.

Several technologies have been studied as candidates for quaternary treatment for the removal of MCs. Despite efficient removal, they showed significant drawbacks, such as the generation of toxic by-products, the production of high-concentration brines, high space requirements [8,9], etc. In order to address this issue in a real scenario, some works have recognised the coupling of technologies of different natures as a suitable solution [10].

Advanced oxidation processes (AOPs) are very interesting technologies, as they mineralise the MCs instead of separating them from one phase to another or concentrating them as the membranes do. They are based on the production of hydroxyl radicals ($^{\bullet}OH$) [11], which react non-selectively with the pollutants and are very powerful for MC degradation. AOPs are highly efficient in producing $^{\bullet}OH$ and do not generate bromates and carbonyl-related side products, in contrast to ozonation. One of the most studied AOP relies on UV/H_2O_2, which is based on the absorption of UV-C light (254 nm) by hydrogen peroxide (H_2O_2), which then generates $^{\bullet}OH$ [12]. The photo-Fenton process poses a very promising alternative to conventional AOPs due to its higher capacity to remove MCs, which is, in theory, even higher than the UV/H_2O_2 process [13]. The photo-Fenton process is based on the Fenton reaction in which iron and H_2O_2 are in contact, producing $^{\bullet}OH$. In the photo-Fenton process, UV radiation is applied to regenerate iron (III) to iron (II) and increment $^{\bullet}OH$ production [14]. The pH strongly affects the efficiency of the process, with the optimal pH being 2.8.; working at natural pH is not possible, as the iron is known to precipitate. Operating at acidic pH in a WWTP is not feasible due to the complexity associated with the acidification and neutralisation steps required. Thus, chelating agents are presented as the key to keeping the iron in solution at natural pH. The chelating agents are organic molecules, such as ethylenediamine-N,N-disuccinic acid (EDDS), nitrilotriacetic acid (NTA), oxalic acid, or citric acid, that can form bonds with iron and avoid its precipitation. The stability and efficiency of the chelating agents at neutral pH allow for the removal of up to 80% of carbamazepine when working with NTA [15] or EDDS [16]. Moreover, the coupling of the photo-Fenton process at neutral pH with adsorption to remove MCs has rarely been performed [17,18]. To the best of our knowledge, all the works analysing MCs removal were carried out under solar irradiance (solar photo-Fenton). Still, this condition limits the application of the photo-Fenton process to sunny hours; thus, water can only be treated for a few hours per day. This means that the technology is not feasible in a WWTP that receives water all day long or in areas with low solar irradiation. Considering these climatic conditions, medium-pressure (MP) UV lamps (instead of sunlight) were used as a light source.

Adsorption processes are consolidated technologies that transfer MCs from the water matrix to an adsorbent material. Granular activated carbon (GAC) filtration is a robust technology that is easy to implement and operate. The capacity to remove MCs has been widely demonstrated, up to 80% of carbamazepine or diclofenac can be adsorbed in GAC filtration [19]. The removal percentage by GAC filtration lies in the affinity of the MC with the selected GAC [19,20]. The main drawback is the necessity to replace it once the exhaustion is achieved. In recent years, regenerated GAC has been shown to be an alternative to reduce waste, energy consumption, and carbon footprint [13,21]. Thus, in this study, a comparison between commercial fresh and regenerated GAC is also performed.

The GAC filtration test needs a long operation time of several months to reach the breakthrough and achieve exhaustion. In order to address this problem, Crittenden et al. [22] proposed the use of a rapid small-scale column test (RSSCT) as a bench-scale alternative that can reach the breakthrough in days, reducing the operation time and the amount of water required for the experiments. The results obtained using the RSSCT are not directly comparable to the full-scale treatment results, but they allow for a comparison between the performance of different GACs [23]. Despite its limitations, this test is the preferred process when selecting the best GAC and indicatively predicting pollution behaviour in GAC columns.

This work studied the coupling of two processes: AOPs and GAC filtration. Despite previous research on the coupling of the photo-Fenton process and adsorption tested it with dyes [18], has been rarely proposed to remove MCs with photo-Fenton at natural pH [17]. Specifically, to the best of our knowledge, this work studies the coupling of both technologies to remove MCs at natural pH, using lamps as a source of irradiation for the very first time. The use of the solar photo-Fenton process has been deeply evaluated, but the use of lamps is mostly studied at the lab scale [24–26], in contrast to the pilot scale of this study. Furthermore, operating at natural pH (7–7.5) was a requirement to avoid initial acidification, later neutralisation, and a consequent increase in effluent salinity. In addition, RSSCTs were performed as stand-alone and coupled systems, which allowed for a comparison of GAC lifetime, in contrast to the general approach of performing batch tests that are far from representing the real operation.

The main target and novelty of this work is the focus on exploring the synergies and benefits of the combination, unlike other studies that combine different technologies to improve removal efficiency [27–29]. The potential benefits are mainly associated with cost reduction, GAC lifetime extension, exploring the possibilities of using regenerated GAC (lowering carbon footprint), a reduction in the reagents in the AOPs, or a reduction in treatment time, thus reducing energy consumption and coming closer to energy neutrality. The possible synergies are related to the most recalcitrant MC, which are different for each technology. The first step was the AOP, which was applied as a polishing step to reduce the MCs' concentration but not completely eliminate them. Thus, it can be operated with lower reagent doses and treatment times. This approach will advance the scaling up of the photo-Fenton process. Additionally, technical-grade reagents at natural pH and lamps for UV irradiation were used to study the most realistic scenario in Luxembourg.

Therefore, the main objective of the present work was to reach 80% removal of benzotriazole, carbamazepine, clarithromycin, and diclofenac via (1) photo-Fenton, (2) UV/H_2O_2, (3) GAC filtration, and (4) the coupling of GAC with the two previous AOPs. Different photo-Fenton scenarios were tested for the best operational conditions, modifying diverse parameters, e.g., type of lamps, reagent dosage, or pH. Additionally, the feasibility of the regenerated GAC was compared to fresh. The best scenarios from AOPs were performed on the coupling of technologies. Toxicity was analysed (as acute toxicity and phototoxicity) to assess the suitability of the technologies and select the best technology not only in terms of MCs removal but also in considering the quality of the influent of the treatments.

2. Materials and Methods
2.1. Chemicals

Benzotriazole, carbamazepine, clarithromycin, and diclofenac (purity ≥99% HPLC Standards GmbH, Borsdorf, Germany) were spiked in the water matrix to increase the concentration to 1 µg L^{-1} for benzotriazole and diclofenac and 2 µg L^{-1} for carbamazepine and clarithromycin. Two sources of iron were used: ferric sulphate heptahydrate (97%, Sigma-Aldrich, Hamburg, Geramany (analytic grade)) and iron sulphate solution (12%, Dr Paul Lohman GmbH, Emmerthal, Germanu (technical grade)). The iron was initially chelated with EDDS (35%, Sigma Aldrich) and subsequently with analytic citric acid (99%, CarlRoth GmbH., Karlsruhe, Germany) and technical citric acid (StockMeier Chemie Dillenburg GmbH., Dillenburg, Germany). Ammonium acetate (98%), ortho-phenanthroline (99%), ascorbic acid (99%), and sodium thiosulfate (98%) from Sigma-Aldrich were used to measure iron and H_2O_2 concentration. Bovine catalase (Sigma-Aldirch, Hamburg, Germany) was added to all the samples to consume the H_2O_2 and quench the Fenton reaction.

2.2. Water Matrix

MC removal was tested in municipal wastewater effluent from the Heiderscheidergrund municipal WWTP (12,000 p.e., Luxembourg), which was collected after the clarifier and thus treated with a conventional activated sludge process. The main characteristics of the wastewater were the following: pH: 6.9–7.7, 400–640 µS cm^{-1} of conductivity, 80–150 mg HCO_3 L^{-1}, 9.7–18 mg L^{-1} of chemical oxygen demand (COD), 1.1–5 mg L^{-1} of total nitrogen (TN), 0.66–2.5 mg L^{-1} of NH_4^+–N, 0.23–2.2 mg L^{-1} of NO_3^-–N, and 0.85–3.3 mg L^{-1} of total phosphorus (yearly range, 2022).

2.3. Analytical Determination

Benzotriazole, carbamazepine, clarithromycin, and diclofenac were analysed externally (Luxembourg Institute of Science and Technology LIST, Luxembourg). The samples were pre-concentrated before the injection by solid phase extraction. Then, they were injected in liquid chromatography (Agilent 1200 SL LC) coupled in tandem with mass spectrometry (LC-MS/MS), a Hybrid Quadrupole-Linear Ion Trap instrument (Sciex 4500 QTrap) with electrospray ionisation in positive polarity. Venditti et al. [30] previously described the methodology; the recovery rates, relative standard deviations, and limits of quantification (LoQ) of MCs are defined in Table S1 of the supplementary information. The H_2O_2 concentration was monitored using the spectrophotometric method with titanium oxysulfate (DIN 38402 H15). The total and dissolved iron were determined via the ortho-phenanthroline method at 510 nm (ISO 6332 [31]). A UV6300 PC Double Beam spectrophotometer from VWR was used for both analyses.

A pH/conductometer 3320 from WTW was used to monitor pH and conductivity. Hach-Langue kits measured COD (LCK 1414), TN (LCK 238), NH_4^+–N (LCK 304), NO_3^-–N (LCK 339), and total phosphorus (LCK 349).

2.4. Experimental Setup

2.4.1. Photo-Fenton Systems

The photo-Fenton system, previously described by Núñez-Tafalla et al. [18], consisted of three borosilicate reactors in series connected to a tank and working in batch mode (Figure 1a). The flow was 800 L h^{-1}, provided by a centrifugal pump (Schmitt MPN115 (0.25 kW)); this flow ensured a turbulent regime. Each reactor contained a different lamp: a 500 W MP lamp (89 W in UV), a 150 W MP lamp (45 W in UV), and a 40 W low-pressure (LP) lamp (11 W in UV). The MP lamp provided polychromatic irradiation compared to the LP lamp, with two irradiation peaks at 185 and 254 nm. The system allowed for all the lamps to be simultaneously or separately switched on. The reactor's external surface was 0.1 cm^2, and the volume was 1.2 L, meaning a hydraulic retention time (or irradiation time) of 5.5 s. During the process, the temperature was controlled and maintained between 18 and 25 °C to avoid lamp damage.

The tank was filled with 60 L of the water and spiked with benzotriazole, carbamazepine, clarithromycin, and diclofenac (the concentration range was 1.5–1.9, 3.3–4.8, 1.3–2.8, and 1.1–2.5 µg L^{-1}, respectively) and then homogenised by recirculating the solution for 15 min. Then, the preprepared iron complex was added and recirculated for 15 min. Finally, H_2O_2 was added, and the moment the lamps were switched on was considered time zero. The iron complex was prepared by dissolving iron sulphate in acidic water. Once it was totally dissolved, the chelating agent was added and stirred until complete complex formation. The molar ratio concentration was 1:1 for both chelating agents, EDDS, and citric acid, as shown by other authors as the most suitable value [32]. The samples were taken at predetermined times, and the bovine catalase solution (0.1 g L^{-1}) was added to the samples to remove the residual H_2O_2 and stop the Fenton reaction. The results were graphically represented by setting the operating time in minutes on the abscissa axis and the normalised effluent concentration (C/C0) on the ordinate axis, which is calculated as shown in Equation (1):

$$\text{Normalised concentration} = C/C_0 = \frac{C_i}{C_0} \qquad (1)$$

where C is the concentration in the water matrix at time i, and C_0 is the initial concentration.

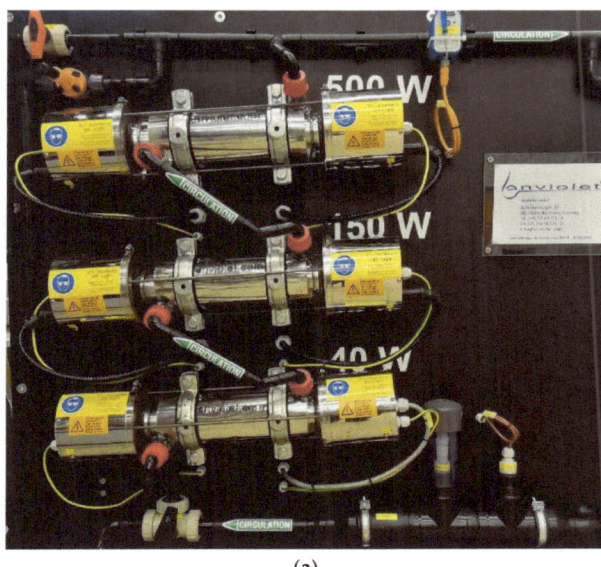 

(a) (b)

Figure 1. (a) Photo-Fenton equipment; (b) RSSC equipment.

2.4.2. Rapid Small-Scale Columns

The RSSC (Figure 1b) system was designed based on the study by Zietzschmann et al. [33]. It consisted of two glass columns operated in parallel, with a diameter of 8 mm and a length of 100 mm; the empty bed volume was adjustable. The water, spiked with the desired MCs concentration, was stirred until complete homogenisation and was stored in a 50 L tank from which the columns were fed simultaneously by a diaphragm pump (KNF Simdos 02, KNF DAC GmbH). An autosampler (AutoSam 2.3, Hitec Zang) allowed for samples to be taken at preprogrammed times. The operational conditions were stabilised in 1 mL of empty bed volume and a flow of 10 mL min^{-1}. A sample was taken from the tank and was considered the initial concentration.

Two commercial GACs were studied within the project: a fresh GAC CarboTech DGF 8 × 30 GL (Carbo Tech Ac Gmbh. Essen, Germany) and a regenerated one, CarboTech Pool W1-3 (CarboTech Ac Gmbh, Essen, Germany) (being CarboTech DGF 8 × 30 GL after reconditioning). Table S2 shows their main physicochemical properties. The GAC was preprocessed to avoid wall effects on the columns. It was crushed with a mortar and pestle, sieved to obtain a uniform size of between 100 and 500 μm, rinsed with deionised water, and dried in the oven. The GAC was brought in the columns, using glass wool as a retainer.

The breakthrough of the adsorption process on the columns was graphically represented, setting the abscissa axis as the bed volumes (BV) and normalised effluent concentration (C/C$_0$, C and C$_0$ are the outlet and inlet concentrations, respectively) as the ordinate axis. BV is a dimensionless parameter that describes the relationship between the volume feed and the volume of the column at a defined operation time. The breakthrough was considered to be the time between C/C$_0$ equal to 0.05 and 0.95 [34]. The breakthrough was modelled and fit according to the hypothesis that the adsorption of the MCs follows a first-order kinetic model, as described in Equation (2) [18]:

$$\frac{C}{C_0} = \frac{qKBV}{1 + KBV} \quad (2)$$

where C is the concentration at time t, C_0 is the initial concentration, q and K are adsorption parameters, and BV is the BV at which the effluent concentration is measured.

When working with an adsorption system, 80% MC removal corresponds to $C/C_0 = 0.20$ (meaning a maximum of 20% of MCs released into the environment), which is defined as 20 BT. The order in which the MCs achieved 20 BT can be used to scale up the process and predict the time of GAC replace. On the contrary, the BV in which 20 BT was achieved in the RSSCT could not directly correlate with a full-scale installation due to various influencing factors, such as the fouling produced by the organic matter [23].

2.4.3. Coupling of Advance Oxidation Processes and Granular Activated Carbon

AOP was selected as the first step when coupling technologies. Four scenarios of AOP were operated and evaluated, varying the selected process (photo-Fenton or UV/H_2O_2) and the chelating agents: EDDS, analytic citric acid, and technical citric acid. The final selection was based on the best performance and also an assessment of practical feasibility at full scale. The AOP was run for 5 min (5.5 s of irradiation time), then the lamps were switched off. Bovine catalase was added to the tank to remove the residual H_2O_2 and stop the Fenton reaction. The AOP effluent was transferred to the RSSCT feeding tank, and the MC adsorption was evaluated under the same conditions as the stand-alone process for both GAC types. In order to determine the possible extension of GAC lifetime, C0 was the concentration of the untreated water before the photo-Fenton process.

2.5. Experimental Design

The tests of this study were performed in five phases:

Phase 1: Assess the best lamp type (monochromatic versus polychromatic irradiation, LP and MP, respectively) to maximise MC removal in the lowest process time (UV/H_2O_2 and photo-Fenton).

Phase 2: Assess the best iron and H_2O_2 dosage, prioritising the minimisation of iron and aiming at the lowest reagent concentration possible, irradiation with the best lamp from Phase 1, and using EDDS as a chelating agent (analytical grade).

Phase 3: Assess the suitability of the best dosage from Phase 1 with citric acid as a technical-grade, commercially available chelating agent.

Phase 4: Removal of the iron (no iron) dosage and assess the best H_2O_2 concentration.

Phase 5: Perform a comparative characterisation of fresh and regenerated GAC (stand-alone).

Phase 6: Use a combination of the technologies, applying the best scenarios of the AOP obtained in the previous phases (Phase 1 to 3) with the referred GACs (Phase 4) to find the best combination process conditions that maximise GAC lifetime within the same process time.

The best scenario was selected based on a removal (%) value obtained by calculating the average of the individual removals for the four mandatory MCs, as defined in the recast of the existing UWWTD [5].

2.6. Toxicity Test

Toxicity and phytotoxicity were analysed in the treated water for the AOPs and the coupling of technologies. For the AOPs, two samples were analysed: untreated water and the sample in which 80% removal was achieved. When the coupling of technologies was performed, four samples were tested: untreated water, water after 5 min of AOP treatment, and 4000 BV for fresh and regenerated GAC.

Acute toxicity was determined by *Daphnia Magna* immobilisation using a DAPH-TOXKIT F kit (MicroBioTest Inc. (Gent, Belgium)). The lethal concentration, 50% (LC50), was evaluated at 24 and 48 h in a dilution row of 100% and 50%. Phytotoxicity was assessed in three species, *Lepidium sativum*, *Sinapis alba*, and *Sorghum saccharatum*, with the PHYTOTOXKIT for Liquid Samples kit (MicroBioTest Inc.). Seed germination and the root and leaf growth were analysed, showing the water matrix's germination rates, as well as the stimulation or inhibition effects on the roots and leaves.

3. Results and Discussion

3.1. AOP Experiments

The UV/H_2O_2 process using LP lamps and an initial H_2O_2 concentration of 40 mg L^{-1} was tested according to previous studies [18]. Figure S1 shows the removal of the average of the 4 MCs. A total of 80% removal of the average was achieved in 60 min. When looking at the performance of individual compounds, diclofenac appeared to be the first removed, which was expected due to its photosensitivity [35], while clarithromycin was the last. this behaviour was observed in all the AOP processes (Table S3). Therefore, clarithromycin was adopted as a control compound since it is the most recalcitrant. In the next experiments, iron was added to speed up the treatment process, resulting in a halved treatment time regarding UV/H_2O_2.

In the photo-Fenton test (with LP lamp), EDDS was chosen as the chelating agent due to its reported stability [36], with an iron concentration of 3 mg L^{-1} (molar ratio with EDDS 1:1). The first tests were performed upon using analytic iron and chelating agents. The average removal of MCs slightly increased, with less positive results than those previously reported [37,38]. This slight improvement allowed us to achieve the objective of removing 80% of the average of the compounds in 30 min. However, 80% clarithromycin removal was achieved in 60 min. The operational duration (minimum 30 min, meaning 33 s or irradiation time) was found to be too long for future full-scale implementation [39]. Therefore, the simultaneous operation of both LP and MP lamps was proposed to enhance the irradiation.

In the following test, the 500 W MP and 40 W LP lamps were used simultaneously with the previous reagent dosage: 3 mg L^{-1} of iron and 40 mg L^{-1} of H_2O_2. The average MC removal reached 60% in 5 min, and the objective was obtained in 10 min (Figure S2). However, 80% clarithromycin removal was achieved in 15 min. Looking at these results the iron dosage was reduced to 1.5 mg L^{-1} in order to mitigate iron release into the environment. The adjustments would align with recommendations in certain German regions, where the concentration limit in surface water bodies ranges between 0.7 and 1.8 mg L^{-1} [40]. The obtained removal rate was similar to the previous dosage, with 80% removal of the MC average in 10 min. The high difference in the treatment time between the 40 W LP lamp and its simultaneous use with the 500 W MP lamp suggested that the 500 W MP lamp was the main contributor; thus, the next tests used this lamp alone, and this reduced the energy applied for the treatment.

According to a previous study with the 500 W MP lamp [41], 5 mg L^{-1} of iron and 40 mg L^{-1} of H_2O_2 were tested, as well as 3 mg L^{-1} of iron and 40 mg L^{-1} of H_2O_2. In Figure 2, the removal of the average of the four compounds is shown to achieve 80% in both cases after 10 min (11 s of irradiance) of the process. In contrast, the results for clarithromycin removal were different, with the process taking 5 min less when 3 mg L^{-1} was dosed. These results could be related to the EDDS increment needed for a higher iron dosage, which raised the organic content of the solution consuming •OH [42].

Once again, iron was reduced to 1.5 mg L^{-1}, and H_2O_2 was maintained at 40 mg L^{-1}. The MC removal rate was lower but close to previous tests, with 75% removal of the average in 10 min, reaching 80% after 15 min (16.5 s of irradiation time); clarithromycin was eliminated after 30 min. Lastly, H_2O_2 was reduced to 20 mg L^{-1}, showing a similar trend of 40 mg L^{-1}. The obtained results confirmed the possibility of working with a low reagent dosage. The effect was more remarkable for iron than for H_2O_2, which was reduced by up to 70%, considering previous works [25,43,44]. Reducing the iron concentration is also crucial for achieving the legal limits in water bodies. Furthermore, the treatment time was also much lower compared with previous work, as it decreased from 10 [45], 30 [25], or 180 min [46] to 16.5 s, which would reduce future costs remarkably on a larger scale.

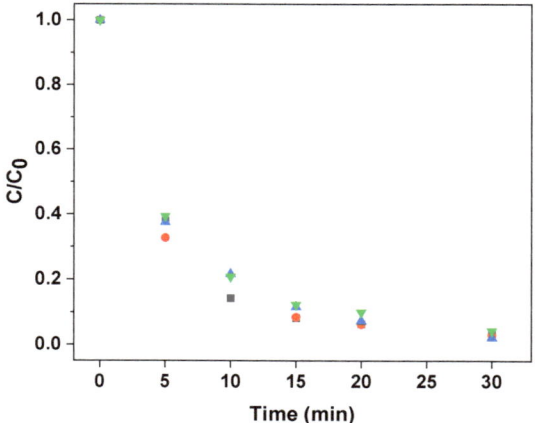

Figure 2. Average removal of four mandatory MCs using the photo-Fenton process, a 500 W MP lamp, and EDDS as a chelating agent. ■: 5 mg L^{-1} iron and 40 mg L^{-1} of H$_2$O$_2$; ●: 3 mg L^{-1} iron and 40 mg L^{-1} of H$_2$O$_2$; ▲: 1.5 mg L^{-1} iron and 40 mg L^{-1} of H$_2$O$_2$; ▼: 1.5 mg L^{-1} iron and 20 mg L^{-1} of H$_2$O$_2$.

As the combined photo-Fenton process aimed to reduce the MC concentration but not mineralize them, the removal of MCs after 5 min was compared (Figure 2), obtaining similar trend results under all conditions. Thus, 1.5 m L^{-1} of iron and 20 mg L^{-1} of H$_2$O$_2$ were selected for the following tests.

When anticipating the scaling up of the process in the future, alternative chelating agents available as industrial products were proposed since EDDS is highly expensive and difficult to find provided in large quantities from the market. Citric acid is a compound that is widely used in industry, is easy to handle (in the form of powder) at an affordable cost and is not toxic. Given its demonstrated suitability by previous authors [47], it was selected as the preferred option for potential future scale-up. Citric acid was used as an iron chelating agent for two iron forms: (i) analytical iron chelated with analytical-grade citric acid and (ii) technical-grade iron using technical-grade citric acid as chelating agents to properly assess the influence of the quality of the reagents in the performance of the process. A total of 80% average MCs removal was achieved with both iron:citric acid qualities in 15 min; a slight removal rate of the analytical iron:citric acid was observed, increasing the process time from 20 to 30 min for clarithromycin degradation. The slower removal when using citric acid as a chelating agent is related to the lower stability of the complex [48]. In contrast, the good performance of the technical-grade product was satisfactory, as previous works did not show the possibility of completely removing all the MCs [17,45]. This difference can be caused by the higher energy irradiated on the system, as previous works were performed under solar light. Figure 3 compares MC degradation with the three chelated iron sources (analytical-grade iron:EDDS, analytical iron:citric acid, and technical-grade iron:citric acid). These three scenarios were further studied on the coupled treatment.

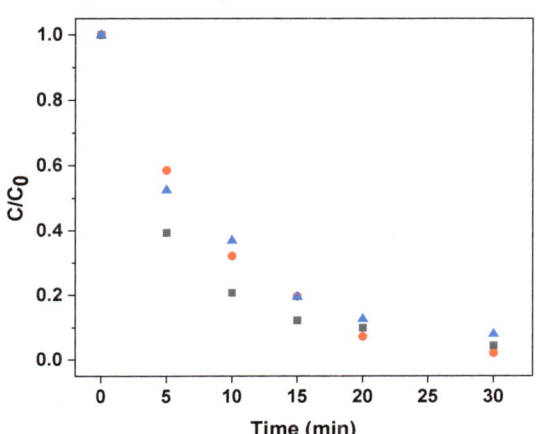

Figure 3. Average removal of four mandatory MCs using the photo-Fenton process, a 500 W MP lamp, 1.5 mg L^{-1} iron, and 20 mg L^{-1} of H$_2$O$_2$. ■: EDDS; ●: analytic citric acid; ▲: technical-grade citric acid.

UV/H$_2$O$_2$ (no iron) was also investigated using the 500 W MP lamp. Figure 4 shows the removal trend using 40 mg L^{-1}. A sharp removal was observed in the first minutes, achieving 80% of the MCs average removal in 10 min (11 s of irradiation time). As in the previous tests, diclofenac was the first to be removed (in 5 min), and clarithromycin was the most recalcitrant, needing 15 min to accomplish the objective (Table S3). The H$_2$O$_2$ dose was increased to 100 mg L^{-1}, with the aim of achieving higher degradation rates. The results were similar to the previous test; the close results can be related to an excess of H$_2$O$_2$ that quenches •OH by producing hydroperoxyl radicals [49], which has been previously reported by other authors [7]. Considering this phenomenon, the H$_2$O$_2$ dosage was reduced to 20 mg L^{-1} of H$_2$O$_2$. Under this condition, removal rates were slower, reaching 80% average removal in 15 min and for clarithromycin in 20 min. Therefore, UV/H$_2$O$_2$ with 40 mg L^{-1} was selected to be coupled with GAC filtration. Table 1 presents a comprehensive summary of the key parameters from the tests, providing clear evidence to support and validate the selected scenario.

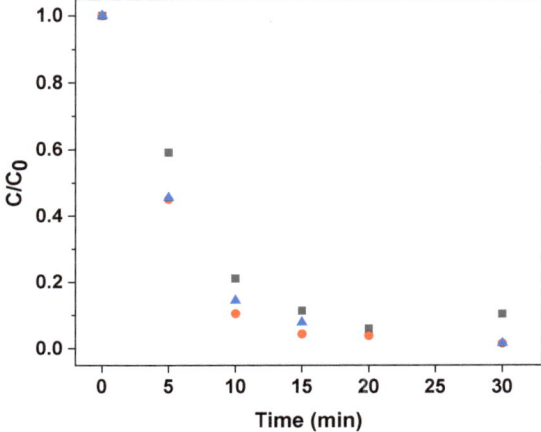

Figure 4. Average removal of four mandatory MCs using the UV/H$_2$O$_2$ process and a 500 W MP lamp. ■: 20 mg L^{-1} of H$_2$O$_2$; ●: 40 mg L^{-1}; ▲: 100 mg L^{-1}.

Table 1. Time, energy consumption, H_2O_2 consumption, and dissolved iron concentration at the point when 80% removal of the average four mandatory MCs was achieved during the AOP tests.

	Time (min)	Energy (kWh)	H_2O_2 (mg L^{-1})	Diss. Iron (mg L^{-1})
UV/H_2O_2: 40 W LP lamp 40 mg L^{-1} of H_2O_2	60	40	5.1	--
Photo-Fenton: 40 W LP lamp 3 mg L^{-1} of iron and 40 mg L^{-1} of H_2O_2	60	40	14.2	2.6
Photo-Fenton: 500 W + 40 W 3 mg L^{-1} of iron 40 mg L^{-1} of H_2O_2	10	90	8.8	3.0
Photo-Fenton: 500 W + 40 W 1.5 mg L^{-1} of iron 40 mg L^{-1} of H_2O_2	10	90	4.8	1.6
Photo-Fenton: Fe:EDDS 5 mg L^{-1} of iron 40 mg L^{-1} of H_2O_2	10	83	9.2	4.7
Photo-Fenton: Fe:EDDS 3 mg L^{-1} of iron 40 mg L^{-1} of H_2O_2	10	83	6.2	3.0
Photo-Fenton: Fe:EDDS 1.5 mg L^{-1} of iron 40 mg L^{-1} of H_2O_2	15	125	5.0	1.4
Photo-Fenton: Fe:EDDS 1.5 mg L^{-1} of iron 20 mg L^{-1} of H_2O_2	15	125	5.2	1.5
Photo-Fenton: Fe:Citric Acid Analytical-grade 1.5 mg L^{-1} of iron 20 mg L^{-1} of H_2O_2	15	125	3.9	1.5
Photo-Fenton: Fe:Citric Acid Technical-grade 1.5 mg L^{-1} of iron 20 mg L^{-1} of H_2O_2	15	125	3.8	1.7
UV/H_2O_2: 500 W 20 mg L^{-1} of H_2O_2	15	125	3.0	--
UV/H_2O_2: 500 W 40 mg L^{-1} of H_2O_2	10	83	5.0	--
UV/H_2O_2: 500 W 100 mg L^{-1} of H_2O_2	10	83	6.7	--

3.2. Adsorption Process

The adsorption capacity depends on a wide range of compound characteristics, such as the polarity or the molecular weight. A determinant characteristic for the adsorption of MCs is their hydrophobic nature, which is usually described by the octanol-water partition coefficient (K_{OW}) represented by the log K_{OW}. The affinity of each pollutant to adsorption can thus be categorised as follows: low adsorption affinity (log K_{OW} < 2.5), medium adsorption affinity (2.5 < log K_{OW} < 4), and high adsorption affinity (log K_{OW} > 4) [50]. The K_{OW} values of the target compounds are shown in Table S4. The water matrix content has also been described as a determinant of the adsorption of MCs, as dissolved organic matter competes with MCs for the adsorption sites [51].

Figure 5 shows the breakthrough of the average of the MCs adsorption for the fresh and the regenerated GAC. The breakthrough was fitted following the model described in Equation (2); the values of the model are shown in Table S5. The model was used to calculate the adsorption capacity for both GACs. The results were similar since the regenerated GAC was produced from the fresh GAC, attaining the average MCs 20 BT in 1080 and 720 BV for fresh and regenerated GAC, respectively. Diclofenac was the first to achieve 20 BT at 400 BV for both GAC types. Diclofenac is a strong competitor for the adsorption sites, quickly occupying the available adsorption sites [2], thus, it has been reported as one of the first compounds to show signs of a breakthrough (Table S6). The second MC to achieve 20 BT was clarithromycin, showing a higher difference between adsorbent materials, reaching 20 BT in 1000 BV and 500 BV for fresh and regenerated GAC, respectively. Next, carbamazepine and benzotriazole achieved the breakthroughs (third and fourth, respectively). The order of reaching 20 BT was directly related to the molecular mass (Table S4). The similar performance of both materials evidenced the possibility of implementing regenerated GAC at full scale, lowering the carbon footprint.

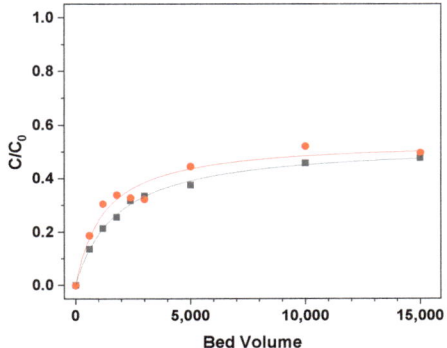

Figure 5. MC average breakthrough of ■ CarboTech DGF 8 × 30 GL (fresh GAC) and ● CarboTech Pool W1-3 (regenerated GAC).

3.3. AOP and Adsorption Process Combination

Four scenarios were selected to study the benefits of the coupling of these technologies. The scenarios were selected to analyse from the most favourable scenarios at the lab scale (EDDS) to the most realistic setup for a future scale-up, such as technical-grade iron and citric acid. All cases showed a GAC lifetime extension for diclofenac, which was the first to achieve 20 BT in the GAC stand-alone filtration and was quickly removed by AOPs. However, the influence on GAC lifetime should be considered due to the presence of technical-grade product excipients, which can block the pores of the GAC.

Figure 6 shows the percentage of removal attained by each process within the 80% target for the average of the MCs in the different scenarios. The BV when 80% was achieved is shown above the bars. The first bar shows the GAC stand-alone, which was used to facilitate a comparison of the extension of the GAC lifetime.

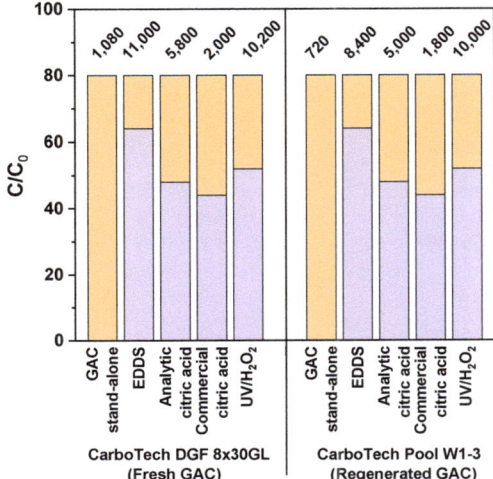

Figure 6. Contribution of each step of the coupled treatment (until 80% target) in the four selected scenarios (orange: removal by adsorption; purple: removal by the AOP). The bed volumes when 80% removal was achieved can be seen above the bars.

When using iron: EDDS, 1.5 mg L^{-1} of iron and 20 mg L^{-1} of H$_2$O$_2$, 64% of the average MC was removed in the photo-Fenton treatment, eliminating up to 72% of diclofenac. Thanks to the high removal of diclofenac, the fresh and regenerated GAC lifetime was extended up to 10 and 11 times, respectively. When the MCs were analysed separately, the first to achieve the minimum removal limit was clarithromycin, and the second was diclofenac. Up to 80% of the benzotriazole was removed in all the tests.

The removal in the first step, AOP, was close to those observed by previous authors [27,29]. Still, in the present work, iron concentration was reduced from 5 mg L^{-1} to 1.5 mg L^{-1}, and H$_2$O$_2$ was reduced from concentrations between 50 and 150 mg L^{-1} to 20 mg L^{-1}. Moreover, the treatment time was reduced since solar irradiation has a lower irradiance than lamps. These results highlight the benefits of working with UV lamps instead of solar irradiation. From the previous investigations, it is highlighted that even when working at acidic pH and high reagent dosage, it is necessary to evaluate the toxicity of the matrix, as several by-products are generated and not removed in the adsorption step. The GAC phase in the previous works was, in general, performed in batch mode, so no comparison was possible.

When using citric acid as a chelating agent, the efficiency of the photo-Fenton process was reduced, as observed in the previous test, with 48% and 44% removal for analytical- and technical-grade chemicals. The GAC lifetime was extended 5 and 7 times for fresh and regenerated GAC, respectively, when analytic iron and citric acid were used in the AOP. When the technical grade chemicals were employed, this extension was reduced to 2 and 2.5 for fresh and regenerated GAC. The difference between analytical-grade and technical-grade products highlights the necessity of testing products similar to those on a large scale. The lower extension of the GAC lifetime when technical-grade chemicals were used could not be exclusively related to the lower removal in the first step, as it was close to the analytical grade. This effect can be caused by competition for the adsorption sites with the excipients contained in the technical reagents.; the iron source contained only 12% iron, meaning 88% constituted other additives. This effect was observed by the authors [17], whose iron concentration was higher; therefore, the GAC lifetime was shorter. Diclofenac was the most affected by this impact, being the first to achieve the breakthrough instead of clarithromycin. There are two possible causes of this effect. On one side, the higher molar mass of diclofenac disallows it from occupying the clogged pores. On the other hand, the presence of additives in the technical-grade chelating agents could provoke a modification in the GAC charge, thus causing a lower affinity between the material and GAC.

In the case of the UV/H$_2$O$_2$ process, using 40 mg L^{-1} of H$_2$O$_2$, 52% of the MCs average was removed in the first stage, which is slightly higher than the tests in which citric acid was selected. However, the GAC lifetime extension was closer to using EDDS as a chelating agent, being extended by up to 9 and 13 times for fresh and regenerated GAC, respectively. The increment in GAC lifetime can be caused by the lower COD content from the chelating agent in the water matrix. As reported in the literature, COD competes with MCs in the adsorption process [19,52], reducing the lifetime. In addition to the COD, the presence of iron can also compete with MCs for the adsorption sites, being less adsorbed.

The coupling of processes strengthened the use of regenerated GAC to remove MCs. In almost all the scenarios, the lifetime of regenerated GAC was close to the fresh one. In addition, the regenerated GAC lifetime extension was higher than the fresh GAC in all the scenarios, making it possible to extend it between 2.5 and 13 times. Thus, the approach of coupling the technologies greatly promotes the use of regenerated GAC on a larger scale.

These results, when linked to the previous test with the stand-alone AOP, show that the UV/H$_2$O$_2$ process is more suitable than the photo-Fenton process for removing MCs. However, toxicity tests were required to study the possible presence of by-products in the treated water, which could produce a potential toxic release into the environment.

3.4. Toxicity Tests

The toxicity evaluation is highly relevant when working with the AOP, since by-products can be generated and not mineralised in the process. These by-products can be more toxic than the original pollutants [53], obtaining a final effluent with a lower concentration of the selected MCs but more harmful than the untreated water. During the AOPs, three different responses to toxicity can occur: a general decrease in toxicity, an increase in toxicity in the first minutes but a decrease after a determined time, or a general increase in toxicity [54]. Furthermore, it has been demonstrated that the toxicity is not the result of the average of the MC's toxicities; rather, synergetic effects with the matrix take place, affecting the general toxicity [55,56].

Acute toxicity was analysed using the *Daphnia Magna* test; an immobilisation lower than 10% was considered a non-acute toxic sample [55]. No acute toxicity, neither in 24 h nor in 48 h of sample contact, was detected in any of the stand-alone AOP scenarios (Table S7). In the combination of processes, the untreated water showed slight toxicity after 24 h of sample contact in the scenario where EDDS was used as a chelating agent (Table S8), disappearing after 5 min of the photo-Fenton process. In the analysis, after 48 h of sample contact, 20% of immobilised *Daphnia Magna* was reported after 5 min of the photo-Fenton process using technical-grade citric acid. The lower oxidative capacity of the commercial citric acid shown in MCs removal results in a lower capacity for mineralising MCs and releasing degradation by-products. However, the following GAC step significantly reduced that toxicity, with even a value of 0 for the regenerated GAC.

Phytotoxicity was evaluated in the same scenarios and samples as for *Daphnia Magna*. The most sensitive species, principally in germination but also in the root and shoot length, was *Sorghum saccharatum*.

The use of the 40 W LP lamp showed root and shoot length inhibition for *Sorghum saccharatum* (Table S9). The negative effect might be caused by the lower UV dose, which slows down the process, removing partially the MCs, with the possible presence of by-products. Iron:EDDS addition when operating this lamp increased the removal of MCs but had a negative effect on the root and shoot length growth of *Lepidium sativum*. The negative effect of iron:EDDS addition can be caused by the use of EDDS as a chelating agent, which is in agreement with Ahile et al. [47].

The simultaneous use of MP and LP lamps dosing 3 mg L^{-1} of iron (chelated with EDDS) showed no phytotoxic effect in any of the studied species. However, the treatment did not solve the growth inhibition observed in the untreated water. In the case of dosing 1.5 mg L^{-1} of iron (chelated with EDDS), a negative effect was observed in the root and shoot length of *Lepidium sativum* and *Sorghum saccharatum*. The incomplete mineralisation of MCs can explain this due effect to the lower •OH production related to the low concentration of iron:EDDS.

Using 500 W MP lamps with EDDS as a chelating agent achieved satisfactory MC removal results. Regarding toxicity, the best dosage was 5 mg L^{-1} of iron, which generally had a positive effect on the phytotoxicity of all the plant species. The reduction in iron to 3 mg L^{-1} had a slight negative impact on phytotoxicity despite good results for removing MCs. The difference between both scenarios could be associated with the higher amount of iron in the first scenario, entailing a higher amount of EDDS, which was mineralised by •OH [56]. A similar phytotoxicity increment was observed when iron was decreased to 1.5 mg L^{-1} and when H_2O_2 was maintained at 40 mg L^{-1}. In contrast, reducing the H_2O_2 dose to 20 mg L^{-1} did not affect toxicity despite the slowing down of the removal of MCs, which was also observed in the UV/H_2O_2 tests dosing different H_2O_2 concentrations.

The use of analytical citric acid had a positive effect on the stimulation of root and shoot length growth in *Sorghum saccharatum*. This stimulation was higher than when working with EDDS, which is in accordance with Ahile et al. [47], who demonstrated lower phytotoxicity for citric acid. The use of technical-grade iron:citric acid had a negative effect, reducing the stimulation of the growth of *Sorghum saccharatum* and *Sinapis alba*. A possible explanation might be that there is lower •OH production; thus, more by-products were not removed; it could also be because the additives of technical-grade iron which can have phytotoxic effects.

In the tests with UV/H_2O_2, no phytotoxicity effect was observed for *Sorghum saccharatum*. Root and shoot growth were stimulated in the three scenarios for *Lepidium sativum* and *Sinapis alba*, increasing the stimulation percentage as the H_2O_2 dose increased. These results support the previous conclusion, which is that the UV/H$_2$O$_2$ process is preferred vs. the photo-Fenton process.

On the coupling of processes, using EDDS as a chelating agent had a negligible effect on the phytotoxicity of *Sorghum saccharatum* (Figure 7). The impact on the other two species was positive, stimulating growth. However, the use of citric acid had a negative effect in both cases, analytical grade or technical grade. The germination rate was reduced, as well as root and shoot length growth. However, only the technical grade had a negative effect on *Lepidium sativum* growth, and it had no effect on *Sinapis alba*. The operation during 5 min of the UV/H$_2$O$_2$ process also showed negative effects on *Sorghum saccharatum* root and shoot growth. Thus, the photo-Fenton process using EDDS as a chelating agent was the best scenario when it was operated for only 5 min, and no presence of phytotoxic by-products was observed. This supports the high •OH production that not only mineralises the MCs but also degrades the EDDS, removing the toxicity reported by other authors. Nevertheless, the use of EDDS was not an alternative, as it had previously been discarded for scaling up due to the lack of technical-grade products. The GAC filtration showed a positive effect on *Lepidium sativum* and *Sinapis alba* growth, obtaining higher-quality water in all cases. In the case of *Sorghum saccharatum*, no effect was observed for growth. The present results show the usefulness of bioanalytical monitoring with different species in the coupling of processes.

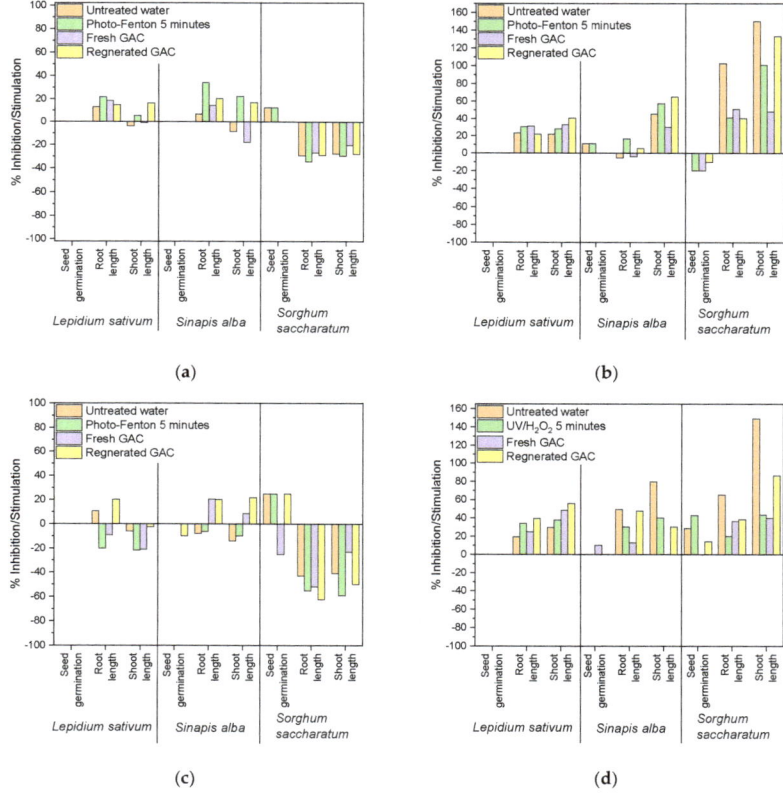

Figure 7. Phytotoxicity inhibition (negative values) and stimulation (positive values) of seed germination and root and shoot length for the three plant species. (**a**) Coupling of the technologies using EDDS as a chelating agent. (**b**) Coupling of the technologies using analytic citric acid as a chelating agent. (**c**) Coupling of the technologies using technical citric acid as a chelating agent. (**d**) Coupling of the technologies using the UV/H$_2$O$_2$ process.

4. Conclusions

In this study, the synergies and advantages of coupling different AOPs using lamps as UV sources with GAC filtration was studied. The reduction in reagents for the photo-Fenton process was demonstrated using 1.5 mg L^{-1} iron and 20 mg L^{-1} of H$_2$O$_2$ and achieving satisfactory results, which is considerably lower than the reported value for a similar process in the literature. Despite high removal when using the photo-Fenton process, the UV/H$_2$O$_2$ process showed slightly lower removal, but the use of iron was avoided. Before the coupling of the technologies, a synergy effect was observed between the AOPs and GAC filtration. Diclofenac was the most recalcitrant MC for GAC filtration but the best removed in the photo-Fenton process. This shows evidence of the advantage of coupling two different quaternary treatments where the most recalcitrant compounds are different for each. Moreover, the coupling reduced the AOP treatment time by between 50 and 66%, consequently reducing energy consumption. Additionally, depending on the scenario, GAC lifetime extension varied between 2 and 11 times, being GAC lifetime comparable when applying the photo-Fenton or UV/H$_2$O$_2$ processes. This result showed that UV/H$_2$O$_2$ avoided the addition of iron, simplifying the process and operation at natural pH in any case. The coupling of technologies produced a higher positive effect in the regenerated GAC than in the fresh version, obtaining a similar GAC lifetime for both types. Thus, regenerated GAC could be adopted, reducing the carbon footprint, energy consumption, and waste management costs for its disposal. The lack of toxicity highlighted the absence of toxic by-products, as can appear when working with ozone. Further studies will implement the coupling of technologies in an on-site pilot-scale plant in the WWTP of Heiderscheidergrund (north of Luxembourg).

Supplementary Materials: The following supporting information can be downloaded at: https://www.mdpi.com/article/10.3390/w16192824/s1. Table S1: MC recovery rates, relative standard deviations, and limits of quantification; Table S2: Selected GAC main physicochemical properties; Table S3: C/C$_0$ of the independent MCs in the AOP tests; Table S4: Adsorption characteristics of the selected MCs; Table S5: First-order kinetic model values of stand-alone GAC adsorption; Table S6: 1-(C/C0) of the independent MCs in the adsorption tests; Table S7: Acute toxicity values in AOPs based on *Daphnia Magna* immobilisation; Table S8: Acute toxicity values in the coupling of processes based on *Daphnia Magna* immobilisation; Table S9: Phytotoxicity inhibition (negative values) and stimulation (positive values) of seed germination and the root and shoot length of the three plant species; Figure S1: Microcontaminant removal using the AOP; Figure S2: MC removal using the photo-Fenton process and a 40W LP lamp and 500W MP lamp simultaneously.

Author Contributions: P.N.-T.: Conceptualisation, investigation, methodology, visualization, writing—original draft preparation. I.S.: Conceptualisation, methodology, validation, writing—review and editing. S.V.: Conceptualisation, writing—review and editing, project administration. J.H.: Writing—review and editing, supervision, project administration, funding acquisition. All authors have read and agreed to the published version of the manuscript.

Funding: This research received no external funding.

Data Availability Statement: The original contributions presented in this study are included in the article and Supplementary Materials. Further inquiries can be directed to the corresponding author.

Acknowledgments: The authors wish to thank the Administration de la gestion de l'eau—Ministère du développement durable et des infrastructures of Luxembourg for funding this research under the Fentastic Project, as well as our project partners SIDEN (Syndicat intercommunal de dépollution des eaux résiduaires du Nord), Hydro-Ingenieure, TR-Engineering, and WiW (Wupperverbandsgesellschaft für integrale Wasserwirtschaft) mbh. The authors wish to thank Adnan Skrijelj for his technical support.

Conflicts of Interest: The authors declare no conflicts of interest.

References

1. Choi, S.; Lee, W.; Son, H.; Lee, W.; Choi, Y.; Yeom, H.; Seo, C.; Lee, H.; Lee, Y.; Lim, S.J.; et al. Occurrence, removal, and prioritization of organic micropollutants in four full-scale wastewater treatment plants in Korea. *Chemosphere* **2024**, *361*, 142460. [CrossRef] [PubMed]
2. Svahn, O.; Borg, S. Assessment of full-scale 4th treatment step for micro pollutant removal in Sweden: Sand and GAC filter combo. *Sci. Total Environ.* **2024**, *906*, 167424. [CrossRef] [PubMed]
3. Miyawaki, T.; Nishino, T.; Asakawa, D.; Haga, Y.; Hasegawa, H.; Kadokami, K. Development of a rapid and comprehensive method for identifying organic micropollutants with high ecological risk to the aquatic environment. *Chemosphere* **2021**, *263*, 128258. [CrossRef] [PubMed]
4. Chen, M.; Jin, X.; Guo, C.; Liu, Y.; Zhang, H.; Wang, J.; Dong, G.; Liu, N.; Guo, W.; Giesy, J.P.; et al. Micropollutants but high risks: Human multiple stressors increase risks of freshwater ecosystems at the megacity-scale. *J. Hazard. Mater.* **2023**, *460*, 132497. [CrossRef]
5. Council of the European Union. Proposal for a Directive of the European Parliament and of the Council Concerning Urban Wastewater Treatment (Recast). 2024; pp. 1–23. Available online: https://data.consilium.europa.eu/doc/document/ST-7108-2024-INIT/en/pdf (accessed on 15 June 2024).
6. AGE-Luxembourg Gouvernement. Directive-Cadre sur L'eau, Administration de Gestion de L'eau-Luxembourg Gouvernement. 2000. Available online: https://eau.gouvernement.lu/fr/administration/directives/Directive-cadre-sur-leau.html (accessed on 15 June 2023).
7. Sánchez-Montes, I.; Salmerón García, I.; Rivas Ibañez, G.; Aquino, J.M.; Polo-López, M.I.; Malato, S.; Oller, I. UVC-based advanced oxidation processes for simultaneous removal of microcontaminants and pathogens from simulated municipal wastewater at pilot plant scale. *Environ. Sci. Water Res. Technol.* **2020**, *6*, 2553–2566. [CrossRef]
8. Soltermann, F.; Abegglen, C.; Tschui, M.; Stahel, S.; von Gunten, U. Options and limitations for bromate control during ozonation of wastewater. *Water Res.* **2017**, *116*, 76–85. [CrossRef]
9. Krahnstöver, T.; Wintgens, T. Separating powdered activated carbon (PAC) from wastewater—Technical process options and assessment of removal efficiency. *J. Environ. Chem. Eng.* **2018**, *6*, 5744–5762. [CrossRef]
10. Rizzo, L.; Gernjak, W.; Krzeminski, P.; Malato, S.; McArdell, C.S.; Perez, J.A.S.; Schaar, H.; Fatta-Kassinos, D. Best available technologies and treatment trains to address current challenges in urban wastewater reuse for irrigation of crops in EU countries. *Sci. Total Environ.* **2020**, *710*, 136312. [CrossRef]
11. Rueda-Márquez, J.J.; Levchuk, I.; Manzano, M.; Sillanpää, M. Toxicity reduction of industrial and municipal wastewater by advanced oxidation processes (Photo-fenton, UVC/H2O2, electro-fenton and galvanic fenton): A review. *Catalysts* **2020**, *10*, 612. [CrossRef]
12. Mukherjee, J.; Lodh, B.K.; Sharma, R.; Mahata, N.; Shah, M.P.; Mandal, S.; Ghanta, S.; Bhunia, B. Advanced oxidation process for the treatment of industrial wastewater: A review on strategies, mechanisms, bottlenecks and prospects. *Chemosphere* **2023**, *345*, 140473. [CrossRef]
13. Rizzo, L.; Malato, S.; Antakyali, D.; Beretsou, V.G.; Đolić, M.B.; Gernjak, W.; Heath, E.; Ivancev-Tumbas, I.; Karaolia, P.; Lado Ribeiro, A.R.; et al. Consolidated vs new advanced treatment methods for the removal of contaminants of emerging concern from urban wastewater. *Sci. Total Environ.* **2019**, *655*, 986–1008. [CrossRef] [PubMed]
14. Ameta, R.; Chohadia, A.K.; Jain, A.; Punjabi, P.B. *Fenton and Photo-Fenton Processes*; Academic Press: Cambridge, MA, USA, 2018. [CrossRef]
15. Dong, W.; Jin, Y.; Zhou, K.; Sun, S.P.; Li, Y.; Chen, X.D. Efficient degradation of pharmaceutical micropollutants in water and wastewater by FeIII-NTA-catalyzed neutral photo-Fenton process. *Sci. Total Environ.* **2019**, *688*, 513–520. [CrossRef] [PubMed]
16. Maniakova, G.; Salmerón, I.; Aliste, M.; Inmaculada Polo-López, M.; Oller, I.; Malato, S.; Rizzo, L. Solar photo-Fenton at circumneutral pH using Fe(III)-EDDS compared to ozonation for tertiary treatment of urban wastewater: Contaminants of emerging concern removal and toxicity assessment. *Chem. Eng. J.* **2022**, *431*, 133474. [CrossRef]
17. Núñez-Tafalla, P.; Salmerón, I.; Oller, I.; Venditti, S.; Malato, S.; Hansen, J. Micropollutant elimination by sustainable technologies: Coupling activated carbon with solar photo-Fenton as pre-oxydation step. *J. Environ. Chem. Eng.* **2024**, *12*, 113305. [CrossRef]
18. Núñez-Tafalla, P.; Salmerón, I.; Venditti, S.; Hansen, J. Combination of Photo-Fenton and Granular Activated Carbon for the Removal of Microcontaminants from Municipal Wastewater via an Acidic Dye. *Sustainability* **2024**, *16*, 1605. [CrossRef]
19. Gidstedt, S.; Betsholtz, A.; Falås, P.; Cimbritz, M.; Davidsson, Å.; Micolucci, F.; Svahn, O. A comparison of adsorption of organic micropollutants onto activated carbon following chemically enhanced primary treatment with microsieving, direct membrane filtration and tertiary treatment of municipal wastewater. *Sci. Total Environ.* **2022**, *811*, 152225. [CrossRef]
20. Rattier, M.; Reungoat, J.; Keller, J.; Gernjak, W. Removal of micropollutants during tertiary wastewater treatment by biofiltration: Role of nitrifiers and removal mechanisms. *Water Res.* **2014**, *54*, 89–99. [CrossRef]
21. Mo, W.; Cornejo, P.K.; Malley, J.P.; Kane, T.E.; Collins, M.R. Life cycle environmental and economic implications of small drinking water system upgrades to reduce disinfection byproducts. *Water Res.* **2018**, *143*, 155–164. [CrossRef]
22. Crittenden, J.C.; Reddy, P.S.; Arora, H.; Trynoski, J.; Hand, W.; Perram, D.L.; Summers, R.S.; Crittenden, J.C.; Reddy, P.S.; Arora, H.; et al. Rapid Small-Scale Column Tests. *Am. Water Work. Assoc.* **1991**, *83*, 77–87. [CrossRef]
23. Freihardt, J.; Jekel, M.; Ruhl, A.S. Comparing test methods for granular activated carbon for organic micropollutant elimination. *J. Environ. Chem. Eng.* **2017**, *5*, 2542–2551. [CrossRef]

24. Bhanot, P.; Celin, S.M.; Sharma, P.; Sahai, S.K.; Kalsi, A. Comparative Evaluation of Low- and Medium-Pressure UV Lamps for Photo-degradation of RDX Wastewater. *Water Air Soil Pollut.* **2023**, *234*, 587. [CrossRef]
25. Martínez-Escudero, C.M.; Garrido, I.; Contreras, F.; Hellín, P.; Flores, P.; León-Morán, L.O.; Arroyo-Manzanares, N.; Campillo, N.; Pastor, M.; Viñas, P.; et al. Degradation of imidacloprid in water by photo-Fenton process using UV-LED lamps at neutral pH: Study of intermediate products by liquid chromatography mass spectrometry after dispersive liquid–liquid microextraction. *Catal. Today* **2024**, *429*, 114501. [CrossRef]
26. Gamarra-Güere, C.D.; Dionisio, D.; Santos, G.O.S.; Vasconcelos Lanza, M.R.; de Jesus Motheo, A. Application of Fenton, photo-Fenton and electro-Fenton processes for the methylparaben degradation: A comparative study. *J. Environ. Chem. Eng.* **2022**, *10*, 106992. [CrossRef]
27. Della-Flora, A.; Wilde, M.L.; Thue, P.S.; Lima, D.; Lima, E.C.; Sirtori, C. Combination of solar photo-Fenton and adsorption process for removal of the anticancer drug Flutamide and its transformation products from hospital wastewater. *J. Hazard. Mater.* **2020**, *396*, 122699. [CrossRef] [PubMed]
28. Della-Flora, A.; Wilde, M.L.; Lima, D.; Lima, E.C.; Sirtori, C. Combination of tertiary solar photo-Fenton and adsorption processes in the treatment of hospital wastewater: The removal of pharmaceuticals and their transformation products. *J. Environ. Chem. Eng.* **2021**, *9*, 105666. [CrossRef]
29. Michael, S.G.; Michael-Kordatou, I.; Beretsou, V.G.; Jäger, T.; Michael, C.; Schwartz, T.; Fatta-Kassinos, D. Solar photo-Fenton oxidation followed by adsorption on activated carbon for the minimisation of antibiotic resistance determinants and toxicity present in urban wastewater. *Appl. Catal. B Environ.* **2019**, *244*, 871–880. [CrossRef]
30. Venditti, S.; Brunhoferova, H.; Hansen, J. Behaviour of 27 selected emerging contaminants in vertical flow constructed wetlands as post-treatment for municipal wastewater. *Sci. Total Environ.* **2022**, *819*, 153234. [CrossRef]
31. Mejri, A.; Soriano-Molina, P.; Miralles-Cuevas, S.; Trabelsi, I.; Sánchez Pérez, J.A. Effect of liquid depth on microcontaminant removal by solar photo-Fenton with Fe(III):EDDS at neutral pH in high salinity wastewater. *Environ. Sci. Pollut. Res.* **2019**, *26*, 28071–28079. [CrossRef]
32. Zietzschmann, F.; Müller, J.; Sperlich, A.; Ruhl, A.S.; Meinel, F.; Altmann, J.; Jekel, M. Rapid small-scale column testing of granular activated carbon for organic micro-pollutant removal in treated domestic wastewater. *Water Sci. Technol.* **2014**, *70*, 1271–1278. [CrossRef]
33. Worch, E. *Adsorption Technology in Water Treatment: Fundamentals, Processes, and Modeling*, 2nd ed.; De Gruyter: Berlin, Germany, 2021. [CrossRef]
34. Keen, O.S.; Thurman, E.M.; Ferrer, I.; Dotson, A.D.; Linden, K.G. Dimer formation during UV photolysis of diclofenac. *Chemosphere* **2013**, *93*, 1948–1956. [CrossRef]
35. Bello, M.M.; Abdul Raman, A.A.; Asghar, A. A review on approaches for addressing the limitations of Fenton oxidation for recalcitrant wastewater treatment. *Process Saf. Environ. Prot.* **2019**, *126*, 119–140. [CrossRef]
36. Mandavgane, S.A. Study of degradation of p-toluic acid by photo-oxidation, peroxidation, photo-peroxidation and photo-fenton processes. *Mater. Today Proc.* **2020**, *29*, 1213–1216. [CrossRef]
37. Adityosulindro, S.; Julcour, C.; Riboul, D.; Barthe, L. Degradation of ibuprofen by photo-based advanced oxidation processes: Exploring methods of activation and related reaction routes. *Int. J. Environ. Sci. Technol.* **2022**, *19*, 3247–3260. [CrossRef]
38. De la Cruz, N.; Esquius, L.; Grandjean, D.; Magnet, A.; Tungler, A.; de Alencastro, L.F.; Pulgarín, C. Degradation of emergent contaminants by UV, UV/H2O2 and neutral photo-Fenton at pilot scale in a domestic wastewater treatment plant. *Water Res.* **2013**, *47*, 5836–5845. [CrossRef]
39. Land Brandenburg. Strategischen Gesamtplan zur Senkung der bergbaubedingten StoffeintStrategischen Gesamtplan zur Senkung der Bergbaubedingten Stoffeinträge in die Spree und deren Zuflüsse in der Lausitz. 2015. Available online: https://mluk.brandenburg.de/mluk/de/umwelt/wasser/bergbaufolgen-fuer-den-wasserhaushalt/# (accessed on 15 June 2024).
40. Núñez-Tafalla, P.; Salmerón, I.; Venditti, S.; Hansen, J. Treatment of wastewater effluent using photo-Fenton combined with granular activated carbon enhancing microcontaminants removal. In Proceedings of the 39th IAHR World Congress, Granada, Spain, 19–24 June 2022; pp. 2041–2046. [CrossRef]
41. Cruz, A.; Couto, L.; Esplugas, S.; Sans, C. Study of the contribution of homogeneous catalysis on heterogeneous Fe(III)/alginate mediated photo-Fenton process. *Chem. Eng. J.* **2017**, *318*, 272–280. [CrossRef]
42. Maniakova, G.; Rizzo, L. Pharmaceuticals degradation and pathogens inactivation in municipal wastewater: A comparison among UVC photo-Fenton with chelating agents, UVC/H2O2 and ozonation. *J. Environ. Chem. Eng.* **2023**, *11*, 111356. [CrossRef]
43. Ricardo, I.A.; Marson, E.O.; Paniagua, C.E.S.; Macuvele, D.L.P.; Starling, M.C.V.M.; Pérez, J.A.S.; Trovó, A.G. Microcontaminants degradation in tertiary effluent by modified solar photo-Fenton process at neutral pH using organic iron complexes: Influence of the peroxide source and matrix composition. *Chem. Eng. J.* **2024**, *487*, 150505. [CrossRef]
44. Nahim-Granados, S.; Berruti, I.; Oller, I.; Polo-López, M.I.; Malato, S. Assessment of a commercial biodegradable iron fertilizer (Fe^{3+}-IDS) for water treatment by solar photo-Fenton at near-neutral pH. *Catal. Today* **2024**, *434*, 114639. [CrossRef]
45. Giménez, B.N.; Conte, L.O.; Duarte, S.A.; Schenone, A.V. Improvement of ferrioxalate assisted Fenton and photo-Fenton processes for paracetamol degradation by hydrogen peroxide dosage. *Environ. Sci. Pollut. Res.* **2024**, *31*, 13489–13500. [CrossRef]
46. Ahile, U.J.; Wuana, R.A.; Itodo, A.U.; Sha'Ato, R.; Malvestiti, J.A.; Dantas, R.F. Are iron chelates suitable to perform photo-Fenton at neutral pH for secondary effluent treatment? *J. Environ. Manag.* **2021**, *278*, 111566. [CrossRef]

47. Ahile, U.J.; Wuana, R.A.; Itodo, A.U.; Sha'Ato, R.; Dantas, R.F. Stability of iron chelates during photo-Fenton process: The role of pH, hydroxyl radical attack and temperature. *J. Water Process Eng.* **2020**, *36*, 101320. [CrossRef]
48. Hong, A.J.; Lee, J.; Cha, Y.; Zoh, K.D. Propiconazole degradation and its toxicity removal during UV/H_2O_2 and UV photolysis processes. *Chemosphere* **2022**, *302*, 134876. [CrossRef] [PubMed]
49. García, L.; Leyva-Díaz, J.C.; Díaz, E.; Ordóñez, S. A review of the adsorption-biological hybrid processes for the abatement of emerging pollutants: Removal efficiencies, physicochemical analysis, and economic evaluation. *Sci. Total Environ.* **2021**, *780*, 146554. [CrossRef] [PubMed]
50. Real, F.J.; Benitez, F.J.; Acero, J.L.; Casas, F. Adsorption of selected emerging contaminants onto PAC and GAC: Equilibrium isotherms, kinetics, and effect of the water matrix. *J. Environ. Sci. Health-Part A Toxic/Hazard. Subst. Environ. Eng.* **2017**, *52*, 727–734. [CrossRef]
51. Gutiérrez, M.; Grillini, V.; Mutavdžić Pavlović, D.; Verlicchi, P. Activated carbon coupled with advanced biological wastewater treatment: A review of the enhancement in micropollutant removal. *Sci. Total Environ.* **2021**, *790*, 148050. [CrossRef]
52. Calvalheri, P.S.; Santos Machado, B.; Ferreira da Silva, T.; Warszawski De Oliveira, K.R.; Magalh Correa, F.J.; Nazário, C.E.; Pereira Cavalcante, R.; De Oliveira, S.C.; Machulek, A.J. Ketoprofen and diclofenac removal and toxicity abatement in a real scale sewage treatment Plant by photo-Fenton Process with design of experiments. *J. Environ. Chem. Eng.* **2023**, *11*, 110699. [CrossRef]
53. Rodrigues-Silva, F.; Lemos, C.R.; Naico, A.A.; Fachi, M.M.; do Amaral, B.; de Paula, V.C.S.; Rampon, D.S.; Beraldi-Magalhães, F.; Prola, L.D.T.; Pontarolo, R.; et al. Study of isoniazid degradation by Fenton and photo-Fenton processes, by-products analysis and toxicity evaluation. *J. Photochem. Photobiol. A Chem.* **2022**, *425*, 113671. [CrossRef]
54. Davis Sá, R.; Patricia Rodríguez-Pérez, A.; Rodrigues-Silva, F.; de Carvalho Soares de Paula, V.; Daniela Tentler Prola, L.; Martins de Freitas, A.; Querne de Carvalho, K.; Vinicius de Liz, M. Treatment of a clinical analysis laboratory wastewater from a hospital by photo-Fenton process at four radiation settings and toxicity response. *Environ. Sci. Pollut. Res.* **2021**, *28*, 24180–24190. [CrossRef]
55. Freitas, A.M.; Rivas, G.; Campos-Mañas, M.C.; Casas López, J.L.; Agüera, A.; Sánchez Pérez, J.A. Ecotoxicity evaluation of a WWTP effluent treated by solar photo-Fenton at neutral pH in a raceway pond reactor. *Environ. Sci. Pollut. Res.* **2017**, *24*, 1093–1104. [CrossRef]
56. Soriano-Molina, P.; García Sánchez, J.L.; Alfano, O.M.; Conte, L.O.; Malato, S.; Sánchez Pérez, J.A. Mechanistic modeling of solar photo-Fenton process with Fe^{3+}-EDDS at neutral pH. *Appl. Catal. B Environ.* **2018**, *233*, 234–242. [CrossRef]

Disclaimer/Publisher's Note: The statements, opinions and data contained in all publications are solely those of the individual author(s) and contributor(s) and not of MDPI and/or the editor(s). MDPI and/or the editor(s) disclaim responsibility for any injury to people or property resulting from any ideas, methods, instructions or products referred to in the content.

Article

Methyl Red Adsorption from Aqueous Solution Using Rumex Abyssinicus-Derived Biochar: Studies of Kinetics and Isotherm

Meseret Dawit Teweldebrihan * and Megersa Olumana Dinka

Department of Civil Engineering Sciences, Faculty of Engineering and the Built Environment, University of Johannesburg APK Campus, Auckland Park 2006, Johannesburg P.O. Box 524, South Africa
* Correspondence: mesidawit1@gmail.com

Abstract: This work focused on the decolorization of methyl red (MR) from an aqueous solution utilizing Rumex abyssinicus-derived biochar (RAB). RAB was prepared to involve unit operations such as size reduction, drying, and carbonization. The pyrolysis of the precursor material was carried out at a temperature of 500 °C for two hours. After that, the prepared RAB was characterized by the pH point of zero charge (pHpzc), the Brunauer–Emmett–Teller (BET) method, Scanning Electron Microscopy (SEM) and Fourier-Transform Infrared (FTIR) spectroscopy. On the other hand, a batch adsorption experiment of MR removal onto RAB was conducted, considering four operating parameters: pH, contact time, adsorbent dose, and initial dye concentration. The characterization of the adsorbent material revealed a porous and heterogeneous surface morphology during SEM, a specific surface area of 45.8 m^2/g during the BET method, the presence of various functional groups during FTIR, and a pHpzc of 6.2. The batch adsorption experiment analysis results revealed that a maximum removal efficiency of 99.2% was attained at an optimum working condition of pH 6, contact time of 40 min, initial dye concentration of 70 mg/L and adsorbent dosage of 0.2 g/100 mL. Furthermore, Freundlich isotherm (R^2 = 0.99) and pseudo-second-order kinetics (R^2 = 0.99) models confirmed the heterogeneous surface interaction and chemisorption nature. Generally, this study highlighted that RAB could be a potential adsorbent for the detoxification of MR-containing industrial effluents.

Keywords: adsorption; methyl red; *Rumex abyssinicus*; textile wastewater; water quality

Citation: Teweldebrihan, M.D.; Dinka, M.O. Methyl Red Adsorption from Aqueous Solution Using Rumex Abyssinicus-Derived Biochar: Studies of Kinetics and Isotherm. *Water* **2024**, *16*, 2237. https://doi.org/10.3390/w16162237

Academic Editor: Alessandro Erto

Received: 28 April 2024
Revised: 20 May 2024
Accepted: 20 May 2024
Published: 8 August 2024

Copyright: © 2024 by the authors. Licensee MDPI, Basel, Switzerland. This article is an open access article distributed under the terms and conditions of the Creative Commons Attribution (CC BY) license (https://creativecommons.org/licenses/by/4.0/).

1. Introduction

Meeting goal 6 of the sustainable development agenda (clean, safe, sufficient, and affordable water for all of humanity) has remained challenging due to various anthropogenic and natural factors [1]. Human-induced activities, in particular, have become serious sources of water pollution, affecting the quality and quantity of freshwater [2]. According to a European investment bank report, about 380 billion cubic meters (BCM) of wastewater is generated annually worldwide, increasing alarmingly, and it is estimated to reach 470 and 574 BCM by 2030 and 2050, respectively [3]. This condition causes water quality deterioration and a reduction in its quantity, as most untreated and inadequately treated wastewater is discharged into nearby water bodies [4,5]. In line with this, industrialization has contributed profoundly to wastewater generation, and about 80% of industrial effluents are emitted into the nearby water channels [6].

Regarding wastewater generation, the textile industry is among the frontline contributors of wastewater, because about 93 BCM of fresh water is utilized in different stages of textile processing, which is estimated to be 4% of freshwater extraction globally [7]. Furthermore, industrial textile processing, such as the dyeing and finishing stages, consumes 200 L of water per kg of fabric processed, resulting in 17–20% wastewater [8]. In line with this, textile industry-based wastewater is characterized by high amounts of biochemical oxygen demand (BOD), chemical oxygen demand (COD), total solids and massive dyes [9], posing serious risks to human health and aquatic biota [10]. In particular, dye-saturated textile wastewater can potentially cause diseases like cancer and negatively affect aquatic biota [11].

Methyl red (MR) is a synthetic, anionic mono-azo dye with a chemical formula of $C_{15}H_{15}N_3O_2$ and a molecular weight of 269.30 g/mol that is almost insoluble in water, but more soluble in other solvents like glacial acetic acid, ether and methanol [12–14]. It is a reactive dye widely used as an acid–base indicator in laboratories [15]. Additionally, methyl red is applied in the textile dyeing and finishing processes [16] due to its high color-fixing performance and mild fading [16]. However, MR-saturated effluents raise concerns regarding human health and environment-related problems [17,18]. Long-term exposure to MR has been found to pose health risks like skin damage, and it can irritate the digestive system and eyes [19].

Furthermore, MR is known to be mutagenic, toxic, and carcinogenic, and its complex decomposition in water bodies makes it harmful to aquatic biota [20,21]. MR is considered mutagenic, toxic, and carcinogenic due to its complex aromatic molecular structure, which makes it more stable and difficult to biodegrade [22]. The presence of azo groups in the chemical structure of MR is responsible for its mutagenic and carcinogenic properties [23]. The azo group (–N=N–) is known to undergo reductive cleavage, leading to the formation of toxic aromatic amines [24]. The presence of these aromatic amines in MR contributes to its toxicity and potential carcinogenicity [25]. The chemical structure of MR consists of a benzoic acid moiety attached to a dimethylamino group through an azo linkage [26]. The azo linkage (–N=N–) is the functional group responsible for the color and reactivity of MR. The chemical structure of MR is shown in Figure 1.

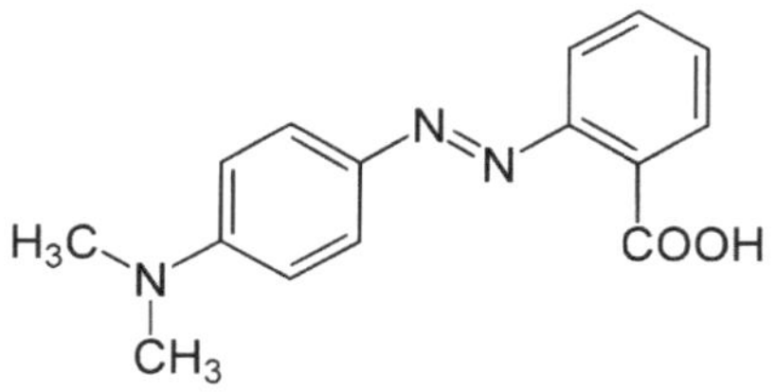

Figure 1. Chemical structure of methyl red.

Hence, treatment interventions are critical before discharging MR-saturated wastewater into nearby water bodies. Various MR decolorization techniques have been explored, aiming to mitigate its environmental and health impacts. These include traditional wastewater treatment techniques such as coagulation/flocculation [27], neutralization [28] and precipitation. However, these conventional methods face challenges in effectively removing MR from wastewater due to the persistent nature of the dye [29]. Hence, advanced wastewater treatment techniques such as the electrochemical process [14], photocatalytic degradation [30], membrane filtration [31], electron photo-Fenton oxidation [32], and reverse osmosis [33] are widely used for MR decontamination. These techniques effectively remove MR from aqueous or textile wastewater [34]. However, certain drawbacks such as their massive chemical consumption, high capital requirement, energy intensiveness, need for skilled human resources and formation of secondary pollutants [35] necessitate the search for environmentally benign, socially acceptable, low-cost and easily accessible treatment methods such as adsorption [36].

Adsorption, defined as the deposition of molecular species onto a solid surface [37], is regarded as a green, clean and versatile method for water and wastewater treatment [37]. Herein, the molecular species that gets adsorbed on the surface is known as the adsorbate, and the surface on which adsorption occurs is known as the adsorbent [38]. Hence, adsorption is a surface phenomenon affected by various parameters such as the temperature, concentration, contact time, pH and adsorbent dosage [39]. Adsorption techniques are classified into physisorption and chemisorption. Physisorption (physical adsorption) is reversible, does not need activation energy, is characterized by non-specificity and occurs via weak Van der Waals forces.

In contrast, chemisorption (chemical adsorption) is known for its irreversibility, specificity, formation of chemical bonds and need for activation energy [40]. Various adsorbent materials are utilized to remove contaminants from water and wastewater. These include natural minerals such as rocks and soils [41], coal-based adsorbents [42] and biomass-derived adsorbents [43]. Numerous selection criteria are considered when choosing the best-performing adsorbent materials. These include local availability, a large specific surface area, high chemical reactivity, porosity, cost-effectiveness, renewability and removal efficiency [44].

Adsorption has been extensively studied for MR removal from textile wastewater and the aqueous environment. Previous research has demonstrated the efficiency of various adsorbents, such as commercial activated carbon [45], zeolites [46], and biosorbents derived from agricultural waste [47]. Among these, commercial activated carbon has gained considerable attention due to its high surface area, pore structure, and adsorption capacity [48]. However, the sustainable development of an adsorbent is crucial to ensure its long-term viability as an effective adsorbent. The production of an adsorbent from non-renewable resources raises concerns regarding its environmental impact and long-term availability [49]. Therefore, researchers have focused on developing biobased activated carbon/biochar derived from biomass waste materials, such as agricultural residues, wood waste, and biochar. These biobased activated carbons offer a sustainable alternative and economic benefits by utilizing waste materials [50]

Various biomass-derived adsorbents have been applied for the removal of methyl red. For instance, a sorbent made of Annona squamosa leaves and barks achieved a maximum adsorption capacity of 7.2 mg/g [51]. Additionally, Bali cow bones-based hydrochar material reached a capacity of 7.2 mg/g [52]. Raphanus caudatus powdered leaves biomass exhibited a capacity of 30.86 mg/g [53]. These studies showcase the diverse adsorption capacities of different biomass-derived adsorbents for methyl red, highlighting their potential in wastewater treatment applications. However, the low adsorption capacities recorded in these studies necessitate the exploration of new locally available precursor materials, such as Rumex abyssinicus, which exhibit greater adsorption capacity.

Rumex abyssinicus, a 3–4 m tall herb found in tropical Africa, North Africa, and Ethiopia [54], is a traditional medicine used for treating diseases like sexually transmitted diseases, fungal infections, diabetes, lung tuberculosis, and leprosy [55,56]. Its roots and bark lower blood pressure, heal wounds and treat stomachaches [57]. However, the remaining parts are often discarded. This study aims to use biochar synthesized from the Rumex abyssinicus stem as an effective adsorbent for pollutant detoxification. Previous research focused on producing activated carbon from Rumex abyssinicus and its application in dye decolorization. However, preparing activated carbon from biomaterials necessitates activating agents such as chemicals and carbonization at an elevated temperature.

Furthermore, the tedious steps required to wash the acid/base used for neutralization, in addition to the massive chemical consumption and energy intensiveness, make the bio-based activated carbon preparation process uneconomical and time-consuming. Hence, this work focused on producing and applying Rumex abyssinicus-based biochar (RAB) for MR removal from aqueous solutions. To the best of the authors' knowledge, no research has been conducted on the preparation of biochar from the Rumex abyssinicus plant. Additionally, no study has addressed the removal of MR from wastewater or aqueous solutions using RAB, which highlights the presence of a research gap. Hence, this study addresses the lack of research on the production and application of Rumex abyssinicus-based biochar for MR removal from aqueous solutions.

2. Material and Methods

2.1. Biochar Preparation

The precursor material (Rumex abyssinicus) utilized in this work was collected from Addis Ababa, Ethiopia. The sun-dried stem of Rumex abyssiniccus was thoroughly washed with tap water before being rinsed with distilled water to eliminate dirty materials that may affect the potential of the adsorbent material. After that, the dust-free sample of Rumex abyssinicus was oven-dried (model BOV-T50F, BIOBASE, China) at a temperature and time of 105 °C and 24 h, respectively. The completely dried precursor material was then pulverized into smaller pieces using a mechanical granulator (Hummer mill-008, STEDMAN, USA). The pyrolysis of the precursor material was carried out using a muffle furnace (Model Nabertherm F 330, Cole Parmer, Canada) at a carbonization temperature of 500 °C for two hours. During the carbonization process, a heating rate of 25 °C/min was used. Figure 2 depicts the major steps followed for the development of biochar from Rumex abyssinicus biomass. Finally, the prepared adsorbent material was cooled in a desiccator, reduced to a particle size of 250 µm and stored in a zip locker for subsequent batch adsorption and adsorbent characterization experiments [58–60].

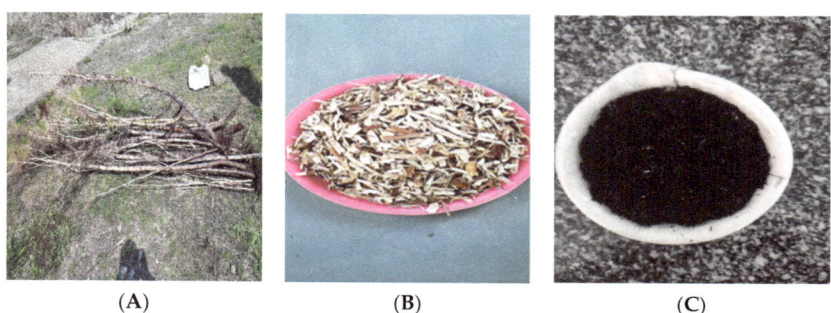

Figure 2. RAB preparation stages: (**A**) stem of Rumex abyssinicus, (**B**) partially size-reduced sample and (**C**) ready-made adsorbent.

2.2. Adsorbent Characterization

2.2.1. pH Point of Zero Charge

Techniques such as acid titration and salt addition are widely used to determine the pH point of zero charges (pHpzc), so the mass addition method for pHpzc determination was used in this study. Precisely, six 250 mL Erlenmeyer flasks were used to prepare 50 mL of 0.1 M NaCl, and the pH variation between 2 and 12 was adjusted using 0.1 M HCl or 0.1 M NaOH. After adjusting the pH, 1 g of the sample was added to each of the flasks and the sample was shaken at 125 rpm for 48 h. In the end, the pHpzc for the sample was determined by drawing curves relating to the initial pH and ΔpH (pH initial–pH final) [61].

2.2.2. The Brunauer–Emmett–Teller Surface Area Analysis

The Brunauer–Emmett–Teller (BET) method was employed in order to ascertain the specific surface area of the adsorbent material. Within this methodology, a BET sampler was utilized to prepare 0.4 g of the adsorbent, which was subsequently degassed under vacuum conditions at a temperature of 200 °C for a duration of 1 h. The adsorption–desorption isotherm of nitrogen at an atmospheric pressure of 700 mm was employed to determine the specific surface area of the adsorbent. The adsorption of nitrogen on the adsorbent's surface was intensified through the utilization of liquid nitrogen at a temperature of −196.5 °C. Ultimately, the BET-specific surface area of the biochar was assessed on the basis of the partial pressure to saturated vapor pressure ratio [62].

2.2.3. Surface Morphology Analysis

Scanning Electron Microscopy (SEM) analysis was performed in order to examine the various surface morphologies and porosity formation that characterized the adsorbent. The operating procedures outlined in the SEM machine manual were utilized to prepare the samples and conduct the scanning. In this study, RAB powder was attached to aluminum specimen stubs that had double-sided SEM functionality. These stubs were sputter-coated with a thin layer of gold. Imaging was carried out using a resolution of 10 µm and a guaranteed capacity of 10 kV, using the JCM-6000PLUS BENCHTOP SEM (JOEL), a model from Japan [63,64].

2.2.4. Functional Group Analysis

Functional group analysis was carried out using Fourier-transform infrared (FTIR, (FT-IR spectroscopy, Perkin Elmer, USA)). Recording of the FTIR spectrum was performed in the region of 4000 to 400 cm^{-1}. In this process, the adsorbent was mixed thoroughly with dry KBr at the ratio of 2:200. Formerly, the mixture of the adsorbent and dry KBr was crushed in a mortar to yield a homogenous powder combination. Then, the powder was injected into a molder to produce a very fine plate. Finally, FTIR data graphs were presented using Origin Pro version 22.

2.3. Batch Adsorption Experiments

The stock solution of the target pollutant (MR) was prepared by dissolving 1 g of MR in 1 L of distilled water, giving a dye concentration of 1000 mg/L. Then, serial dilution was thoroughly performed to obtain the intended concentration of MR. The optimization of MR removal was carried out considering the four operating parameters affecting the batch adsorption efficiency. Accordingly, the pH, adsorbent dosage, contact time and initial MR dye concentration were optimized using the one variable at a time (OVAT) approach. Normally, in OVAT, changing one variable is accompanied by fixing other variables at their specified values. Therefore, the effects of the operating parameters were investigated by varying the pH (3–9), contact time (20–60 min), adsorbent dosage (0.1–0.3 g/100 mL) and MR concentration of 50–100 mg/L. The temperature was maintained at 25 °C for all batch adsorption experiments, whereas a shaking speed of 250 rpm was used. The residual MR left unadsorbed was quantified using UV–Vis spectroscopy (JASCO V-770, JASCO, and Oceania) at a maximum wavelength of 520 nm. Precisely, 100 mL of the target pollutant solution was transferred into an Erlenmeyer flask, whose pH was adjusted using 0.1 M NaOH and HCl, depending on the required pH value. A known amount of the adsorbent material was added to the pH-adjusted solution. The adsorption process was then initiated by placing the target pollutant and adsorbent material mixture in an orbital shaker (70–400 LCD Digital Orbital Shaker) at 250 rpm. As the contact time expired, the solution was filtered out using Whatman filter paper and the filtrates were collected using various sampling bottles. Then, the dye-containing solution was transferred into a cuvette, and the absorbance reading was carried out by inserting the residual dye-containing cuvette into the UV–Vis spectroscope (JASCO V-770, JASCO, Oceania). The absorbance read was then used to calculate the final MR concentration. Finally, the adsorption capacity (Q_E) in mg/g and the removal efficiency (R_E) in % were calculated using Equations (1) and (2), respectively.

$$Q_E = (C_O - C_f) \times \frac{V}{m} \quad (1)$$

$$R_E = \frac{C_O - C_f}{C_O} \times 100\% \quad (2)$$

where C_O and C_f, both in mg/L, denote the initial and final dye concentration, respectively, and m and V are the mass of Rumex abyssinicus-derived biochar (g) and the volume of the solution (L), respectively [52].

2.4. Adsorption Isotherm

Langmuir and Freundlich are the most popular adsorption isotherm models used to investigate the nature of the adsorption at equilibrium [65,66]. Fundamentally, these two isotherm models differ because the Langmuir isotherm is assumed to indicate monolayer and homogenous surface interaction. In contrast, the heterogeneous and multilayered interaction was observed when the Freundlich isotherm model was found to fit data better [67,68]. The equations for the Langmuir and Freundlich isotherm models are presented in Table 1.

Table 1. Langmuir and Freundlich isotherm models.

Model	Equation	Parameters	Equation Number
Langmuir	$Q_e = \frac{Q_{max} K_L C_e}{1 + K_L C_e}$	Q_{max} is a maximum adsorption capacity (mg/g) C_e is equilibrium MR concentration mg/L K_L is Langmuir isotherm constant (L/mg)	(3)
Freundlich	$Q_e = K_F C_e^{1/n}$	K_F—Freundlich constant $(mg/g)(mg/L)^n$ $1/n$—Freundlich exponent that describes the nonlinearity of the sorption isotherm	(4)

Varied initial dye concentrations (50, 60, 70, 80, 90 and 100 mg/L) at the constant operating conditions of pH 6, a contact time of 40 min and an adsorbent dosage of 0.2 g/100 mL were used while investigating the adsorption isotherm.

2.5. Adsorption Kinetics

The rate of the target pollutant uptake and the adsorption mechanism were investigated using various kinetics models [69,70]. The pseudo-first-order (PFO), pseudo-second-order (PSO) and intraparticle diffusion (IPD) models are the most widely used kinetic models. During the kinetic study of MR adsorption onto RAB, the contact time was varied from 20 to 60 min while maintaining the other independent variables at their respective values of pH 6, an MR concentration of 70 mg/L and an adsorbent dosage of 0.2 g/100 mL. Finally, the equations for the PFO, PSO, and IPD models are presented in Table 2 [61].

Table 2. Adsorption kinetics models.

Model	Equation	Parameters	Equation Number
PFO	$Q_t = Q_e(1 - e^{-K_1 t})$	K_1—PFO rate constant in (g/(mg min)), t(min)—contact time Q_t (mg/g)—adsorption capacity at time t	(5)
PSO	$Q_t = \frac{Q_e^2 K_2 t}{1 + Q_e K_2 t}$	K_2—PSO rate constant in (g/(mg min))	(6)
IPD	$Q_e = K_d \times t^{0.5} + C$	C (mg/g)—intercept K_d(mg/(g min$^{0.5}$))—IPD rate constant	(7)

3. Results and Discussions

3.1. Adsorbent Characteristics

3.1.1. pH Point of Zero Charge

The pH of zero charges (pHpzc) is the pH value at which the surface density of cations, or positive charges, equals the surface density of anions or negative charges. Its value is paramount since many adsorption processes depend highly on pH. As a result, the pHpzc of the RAB was determined to be 6.2, as shown in Figure 3. This indicated that the surface density of the adsorbent material is negatively and positively charged, above and below the pH value of 6.2, respectively. Hence, anionic dyes like MR are significantly adsorbed when the pH of the solution is less than 6.2. This pHpzc value is comparable with the values reported by the same researchers for Rumex abyssinicus-based adsorbents, such as pHpzc values of 7.9 [71], 5.03 [72], 6.9 [73], 5.1 [74] and 7.2 [75].

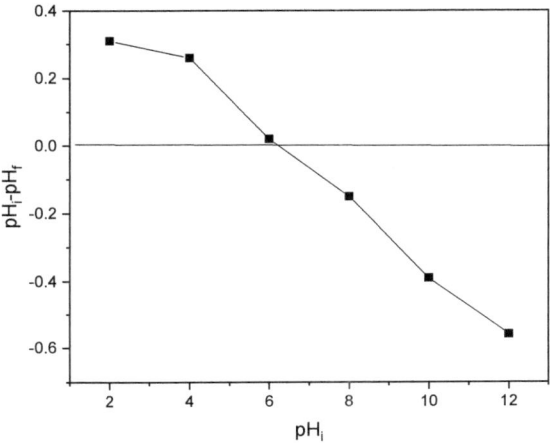

Figure 3. pH point of zero charge for RAB0.

3.1.2. SEM Analysis

Figure 4 displays the RAB SEM image magnified 2000 times at a resolution of 10 μm. The RAB appears in dark and bright colors, as shown in Figure 1. Throughout the carbonization process, Rumex abyssinicus undergoes breakdown, polycondensation, and cyclization, which turn it into a material with a high carbon content. The SEM micrograph shows that surface porosity and roughness developed on the char's smooth surface. This was explained by the carbon structure breaking down during pyrolysis at 500 °C. Fundamentally, biochar with holes, cavities, morphological cracks, pores and rough surfaces is suitable for surface adsorption. As a result, the adsorption capacity and removal efficiency of the target pollutant tend to increase [76–78].

Figure 4. SEM micrograph of Rumex abyssinicus-derived biochar.

3.1.3. FTIR Analysis

The functional group composition of the adsorbent material was investigated using FTIR along a wavelength range of 4000–400 cm^{-1}, as shown in Figure 5. The FTIR spectrum of Rumex abyssinicus-based biochar reveals several prominent peaks corresponding to specific functional groups. The peak at 3367 cm^{-1} indicates the presence of hydroxyl (OH) groups, possibly associated with surface hydroxyls or hydroxylated aromatic structures. Aliphatic C–H stretching vibrations are observed at 2922 cm^{-1}, suggesting the presence of aliphatic hydrocarbons. Carboxylic acid (COOH) functional groups are evident at 1190 cm^{-1}, while the peak at 1417 cm^{-1} signifies aromatic C=C stretching vibrations, indicative of aromatic structures within the biochar. Additionally, the presence of C–O stretching vibrations at 1560 cm^{-1} suggests the presence of ether or ester functional groups, while the peak at 580 cm^{-1} corresponds to alkene (C=C) stretching vibrations, indicating the presence of unsaturated hydrocarbons. Together, these functional groups contribute to the diverse chemical properties of the biochar, potentially influencing its applications in various fields [79].

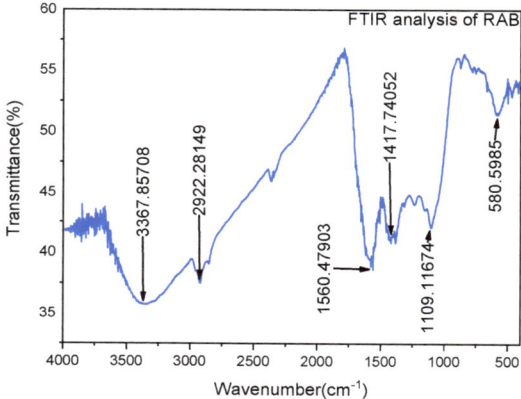

Figure 5. FTIR peaks for Rumex abyssinicus-derived biochar.

3.1.4. BET Analysis

The BET surface area of the adsorbent material was determined to be 45.8 m^2/g. This surface area is lower than that reported for the same precursor material-based activated carbons. Precisely, BET-specific surface areas of 2522 m^2/g [72], 524 m^2/g [73], 962.3 m^2/g [74] and 3619.7 m^2/g [75] were reported in previous research. The significant decrease in the surface area of the biochar compared to the activated carbons of the same material is attributed to the absence of chemical activation. Normally, activated carbons have higher surface areas than biochar because activated carbons are produced at elevated temperatures and necessitate chemical treatments. However, a higher specific surface area alone does not determine the quality of the adsorbent material since the surface morphologies and composition of functional groups are vital for adsorbent development. The specific surface area of this adsorbent material is still higher than that of many unmodified biochars prepared from biomass residue and lignocellulosic materials, with surface areas like 1.71 m^2/g [80], 44.38 m^2/g [81], 17.65 m^2/g, 20.9 m^2/g [82], 9 m^2/g [83] and 17.65 m^2/g [84].

3.2. Effect of Operating Parameters
3.2.1. Effect of Contact Time

The contact time impacts the removal efficiency and capacity of pollutants to be adsorbed onto the adsorbent materials. This study evaluates the impact of contact time on the adsorption capacity and removal percentage when removing methyl red from an aqueous solution using RAB. Different contact times were used, with a constant pH of 6, adsorbent dose of 0.2 g/100 mL, and initial methyl red concentration of 70 mg/L. The removal process occurs in three stages: initial, intermediate, and equilibrium. The initial stage involves rapid uptake due to a higher concentration gradient and increased adsorption capacity. As the contact time increases, the concentration gradient driving force and available active sites decrease, decreasing the adsorption rate and removal percentage [85]. As shown in Figure 6, the equilibrium stage occurs at 40 min, with a maximum removal percentage of 99%.

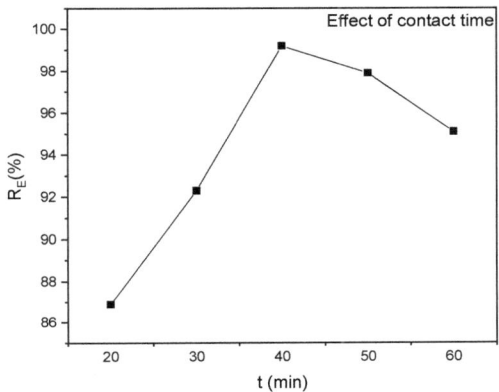

Figure 6. Effect of contact time on MR removal efficiency at pH 6, an adsorbent dosage of 0.2 g/100 mL and an initial dye concentration of 70 mg/L.

3.2.2. Impact of Initial MR Concentration

This study investigated the influence of the initial solution concentration on the uptake of MR on RAB. The results showed a reduction in the MR percentage removal from 91.1 to 82.62% when the concentration of MR was increased from 50 to 100 mg/L, as depicted in Figure 7. This is due to the high surface area and more unoccupied adsorption active sites at the lowest solution concentration. However, there are limited active sites at higher initial concentrations, resulting in a reduction in MR ion removal [86]. The study also found that increasing the solution concentration increases the number of solute molecules dispersed in the solution, allowing for toxic MR uptake onto the RAB surface.

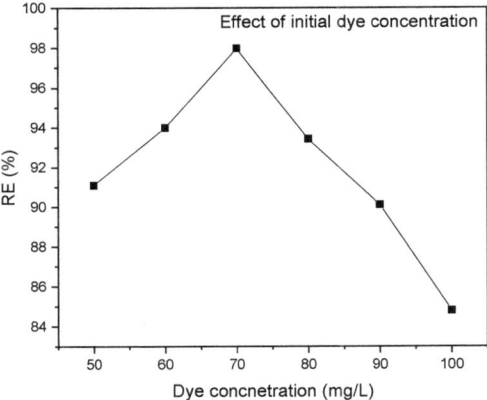

Figure 7. Effect of initial MR concentration on the dye removal efficiency at pH 6, a contact time of 40 min and an adsorbent dosage of 0.2 g/100 mL.

3.2.3. Effect of Adsorbent Dosage

The adsorbent dose is crucial in adsorption studies as it affects the adsorbent–adsorbate equilibrium. This study investigated the effect of the RAB dose on MR adsorption at pH 6, a concentration of 70 mg/L, and a contact time of 40 min, as shown in Figure 8. The results showed that increasing the RAB dose from 0.1 to 0.3 g/100 mL increased the MR adsorption capacity from 28.3 to 33.79 mg/g. Under the same condition, the MR molecule percentage removal increased with increasing the adsorbent dose from 82.4 to 96.46%. A maximum MR adsorption removal of 98.56% and an adsorption capacity Q_e of 34.5 mg/g were achieved at an adsorbent dosage of 0.2 g/100 mL. The higher uptake at a low pH is attributed to the strong attraction between the protonated surface of the RAB and the species of MR in the solution. The optimum adsorbent dosage was 0.2 g; when the biosorbent dosage is increased beyond this point, the adsorption capability decreases because fewer MR ions occupy the active sites [87]. Additionally, the slight decrease in the removal efficiency observed after a 0.20 mg/100 mL adsorbent dosage is likely due to the agglomeration of adsorbent particles at higher dosages, which can block active sites and hinder mass transfer rates. The dense packing of adsorbents on the surface of the adsorption medium also leads to longer diffusion distances and reduced mass transfer rates, resulting in a decrease in the removal efficiency [88].

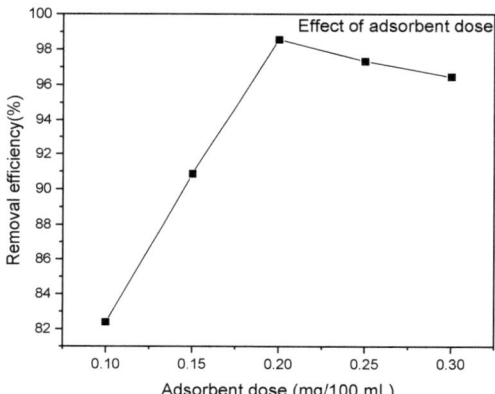

Figure 8. Effect of adsorbent dosage on MR removal efficiency at pH 6, a dye concentration of 70 mg/L, a contact time of 40 min and a temperature of 25 °C.

3.2.4. Effect of pH

This study investigated the effect of the initial solution pH on the adsorption of MR ions onto RAB, as shown in Figure 9. The results show that the adsorption capacity and percentage removal increase with increasing the pH from 3 to 6. However, a further increase in pH from 6 to 9 decreased the removal efficiency and adsorption capacity. This is probably due to the interaction effect among the operating parameters. A maximum MR adsorption removal of 99.17% and adsorption capacity Q_e of 40.9 mg/g were achieved at pH 6. The MR uptake became high at a lower solution pH and decreased as the solution pH increased. This is due to the strong attraction between the RAB adsorbent material's protonated surface and the MR species in the solution [87]. This study selected pH six as the optimum MR uptake due to its highest removal efficiency. The removal efficiency of MR onto RAB decreases at near-neutral and alkaline pH conditions due to the enhanced deprotonation of acid functional groups (AFGs) on the biochar surface. This results in a greater negative surface charge density (SCD), which repulses the methyl red molecules and reduces their sorption onto the biochar. The decrease in sorption is attributed to the repulsive forces between the negatively charged biochar surface and the negatively charged methyl red molecules at higher pH values. Furthermore, the pHpzc of RAB was determined to be

6.2, indicating that the surface charge of the adsorbent is dominated by negative charges above a pH of 6.2. Hence, repulsive forces between negatively charged MR molecules and negatively charged RAB above a pH of 6.2 resulted in a decrease in the removal efficiency at near-neutral and alkaline pH conditions.

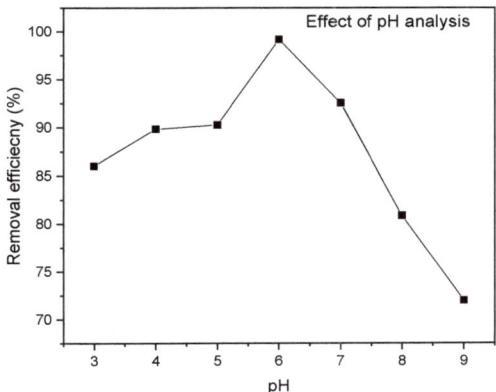

Figure 9. Effect of pH on the MR removal efficiency at a contact time of 40 min, initial dye concentration of 70 mg/L, adsorbent dosage of 0.2 g/100 mL and temperature of 25 °C.

3.3. Adsorption Isotherm

The adsorption isotherm was investigated using Langmuir and Freundlich models, as shown in Figures 10 and 11, respectively. The investigation of methyl red adsorption onto Rumex abyssinicus-derived biochar using Langmuir and Freundlich isotherm models reveals important insights. The Langmuir model suggests a maximum adsorption capacity (Qmax) of 42.34 mg/g, favorable adsorption (RL < 1) and a good fit to the experimental data (R^2 = 0.96), as shown in Table 3. Meanwhile, the Freundlich model also fits well (R^2 = 0.99) and suggests a higher adsorption capacity (K_F) = 19.19 (mg/g)(mg/L)^n and a more heterogeneous adsorption surface (1/n = 0.266). However, the lower Reduced Chi-Square value for the Freundlich model implies a better fit to the data compared to the Langmuir model. Overall, both models provide valuable insights into the adsorption process, with the Freundlich model potentially better capturing the complexities of the system due to its lower Reduced Chi-Square value [89,90].

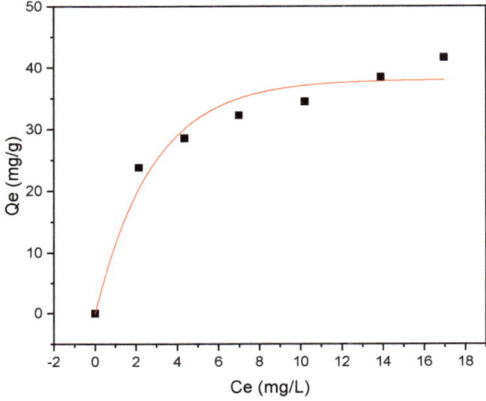

Figure 10. Langmuir isotherm model for MR adsorption onto RAB at pH 6, a contact time of 60 min, an adsorbent dosage of 0.4 g/100 mL and a temperature of 25 °C.

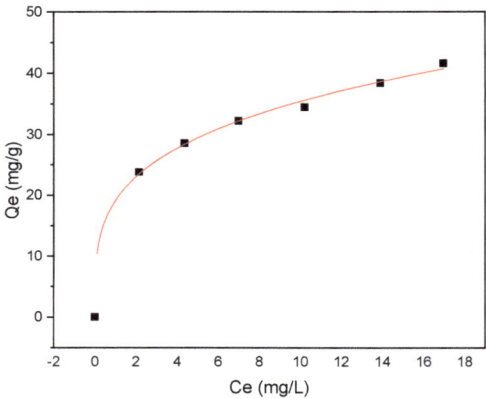

Figure 11. Freundlich isotherm model for MR adsorption onto RAB at pH 6, a temperature of 25 °C, an adsorbent dosage of 0.2 g/100 mL and a contact time of 60 min.

Table 3. Langmuir and Freundlich adsorption isotherm parameters.

Langmuir Isotherm	Freundlich Isotherm
Q_{max} = 42.34 mg/g	K_F = 19.19 (mg/g)(mg/L)^n
R^2 = 0.96	R^2 = 0.99
K_L = 0.359 L/mg	$1/n$ = 0.266
Reduced Chi-Sqr = 1.5	Reduced Chi-Sqr = 0.565

3.4. Adsorption Kinetics

The investigation of methyl red adsorption kinetics onto Rumex abyssinicus-derived biochar reveals significant insights across multiple kinetic models. The kinetic parameters associated with PFO, PSO, and IPD were determined from nonlinear plots of their respective models, as shown in Figures 12–14, respectively. The pseudo-first-order model yielded a rate constant (K_1) of 0.018 min^{-1} and an equilibrium adsorption capacity (Q_e) of 39.74 mg/g, exhibiting a good fit to the experimental data with an R^2 of 0.98 and a Reduced Chi-Square value of 1.01. Meanwhile, the pseudo-second-order model demonstrated a higher rate constant (K_2) of 1.34 g/mg min and a slightly higher equilibrium adsorption capacity (Q_e = 41.86 mg/g), with a superior goodness of fit (R^2 = 0.99) and a lower Reduced Chi-Square value of 0.57, as shown in Table 4. Additionally, the intraparticle diffusion model revealed a diffusion rate constant (K_p) of 4.14 mg/g min$^{0.5}$ and a constant intercept (C) of 5.84 mg/g, indicating that while intraparticle diffusion contributes to the overall adsorption process, it might not be the sole controlling mechanism. Overall, the pseudo-second-order model seems to best describe the experimental data, suggesting that the adsorption of methyl red onto Rumex abyssinicus-derived biochar involves chemisorption and further indicating the complexity of the adsorption kinetics, which are potentially influenced by multiple mechanisms [72,91].

Table 4. Values of parameters for PFO, PSO and IPD adsorption kinetics models.

PFO	PSO	IPD
K_1 = 0.018 min^{-1}	K_2 = 1.34 g/mg/min	K_P = 4.14 mg/g min$^{0.5}$
Q_e = 39.74 mg/g	Q_e = 41.86 mg/g	C = 5.84 mg/g
R^2 = 0.98	R^2 = 0.99	R^2 = 0.96
Reduced Chi-Sqr = 1.01	Reduced Chi-Sqr = 0.57	Reduced Chi-Sqr = 1.91

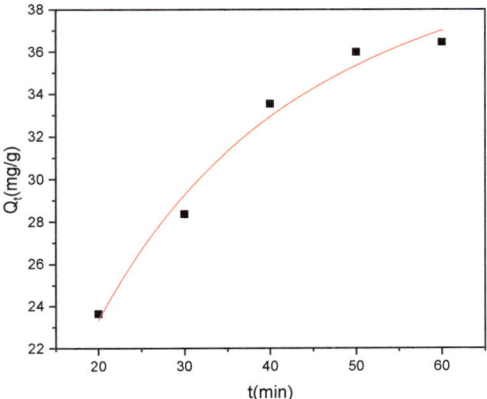

Figure 12. PFO kinetic model for the adsorption of MR onto RAB at pH 6, an adsorbent dosage of 0.4 g/100 mL, a temperature of 25 °C and an initial MR concentration of 70 mg/L.

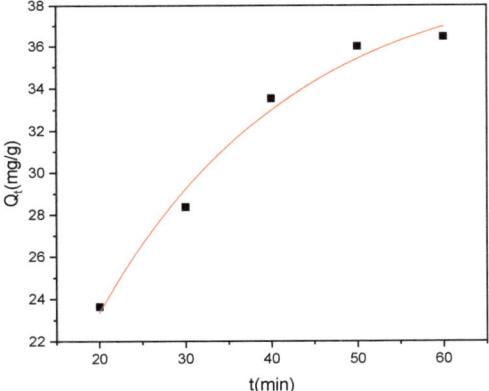

Figure 13. PSO kinetics plot at an adsorbent dosage of 0.2 g/100 mL, dye concentration of 70 mg/L, pH of 6 and temperature of 25 °C.

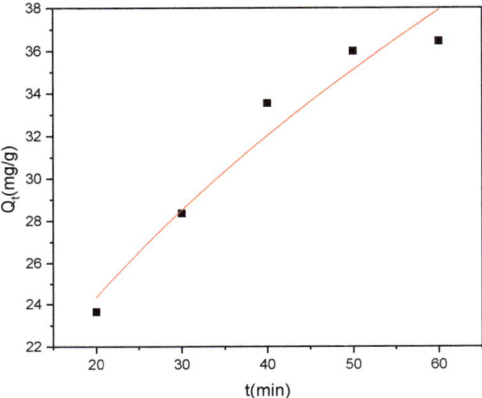

Figure 14. IPD model for adsorption of MR from aqueous solution at pH 6, a dye concentration of 70 mg/L, an adsorbent dosage of 0.2 g/100 mL and a temperature of 25 °C.

3.5. Comparative Analysis

The maximum adsorption capacity (42.34 mg/g) for MR adsorption from aqueous solution obtained using RAB is compared with other biosorbents derived from lignocellulosic, agricultural leftovers and woody materials. As a result, the findings of the comparison with recent works in the literature indicated that RAB is superior in removing MR from aqueous solution. This is probably due to the nature of the precursor material and the affinity of the target pollutant towards forming an interaction with the adsorbent materials. Hence, the prepared biochar can be taken as an alternative adsorbent for the removal of reactive and persistent dyes like MR. Table 5 presents the findings of a literature search where biosorbents prepared from different materials used to remove MR are compared with this research.

Table 5. Comparative analysis of methyl red adsorption using biobased adsorbents.

S.N	Precursor Materials	Maximum Adsorption Capacity (mg/g)	References
1.	Funnel seeds	26	[92]
2.	Caraway seeds	28	[93]
3.	Biogas plant waste	31	[47]
4.	Moringa olifera	28.67	[94]
5.	Rumex abyssinicus	42.34	Current work

3.6. Proposed Adsorption Mechanism

The conceptual illustration of the interaction between methyl red (MR) and the proposed sorbent, based on the characterization and adsorption performance results, reveals several key points, as shown in Figure 15. Firstly, the porous surface morphology of the biochar provides ample surface area, enhancing the contact between dye molecules and the sorbent, supported by a high BET surface area (45.8 m^2/g). Secondly, various functional groups on the biochar surface facilitate interaction between the adsorbent and the adsorbate. Hydroxyl groups (–OH) can participate in hydrogen bonding with the functional groups of methyl red molecules. Aliphatic hydrocarbons can contribute to hydrophobic interactions with the hydrophobic regions of methyl red molecules. Carboxylic acid groups (–COOH) can undergo ion–dipole interactions with the charged regions of methyl red molecules. Aromatic carbon–carbon double bonds can engage in π-π interactions with the aromatic rings of methyl red molecules. Additionally, the C–O groups may participate in hydrogen bonding or other polar interactions with functional groups of methyl red. On the other hand, alkene groups can contribute to hydrophobic interactions with the hydrophobic regions of methyl red molecules. These functional groups collectively play a pivotal role in the adsorption process of methyl red onto Rumex abyssinicus-based biochar, highlighting the diverse chemical interactions involved in this adsorption mechanism. Finally, at pH 6, the maximum removal efficiency of 99.2% suggests the significance of surface charge interactions in the adsorption process [95]. Overall, the interaction between methyl red and the proposed sorbent involves a combination of physical and chemical mechanisms, facilitated by the surface morphology and functional groups of the biochar.

3.7. Scale-Up and Cost Implications, and Environmental Impact Assessment

Scaling up the production of biochar derived from Rumex abyssinicus stems for the removal of methyl red from aqueous solutions needs a comprehensive analysis with significant implications. For adsorbent preparation from Rumex abyssinicus biochar, only 4.64 Ethiopian birr (ETB) was utilized per 1 kg of the product. The total cost includes the estimated costs for size reduction (2.29 ETB/kg) and thermal activation (2.65 ETB/kg). Hence, leveraging the underutilized stems of Rumex abyssinicus as a raw material could potentially provide a cost-effective and sustainable source for the adsorbent. However, the scale-up process necessitates the careful consideration of factors such as an increased production capacity, the efficient collection and processing of raw materials, and market demand. The estimated costs for size reduction and thermal activation are critical components, with opportunities for cost optimization through process efficiency improvements. Moreover, the environmental impacts, including resource utilization, energy consumption, and waste generation, require a thorough assessment to ensure sustainability. Overall, a comprehensive approach that integrates considerations of cost, scalability, and environmental sustainability is essential for realizing the full potential of biochar made from Rumex abyssinicus as an effective adsorbent for methyl red removal on a larger scale.

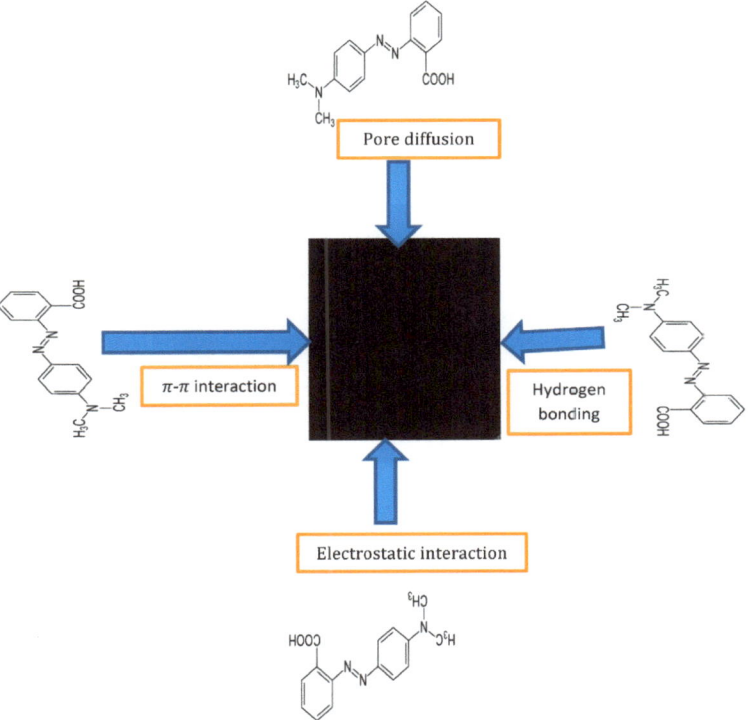

Figure 15. Methyl red adsorption mechanism onto Rumex abyssinicus biochar.

4. Conclusions

This study investigated the removal of methyl red from an aqueous solution using biochar prepared from Rumex abyssinicus stem. Accordingly, the prepared adsorbent was found to reveal characteristics of a 45.8 m²/g BET surface area composed of various functional groups. Additionally, morphological studies showed irregular and heterogeneous surfaces. The effects of various operating parameters such as the contact time, initial methyl red concentration, biochar dosage and pH were investigated by varying the respective parameters while holding the other parameters at constant values. As a result, a maximum methyl red removal efficiency of 99.2% was attained at optimum treatment conditions of pH 6, a contact time of 40 min, an initial dye concentration of 70 mg/L and an adsorbent dosage of 0.2 mg/100 mL. The adsorption isotherm was investigated using Langmuir and Freundlich isotherm models, in which the Freundlich isotherm with a maximum R^2 of 0.99 was determined to give a suitable description of the adsorption process. Additionally, a pseudo-second-order kinetics model was found to express adsorption with an $R^2 = 0.99$. In conclusion, it could be shown that Rumex abyssinicus-based biochar has the potential to effectively remove persistent dyes like methyl red from wastewater.

Author Contributions: Conceptualization, M.D.T. methodology, M.D.T.; software, M.D.T.; validation, M.D.T. and M.O.D.; formal analysis, M.D.T.; investigation, M.D.T.; resources, M.D.T.; data curation, M.D.T.; writing—original draft preparation, M.D.T.; writing—review and editing, M.D.T.; supervision, M.O.D.; project administration, M.D.T.; funding acquisition, M.D.T. All authors have read and agreed to the published version of the manuscript.

Funding: This research received no external funding.

Data Availability Statement: All data generated is included in the main body of the manuscript.

Acknowledgments: We want to express our sincere gratitude to the University of Johannesburg for its dedication to creating a lively and supportive research environment, as their support has given us an invaluable opportunity to delve deeply into the research. We have succeeded in my research endeavors because of its state-of-the-art facilities, access to cutting-edge resources, and collaborative atmosphere. In addition, We would like to express our gratitude to Adama and Addis Ababa Science and Technology Universities for their support in our laboratory facilities.

Conflicts of Interest: The authors declare no conflicts of interest.

References

1. Roy, A.; Pramanick, K. Analysing progress of sustainable development goal 6 in India: Past, present, and future. *J. Environ. Manag.* **2019**, *232*, 1049–1065. [CrossRef]
2. Akhtar, N.; Syakir Ishak, M.I.; Bhawani, S.A.; Umar, K. Various natural and anthropogenic factors responsible for water quality degradation: A review. *Water* **2021**, *13*, 2660. [CrossRef]
3. Environment and Natural Resources Department. *Wastewater as Resources: May 2022*; European Investment Bank: Athens, Greece, 2022; ISBN 9789286153358.
4. Tariq, A.; Mushtaq, A. Untreated Wastewater Reasons and Causes: A Review of Most Affected Areas and Cities. *Int. J. Chem. Biochem. Sci.* **2023**, *23*, 121–143.
5. Bisimwa, A.M.; Amisi, F.M.; Bamawa, C.M.; Muhaya, B.B.; Kankonda, A.B. Water quality assessment and pollution source analysis in Bukavu urban rivers of the Lake Kivu basin (Eastern Democratic Republic of Congo). *Environ. Sustain. Indic.* **2022**, *14*, 100183. [CrossRef]
6. Kishor, R.; Purchase, D.; Saratale, G.D.; Saratale, R.G.; Ferreira, L.F.R.; Bilal, M.; Chandra, R.; Bharagava, R.N. Ecotoxicological and health concerns of persistent coloring pollutants of textile industry wastewater and treatment approaches for environmental safety. *J. Environ. Chem. Eng.* **2021**, *9*, 105012. [CrossRef]
7. Okafor, C.C.; Madu, C.N.; Ajaero, C.C.; Ibekwe, J.C. Sustainable management of textile and clothing. *Clean Technol. Recycl* **2021**, *1*, 70–87. [CrossRef]
8. Christian, D.; Gaekwad, A.; Dani, H.; Shabiimam, M.A.; Kandya, A. Recent techniques of textile industrial wastewater treatment: A review. *Mater. Today Proc.* **2023**, *77*, 277–285. [CrossRef]
9. Bidu, J.M.; Van der Bruggen, B.; Rwiza, M.J.; Njau, K.N. Current status of textile wastewater management practices and effluent characteristics in Tanzania. *Water Sci. Technol.* **2021**, *83*, 2363–2376. [CrossRef]
10. Khan, W.U.; Ahmed, S.; Dhoble, Y.; Madhav, S. A critical review of hazardous waste generation from textile industries and associated ecological impacts. *J. Indian Chem. Soc.* **2023**, *100*, 100829. [CrossRef]
11. Islam, T.; Repon, M.R.; Islam, T.; Sarwar, Z.; Rahman, M.M. Impact of textile dyes on health and ecosystem: A review of structure, causes, and potential solutions. *Environ. Sci. Pollut. Res.* **2023**, *30*, 9207–9242. [CrossRef]
12. de Jesus, S.S.; Ferreira, G.F.; Moreira, L.S.; Maciel Filho, R. Biodiesel production from microalgae by direct transesterification using green solvents. *Renew. Energy* **2020**, *160*, 1283–1294. [CrossRef]
13. Hadi, S.M.; Al-Mashhadani, M.K.H.; Eisa, M.Y. Optimization of dye adsorption process for Albizia lebbeck pods as a biomass using central composite rotatable design model. *Chem. Ind. Chem. Eng. Q.* **2019**, *25*, 39–46. [CrossRef]
14. Sharma, K.; Pandit, S.; Mathuriya, A.S.; Gupta, P.K.; Pant, K.; Jadhav, D.A. Microbial Electrochemical Treatment of Methyl Red Dye Degradation Using Co-Culture Method. *Water* **2022**, *15*, 56. [CrossRef]
15. Ikram, M.; Naeem, M.; Zahoor, M.; Rahim, A.; Hanafiah, M.M.; Oyekanmi, A.A.; Shah, A.B.; Mahnashi, M.H.; Al Ali, A.; Jalal, N.A. Biodegradation of Azo Dye Methyl Red by Pseudomonas aeruginosa: Optimization of Process Conditions. *Int. J. Environ. Res. Public Health* **2022**, *19*, 9962. [CrossRef]
16. Fan, J.; Wu, W.; Liu, Y.; Ji, B.; Xu, H.; Zhong, Y.; Zhang, L.; Mao, Z. Customizable High-Contrast Optical Responses: Dual Photosensitive Colors for Smart Textiles. *ACS Appl. Mater. Interfaces* **2023**, *15*, 54085–54097. [CrossRef] [PubMed]
17. Mahajan, P.; Kaushal, J. Phytoremediation of azo dye methyl red by macroalgae Chara vulgaris L.: Kinetic and equilibrium studies. *Environ. Sci. Pollut. Res.* **2020**, *27*, 26406–26418.
18. Adusei, J.K.; Agorku, E.S.; Voegborlo, R.B.; Ampong, F.K.; Danu, B.Y.; Amarh, F.A. Removal of Methyl red in aqueous systems using synthesized NaAlg-g-CHIT/nZVI adsorbent. *Sci. Afr.* **2022**, *17*, e01273.
19. Waghchaure, R.H.; Adole, V.A.; Jagdale, B.S. Photocatalytic degradation of methylene blue, rhodamine B, methyl orange and Eriochrome black T dyes by modified ZnO nanocatalysts: A concise review. *Inorg. Chem. Commun.* **2022**, *143*, 109764.

20. Hanafi, M.F.; Sapawe, N. A review on the water problem associate with organic pollutants derived from phenol, methyl orange, and remazol brilliant blue dyes. *Mater. Today Proc.* **2020**, *31*, A141–A150. [CrossRef]
21. Saxena, A.; Gupta, S. Toxicological Impact of Azo Dyes Azo dyes and Their Microbial Degraded Byproducts on Flora and Fauna. In *Innovations in Environmental Biotechnology*; Springer: Berlin/Heidelberg, Germany, 2022; pp. 319–343.
22. Olawale, M.D.; Akintemi, E.O.; Agbaffa, B.E.; Obaleye, J.A. Synthesis, characterization, adsorption study, quantum mechanics, monte carlo and molecular dynamics of lead based polymeric compound towards mopping of aqueous methyl red dye. *Results Chem.* **2022**, *4*, 100499. [CrossRef]
23. Takkar, S.; Tyagi, B.; Kumar, N.; Kumari, T.; Iqbal, K.; Varma, A.; Thakur, I.S.; Mishra, A. Biodegradation of methyl red dye by a novel actinobacterium Zhihengliuella sp. ISTPL4: Kinetic studies, isotherm and biodegradation pathway. *Environ. Technol. Innov.* **2022**, *26*, 102348. [CrossRef]
24. Das, D.K.; Goswami, P.; Barman, C.; Das, B. Methyl red: A fluorescent sensor for Hg^{2+} over Na^{+}, K^{+}, Ca^{2+}, Mg^{2+}, Zn^{2+}, and Cd^{2+}. *Environ. Eng. Res.* **2012**, *17*, 75–78.
25. Manjunatha, A.S.; Sukhdev, A. Puttaswamy spectrophotometric oxidative decolorization of methyl red with chloramine-T and bromamine-T: Comparative kinetic modeling and mechanistic study. *Russ. J. Phys. Chem. A* **2018**, *92*, 2647–2655. [CrossRef]
26. Khouri, S.J.; Abdel-Rahim, I.A.; Alshamaileh, E.M.; Altwaiq, A.M. Equilibrium and structural study of m-methyl red in aqueous solutions: Distribution diagram construction. *J. Solut. Chem.* **2013**, *42*, 1844–1853. [CrossRef]
27. Fosso-Kankeu, E.; Webster, A.; Ntwampe, I.O.; Waanders, F.B. Coagulation/flocculation potential of polyaluminium chloride and bentonite clay tested in the removal of methyl red and crystal violet. *Arab. J. Sci. Eng.* **2017**, *42*, 1389–1397. [CrossRef]
28. Bhattacharya, A.; Goyal, N.; Gupta, A. Degradation of azo dye methyl red by alkaliphilic, halotolerant Nesterenkonia lacusekhoensis EMLA3: Application in alkaline and salt-rich dyeing effluent treatment. *Extremophiles* **2017**, *21*, 479–490. [CrossRef]
29. Kokkiligadda, V.R.; Pokala, R.K.V.; Karumuri, A.; Bollikola, H.B. Adsorption Potentialities of Bio-Sorbents Derived from Pomegranate in the Removal of Methyl Red Dye from Polluted Waters. *Caribb. J. Sci. Technol.* **2020**, *8*, 105–118. [CrossRef]
30. Goswami, Y.C.; Kaundal, J.B.; Begzaad, S.; Tiwari, R.K. Photocatalytic degradation of Methyl Red dye using highly efficient ZnO/CdS hierarchical heterostructures under white LED. *J. Iran. Chem. Soc.* **2023**, *20*, 1681–1697 [CrossRef]
31. Hou, T.; Guo, K.; Wang, Z.; Zhang, X.-F.; Feng, Y.; He, M.; Yao, J. Glutaraldehyde and polyvinyl alcohol crosslinked cellulose membranes for efficient methyl orange and Congo red removal. *Cellulose* **2019**, *26*, 5065–5074. [CrossRef]
32. Ebratkhahan, M.; Naghash Hamed, S.; Zarei, M.; Jafarizad, A.; Rostamizadeh, M. Removal of neutral red dye via electro-fenton process: A response surface methodology modeling. *Electrocatalysis* **2021**, *12*, 579–594. [CrossRef]
33. Wen, H.; Huang, W.; Liu, C. Double-barrier forward osmosis membrane for rejection and destruction of bacteria and removal of dyes. *Desalination* **2022**, *529*, 115609. [CrossRef]
34. Kiran, S.; Nosheen, S.; Abrar, S.; Anjum, F.; Gulzar, T.; Naz, S. Advanced approaches for remediation of textile wastewater: A comparative study. In *Advanced Functional Textiles and Polymers: Fabrication, Processing and Applications*; Scrivener Publishing LLC: Beverly, MA, USA, 2019; pp. 201–264.
35. Nidheesh, P.V.; Ravindran, V.; Gopinath, A.; Kumar, M.S. Emerging technologies for mixed industrial wastewater treatment in developing countries: An overview. *Environ. Qual. Manag.* **2022**, *31*, 121–141. [CrossRef]
36. Velusamy, S.; Roy, A.; Sundaram, S.; Kumar Mallick, T. A review on heavy metal ions and containing dyes removal through graphene oxide-based adsorption strategies for textile wastewater treatment. *Chem. Rec.* **2021**, *21*, 1570–1610. [CrossRef]
37. Han, D.; Tu, Y.; Li, X.; Zeng, Z.; Xu, Q.; Zhu, J. Adsorption and reaction of an alkyne molecule on diverse oxygen-reconstructed Cu (110) surfaces. *Surf. Sci.* **2022**, *719*, 122039. [CrossRef]
38. Pellenz, L.; de Oliveira, C.R.S.; da Silva Júnior, A.H.; da Silva, L.J.S.; da Silva, L.; de Souza, A.A.U.; Ulson, S.M.; Borba, F.H.; da Silva, A. A comprehensive guide for characterization of adsorbent materials. *Sep. Purif. Technol.* **2023**, *305*, 122435. [CrossRef]
39. de Farias Silva, C.E.; da Gama, B.M.V.; da Silva Gonçalves, A.H.; Medeiros, J.A.; de Souza Abud, A.K. Basic-dye adsorption in albedo residue: Effect of pH, contact time, temperature, dye concentration, biomass dosage, rotation and ionic strength. *J. King Saud Univ. Sci.* **2020**, *32*, 351–359.
40. Akperov, E.O.; Akperov, O.H. Removal of the basic green 5 dye from aqueous solutions by grape (*Vitis vinifera* L.) bushes wastes. *New Mater. Compd. Appl.* **2019**, *3*, 171.
41. An, N.; Zagorščak, R.; Thomas, H.R. Adsorption characteristics of rocks and soils, and their potential for mitigating the environmental impact of underground coal gasification technology: A review. *J. Environ. Manag.* **2022**, *305*, 114390. [CrossRef]
42. Shaida, M.A.; Dutta, R.K.; Sen, A.K.; Ram, S.S.; Sudarshan, M.; Naushad, M.; Boczkaj, G.; Nawab, M.S. Chemical analysis of low carbon content coals and their applications as dye adsorbent. *Chemosphere* **2022**, *287*, 132286. [CrossRef] [PubMed]
43. Hashmi, Z.; Jatoi, A.S.; Nadeem, S.; Anjum, A.; Imam, S.M.; Jangda, H. Comparative analysis of conventional to biomass-derived adsorbent for wastewater treatment: A review. *Biomass Convers. Biorefinery* **2024**, *14*, 45–76. [CrossRef]
44. Azari, A.; Nabizadeh, R.; Mahvi, A.H.; Nasseri, S. Integrated Fuzzy AHP-TOPSIS for selecting the best color removal process using carbon-based adsorbent materials: Multi-criteria decision making vs. systematic review approaches and modeling of textile wastewater treatment in real conditions. *Int. J. Environ. Anal. Chem.* **2022**, *102*, 7329–7344. [CrossRef]
45. Bouzid, T.; Grich, A.; Naboulsi, A.; Regti, A.; Tahiri, A.A.; El Himri, M.; El Haddad, M. Adsorption of Methyl Red on porous activated carbon from agriculture waste: Characterization and response surface methodology optimization. *Inorg. Chem. Commun.* **2023**, *158*, 111544. [CrossRef]

46. Radoor, S.; Karayil, J.; Jayakumar, A.; Parameswaranpillai, J.; Siengchin, S. Efficient removal of methyl orange from aqueous solution using mesoporous ZSM-5 zeolite: Synthesis, kinetics and isotherm studies. *Colloids Surfaces A Physicochem. Eng. Asp.* **2021**, *611*, 125852. [CrossRef]
47. Wolski, R.; Bazan-Wozniak, A.; Pietrzak, R. Adsorption of methyl red and methylene blue on carbon bioadsorbents obtained from biogas plant waste materials. *Molecules* **2023**, *28*, 6712. [CrossRef]
48. Zhang, K.; Sun, J.; Lei, E.; Ma, C.; Luo, S.; Wu, Z.; Li, W.; Liu, S. Effects of the pore structure of commercial activated carbon on the electrochemical performance of supercapacitors. *J. Energy Storage* **2022**, *45*, 103457. [CrossRef]
49. Selamat, N.A.; Abd Halim Md Ali, M.R.; Yusof, M.; Che, N.W.; Jusoh, N.R.J.; Mail, K.M.P. Development of carbon dioxide adsorbents from renewable and non-renewable sources: A review. *Malays. J. Fundam. Appl. Sci.* **2020**, *16*, 544–556.
50. Adeleye, A.T.; Akande, A.A.; Odoh, C.K.; Philip, M.; Fidelis, T.T.; Amos, P.I.; Banjoko, O.O. Efficient synthesis of bio-based activated carbon (AC) for catalytic systems: A green and sustainable approach. *J. Ind. Eng. Chem.* **2021**, *96*, 59–75. [CrossRef]
51. Ramana, K.V.; Mohan, K.C.; Ravindhranath, K.; Babu, B.H. Bio-Sorbent Derived from Annona Squamosa for the Removal of Methyl Red Dye in Polluted Waters: A Study on Adsorption Potential. *Chem. Chem. Technol* **2020**, *14*, 274–283. [CrossRef]
52. Neolaka, Y.A.B.; Lawa, Y.; Naat, J.; Lalang, A.C.; Widyaningrum, B.A.; Ngasu, G.F.; Niga, K.A.; Darmokoesoemo, H.; Iqbal, M.; Kusuma, H.S. Adsorption of methyl red from aqueous solution using Bali cow bones (*Bos javanicus domesticus*) hydrochar powder. *Results Eng.* **2023**, *17*, 100824. [CrossRef]
53. Balu, P.; Asharani, I.V.; Thirumalai, D. Catalytic degradation of hazardous textile dyes by iron oxide nanoparticles prepared from Raphanus sativus leaves' extract: A greener approach. *J. Mater. Sci. Mater. Electron.* **2020**, *31*, 10669–10676. [CrossRef]
54. Eguale, T.; Tadesse, D.; Giday, M. In vitro anthelmintic activity of crude extracts of five medicinal plants against egg-hatching and larval development of Haemonchus contortus. *J. Ethnopharmacol.* **2011**, *137*, 108–113. [CrossRef] [PubMed]
55. Iriti, M.; Date, P. Bioactive compounds and health benefits of edible Rumex species—A review. *Cell. Mol. Biol.* **2018**, *64*, 27–34.
56. Mekonnen, T.; Urga, K.; Engidawork, E. Evaluation of the diuretic and analgesic activities of the rhizomes of Rumex abyssinicus Jacq in mice. *J. Ethnopharmacol.* **2010**, *127*, 433–439. [CrossRef] [PubMed]
57. Ahmed, S.; Panda, R.C.; Madhan, B. Extraction of bio-active compounds from Ethiopian plant material Rumex abyssinicus (mekmeko) root—A study on kinetics, optimization, antioxidant and antibacterial activity. *J. Taiwan Inst. Chem. Eng.* **2017**, *75*, 228–239. [CrossRef]
58. Wang, S.; Yan, Y.; Hsu, C.; Lin, H. Waste bamboo-derived biochar and multiporous carbon as adsorbents for methyl orange removal. *J. Chinese Chem. Soc.* **2023**, *70*, 1628–1635. [CrossRef]
59. Diaz-Uribe, C.; Ortiz, J.; Duran, F.; Vallejo, M.; Fals, J. Methyl Orange Adsorption on Biochar Obtained from Prosopis juliflora Waste: Thermodynamic and Kinetic Study. *ChemEngineering* **2023**, *7*, 114. [CrossRef]
60. Hou, Y.; Liang, Y.; Hu, H.; Tao, Y.; Zhou, J.; Cai, J. Facile preparation of multi-porous biochar from lotus biomass for methyl orange removal: Kinetics, isotherms, and regeneration studies. *Bioresour. Technol.* **2021**, *329*, 124877. [CrossRef]
61. Kifetew, M.; Prabhu, V.; Worku, Z.; Fito, J.; Alemayehu, E. Adsorptive removal of reactive yellow 145 dye from textile industry effluents using teff straw-activated carbon: RSM-based process optimization. *Water Pract. Technol.* **2024**, *19*, 362–383. [CrossRef]
62. Takele, T.; Angassa, K.; Abewaa, M.; Kebede, A.M.; Tessema, I. Adsorption of methylene blue from textile industrial wastewater using activated carbon developed from H_3PO_4-activated khat stem waste. *Biomass Convers. Biorefinery* **2023**. [CrossRef]
63. Abewaa, M.; Arka, A.; Haddis, T.; Mengistu, A.; Takele, T.; Adino, E.; Abay, Y.; Bekele, N.; Andualem, G.; Girmay, H. Results in Engineering Hexavalent chromium adsorption from aqueous solution utilizing activated carbon developed from Rumex abyssinicus. *Results Eng.* **2024**, *22*, 102274. [CrossRef]
64. Dimbo, D.; Abewaa, M.; Adino, E.; Mengistu, A.; Takele, T.; Oro, A.; Rangaraju, M. Methylene blue adsorption from aqueous solution using activated carbon of spathodea campanulata. *Results Eng.* **2024**, *21*, 101910. [CrossRef]
65. Mabuza, M.; Premlall, K.; Daramola, M.O. Modelling and thermodynamic properties of pure CO_2 and flue gas sorption data on South African coals using Langmuir, Freundlich, Temkin, and extended Langmuir isotherm models. *Int. J. Coal Sci. Technol.* **2022**, *9*, 45. [CrossRef]
66. Pereira, S.K.; Kini, S.; Prabhu, B.; Jeppu, G.P. A simplified modeling procedure for adsorption at varying pH conditions using the modified Langmuir–Freundlich isotherm. *Appl. Water Sci.* **2023**, *13*, 29. [CrossRef]
67. Al-Ghouti, M.A.; Da'ana, D.A. Guidelines for the use and interpretation of adsorption isotherm models: A review. *J. Hazard. Mater.* **2020**, *393*, 122383. [CrossRef]
68. Azizian, S.; Eris, S. Adsorption isotherms and kinetics. In *Interface Science and Technology*; Elsevier: Amsterdam, The Netherlands, 2021; Volume 33, pp. 445–509. ISBN 1573-4285.
69. Sharma, G.; Naushad, M. Adsorptive removal of noxious cadmium ions from aqueous medium using activated carbon/zirconium oxide composite: Isotherm and kinetic modelling. *J. Mol. Liq.* **2020**, *310*, 113025. [CrossRef]
70. Islam, M.A.; Nazal, M.K.; Angove, M.J.; Morton, D.W.; Hoque, K.A.; Reaz, A.H.; Islam, M.T.; Karim, S.M.A.; Chowdhury, A.-N. Emerging iron-based mesoporous materials for adsorptive removal of pollutants: Mechanism, optimization, challenges, and future perspective. *Chemosphere* **2023**, *349*, 140846. [PubMed]
71. Abewaa, M.; Adino, E.; Mengistu, A. Heliyon Preparation of Rumex abyssinicus based biosorbent for the removal of methyl orange from aqueous solution. *Heliyon* **2023**, *9*, e22447. [CrossRef] [PubMed]
72. Fito, J.; Abewaa, M.; Mengistu, A.; Angassa, K.; Ambaye, A.D.; Moyo, W.; Nkambule, T. Adsorption of methylene blue from textile industrial wastewater using activated carbon developed from Rumex abyssinicus plant. *Sci. Rep.* **2023**, *13*, 5427. [CrossRef]

73. Mengistu, A.; Abewaa, M.; Adino, E.; Gizachew, E.; Abdu, J. The application of Rumex abyssinicus based activated carbon for Brilliant Blue Reactive dye adsorption from aqueous solution. *BMC Chem.* **2023**, *17*, 82. [CrossRef]
74. Abewaa, M.; Mengistu, A.; Takele, T.; Fito, J.; Nkambule, T. Adsorptive removal of malachite green dye from aqueous solution using Rumex abyssinicus derived activated carbon. *Sci. Rep.* **2023**, *13*, 14701. [CrossRef]
75. Fito, J.; Mengistu, A.; Abewaa, M.; Angassa, K.; Moyo, W.; Phiri, Z.; Mafa, P.J.; Kuvarega, A.T.; Nkambule, T.T.I. Journal of the Taiwan Institute of Chemical Engineers Adsorption of Black MNN reactive dye from tannery wastewater using activated carbon of Rumex Abysinicus. *J. Taiwan Inst. Chem. Eng.* **2023**, *151*, 105138. [CrossRef]
76. Mo, Z.; Tai, D.; Zhang, H.; Shahab, A. A comprehensive review on the adsorption of heavy metals by zeolite imidazole framework (ZIF-8) based nanocomposite in water. *Chem. Eng. J.* **2022**, *443*, 136320. [CrossRef]
77. Chu, S.; Liu, C.; Feng, X.; Wu, H.; Liu, X. Aromatic polymer dual-confined magnetic metal-organic framework microspheres enable highly efficient removal of dyes, heavy metals, and antibiotics. *Chem. Eng. J.* **2023**, *472*, 145159. [CrossRef]
78. Rathi, B.S.; Kumar, P.S. Application of adsorption process for effective removal of emerging contaminants from water and wastewater. *Environ. Pollut.* **2021**, *280*, 116995. [CrossRef] [PubMed]
79. Demiral, İ.; Samdan, C.; Demiral, H. Enrichment of the surface functional groups of activated carbon by modification method. *Surf. Interfaces* **2021**, *22*, 100873. [CrossRef]
80. Švábová, M.; Bičáková, O.; Vorokhta, M. Biochar as an effective material for acetone sorption and the effect of surface area on the mechanism of sorption. *J. Environ. Manag.* **2023**, *348*, 119205. [CrossRef]
81. Zamani, S.A.; Yunus, R.; Samsuri, A.W.; Salleh, M.A.M.; Asady, B. Removal of Zinc from Aqueous Solution by Optimized Oil Palm Empty Fruit Bunches Biochar as Low Cost Adsorbent. *Bioinorg. Chem. Appl.* **2017**, *2017*, 7914714. [CrossRef] [PubMed]
82. Liu, C.; Ye, J.; Lin, Y.; Wu, J.; Price, G.W.; Burton, D.; Wang, Y. Removal of Cadmium (II) using water hyacinth (Eichhornia crassipes) biochar alginate beads in aqueous solutions. *Environ. Pollut.* **2020**, *264*, 114785. [CrossRef] [PubMed]
83. Esteves, B.M.; Morales-Torres, S.; Maldonado-Hódar, F.J.; Madeira, L.M. Fitting biochars and activated carbons from residues of the olive oil industry as supports of fe-catalysts for the heterogeneous fenton-like treatment of simulated olive mill wastewater. *Nanomaterials* **2020**, *10*, 876. [CrossRef] [PubMed]
84. Wahi, R.; Zuhaidi, N.F.Q.A.; Yusof, Y.; Jamel, J.; Kanakaraju, D.; Ngaini, Z. Chemically treated microwave-derived biochar: An overview. *Biomass Bioenergy* **2017**, *107*, 411–421. [CrossRef]
85. Villabona-Ortíz, Á.; Figueroa-Lopez, K.J.; Ortega-Toro, R. Kinetics and adsorption equilibrium in the removal of azo-anionic dyes by modified cellulose. *Sustainability* **2022**, *14*, 3640. [CrossRef]
86. Masuku, M.; Nure, J.F.; Atagana, H.I.; Hlongwa, N.; Nkambule, T.T.I. Advancing the development of nanocomposite adsorbent through zinc-doped nickel ferrite-pinecone biochar for removal of chromium (VI) from wastewater. *Sci. Total Environ.* **2024**, *908*, 168136. [CrossRef] [PubMed]
87. Tee, G.T.; Gok, X.Y.; Yong, W.F. Adsorption of pollutants in wastewater via biosorbents, nanoparticles and magnetic biosorbents: A review. *Environ. Res.* **2022**, *212*, 113248. [CrossRef] [PubMed]
88. Fa Soliman, M.; Mahrous, M.; Gad, A.; Ali, I. Optimization of total hardness removal efficiency of industrial wastewater using novel adsorbing materials. *Aswan Univ. J. Environ. Stud.* **2023**, *4*, 169–192. [CrossRef]
89. Azmier, M.; Norhidayah, A. Modified durian seed as adsorbent for the removal of methyl red dye from aqueous solutions. *Appl. Water Sci.* **2015**, *5*, 407–423. [CrossRef]
90. Solution, A.; White, U.; Peel, P.; Enenebeaku, C.K.; Okorocha, N.J. Adsorption and Equilibrium Studies on the Removal of Methyl Red from Adsorption and Equilibrium Studies on the Removal of Methyl Red from Aqueous Solution Using White Potato Peel Powder. *Int. Lett. Chem. Phys. Astron.* **2017**, *72*, 52. [CrossRef]
91. Khomeyrani, S.F.N.; Azghandi, M.H.A.; Ghalami-Choobar, B. Rapid and efficient ultrasonic assisted adsorption of PNP onto LDH-GO-CNTs: ANFIS, GRNN and RSM modeling, optimization, isotherm, kinetic, and thermodynamic study. *J. Mol. Liq.* **2021**, *333*, 115917. [CrossRef]
92. Paluch, D.; Bazan-Wozniak, A.; Wolski, R.; Nosal-Wiercińska, A.; Pietrzak, R. Removal of Methyl Red from Aqueous Solution Using Biochar Derived from Fennel Seeds. *Molecules* **2023**, *28*, 7786. [CrossRef]
93. Paluch, D.; Bazan-Wozniak, A.; Nosal-Wiercińska, A.; Pietrzak, R. Removal of Methylene Blue and Methyl Red from Aqueous Solutions Using Activated Carbons Obtained by Chemical Activation of Caraway Seed. *Molecules* **2023**, *28*, 6306. [CrossRef]
94. Khalfaoui, A.; Bouchareb, E.M.; Derbal, K.; Boukhaloua, S.; Chahbouni, B.; Bouchareb, R. Uptake of Methyl Red dye from aqueous solution using activated carbons prepared from Moringa Oleifera shells. *Clean. Chem. Eng.* **2022**, *4*, 100069. [CrossRef]
95. Amari, A.; Yadav, V.K.; Pathan, S.K.; Singh, B.; Osman, H.; Choudhary, N.; Khedher, K.M.; Basnet, A. Remediation of Methyl Red Dye from Aqueous Solutions by Using Biosorbents Developed from Floral Waste. *Adsorpt. Sci. Technol.* **2023**, *2023*, 1532660. [CrossRef]

Disclaimer/Publisher's Note: The statements, opinions and data contained in all publications are solely those of the individual author(s) and contributor(s) and not of MDPI and/or the editor(s). MDPI and/or the editor(s) disclaim responsibility for any injury to people or property resulting from any ideas, methods, instructions or products referred to in the content.

Article

Impact of Design and Mixing Strategies on Biogas Production in Anaerobic Digesters

Thomas Neuner [1,*], Michael Meister [2], Martin Pillei [1], Thomas Senfter [1], Simon Draxl-Weiskopf [3,4], Christian Ebner [3], Jacqueline Winkler [3] and Wolfgang Rauch [3]

[1] Department of Industrial Engineering, MCI—The Entrepreneurial School, 6020 Innsbruck, Austria; martin.pillei@mci.edu (M.P.); thomas.senfter@mci.edu (T.S.)

[2] Department of Environmental, Process & Energy Engineering, MCI—The Entrepreneurial School, 6020 Innsbruck, Austria; michael.meister@mci.edu

[3] Unit of Environmental Engineering, University of Innsbruck, Technikerstraße 13, 6020 Innsbruck, Austria; s.draxl-weiskopf@biotreat.at (S.D.-W.); christian.ebner@uibk.ac.at (C.E.); jacqueline.winkler@uibk.ac.at (J.W.); wolfgang.rauch@uibk.ac.at (W.R.)

[4] BioTreaT GmbH, Technikerstraße 21a, 6020 Innsbruck, Austria

* Correspondence: thomas.neuner@mci.edu

Citation: Neuner, T.; Meister, M.; Pillei, M.; Senfter, T.; Draxl-Weiskopf, S.; Ebner, C.; Winkler, J.; Rauch, W. Impact of Design and Mixing Strategies on Biogas Production in Anaerobic Digesters. *Water* **2024**, *16*, 2205. https://doi.org/10.3390/w16152205

Academic Editors: Christos S. Akratos and Carmen Teodosiu

Received: 25 June 2024
Revised: 1 August 2024
Accepted: 2 August 2024
Published: 4 August 2024

Copyright: © 2024 by the authors. Licensee MDPI, Basel, Switzerland. This article is an open access article distributed under the terms and conditions of the Creative Commons Attribution (CC BY) license (https:// creativecommons.org/licenses/by/ 4.0/).

Abstract: Anaerobic digestion (AD) is a biological process that breaks down organic matter in the absence of oxygen, producing biogas and nutrient-rich digestate. Various reactor designs and mixing strategies are well-established in AD processes, each with their own advantages and benefits. The presented study summarizes and investigates the state of the art of AD in domestic wastewater treatment plants (WWTPs) in an Austrian alpine region, with a primary focus on finding similarities among the most efficient plants regarding digester design, mixing approaches, and biogas production. By combining surveys and detailed field studies in cooperation with 34 WWTPs, the study provides a comprehensive overview of common AD practices, reactor shapes, and inherent mixing methods, highlighting their potential regarding energetic efficiency and biogas production. The results of the survey reveal qualitative trends in efficient AD design alongside detailed quantitative data derived from the supervised in-field optimization studies. Notably, one of the studies demonstrated energetic savings of 52% with no decrease in biogas production, achieved by transitioning from gas injection to mechanical agitation. Redundant impeller-based overmixing was also practically investigated and demonstrated in another field study. After optimization, the adaptations also resulted in energy savings of 30%, still proving sufficient substrate mixing with biomethane potential analysis. In conclusion, this research emphasizes the economic and environmental importance of energy-refined practices and optimized processes while highlighting the sustainability of AD, particularly for large domestic WWTPs but also for different comparable applications.

Keywords: anaerobic digestion; biogas; mixing; process design; wastewater

1. Introduction

The transition to renewable energy sources and the optimization of energy efficiency in existing processes are becoming increasingly important due to the rising levels of greenhouse gas emissions, namely carbon dioxide (CO_2) [1]. Among today's various green and renewable methods, anaerobic digestion (AD) of organic matter represents a significant and important energy source both because of environmental and economic reasons. Biogas, the end product of the complex decomposition processes in AD, can be utilized in multiple ways: it can be combusted onsite as a short-term energy source, refined and injected into the gas grid, or used as an efficient long-term energy storage solution [2]. Consequently, AD plays an increasingly important role in the context of climate-neutral energy management. This is also reflected in the fact that AD is integrated in in almost every wastewater treatment plant (WWTP). However, while certain AD designs and mixing strategies have been well-established within the AD context, the processes are often prone to inefficient operation due to the difficulties in process monitoring. For evaluation of the AD process, laboratory experiments, computer-aided simulations, and in-field experiments are promising tools to help in understanding and optimizing the individual AD processes [3–12]. However, besides these supportive tools, a pivotal share in efficient AD operation is often contributed by the long-time experience of plant operators. Hence, to harness the full potential of AD for energy transition, a comprehensive theoretical understanding of the processes and parameters involved is as essential as the consideration and implementation of practical recommendations provided by plant supervisors.

In detail, AD involves the generation of biogas from biodegradable substances within large-scale reactors. AD requires a specific environment for the utilized microorganisms and thorough mixing to ensure continuous and efficient biogas production [13]. The resulting biogas typically consists of approximately 60% methane (CH_4), 40% CO_2, and small amounts of hydrogen (H_2), sulfide, and other trace gasses [14]. Industrial large-scale AD plants are capable of generating up to 5000 m^3 biogas per day [15–17], as also presented within this study. The exothermic combustion of CH_4 converts it into energy and less climate-damaging CO_2, providing a dual benefit of energy recovery from organic residues and reduced CH_4 emissions [18]. However, AD can also contribute to unwanted CH_4 emissions due to gas losses. While CH_4 emissions may occur along the WWTPs, the majority originate from the sludge line [19,20]. In preliminary studies to this work, CH_4 emissions from an Austrian WWTP serving 260,000 population equivalents (PEs) were estimated to be approximately 25 g CH_4 per PEs per year, accounting for over a quarter of the plant's emission footprint. Therefore, optimizing AD is necessary to maximize the potential of organic waste disposal and energy recovery, thereby reducing greenhouse gas emissions.

The fundamental operational design of AD towers is usually similar to continuously operated and stirred large-scale bioreactors [21]. These continuous stirred tank reactors (CSTRs) are simple and efficient in both design and operation, enabling continuous feeding and withdrawal of digestate. CSTRs are well-suited for treating homogeneous substrates and maintaining stable operating conditions. However, they require relatively long hydraulic retention times (HRTs) to achieve sufficient organic matter degradation [22]. Consequently, volumetric biogas production rates are generally lower compared to other designs, such as anaerobic sequencing batch reactors (ASBRs). Batch cycle-based AD systems, such as ASBR, offer higher flexibility in feeding and substrate types [23,24]. While these systems achieve high levels of organic matter removal and biogas production efficiency, they are associated with higher energy requirements and more complex system control. Anaerobic fixed film reactors (AFFRs) utilize fixed media (e.g., plastic or ceramic materials) to support microbial growth, thereby increasing the surface area available for digestion [25]. AFFRs have shorter HRT and higher organic loading rates compared to suspended growth systems but may become clogged or require periodic replacement. Up-flow anaerobic sludge blanket (UASB) reactors employ dense sludge blankets to retain microbial biomass and facilitate AD [26]. UASB reactors exhibit relatively high volumetric biogas production rates and can tolerate high organic loading rates [27]. However, these reactors require a careful control of influent characteristics and hydraulic conditions to prevent sludge washout or stratification. Additionally, UASB reactors are sensitive to temperature fluctuations and variations in substrate composition.

The geometric design of AD towers is influenced by technical, economic, and practical considerations and aims to achieve efficient substrate digestion, biogas production, and optimal system performance [7,28,29]. The structure of AD towers depends on several factors, including specific design requirements, available space, and operational influencing factors, resulting in significant variability in physical configurations [30]. Common AD system designs typically employ vertical or horizontal cylindrical configurations due to their high structural stability, simple construction, and efficient space utilization [7,31–33]. Vertical designs are prevalent due to their capacity and ease of access for maintenance, though this often requires specialized equipment. Horizontal designs are often partially installed underground for thermal insulation. Cylindrical AD reactors usually feature a conical tapering at the bottom to facilitate sludge removal [17]. Egg-shaped reactors are also common in AD, most often combined with mechanical agitators or impellers inside draft tubes [34–36]. More complex shapes, such as quadratic, rectangular, or spherical, are less common due to distinct disadvantages [16,17]. Rectangular tanks, for example, allow for a better utilization of construction space and simpler construction but suffer from constricted hydrodynamics, creating dead zones and uneven flow distribution that necessitate additional mixing equipment [37]. The corners are especially prone to decreased hydrodynamics, resulting in low mixing and potentially unprocessed organic matter [38]. In some designs, quadratic AD towers can be partitioned for multi-stage digestion processes [39]. Spherically designed AD towers, while less optimal for space utilization, offer excellent volume-to-surface area ratios, promoting uniform temperature distribution and efficient mixing [16,17]. Their structural integrity allows them to withstand internal pressures effectively. Custom or hybrid designs combining elements of different geometric shapes could be employed to meet specific site requirements and operational needs, though these are associated with higher upfront and potential operational costs. For structural reasons, AD towers are almost exclusively constructed of concrete [40]. To enhance operational control, AD towers can be designed in pairs, which provides the possibility of serial or parallel operation depending on the situation [41]. When possible, serial operation can increase gas output; however, parallel operation may be beneficial when high amounts of substrate or external influences (e.g., low temperature) are overloading one single AD tower.

AD systems encompass various technical processes, and the overall energy requirement for an AD tower arises from the sum of the individual operational units. The primary energy-consuming aspects of AD operation include mixing, feedstock pumping, temperature control, and auxiliary processes such as monitoring and control equipment [16,42,43]. Maintaining proper feedstock pumping and recirculation, as well as mixing and agitation within the AD tower, is crucial for supporting microbial activity, ensuring uniform substrate distribution, and preventing solids settlement [6]. Besides mixing, AD processes are temperature-sensitive, with optimal microbial activity occurring within a specific temperature range. Energy is required to maintain consistent temperatures, particularly in colder climates or during cold seasons. The ideal temperature for AD depends on the specific types of microorganisms involved in the process and the substrate [24,44,45]. AD is typically operated within the mesophilic temperature range of 25 °C to 40 °C, which is most suitable for a wide range of common anaerobic bacteria and archaea. Thermophilic AD (50 °C to 65 °C) accelerates the digestion process, resulting in faster degradation of organic matter and higher biogas production rates compared to mesophilic digestion [22,46]. However, thermophilic AD requires more energy for heating and is more sensitive to temperature fluctuations. Psychrophilic AD operates at temperatures below 25 °C and is suitable for certain low-temperature environments or feedstocks, though it has slower reaction rates and lower biogas yields [47].

Optimizing mixing in AD is crucial for maintaining uniform conditions within the reactor, enhancing biogas production through increased microbial activity, and, ideally, decreasing energy demand [5,8,9,48,49]. Several approaches to improve mixing have been investigated in recent studies and are applied in new well-planned AD systems. However, implementation in existing AD systems is difficult since most systems are continuously operated. Uninterrupted operation is often essential, and adjustments or maintenance are associated with significant structural and economic challenges. Hydrodynamics and energy demand for mixing can be assessed through experiments in existing plants, laboratory trials, or numerical simulations [3–10]. The choice of mixing method, which significantly impacts energy consumption, should be based on the substrate and AD tower geometry. The energy consumption of the mixing stage can account for up to 50% of the overall energy balance of an AD plant [50,51]. Each mixing approach offers individual advantages and disadvantages dependent on the application and the AD design. Hence, various distinct mixing approaches are established for AD mixing [15–17,52]. The main mixing approaches include pumped recirculation, mechanically induced mixing with an agitator or mixing with an impeller inside a draft-tube, and gas injection-based designs [4,8,17,28,35,40,52]. Internal hydraulic mixing via pumped recirculation, the most fundamental approach, is typically achieved using the inlet feed. In CSTR, the inlet feed supplies necessary nutrients and manages the reactor concentration and temperature. Within the AD context, pumped recirculation is primarily utilized for providing fresh substrate, enhancing sludge mixing, and maintaining mesophilic temperature conditions within the reactor. Depending on the fluid properties as well as the reactor design, a powerful sludge recirculation can be sufficient for fluid intermixing throughout the reactors [35] However, AD mixing that relies solely on sludge recirculation is seldom and rather an exception. More commonly, a combination of pumped recirculation and additional external mixing is applied. External mixing methods in AD primarily involve mechanical agitation or gas-induced mixing. Mechanical mixing devices vary significantly in design. Multi-stage propeller devices induce mixing near the mixing segments, breaking up solid clusters with increased shear [6], while helical, slower-rotating devices provide more uniform and gentle mixing throughout the reactor [43,53]. High agitation diameters are often preferred due to the increased mixing associated with higher circumferential velocity but are linked to higher physical strain in the devices. While higher shear forces improve sludge dewaterability, they can create a non-ideal environment for sensitive microorganisms [6]. Especially in the context of large-scale AD mixing, the thorough and often-cited studies of Wu et al. have demonstrated that mechanical mixing offers the best ratio of mixing intensity to power consumption [3,54]. Besides large agitation devices, fast rotating impellers with a small diameters are also used in combination with draft tubes to induce sludge mixing [35,55]. Although impeller mixing inside draft tubes is a form of mechanical agitation, it is often considered a distinct mixing approach. Impellers lift sludge through a draft tube from the bottom to the top of the reactor, a method often employed in reactors with tapering bottoms, such as egg-shaped geometries. Draft tube mixing with impellers in reactors with a wide bottom diameter may lead to increased dead volumes near the wall regions [35]. However, when specifically designed and harmonized with the reactor geometry, circulation patterns can result in energy-efficient and uniform mixing [34,36]. This effect is pronounced in egg-shaped designs due to the curvature of the wall and can be further enhanced by aligning the sludge recirculation inlet with the reactor wall curvature [35,36]. Besides mechanical mixing, gas-induced mixing is another major approach in AD reactors. Given the anoxic conditions necessary for biogas production, the induced gas is typically biogas, taken from within the AD system [3,56,57]. High nozzle-driven velocities lift the gas to the top, resulting in sludge mixing and a dispersion of solids and particle clusters. The energy required for gas induction depends on factors such as AD tower height, sludge viscosity, and density. However, gas nozzles are prone to fouling and require more maintenance compared to mechanical agitators [58]. Mechanical agitation offers cost-effective and consistent mixing, ensuring a uniform distribution of microorganisms and nutrients throughout the reactor [16]. In contrast, gas-induced mixing is stated to provide aided microbiological activity due to more thorough mixing, but is generally less uniform and has higher operational costs [56,59]. This leads to a consideration of operating costs and the sufficiency of the induced mixing.

To estimate mixing efficiency in existing systems or during the planning stages of new plants, experimental tests on reference plants can help in predicting hydrodynamics. Furthermore, computational fluid dynamics (CFD) modeling is a promising tool for optimizing reactor design and mixing strategies, as it can predict flow patterns and identify poorly mixed regions without implementing changes in the real plant [4,7,8,32,38,56,60,61]. Computer-aided simulations are also very useful for estimating the required mixing power and energetic demand. The overall energy demand in AD towers can be reduced with energy recovery systems such as heat exchangers to prevent energy losses [62]. Monitoring and controlling operating variables like power consumption, rotational speed, and fluid velocities can optimize mixing efficiency in real-time. Overmixing, which is often applied in good faith to ensure a sufficient distribution of components, can negatively impact biogas production and overall plant efficiency. Depending on the substrate composition, mixing too vigorously can lead to decreased biogas production [63]. Adjusting mixing intensity through the rotational speed, mixing depth, and dynamic operation mode can prevent overmixing and match the characteristics of the feedstock [51,64]. However, too low and insufficient mixing is also counterproductive and can lead to decreased microbial activity and the settlement of solids, reducing the usable volume and necessitating costly cleanouts [65]. It can also cause short-circuit flows and dead zones, compromising CH_4 yield and sludge digestion, resulting in undigested organic matter and therefore reduced gas production. This is stated in a study showing that biogas can be trapped in low mixed zones [66]. Since dynamic velocity control is often linked with operational difficulties, intermittent mixing can be applied to control the induced amount of mixing. For individual cases, intermittent mixing has been reported to reduce the energy demand for mixing by up to 30% while still maintaining constant biogas production rates [51,67,68]. Regarding the impact of intermittent mixing, it has been proven that different mixing intensities promote different methogenesis. In detail, intermittent mixing results in a beneficial balance of mixing-dependent acetotrophic and hydrogenotrophic methanogenesis, leading to overall increased microbial activity [69]. Subsequently, intermitted mixing is reported to both reduce the energetic demand as well as improve biogas production when implemented properly [51]. This results in a more economic AD operation with minimized potential dead zones, promoted fluid circulation, and reduced energy waste.

All the mentioned mixing strategies and approaches in reactor design are common and well-established in anaerobic AD towers. However, for today's standards, some of the static and conventional methods may be less efficient compared to modern, study-refined techniques and mixing strategies. Specifically, unnecessary and avoidable overmixing, which can be attributed to an inefficient operation of AD towers, must be critically examined as it leads to increased costs and partially unutilized methane potential. To assess the integral efficiency of AD systems, both the output in terms of gas quality and production as well as the input in terms of operational costs, maintenance (e.g., evacuation to remove sediments), and required investments need to be considered. In order to evaluate certain trends in the complex AD framework, this study investigates a wide array of differently designed AD towers in existing WWTPs in an alpine region in Austria. Both with a wide-ranging questionnaire-based survey, as well as with direct cooperation with the WWTPs, a comprehensive summarization of AD data is collected and provided within this study. The novel characteristics and distinguishing features of this study lie in the extensive data collection achieved through a combined approach of in-depth field studies, surveys, and the direct exchange of experience and information with plant operators, encompassing a wide range of plant dimensions, reactor geometries, and mixing methods. Because of the wide-spanning and thorough data from field investigations and experimental trials, this work offers significant practical relevance and can serve as a basis and benchmark for more in-depth analyses, such as numerical studies. In detail, AD specific data such as plant characteristics, gas production, energy demand and additional advantages, and disadvantages of specific plants and approaches are carefully summarized, curated, and presented in an anonymous form, ensuring data protection. This study covers a wide range of well-established and state-of-the-art reactor geometries, inherent mixing approaches, sludge rates, and an overall scale of plants ranging from small local communities to larger cities. The primary focus is to highlight similarities between efficient plants in order to provide recommendations for either optimizing existing plants or for the planning stage of new plants. By combining the conducted survey with supervised field studies and laboratory analyses, the outcoming results offer valuable insights into the interplay between mixing strategies and biogas production efficiency, with the decisive factor of direct practical recommendations and feedback of plant operators. Besides qualitative guidelines on efficient plant designs, the conducted field studies also underline the potential energetic savings of optimization studies. Thus, this study can be seen as a broad set of recommendations for plant operators, applicable not only in the field of AD but also in comparable technical processes where hydrodynamics and particularly complex large-scale mixing tasks are important.

2. Materials and Methods

2.1. Operator Survey with AD-Specific Questionnaire

To evaluate AD processes in the studied WWTPs, a specifically tailored questionnaire was developed and employed to gather relevant integral data. Key aspects covered in the questionnaire included geometric design, size (AD volume V_{AD}) and dimensions, mixing strategies, operational parameters, and fluid characteristics specific to the AD process. Additionally, biogas production rates, energy requirements, and optional plant-specific details were among the targeted aspects of inquiry. Since the survey responses highlighted that the AD sludge volume and biogas rate fluctuate over time, operators were requested to provide averaged AD volume and gas rate values that approximate the actual parameters for plant classification. With a response rate of 76%, questionnaire data from 34 plants were gathered and are presented in an anonymized form to ensure data privacy for the individual plants included in this study. Within the 34 plants, additional and more detailed field studies were conducted in cooperation with the plant operators. However, not all returned questionnaires were fully completed, as some parameters, such as specific power and energetic requirements, were not consistently monitored by all plant operators—or they were subject to data protection regulations (e.g., specific geometric dimensions such as diameter and height, as well as biogas utilization) and therefore cannot be published. The detailed questions included the following:

- Name
- City
- Structural information: amount of AD towers, AD volume (m^3), geometric shape, bottom geometry
- Operational information: serial, parallel or dynamic control
- Substrate information: %TS, organic dry matter (ODM), co-Substrate (yes/no)
- Mixing information: Approach and required power and energetic demand
- Biogas production (m^3)
- Energetic processing of the produced biogas

2.2. Investigation of AD Parameters with In-Depth Field Studies

In addition to the wide-ranging comprehensive data collection through the questionnaire, specific plants undergo detailed investigation and monitoring in this study. Energy, biogas rate, and additional sludge-related data are collected through monitoring and experimental investigations. Biogas production is indicated with the CH$_4$ yield, which is the amount of produced CH$_4$ given in normed cubic meters per year (a) (Nm^3 CH$_4$/a). The performance of the digester is monitored with specific CH$_4$ productivity and expressed with gas volume per day (d) and AD volume (Nm^3 CH$_4$/d m^3). It reflects the efficiency in producing CH$_4$. The extent to which the organic material in the substrate is broken down during the AD process is given with the percentage of degree of organic degradation (DOD). The amount of organic material fed into the AD towers is defined as the organic loading rate (OLR) and monitored as kilograms of ODM per AD volume per day (kg ODM/m^3 d). It is a crucial parameter for optimizing the digestion process and preventing overloading of the digester. With the OLR and the AD volume, the overall annual amount of organic matter (OM) can be derived and is expressed in kg ODM/a. Specific gas production (SGP) is expressed using the CH$_4$ yield divided by the OM (Nm^3 CH$_4$/kg ODM). Regarding sludge properties, key parameters such as sludge density ρ (kg m^{-3}), dynamic viscosity η (Pa s), dry matter DM (%), and ODM are determined. ODM represents the dried organic portion of substrate that is available for the microorganisms to decompose during the AD process, crucial for biogas production and yield. Higher ODM levels generally indicate greater gas potential but require specific handling for mixing and logistics caused by specific rheology (e.g., high viscosity of the sludge). ODM quantification involves drying and oxidizing the substrate, with the resulting ODM calculated by subtracting the remaining inorganic material from the dried sample. Parameters such as overall DM and total solids concentration (%TS) are also important for sludge classification in AD, influencing gas production and especially hydrodynamics. Finally, HRT (in d) indicates the duration that the substrate remains in the digesters, is critical for determining the efficiency of the digestion process, and was also collected for the individual plants.

2.3. Measurement of Biomethane and Residual Gas Potential

In the experimental setup focusing on mixing studies, both the biomethane potential (BMP) and residual gas potential (RGP) were assessed as they represent key indicators in the determination of AD efficiency. BMP indicates the maximum CH_4 yield achievable from a specific organic substrate under complete anaerobic conditions, whereas RGP evaluates the remaining biodegradability of partially pre-degraded samples containing unprocessed organic matter [47,70,71]. BMP, RGP as well as SGP are usually given in normed cubic meter per kg of ODM (Nm^3 CH_4/kg ODM) or per ton (t) ODM for large-scale applications (Nm^3 CH_4/t ODM). Conducted under controlled laboratory conditions, the assessments involve placing digestate, supplemented with excess nutrients and microorganisms, into laboratory reactors. The subsequent measurement of biogas production over a predefined period occurs at an average mesophilic temperature (37 °C ± 1 °C). BMP determination employs the state-of-the-art automatic methane potential test system (AMPTS) over a 21-day duration, while RGP is assessed using eudiometer tubes over 10 days (Figure 1).

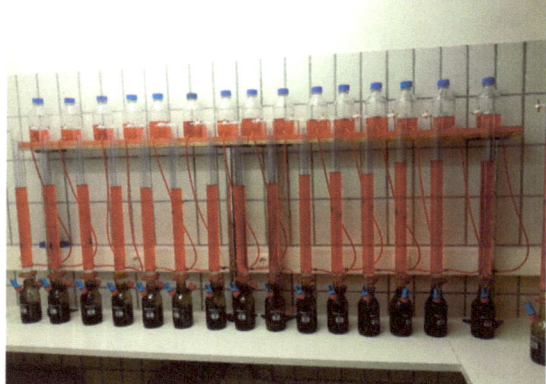

Figure 1. Laboratory setup for evaluation of the RGP in specific sludge samples using eudiometric tubes.

To ensure the integrity of experimental evaluations, samples are extracted either from the fresh substrate or the recirculation pipe, thereby mitigating potential falsification from fresh substrate influence. Laboratory analysis includes determining the DM and ODM content of these samples. Subsequently, 600 g of sludge undergo AD for 10 days at 37 °C, with the produced gas collected, measured for volume, and filtered to remove CO_2 and other trace gasses. Gas volumes are then standardized into norm-gas volumes (norm cubic meter: Nm^3) using Equation (1) (while V_0 represents the normed gas volume (Nm^3), V the measured gas volume (m^3), p the measured pressure (Pa), p_0 the Norm-pressure of 101,300 Pa, p_w vapor-pressure of 5622 Pa of water at 308.15 K, T the gas temperature of 308.15 K, and T_0 the norm temperature of 273.15 K) [72]. This methodology enables the precise assessment of both BMP and RGP, providing valuable insights into both the AD efficiency and mixing evaluation.

$$V_0 = V \frac{(p - p_\mathrm{w})(T_0)}{p_0(T)} \tag{1}$$

2.4. Measurement of the Power Demand for Mixing

Within the detailed in-field experiments, the energy demand of 8 operational plants was experimentally monitored using a Fluke Energy Logger 1732 v2.3. Three-phase current voltage (V), current (I), apparent (S), reactive (Q), and active power (P) were measured. Data post-processing was conducted using Fluke Energy Analyze Plus v3.11.2 software. The energy demand of motor engines driving mechanical agitation, impellers, and gas-inducing units was monitored continuously for up to 24 h. Additionally, the energy demand of recirculation pumps was recorded. Measurements were taken at electric distribution boards or directly at technical installations. Linear averaged active power was used for comparing the energetic requirements of the plants. The specific power demand Φ was calculated using the required power demand P (W) and the overall AD volume V_{AD} (m^3) according to Equation (2) and is a common indicator used to describe the required volumetric power in W m^{-3} for the individual mixing approaches in relation to the AD volume [36,69,73,74]. Additional information such as the associated mixing method and origin of the data are described in the indices of Φ (e.g., $\Phi_{\text{Mixing,Survey}}$). Moreover, energy consumption trends over different operational periods are analyzed to identify potential optimization opportunities. The collected data provides valuable insights into the energy performance of AD systems, aiding in the development of energy-efficient designs and operational strategies.

$$\Phi = \frac{P}{V_{AD}} \tag{2}$$

2.5. Comparison of Gas Injection and Mechanical Agitation in a Cylindrical AD Tower

To highlight differences in the applicability of gas injection and mechanical agitation in cylindrical AD towers, both mixing approaches are investigated in detail in a real-life situation in cooperation with Plant 1. Energetic demand and biogas production were monitored after (Plant 1A) and before (Plant 1B) the conversion from gas injection to mechanical agitation using a propeller-based design. The experimentally measured power demand of the individual mixing approaches was compared to the overall gas composition and production as well as to the SGP, considering the additional impact of season-dependent variables such sludge composition and ODM.

2.6. Investigation of Impeller-Induced Draft Tube Mixing in an Egg-Shaped AD Tower

To highlight the hydrodynamic and energetic impact of additional impeller mixing, coupled with pumped recirculation, a detailed investigation in cooperation with Plant 2 was conducted. The study examined the plant's performance both with and without additional impeller-induced draft tube mixing. Throughout two distinct representative cases, both power consumption and biogas production were monitored, providing comprehensive data on the system's efficiency before and after terminating impeller operation. Moreover, to gauge the efficacy of the remaining pumped recirculation-induced mixing in isolation, BMP was assessed in a laboratory setting for the used substrate and compared to the produced gas of the AD towers. This allowed for an evaluation of whether mixing achieved solely through recirculation mechanisms suffices for optimal biogas production.

3. Results

3.1. Data Evaluation of Survey and Field Studies

The results of the conducted survey and field studies were curated and presented in an anonymized form, providing a plant and mixing specific summarization. The prevalence of different reactor designs and mixing approaches derived from the survey are summarized in Table 1, along with their associated AD volume, operation mode, biogas production, and required specific power for external mixing and internal pumped recirculation.

Table 1. Survey data set collected with an AD-specific questionnaire of 34 individual plants. Frequency of reactor geometry and mixing approach are displayed in Figure 2.

ID	Geometry	Vol. AD1	Vol. AD2	Operation	Mixing	Biogas	$\Phi_{\text{Mixing, Survey}}$	$\Phi_{\text{Pumped Recirculation, Survey}}$	$\sum \Phi_{\text{Survey}}$
(-)	(-)	(m^3)	(m^3)	(-)	(-)	(Nm^3 d^{-1})	(W m^{-3})	(W m^{-3})	(W m^{-3})
1A	Cylindrical	4600	4600	Serial	M.A.	6160	0.4	4.5	4.9
1B	Cylindrical	4600	4600	Parallel	G.I.	4600	6	4.2	10.2
2A	Egg-shaped	2500	2500	Parallel	P.R.	5000	-	4.2	4.2
2B	Egg-shaped	2500	2500	Parallel	I.	5000	1.2	4.2	5.4
3	Egg-shaped	2200	-	Single AD	M.A.	1073	1.4	1.5	2.9
4	Cylindrical	6000	6000	Parallel	G.I.	4027	1.9	1	2.9
5	Truncated cone	1400	-	Single AD	G.I.	431	3.6	3.2	6.8
6	Truncated cone	400	400	Parallel	P.R.	342	-	5.6	5.6
7	Cylindrical	1400	-	Single AD	BIMA	307	-	1.6	1.6
8	Egg-shaped	1700	1700	Serial	M.A.	1064	1.5	2.3	3.8
9	Cylindrical	500	200	Serial	M.A.	53	-	-	-
10	Egg-shaped	1600	-	Single AD	I	938	0.9	2.0	2.9
11	Quadratic	440	440	Parallel	G.I.	100	-	-	-
12	Egg-shaped	2200	-	Single AD	M.A.	1486	1.6	2.2	3.8
13	Quadratic	560	560	Parallel	G.I.	236	1.9	2.5	4.4
14	Egg-shaped	760	-	Single AD	M.A.	628	2.5	4.1	6.6
15	Cylindrical	1750	-	Single AD	M.A.	1226	3.7	5.1	8.8
16	Cylindrical	1050	1050	Serial	G.I.	465	0.5	3.0	3.5
17	Quadratic	430	430	Parallel	G.I.	220	0.8	2.4	3.2
18	Egg-shaped	2200	-	Single AD	G.I.	1134	-	-	-
19	Cylindrical	1200	-	Single AD	M.A.	510	-	-	-
20	Cylindrical	584	584	Parallel	P.R.	661	-	9.4	9.4
21	Egg-shaped	1200	-	Single AD	G.I.	401	-	-	-
22	Rectangular	1215	1215	Parallel	M.A.	697	3.0	2.5	5.5
23	Cylindrical	413	413	Serial	G.I.	400	1.0	1.5	2.5
24	Cylindrical	413	413	Parallel	G.I.	-	-	-	-
25	Egg-shaped	911	-	Single AD	P.R.	840	-	-	-
26	Cylindrical	860	-	Single AD	P.R.	345	-	1.9	1.9
27	Cylindrical	350	350	Parallel	M.A.	150	1.8	4.2	6.0
28	Cylindrical	940	-	Single AD	P.R.	207	-	8.0	8.0
29	Cylindrical	1200	-	Single AD	G.I.	1000	2.4	1.7	4.1
30	Cylindrical	1275	-	Single AD	G.I.	540	7.1	2.6	9.7
31	Cylindrical	790	-	Single AD	P.R.	248	-	-	-
32	Cylindrical	1700	-	Single AD	M.A.	1470	-	-	-
33	Cylindrical	1800	-	Single AD	G.I.	669	0.7	2.0	2.7
34	Cylindrical	2200	-	Single AD	M.A.	1600	-	2.0	2.0

Note(s): '-' indicates missing provided data by the AD operators collected with questionnaire. Abbreviations: P.R.: pumped recirculation, M.A.: mechanical agitation, G.I.: gas-induced mixing, I.: impeller and draft tube.

A graphical summarization is displayed in Figure 2a,b for both the frequency of the geometric reactor shape and the AD mixing approaches, respectively. It is noted that the majority of the investigated plants (55%) employ a cylindrical-based reactor geometry. While evaluating the investigated cylindrical AD towers in this manuscript and in the recent literature, no specific recommended diameter–height ratio can be generalized, but in general AD towers tend to be designed vertically with a larger height than diameter. Besides the simple and well-tested cylindrical shapes, egg-shaped AD designs (26%) were also common and widely used as the state-of-the-art in AD design within this study. While again no specific ratio in geometry is widely applied, most egg-shaped reactors are used in combination with draft tube mixing or aligned inlet configuration [3,35]. Quadratic (9%) and rectangular tanks (3%) are more commonly utilized for smaller AD volumes. While truncated cone geometries (7%) relate to a distinct tapered reactor design, most cylindrical AD towers also offer a tapered contour at the bottom to decrease dead zones.

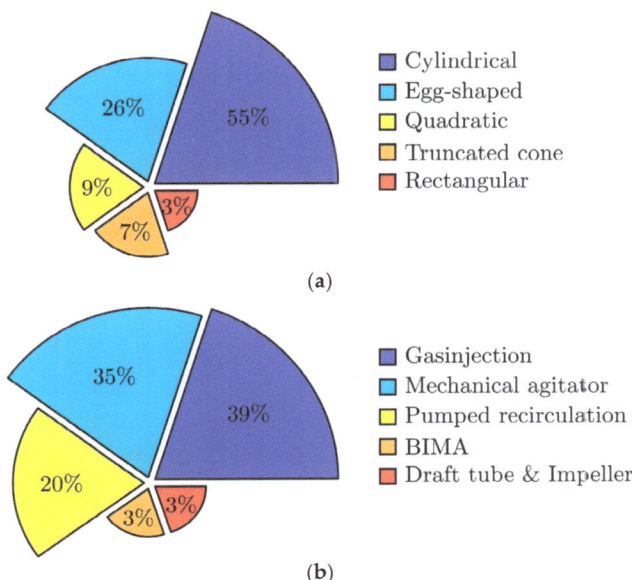

Figure 2. Approaches in state-of-the-art AD design for reactor design (**a**) and mixing (**b**).

Regarding mixing, the frequency of the most common approaches is highlighted in Figure 2b. Mechanical agitation as well as gas injection in conjunction with additional pumped recirculation are the most prominent mixing approaches with 35% and 39%, respectively. While impeller mixing with draft tubes, in general, is a well-established approach [35,36,75], it was less commonly observed in the investigated geographic area of WWTPs, with only two plants employing this approach within this study. Furthermore, within this study, one of the two plants employing draft tube mixing in combination with an impeller terminated its operation to assess its impact on overall mixing. Notably, one of the investigated plants was using a sophisticated biogas-induced mixing arrangement (BIMA). While BIMA is similar to gas-induced mixing methods, no additional energy is required for gas pumping, since the intermixing of the sludge is achieved solely with the pressure formed during the AD process. Both the serial and parallel modes of operation for two or more AD towers are common depending on both the substrate and operational properties. The possibility of dynamic operation was underlined as an important consideration in the overall design of AD plants by direct feedback in the field studies.

The curation of energetic data has highlighted that most energy is required in the pumped recirculation, followed by the internal mixing approaches. The pumped recirculation is used for pumping of the feedstock, temperature control, and also for mixing. As displayed in Figure 3, for 20 of 26 plants that have provided energetic information, the specific energy required for recirculation is higher than for the internal mixing approach.

This is related to the rheology of the sludge as well as to maintaining a certain substrate level and mesophilic temperature inside the AD towers. Regarding the specific power of internal mixing, gas injection and mechanical agitation do not show a specific trend that can be linked to the AD volume or to the mixing method itself. It has to be noted that, especially with gas-induced mixing, plant operators have stated a tendency to using an intermittent operation of the gas injection in order to save energy. In particular, the detailed investigation of Plant 1 shows a severe reduction in energetic demand after the transition to the more cost-effective mechanical agitation (1A), as compared to the prior utilized gas injection (1B). It is also highlighted that the rather seldom-used rectangular and especially quadratic-shaped AD towers are predominantly used in combination with gas-induced mixing and a lower AD volume. This is reasoned with a better applicability of gas-inducing segments at the bottom of the AD towers. Specifically, the corners in the square geometry can lead to increased dead zones, which are particularly poorly mixed, especially with mechanical agitators. The more detailed data gathered in the field studies conducted in cooperation with plant operators are summarized in Table 2.

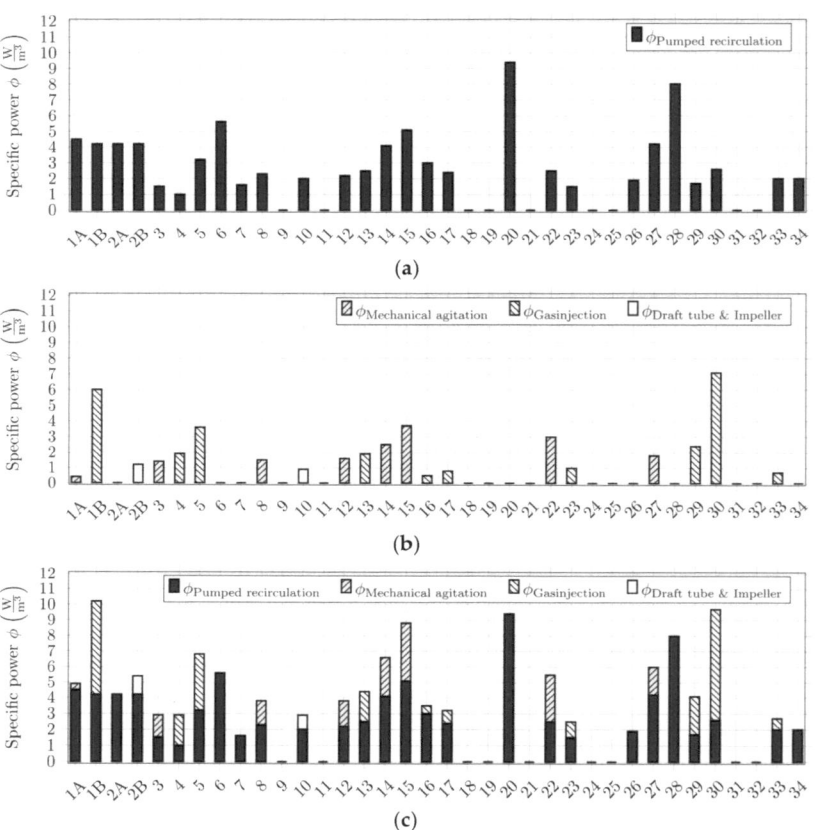

Figure 3. Breakdown of specific power demand for the individual plants required for pumped recirculation (**a**), additional external mixing (**b**), and the sum of both if a combined approach is used (**c**). Plants 2A, 6, 7, 20, 26, 28, and 34 are exclusively mixed with pumped recirculation.

The AD temperature (T_{AD}), HRT, OLR, DOD, and RGP are displayed for the individual plants. It is shown that all plants are within the mesophilic temperature range. Plant 8 and Plant 5 show the lowest RGP while also having the highest HRT. In comparison to energetic data provided by the plant operators in Table 1, Table 2 shows small deviations in the experimentally measured required power. The deviations as well as the DOD are displayed in Figure 4. In addition to determining the DOD of the organic dry matter, the RGP of the digested sludge derived from the field studies in cooperation with the different plants is highlighted in Figure 5. The RGP indicates how much biogas can still be produced after further incubation of the partially digested sludge and can be seen as a measure of the remaining degradable substances in the sludge (thus, the degree of stabilization). The investigated RGP values generally ranged between 50 and 80 Nm^3 CH_4/t ODM. In this range, similar to the case of DOD rates, no correlation between the mixing energy input and the RGP was observed. At Plant 8, a significantly lower RGP is documented compared to the other plants. Besides Plant 5 with a single AD, Plant 1 also showed a lower RGP. These digesters were also operated in serial mode at the time of sampling, but periodically switched to parallel operation due to foaming problems. The sludge age at Plant 1 was also significantly lower compared to Plant 8 at the time of sampling for RGF determination. These data suggest that serial operation with a sufficiently high sludge age can lead to optimal degradation results.

Table 2. Data set collected during field studies in cooperation with individual plant operators. RGP is derived from laboratory experiments. Required specific power demand is determined experimentally using a Fluke Energy Logger 1732 v2.3 and may deviate from survey-collected data displayed in Table 1.

ID	AD	Vol.	Type	Geometry	Mixing	T_{AD}	HRT	OLR	DOD	RGP	$\Phi_{Mixing\,F.E.}$	$\Phi_{Pumped\,Recirculation\,F.E.}$	$\sum \Phi_{F.E.}$
(-)	(-)	(m³)	(-)	(-)	(-)	(°C)	(d)	(kg ODM/ m³ d)	(%)	(Nm³ CH₄/t ODM)	(W m⁻³)	(W m⁻³)	(W m⁻³)
1A	AD 1 AD 2	4600 4600	Serial	Cylindrical	M.A.	37	24	2.3	67	57 ± 3.0	0.7 0.7	4.1 4.9	4.8 5.8
1B	AD 1 AD 2	4600 4600	Parallel	Cylindrical	G.I.	37	25	1.9	55	-	6 6	4.1 4.3	10.1 10.3
2A	AD 1 AD 2	2500 2500	Parallel	Egg-shaped	P.R	39	22	2.1	69	76 ± 5.6 80 ± 1.5	- -	2.9 2.9	2.9 2.9
2B	AD 1 AD 2	2500 2500	Parallel	Egg-shaped	I.	39	22	-	-	-	1.2 1.2	2.9 2.9	4.1 4.1
3	AD 1	2200	Single AD	Egg-shaped	M.A.	34	31	1	53	84 ± 2.8	1.7	1.5	3.2
4	AD 1 AD 2	6000 6000	Parallel	Cylindrical	G.I.	36	34	1.4	63	61 ± 0.8 65 ± 1.7	1.9 1.9	1.1 0.8	3.0 2.7
5	AD 1	1400	Single AD	Truncated cone	G.I.	39	49	1.1	59	48 ± 2.4	5.0	3.2	8.2
6	AD 1 AD 2	400 400	Parallel	Truncated cone	P.R	40	28	1.4	59	60 ± 1.4 58 ± 2.2	- -	4.7 6.4	4.7 6.4
7	AD 1	1400	Single AD	Cylindrical	BIMA	39	39	1	61	63 ± 1.6	-	3.5	3.5
8	AD 1 AD 2	1700 1700	Serial	Egg-Shaped	M.A.	39	39	2.3	71	32 ± 1.5	1.6 1.5	2.3 2.3	3.9 3.8

Note(s): '-' indicates missing provided data by the AD operators collected. Abbreviations: P.R.: pumped recirculation, M.A.: mechanical agitation, G.I.: gas-induced mixing, F.E.: derived from field experiments, I.: impeller and draft tube.

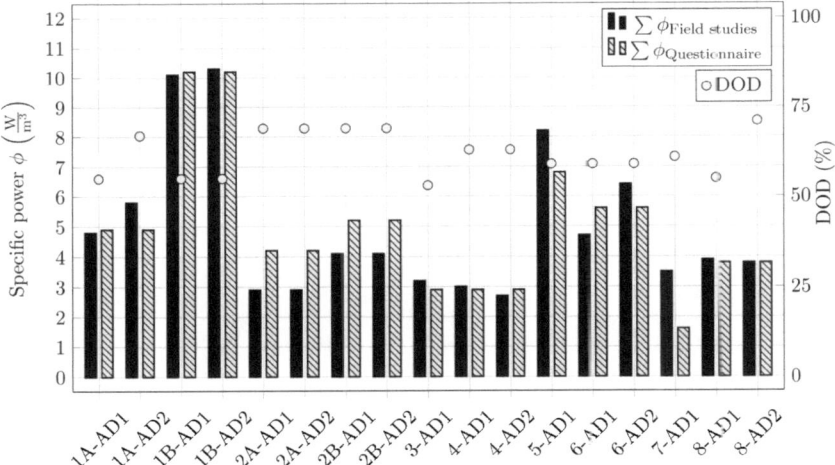

Figure 4. Comparison of the power rating provided by plant operators and of the measured in-field power demand using a Fluke Energy Logger 1732 v2.3. DOD is also highlighted for the specific AD towers.

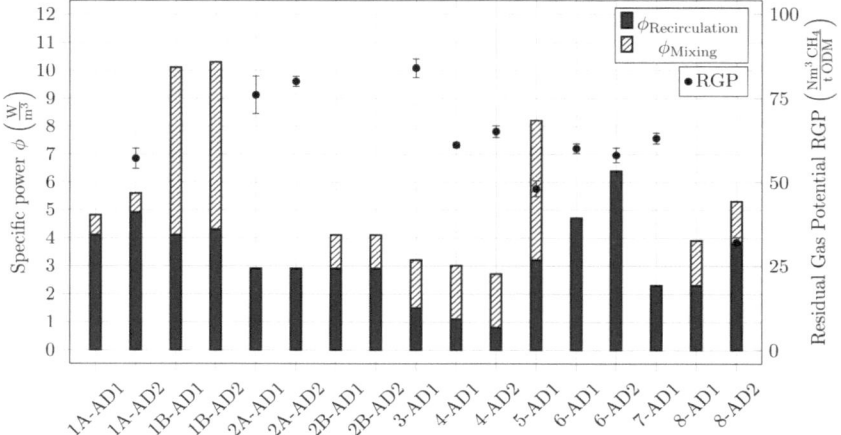

Figure 5. Comparison of specific power demand required for mixing and recirculation using experimentally determined data displayed in Table 2. RGP is additionally displayed for individual AD towers. Note that for serial operation mode, RGP is only given for final AD stage.

3.2. Field Comparison of Gas Injection and Mechanical Agitation in a Cylindrical AD

To assess the suitability of different mixing approaches, specifically regarding the power consumption, energetic demand, and biogas production, Plant 1 was investigated both during operation with mechanical agitation and previously with gas-induced mixing. To determine the required energetic demand, recorded energy data after transitioning to mechanical agitation (1A) were compared to the energetic data of the previous gas injection method (1B). Table 3 illustrates that the energetic demand in Plant 1 was significantly higher when using gas-injected mixing compared to mechanical agitation. Additionally, the comparison of biogas production rates revealed a notable difference after the conversion to mechanical agitation.

Table 3. Detailed data of in-depth field study regarding transition from gas injection to mechanical agitation in cooperation with Plant 1.

	Unit	Case A (Mechanical Agitation)	Case B (Gas Injection)
AD volume	(m^3)	8552	8217
HRT	(d)	24	25
T_{AD}	(°C)	37	37
OLR	(kg ODM/m^3 d)	2.3	1.9
OM	(kg ODM/a)	7,179,404	5,698,489
DOD	(%)	56	55
CH_4 concentration	(%)	60	60
CH_4 yield	(Nm^3/a)	2,248,200	1,680,250
CH_4 productivity	(Nm^3/m^3 d)	0.72	0.56
SGP	(Nm^3 CH_4/kg ODM)	0.313	0.295
Energy demand of external mixing	(MWh/a)	30	485

Note(s): Case A and case B data are derived from 2 years after and 2 before after the transition from gas injection to mechanical agitation, respectively.

In detail, the conducted transition increased the CH_4 yield by 33.8% while reducing the required energy demand for external mixing by 93.79%. However, it is crucial to note that the associated AD volume, OLR, and CH_4 productivity were also higher under mechanical agitation conditions (case A). Therefore, for the comparison of biogas, the SGP should be considered. Furthermore, measurement insecurity, especially in large-scale applications, need to be considered. Nevertheless, energy-related findings hold huge potential for optimizing energy consumption in similar plant setups.

3.3. Field Study Regarding Impact of Impeller Mixing in an Egg-Shaped AD

Evaluation of the survey has indicated that the utilization of impeller mixing within a draft tube is relatively uncommon within the investigated region. Further investigation and exchange of the experience with plant operators has revealed that impeller mixing is indeed less favored due to mechanical abrasion associated with the high velocities generated by small-diameter impellers and solids in the sludge. Additionally, the intense shear forces within the draft tube can lead to foam formation, thereby reducing gas efficiency. The hydrodynamics of impeller mixing necessitate specific reactor geometries, such as a truncated cone or egg-shaped designs, to enhance mixing due to the vertical circulation flow fields [34,35]. The long-term study of impeller operation inside a draft tube within this study has highlighted the disadvantages regarding mechanical durability. The high rotational speed associated with impeller mixing leads to pronounced abrasive effects that potentially damage the impeller blades, especially in comparison to slow rotating mechanical agitators with a larger diameter. It was observed that the mechanical stress on the impeller, caused by solids within the AD tower, resulted in severe abrasive effects on the impeller geometry. Subsequently, the maintenance and repair of the mixing equipment represent substantial costs and prolonged downtime for the AD tower.

To examine the necessity of additional impeller mixing, a detailed investigation was conducted at Plant 2 to assess the potential issue of overmixing. As demonstrated in Table 2 (data of 2A and 2B), the required specific power for mixing $\sum \Phi_{F,E}$ decreased per AD from 4.1 W m^{-3} to 2.9 W m^{-3} due to the termination of the additional impeller mixing through a draft tube. This results in a short-term reduction in the required mixing power demand of approximately 30%. The monitored biogas remained constant before and after the transition, with no significant fluctuations throughout the period of one year. The SGP provided by the plant was 445 Nm^3 CH$_4$/t ODM. To ensure the sufficiency of the AD mixing induced solely via pumped recirculation, the BMP determined in laboratory experiments compared to the plant's gas production is displayed in Figure 6. Evaluation of the gas production from the AD mixed solely with recirculation is displayed for 11 samples that were taken over a span of 16 weeks and digested under ideal laboratory conditions to estimate the BMP. Fluctuations in the samples, as highlighted in Figure 6, are attributed to a less ideal intermixing of primary and excess sludge when the samples were taken.

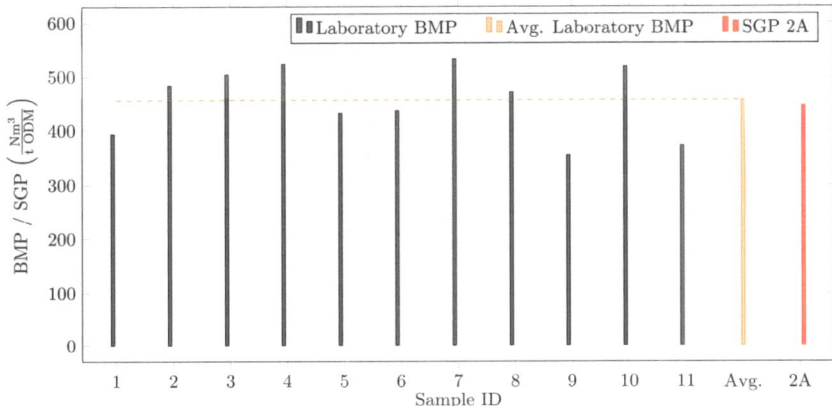

Figure 6. Laboratory evaluation of the BMP and SGP for Plant 2 after termination of impeller mixing.

The comparison shows good agreement between the BMP derived through the ideal lab-scale AD (avg. laboratory BMP = 456 Nm^3 CH$_4$/t ODM) and the SGP of the real

AD towers (SGP of Plant 2A = 445 Nm^3 CH$_4$/t ODM). Small deviations between the gas potentials highlight sufficient intermixing and thorough processing of the organic matter. Regarding biogas production, sludge mixing based solely on recirculation was proven to be sufficient, and additional impeller mixing was deemed unneeded for this specific case. Regarding settlement and the long-term maintenance of the AD towers, the ongoing study did not reveal significant alterations in the hydrodynamic behavior of solid particles within the investigated period. The available AD volume is regularly determined using tracer-based tests. As highlighted in Figure 7, the available AD volume is shown for specific time points before and after impeller shutdown, as well as after complete AD evacuation.

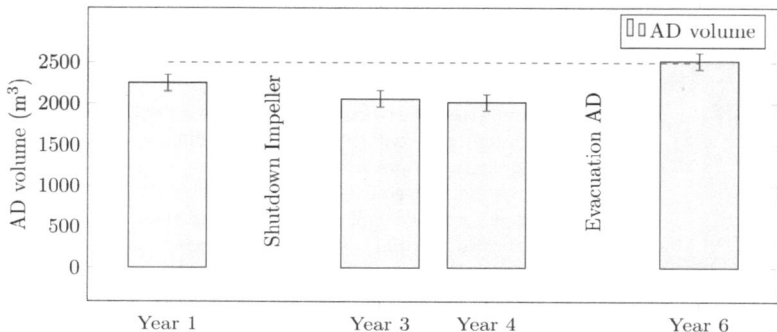

Figure 7. Investigation of sediment behavior after impeller termination. While no significant change in available AD volume was detected while terminating impeller, complete regeneration of AD volume is shown after complete sediment evacuation.

The results indicate that the steady and continuous decrease in usable AD volume was not impacted significantly by the termination of the impeller. These findings suggest that overmixing mitigation strategies, such as terminating the draft tube, can effectively reduce energy demand without adversely affecting biogas production or the hydrodynamics. The results of this in-field study highlight the potential benefits of re-evaluating mixing strategies in WWTPs to optimize both operational efficiency and cost-effectiveness. Further research is warranted to generalize these findings across different plant configurations and operational conditions as well as to assess the long-term implications on system performance and maintenance. Therefore, the investigated plant has terminated the impeller mixing and solely relies on a pumped recirculation after the hydrodynamics and potential overmixing were highlighted.

4. Discussion

The results of the field investigations and survey-based AD studies highlight that while there is a general trend in geometric AD designs and mixing approaches, the operational strategies of most plants investigated in this study rely on individual operators' expertise and manual intervention when specific parameters change during the AD process. While the structural AD design of new plants—regarding the overall volume, shape, and operation mode—is generally based on existing reference plants with a similar PE, the task of digester mixing is often outsourced and planned by external parties. Despite the proven significance of dynamic operation (e.g., intermittent mixing) in several studies [43,67,68,76,77], the majority of the implemented mixing approaches of the investigated plants are designed statically for one single operating point, providing limited flexibility to adapt to changing process properties. This leads to a less ideal harmonization of the mixing approach with the constantly changing operational conditions of the specific plants and subsequently to either overmixing or insufficient mixing. While the specific design of AD has been the subject of many scientific optimization studies [8,12,49,76,78,79], practical AD design and operation are often constricted on the operators' experience and prevailing habits. This can lead to black-box thinking and subsequently inefficient operation and increased expenses, as stated by the operators of several of the investigated plants.

Regarding the structural design of AD, this survey-based study has highlighted a predominance in cylindrical design followed by egg-shaped AD towers. This is reasoned with less ideal hydrodynamics or space efficiency when using rectangular or spherical AD designs, respectively. Scientific investigations regarding optimized hydrodynamics in the context of AD also show a predominantly preferred utilization of cylindrical AD geometries [7,17,32,33,53,80]. The cylindrical shape is favored because of simple construction, good hydrodynamics, and reasonable space utilization. Deeper analyses of scientific publications in recent decades have also shown a continuous growth in publications regarding egg-shaped AD designs [17,31,34–36,81,82]. This is due to the fact that the curved geometry offers a promising investigation template when harmonized with specific mixing approaches and feedstock inlet configurations [34,35]. The pumped recirculation can especially offer a beneficial impact on hydrodynamics when properly aligned with the reactors' curvature.

As highlighted both in the study results and in feedback by the plant operators, the possibility of a dynamic operation of multiple AD towers can significantly help to increase the overall system efficiency by enabling targeted responses to changes in fluid properties or external influences on the AD process. The serial mode of operation is stated to lead to a higher gas production rate by providing different and specifically tailored environmental conditions in each tower [41,43]. The resulting optimized activity of specific microorganisms at different stages of digestion has a beneficial impact on the gas rate. However, the first-stage digester needs to be capable of handling the amount of substrate. According to the feedback of the AD operators, cold temperatures combined with high amounts of co-substrate can result in foaming layers, decreasing the efficiency of AD towers by inhibiting gas release and disrupting the microbial community. Parallel operation is favored when processing large amounts of sludge or different types of substrate that require individual mixing strategies [83].

Regarding mixing, gas injection and mechanical agitation are demonstrated to be the most frequently implemented approaches according to the survey (Figure 2b). This is consistent with the trend observed in scientific studies, which report an increased frequency of 62% for mechanical agitation and 19% for both gas injection and pumped recirculation, as summarized by Caillet et al. [12]. According to Table 1, the sum of the specific power $\sum \Phi_{Survey}$ required for pumped recirculation and optional additional mixing spans from 1.6 W m^{-3} to 10.2 W m^{-3}, with the share of $\Phi_{Pumped\ Recirculation}$ being higher than Φ_{Mixing} for 20 of 34 plants. Therefore, the investigated plants are in the range of the general recommendation of the United States Environmental Protection Agency [74], with 5–8 W m^{-3} for digester mixing, which is often cited in studies addressing AD mixing [36,69,73]. However, this recommendation does not account for fluid properties (e.g., non-Newtonian behavior) or the digester design. In particular, the comprehensive studies of Wu et al. [3] report similar ranges in specific power, with 4.11 W m^{-3}, 5.0 W m^{-3}, and 5.9 W m^{-3} for mechanical mixing, gas-induced mixing, and a pumped recirculation, respectively. Regarding draft tube mixing, similar values are reported between approximately 4 W m^{-3} and 12 W m^{-3} [55,75,84]. The inserted specific power significantly relies on the basic AD design and fluid properties. Lower values are also reported as sufficient for mixing in specific cases without compromising hydrodynamics, as shown by Grazia Leonzio [85] with 1.9 W m^{-3} to 2.7 W m^{-3} exclusively using a pumped recirculation, Xinxin et al. [86] with 0.5 W m^{-3} and Oates et al. [7] with 0.3 W m^{-3} for mechanical agitation, or by Dapelo and Bridgeman [58] with 1 W m^{-3} for gas-induced mixing. However, certain process variables can have a drastic impact on the calculated specific power, as exemplified in the studies of Soroush et al. [80] using 50 W m^{-3} due to a relatively low AD volume and Li et al. [56], reporting ranges from 21 to 131 W m^{-3} due to very high %TS concentrations. This underlines the fact that the recommendation of 5–8 W m^{-3} is only a rough guideline, and individual AD specific consideration of the induced specific mixing power is highly important.

It was noted that some of the provided energetic data of the answered surveys deviates from the experimentally measured energetic demands derived from the field studies, as highlighted in Figure 4. This can be mainly attributed to the fact that the plant operators provide annual averaged data, while the experimental measurements were conducted for a maximum of 24 h.

Direct contact with the surveyed plants has revealed that there is a tendency to transition from gas-inducing methods to mechanical agitation due to a potential decrease in maintenance and operational costs. This was furthermore reasoned with a better dynamic control regarding changing sludge properties and hydrodynamics. This, however, necessitates the implementation of dynamic motor control for the respective mixing approach to specifically adapt to the continuously changing operational conditions and, therefore, prevents insufficient or too-intense mixing. The comparison between gas mixing and mechanical mixing is also a continuously debated topic in scientific studies [11,12,87]. Most plants investigated rely on static speed control, especially when operating mechanical agitation devices. Retrofitting a dynamic control of the rotational speed is often not possible because of static gear shifts. Therefore, the exclusive implementation of a new dynamic mixing strategy is often associated with problematic downtimes and high expenses. Static motor controls usually only allow for changing the direction of rotation, which is useful for preventing clogging of the rotating stirring devices. Dynamic regulation of the rotational speed would allow for optimizing the induced mixing depending on changes in the substrate or reactor volume. It was noted that when a dynamic velocity control is not possible, plant operators tend to implement intermittent mixing as also implemented in several optimization studies [51,67,68,77]. The short-term result is a decrease in operational costs due to lower required mixing power, but it also can increase the biogas production, since different bacteria show specific microbial activity at certain mixing intensities. The potential negative effects of overmixing were also proven in laboratory experiments in previous studies [51,64,88].

Within this study, inefficient mixing was investigated and subsequently confirmed for two individual cases. Changes in the mixing stage were implemented for Plants 1 and 2 during the field study. As presented in Table 3, the transition from gas injection to mechanical agitation decreased the required energetic demand for external mixing by almost 94%. The specific mixing power required for gas injection before the transition is documented in Table 1 with $\Phi_{\text{Mixing, Survey}} = 6$ W m^{-3}. This is consistent with the aforementioned typical values recommended for thorough digester mixing [3,69,73,74]. The combined required specific power demand of mixing and pumped recirculation $\sum \Phi_{\text{Survey}}$ is subsequently reduced by 52% from 10.2 W m^{-3} to 4.9 W m^{-3}. At the same time, according to the data in Table 3, gas production was also affected. The gas production between both cases increased significantly from 1,680,250 m^3 to 2,248,200 m^3 for case A. However, both the OLR and the AD volume were higher in case A, with 2.3 kg ODM/m^3 d and 8552 m^3 compared to 1.9 kg ODM/m^3 d and 8217 m^3 in the previous case B. Even after considering the change in OLR and AD volume, this still results in a slightly increased SGP of 0.313 Nm^3 CH$_4$/kg ODM for case A compared to 0.295 Nm^3 CH$_4$/kg ODM for case B. The comparison of the associated SGPs shows good accordance with values found in the literature ranging between 0.1 and 0.5 Nm^3 CH$_4$/kg ODM [89,90]. Although an increase in gas production can be derived from the data, it is necessary to critically assess this due to large-scale measurements and associated uncertainties in the determination of gas parameters. However, the transition from gas injection to mechanical agitation, which subsequently decreased the required energetic demand, definitely did not impact the gas production negatively. This indicates that for this specific case, the gas injection additional to the pumped recirculation was unneeded and a mechanical agitator with lower power consumption was sufficient. Therefore, it can be stated that the efficiency of the plant was significantly increased.

For Plant 2, the redundancy of impeller mixing inside the egg-shaped reactors was confirmed by the field experiments, since the used pumped recirculation was demonstrated to be sufficient for complete digestion of the organic matter without increasing RGP values. According to Table 2, the specific power demand $\sum \Phi_{FE}$ between cases 2A and 2B was reduced by 30% by terminating the impeller-induced mixing. The $\Phi_{\text{Pumped Recirculation F.E.}}$ of 2.9 W m^{-3}, again, in the common range associated with specific power required for pumped recirculation, is demonstrated to be sufficient according to the absence of observed changes in biogas production, as indicated by the BMP comparison of ideally processed substrate derived from laboratory BMP trials with the monitored SGP of Plant 2A (Figure 6). Notably, the comparison of the SGP of Plant 2A, employing only pumped recirculation, with the BMP derived under ideal conditions shows nearly identical values (avg. laboratory BMP = 456 Nm^3 CH$_4$/t ODM, SGP of 2A = 445 Nm^3 CH$_4$/t ODM). This indicates a very efficient AD as compared to common values found in the literature, where the SGP is usually only 85–95% of the BMP [71,91]. The range is also very comparable to common BMP values, which are strongly dependent on the used sludge composition [92–96]. While impeller mixing has been reported to be efficient, especially in combination with curved egg-shaped geometries [35,36], the presented results demonstrate that for this individual case with pumped recirculation, additional mixing is deemed as redundant. Therefore, the plant's operators ultimately decided to terminate the impeller-induced mixing and since then only rely on mixing through pumped recirculation. The efficiency of mixing with pumped recirculation could further be improved by a reduction in dead zones by dynamically changing the position of the suction of the sludge intake and occasional drainage of the bottom sludge [35,36,81].

The optimization of the mixing stage was successful in both cases, leading to the assumption that there is much more potential for optimizing each of the remaining plants. However, AD in domestic WWTPs is a continuous process that is very difficult to shut down, especially if plant operators run on full capacity and do not have the opportunity to temporarily terminate one of multiple available digesters. This underlines the importance of careful planning when designing new AD towers or implementing changes in existing plants. In particular, a dynamic control of the mixing approach [43,67,68,76,77] as well as the possibility to easily switch between parallel and serial operation modes [41,43] was a reoccurring recommendation by the plant operators in the direct exchange of know-how and operational experience. The possibility of dynamic process control by manipulating mixing variables can improve the plant's efficiency and prevent overmixing. While no correlation between the mixing approach and RGP was proven, the operation mode seems to impact the overall RGP. When using serial operation, each AD tower can be tailored specifically to the requirements for optimal AD and gas production. Particularly, Plant 8 showed a decreased RGP when operated in serial. The reduction in RGP is substantial both for the plants efficiency and sustainability in operation, since both CO_2 and CH_4 are greenhouse gases.

In summary, the investigations and the direct contact with AD operators have revealed that while AD operation and mixing are extensively studied in scientific research, there remains a stigma around overmixing AD towers, which leads to the acceptance of disadvantageous results, such as higher operational costs or decreased microbial activity [51,63,64]. This is related to the fact that most of the time there is no possibility to monitor the mixing of opaque sludge in optically non-accessible AD towers. While real-time mixing monitoring is possible at laboratory scales with sophisticated measurement approaches [5,10,97] which are not practical for constant real-scale live-monitoring, the implementation of a simple dynamic mixing control would be sufficient to enable the possibility of alternating the mixing intensity by simultaneously evaluating the gas production and process-related AD phenomena such as foaming, sedimentation, and RGP. On the basis of Plant A and B, it was shown that even little changes in the mixing stage can drastically decrease energy requirements and subsequently increase the plants' energetic efficiency. This, however, requires individual consideration of the investigated plants' design and operation conditions. In general, cylindrical shapes are favored because of their simplicity in both operation and construction [31–33]. Coupled with dynamic internal pumped recirculation and also an additional dynamic mechanical agitation, the complex mixing process can be adapted to changing process or fluid properties if necessary. Depending on the volume and space, a parallel or serial AD setup can increase capacity or gas production [41,43]. The final determination of the optimal mixing intensity remains a crucial objective, which can be achieved through the implementation of dynamic mixing control, which subsequently helps to reduce costs and minimize the potential for residual gas.

5. Conclusions

The presented combined evaluation of AD plants using surveys and field studies provides a comprehensive summary of the state-of-the-art in AD design and mixing strategies, and it highlights potential optimization points for similar existing systems and comparable technical applications. It was highlighted that besides cylindrical reactor geometries, the variable operation of two or more AD towers with a dynamic mixing strategy is the most favored approach within the investigated region. Regarding mixing, a tendency towards mechanical agitation was noticed, which is related to better control, lower operational costs, and easier maintenance compared to gas injection and especially impeller-driven draft tube mixing. Two specific field studies demonstrated practically that AD systems might often be operated in a habit-driven and potentially inefficient manner by evaluating induced mixing and optimizing it regarding energetic demand and gas-related efficiency. In both cases, the operational power demands for mixing were reduced by 52% and 30%, respectively. Besides plant-specific optimization, the survey, and especially the direct contact with plant operators, highlighted which AD designs are frequently used and identified common operational problems. Furthermore, prevention methods and practical recommendations for common AD phenomena are presented. It is notable that the inefficient operation of AD most often contributed to difficulties in the process monitoring, especially regarding the hydrodynamics. Therefore, there is significant potential for optimization within the AD context, which can reduce operational and maintenance costs, as well as investment costs, with proper planning in the design phase. Applied refinement of AD processes, as presented in this study, can help increase gas production and decrease gas losses and operational expenses, positioning AD as an indispensable green energy source. The conducted optimizations not only hold economic value because of decreased energetic costs, but also have environmental impact because of the reduction in RGP associated with greenhouse gas emissions. In conclusion, this study underscores the importance of continuous improvement and optimization in AD systems to enhance their efficiency and sustainability as a green energy source.

Author Contributions: Conceptualization, T.N., M.P., M.M. and W.R.; methodology, T.N., S.D.-W., C.E., M.P., M.M., T.S. and J.W.; software, S.D.-W. and T.N.; validation, T.N., S.D.-W. and C.E.; formal analysis, T.N., S.D.-W. and C.E.; investigation, T.N., W.R., M.M. and M.P.; resources, M.M., M.P. and W.R.; data curation, T.N., T.S. and J.W.; writing—original draft preparation, T.N., M.M., T.S. and J.W.; writing—review and editing, T.N., T.S., M.M., J.W., M.P. and W.R.; visualization, T.S., T.N. and J.W.; supervision, W.R., M.P. and M.M.; project administration, W.R. and M.M.; funding acquisition, W.R. and M.M. All authors have read and agreed to the published version of the manuscript.

Funding: This research was funded by Austrian Federal Ministry of Agriculture, Forestry, Regions and Water Management, grant number B801259.

Data Availability Statement: Due to data protection, data will be made available in an anonymized form upon request.

Conflicts of Interest: Author Simon Draxl-Weiskopf was employed by the company BioTreaT GmbH. The remaining authors declare that the research was conducted in the absence of any commercial or financial relationships that could be construed as a potential conflict of interest.

References

1. Letcher, T. Global Warming, Greenhouse Gases, Renewable Energy, and Storing Energy. In *Storing Energy*; Elsevier: Amsterdam, The Netherlands, 2022; pp. 2–3.
2. Grangeiro, L.C.; de Almeida, S.G.C.; de Mello, B.S.; Fuess, L.T.; Sarti, A.; Dussán, K.J. Chapter 7-New Trends in Biogas Production and Utilization. In *Sustainable Bioenergy*; Rai, M., Ingle, A.P., Eds.; Elsevier: Amsterdam, The Netherlands, 2019; pp. 199–223, ISBN 978-0-12-817654-2.
3. Wu, B. CFD Simulation of Gas and Non-Newtonian Fluid Two-Phase Flow in Anaerobic Digesters. *Water Res.* 2010, 44, 3861–3874. [CrossRef]
4. Wu, B. CFD Investigation of Turbulence Models for Mechanical Agitation of Non-Newtonian Fluids in Anaerobic Digesters. *Water Res.* 2011, 45, 2082–2094. [CrossRef] [PubMed]
5. Neuner, T.; Meister, M.; Pillei, M.; Koch, M.; Rauch, W. Numerical and Experimental Flow Investigation Using Ultrasonic PIV for Optimizing Mechanically Agitated Lab-Scale Anaerobic Digesters. *Chem. Eng. Sci.* 2022, 264, 118129. [CrossRef]
6. Winkler, J.; Neuner, T.; Hupfauf, S.; Arthofer, A.; Ebner, C.; Rauch, W.; Bockreis, A. Impact of Impeller Design on Anaerobic Digestion: Assessment of Mixing Dynamics, Methane Yield, Microbial Communities and Digestate Dewaterability. *Bioresour. Technol.* 2024, 406, 131095. [CrossRef]
7. Oates, A.; Neuner, T.; Meister, M.; Borman, D.; Camargo-Valero, M.; Sleigh, A.; Fischer, P. Modelling Mechanically Induced Non-Newtonian Flows to Improve the Energy Efficiency of Anaerobic Digesters. *Water* 2020, 12, 2995. [CrossRef]
8. Neuner, T.; Meister, M.; Pillei, M.; Rauch, W. Optimizing Mixing Efficiency of Anaerobic Digesters with High Total Solids Concentrations Using Validated CFD Simulations. *Biochem. Eng. J.* 2024, 208, 109320. [CrossRef]
9. Lebranchu, A.; Delaunay, S.; Marchal, P.; Blanchard, F.; Pacaud, S.; Fick, M.; Olmos, E. Impact of Shear Stress and Impeller Design on the Production of Biogas in Anaerobic Digesters. *Bioresour. Technol.* 2017, 245, 1139–1147. [CrossRef]
10. Tanguy, P.A.; Lacroix, R.; Bertrand, F.; Choplin, L.; Brito-De La Fuente, E. Mixing of Non-Newtonian Viscous Fluids with Helical Impellers: Experimental and Three-Dimensional Numerical Studies. *Am. Inst. Chem. Eng.* 1992, 286, 33–37.
11. Sasidhar, K.B.; Somasundaram, M.; Ekambaram, P.; Arumugam, S.K.; Nataraj, G.; Murugan, M.A. A Critical Review on the Effects of Pneumatic Mixing in Anaerobic Digestion Process. *J. Clean. Prod.* 2022, 378, 134513. [CrossRef]
12. Caillet, H.; Bastide, A.; Adelard, L. Advances in Computational Fluid Dynamics Modeling of Anaerobic Digestion Process for Renewable Energy Production: A Review. *Clean. Waste Syst.* 2023, 6, 100124. [CrossRef]
13. Appels, L.; Lauwers, J.; Degreve, J.; Helsen, L.; Lievens, B.; Willems, K.; Dewil, R. Anaerobic Digestion in Global Bio-Energy Production-Potential and Research. *Renew. Sustain. Energy Rev.* 2011, 15, 4295–4301. [CrossRef]
14. Vasan, V.; Sridharan, N.V.; Feroskhan, M.; Vaithiyanathan, S.; Subramanian, B.; Tsai, P.C.; Lin, Y.C.; Lay, C.H.; Wang, C.T.; Ponnusamy, V.K. Biogas Production and Its Utilization in Internal Combustion Engines—A Review. *Process Saf. Environ. Prot.* 2024, 186, 518–539. [CrossRef]
15. Deublein, D.; Steinhauser, A. *Biogas from Waste and Renewable Resources*; Wiley-VCH: Weinheim, Germany, 2008.
16. Radetic, B. Anaerobic Digestion, Important Aspects Regarding Digester Design and Sludge Mixing Systems. In *Handbook of Water and Used Water Purification*; Lahnsteiner, J., Ed.; Springer International Publishing: Cham, Germany, 2024; pp. 737–749, ISBN 978-3-319-78000-9.
17. Yagna Prasad, K. Gas Mixing in Anaerobic Sludge Digestion. In *Handbook of Water and Used Water Purification*; Lahnsteiner, J., Ed.; Springer International Publishing: Cham, Germany, 2020; pp. 1–16, ISBN 978-3-319-66382-1.
18. Hosseini, S.E.; Wahid, M.A. Development of Biogas Combustion in Combined Heat and Power Generation. *Renew. Sustain. Energy Rev.* 2014, 40, 868–875. [CrossRef]
19. Marañón, E.; Salter, A.M.; Castrillón, L.; Heaven, S.; Fernández-Nava, Y. Reducing the Environmental Impact of Methane Emissions from Dairy Farms by Anaerobic Digestion of Cattle Waste. *Waste Manag.* 2011, 31, 1745–1751. [CrossRef]
20. Ólafsdóttir, S.S.; Jensen, C.D.; Lymperatou, A.; Henriksen, U.B.; Gavala, H.N. Effects of Different Treatments of Manure on Mitigating Methane Emissions during Storage and Preserving the Methane Potential for Anaerobic Digestion. *J. Environ. Manag.* 2023, 325, 116456. [CrossRef] [PubMed]
21. Shah, S.V.; Yadav Lamba, B.; Tiwari, A.K.; Chen, W.-H. Sustainable Biogas Production via Anaerobic Digestion with Focus on CSTR Technology: A Review. *J. Taiwan Inst. Chem. Eng.* 2024, 162, 105575. [CrossRef]
22. Hu, Y.; Ma, H.; Wu, J.; Kobayashi, T.; Xu, K.-Q. Performance Comparison of CSTR and CSFBR in Anaerobic Co-Digestion of Food Waste with Grease Trap Waste. *Energies* 2022, 15, 8929. [CrossRef]
23. Jiraprasertwong, A.; Vichaitanapat, K.; Leethochawalit, M.; Chavadej, S. Three-Stage Anaerobic Sequencing Batch Reactor (ASBR) for Maximum Methane Production: Effects of COD Loading Rate and Reactor Volumetric Ratio. *Energies* 2018, 11, 1543. [CrossRef]
24. Khumalo, S.M.; Bakare, B.F.; Tetteh, E.K.; Rathilal, S. Sequencing Batch Reactor Performance Evaluation on Orthophosphates and COD Removal from Brewery Wastewater. *Fermentation* 2022, 8, 296. [CrossRef]
25. Switzenbaum, M.S. Anaerobic Fixed Film Wastewater Treatment. *Enzyme Microb. Technol.* 1983, 5, 242–250. [CrossRef]
26. Pererva, Y.; Miller, C.D.; Sims, R.C. Approaches in Design of Laboratory-Scale UASB Reactors. *Processes* 2020, 8, 734. [CrossRef]
27. Mainardis, M.; Buttazzoni, M.; Goi, D. Up-Flow Anaerobic Sludge Blanket (UASB) Technology for Energy Recovery: A Review on State-of-the-Art and Recent Technological Advances. *Bioengineering* 2020, 7, 43. [CrossRef] [PubMed]
28. Ward, A.J.; Hobbs, P.J.; Holliman, P.J.; Jones, D.L. Optimisation of the Anaerobic Digestion of Agricultural Resources. *Bioresour. Technol.* 2008, 99, 7928–7940. [CrossRef] [PubMed]

29. Hamawand, I. Anaerobic Digestion Process and Bio-Energy in Meat Industry: A Review and a Potential. *Renew. Sustain. Energy Rev.* **2015**, *44*, 37–51. [CrossRef]
30. Nasir, I.M.; Mohd Ghazi, T.I.; Omar, R. Anaerobic Digestion Technology in Livestock Manure Treatment for Biogas Production: A Review. *Eng. Life Sci.* **2012**, *12*, 258–269. [CrossRef]
31. Fagbohungbe, M.O.; Dodd, I.C.; Herbert, B.M.J.; Li, H.; Ricketts, L.; Semple, K.T. High Solid Anaerobic Digestion: Operational Challenges and Possibilities. *Environ. Technol. Innov.* **2015**, *4*, 268–284. [CrossRef]
32. Subhaschandra Singh, T.; Nath Verma, T.; Nashine, P. Analysis of an Anaerobic Digester Using Numerical and Experimental Method for Biogas Production. *Mater. Today Proc.* **2018**, *5*, 5202–5207. [CrossRef]
33. Subramanian, B.; Pagilla, K.R. Anaerobic Digester Foaming in Full-Scale Cylindrical Digesters–Effects of Organic Loading Rate, Feed Characteristics, and Mixing. *Bioresour. Technol.* **2014**, *159*, 182–192. [CrossRef]
34. Dabiri, S.; Sappl, J.; Kumar, P.; Meister, M.; Rauch, W. On the Effect of the Inlet Configuration for Anaerobic Digester Mixing. *Bioprocess Biosyst. Eng.* **2021**, *44*, 2455–2468. [CrossRef]
35. Meister, M.; Rezavand, M.; Ebner, C.; Pümpel, T.; Rauch, W. Mixing Non-Newtonian Flows in Anaerobic Digesters by Impellers and Pumped Recirculation. *Adv. Eng. Softw.* **2018**, *115*, 194–203. [CrossRef]
36. Wu, B. CFD Simulation of Mixing in Egg-Shaped Anaerobic Digesters. *Water Res.* **2010**, *44*, 1507–1519. [CrossRef] [PubMed]
37. Mendieta-Pino, C.A.; Garcia-Ramirez, T.; Ramos-Martin, A.; Perez-Baez, S.O. Experience of Application of Natural Treatment Systems for Wastewater (NTSW) in Livestock Farms in Canary Islands. *Water* **2022**, *14*, 2279. [CrossRef]
38. Siswantara, A.I.; Daryus, A.; Darmawan, S.; Gunadi, G.G.; Camalia, R. CFD Analysis of Slurry Flow in an Anaerobic Digester. *Int. J. Technol.* **2016**, *7*, 197. [CrossRef]
39. Park, Y.; Khim, J.; Kim, J.D. Application of a Full-Scale Horizontal Anaerobic Digester for the Co-Digestion of Pig Manure, Food Waste, Excretion, and Thickened Sewage Sludge. *Processes* **2023**, *11*, 1294. [CrossRef]
40. Deng, L.; Liu, Y.; Wang, W. Construction Materials and Structures of Digesters. In *Biogas Technology*; Springer: Singapore, 2020; pp. 157–199, ISBN 978-981-15-4940-3.
41. Walter, A.; Hanser, M.; Ebner, C.; Insam, H.; Markt, R.; Hupfauf, S.; Probst, M. Stability of the Anaerobic Digestion Process during Switch from Parallel to Serial Operation—A Microbiome Study. *Sustainability* **2022**, *14*, 7161. [CrossRef]
42. Corigliano, O.; Iannuzzi, M.; Pellegrino, C.; D'Amico, F.; Pagnotta, L.; Fragiacomo, P. Enhancing Energy Processes and Facilities Redesign in an Anaerobic Digestion Plant for Biomethane Production. *Energies* **2023**, *16*, 5782. [CrossRef]
43. Singh, B.; Kovács, K.L.; Bagi, Z.; Nyári, J.; Szepesi, G.L.; Petrik, M.; Siménfalvi, Z.; Szamosi, Z. Enhancing Efficiency of Anaerobic Digestion by Optimization of Mixing Regimes Using Helical Ribbon Impeller. *Fermentation* **2021**, *7*, 251. [CrossRef]
44. Steiniger, B.; Hupfauf, S.; Insam, H.; Schaum, C. Exploring Anaerobic Digestion from Mesophilic to Thermophilic Temperatures—Operational and Microbial Aspects. *Fermentation* **2023**, *9*, 798. [CrossRef]
45. El Ibrahimi, M.; Khay, I.; El Maakoul, A.; Bakhouya, M. Food Waste Treatment through Anaerobic Co-Digestion: Effects of Mixing Intensity on the Thermohydraulic Performance and Methane Production of a Liquid Recirculation Digester. *Process Saf. Environ. Prot.* **2021**, *147*, 1171–1184. [CrossRef]
46. Singh, R.; Hans, M.; Kumar, S.; Yadav, Y.K. Thermophilic Anaerobic Digestion: An Advancement towards Enhanced Biogas Production from Lignocellulosic Biomass. *Sustainability* **2023**, *15*, 1859. [CrossRef]
47. Yang, T.S.; Flores-Rodriguez, C.; Torres-Albarracin, L.; da Silva, A.J. Thermochemical Pretreatment for Improving the Psychrophilic Anaerobic Digestion of Coffee Husks. *Methane* **2024**, *3*, 214–226. [CrossRef]
48. Karim, K.; Varma, R.; Vesvikar, M.; Al-Dahhan, M.H. Flow Pattern Visualization of a Simulated Digester. *Water Res.* **2004**, *38*, 3659–3670. [CrossRef] [PubMed]
49. Terashima, M.; Goel, R.; Komatsu, K.; Yasui, H.; Takahashi, H.; Li, Y.Y.; Noike, T. CFD Simulation of Mixing in Anaerobic Digesters. *Bioresour. Technol.* **2009**, *100*, 2228–2233. [CrossRef]
50. Wiedemann, L.; Conti, F.; Saidi, A.; Sonnleitner, M.; Goldbrunner, M. Modeling Mixing in Anaerobic Digesters with Computational Fluid Dynamics Validated by Experiments. *Chem. Eng. Technol.* **2018**, *41*, 2101–2110. [CrossRef]
51. Kowalczyk, A.; Harnisch, E.; Schwede, S.; Gerber, M.; Span, R. Different Mixing Modes for Biogas Plants Using Energy Crops. *Appl. Energy* **2013**, *112*, 465–472. [CrossRef]
52. Chang, C.-C.; Kuo-Dahab, C.; Chapman, T.; Mei, Y. Anaerobic Digestion, Mixing, Environmental Fate, and Transport. *Water Environ. Res.* **2019**, *91*, 1210–1222. [CrossRef] [PubMed]
53. Ameur, H.; Kamla, Y.; Sahel, D. Performance of Helical Ribbon and Screw Impellers for Mixing Viscous Fluids in Cylindrical Reactors. *ChemEngineering* **2018**, *2*, 26. [CrossRef]
54. Wu, B. CFD Analysis of Mechanical Mixing in Anaerobic Digesters. *Trans. ASABE* **2009**, *52*, 1371–1382. [CrossRef]
55. Meroney, R.N.; Colorado, P.E. CFD Simulation of Mechanical Draft Tube Mixing in Anaerobic Digester Tanks. *Water Res.* **2009**, *43*, 1040–1050. [CrossRef]
56. Li, L.; Wang, K.; Wei, L.; Zhao, Q.; Zhou, H.; Jiang, J. CFD Simulation and Performance Evaluation of Gas Mixing during High Solids Anaerobic Digestion of Food Waste. *Biochem. Eng. J.* **2022**, *178*, 108279. [CrossRef]
57. Vesvikar, M.S.; Al-Dahhan, M. Flow Pattern Visualization in a Mimic Anaerobic Digester Using CFD. *Biotechnol. Bioeng.* **2005**, *89*, 719–732. [CrossRef] [PubMed]
58. Dapelo, D.; Bridgeman, J. Euler-Lagrange Computational Fluid Dynamics Simulation of a Full-Scale Unconfined Anaerobic Digester for Wastewater Sludge Treatment. *Adv. Eng. Softw.* **2018**, *117*, 153–169. [CrossRef]

59. Wellinger, A.; Lindberg, A. Biogas Upgrading and Utilisation. *IEA Bioenergy Task* **2000**, *24*, 3–17.
60. Bridgeman, J. Computational Fluid Dynamics Modelling of Sewage Sludge Mixing in an Anaerobic Digester. *Adv. Eng. Softw.* **2012**, *44*, 54–62. [CrossRef]
61. Morchain, J.; Gabelle, J.-C.; Cockx, A. A Coupled Population Balance Model and CFD Approach for the Simulation of Mixing Issues in Lab-Scale and Industrial Bioreactors. *AIChE J.* **2013**. Early View. [CrossRef]
62. Ahmad, A.; Ghufran, R.; Nasir, Q.; Shahitha, F.; Al-Sibani, M.; Al-Rahbi, A.S. Enhanced Anaerobic Co-Digestion of Food Waste and Solid Poultry Slaughterhouse Waste Using Fixed Bed Digester: Performance and Energy Recovery. *Environ. Technol. Innov.* **2023**, *30*, 103099. [CrossRef]
63. Kaparaju, P.; Buendia, I.; Ellegaard, L.; Angelidakia, I. Effects of Mixing on Methane Production during Thermophilic Anaerobic Digestion of Manure: Lab-Scale and Pilot-Scale Studies. *Bioresour. Technol.* **2008**, *99*, 4919–4928. [CrossRef]
64. Amani, T.; Nosrati, M.; Sreekrishnan, T.R. Anaerobic Digestion from the Viewpoint of Microbiological, Chemical, and Operational Aspects—A Review. *Environ. Rev.* **2010**, *18*, 255–278. [CrossRef]
65. Ning, X.; Huang, Y.; Huang, P.; Ou, X.; Luo, H.; Bai, Z.; Chen, H.; Ge, X.; Li, L. Effect of the Inoculum Mixing Ratio on the Anaerobic Digestion of Food Waste: Reactor Performance and Microbial Community. *Environ. Technol. Innov.* **2024**, *35*, 103680. [CrossRef]
66. Mills, P.J. Minimisation of Energy Input Requirements of an Anaerobic Digester. *Agric. Wastes* **1979**, *1*, 57–66. [CrossRef]
67. Wang, Y.; Zhang, J.; Sun, Y.; Yu, J.; Zheng, Z.; Li, S.; Cui, Z.; Hao, J.; Li, G. Effects of Intermittent Mixing Mode on Solid State Anaerobic Digestion of Agricultural Wastes. *Chemosphere* **2020**, *248*, 126055. [CrossRef]
68. Li, L.; Wang, K.; Sun, Z.; Zhao, Q.; Zhou, H.; Gao, Q.; Jiang, J.; Mei, W. Effect of Optimized Intermittent Mixing during High-Solids Anaerobic Co-Digestion of Food Waste and Sewage Sludge: Simulation, Performance, and Mechanisms. *Sci. Total Environ.* **2022**, *842*, 156882. [CrossRef] [PubMed]
69. Lindmark, J.; Thorin, E.; Bel Fdhila, R.; Dahlquist, E. Effects of Mixing on the Result of Anaerobic Digestion: Review. *Renew. Sustain. Energy Rev.* **2014**, *40*, 1030–1047. [CrossRef]
70. Achinas, S.; Achinas, V.; Euverink, G.J.W. Chapter 2—Microbiology and Biochemistry of Anaerobic Digesters: An Overview. In *Bioreactors*; Singh, L., Yousuf, A., Mahapatra, D.M., Eds.; Elsevier: Amsterdam, The Netherlands, 2020; pp. 17–26, ISBN 978-0-12-821264-6.
71. Cho, J.K.; Park, S.C.; Chang, H.N. Biochemical Methane Potential and Solid State Anaerobic Digestion of Korean Food Wastes. *Bioresour. Technol.* **1995**, *52*, 245–253. [CrossRef]
72. Green, D.W.; Perry, R.H. *Perry's Chemical Engineers' Handbook*, 8th ed.; McGraw-Hill Education: New York, NY, USA, 2008; ISBN 9780071422949.
73. Stukenberg, J.R.; Clark, J.H.; Sandino, J.; Naydo, W.R. Egg-Shaped Digesters: From Germany to the US. *Water Environ. Technol.* **1992**, *4*, 42–51.
74. U.S. Environmental Protection Agency. *Process Design Manual for Sludge Treatment and Disposal*; U.S. Environmental Protection Agency: Washington, DC, USA, 1974.
75. Craig, K.J.; Nieuwoudt, M.N.; Niemand, L.J. CFD Simulation of Anaerobic Digester with Variable Sewage Sludge Rheology. *Water Res.* **2013**, *47*, 4485–4497. [CrossRef]
76. Liu, Y.; Wang, X.; Sun, Y. Optimization and Experimental Study of Variable Frequency Intermittent Stirring Strategy for Anaerobic Digestion Based on CFD. *Fuel* **2024**, *355*, 129371. [CrossRef]
77. Singh, B.; Kovács, K.L.; Bagi, Z.; Petrik, M.; Szepesi, G.L.; Siménfalvi, Z.; Szamosi, Z. Significance of Intermittent Mixing in Mesophilic Anaerobic Digester. *Fermentation* **2022**, *8*, 518. [CrossRef]
78. Weldehans, M.G. Optimization of Distillery-Sourced Wastewater Anaerobic Digestion for Biogas Production. *Clean. Waste Syst.* **2023**, *6*, 100118. [CrossRef]
79. Zamani Abyaneh, E.; Zarghami, R.; Krühne, U.; Rosinha Grundtvig, I.P.; Ramin, P.; Mostoufi, N. Mixing Assessment of an Industrial Anaerobic Digestion Reactor Using CFD. *Renew. Energy* **2022**, *192*, 537–549. [CrossRef]
80. Dabiri, S.; Noorpoor, A.; Arfaee, M.; Kumar, P.; Rauch, W. CFD Modeling of a Stirred Anaerobic Digestion Tank for Evaluating Energy Consumption through Mixing. *Water* **2021**, *13*, 1629. [CrossRef]
81. Subramanian, B.; Miot, A.; Jones, B.; Klibert, C.; Pagilla, K.R. A Full-Scale Study of Mixing and Foaming in Egg-Shaped Anaerobic Digesters. *Bioresour. Technol.* **2015**, *192*, 461–470. [CrossRef] [PubMed]
82. Singh, B.; Singh, N.; Čonka, Z.; Kolcun, M.; Siménfalvi, Z.; Péter, Z.; Szamosi, Z. Critical Analysis of Methods Adopted for Evaluation of Mixing Efficiency in an Anaerobic Digester. *Sustainability* **2021**, *13*, 6668. [CrossRef]
83. Tamborrino, A.; Catalano, F.; Leone, A.; Bianchi, B. A Real Case Study of a Full-Scale Anaerobic Digestion Plant Powered by Olive By-Products. *Foods* **2021**, *10*, 1946. [CrossRef]
84. Kariyama, I.D.; Zhai, X.; Wu, B. Influence of Mixing on Anaerobic Digestion Efficiency in Stirred Tank Digesters: A Review. *Water Res.* **2018**, *143*, 503–517. [CrossRef]
85. Leonzio, G. Study of Mixing Systems and Geometric Configurations for Anaerobic Digesters Using CFD Analysis. *Renew. Energy* **2018**, *123*, 578–589. [CrossRef]
86. Li, X.; Cao, Y.; Huang, Z.; Liang, J.; Cheng, G.; Pan, R. Study on Side-Entering Agitator Flow Field Simulation in Large Scale Biogas Digester. *MATEC Web Conf.* **2018**, *153*, 5003. [CrossRef]

87. Bergamo, U.; Viccione, G.; Coppola, S.; Landi, A.; Meda, A.; Gualtieri, C. Analysis of Anaerobic Digester Mixing: Comparison of Long Shafted Paddle Mixing vs Gas Mixing. *Water Sci. Technol.* **2020**, *81*, 1406–1419. [CrossRef]
88. Kim, M.; Ahn, Y.-H.; Speece, R.E. Comparative Process Stability and Efficiency of Anaerobic Digestion; Mesophilic vs. Thermophilic. *Water Res.* **2002**, *36*, 4369–4385. [CrossRef] [PubMed]
89. Bolzonella, D.; Pavan, P.; Battistoni, P.; Cecchi, F. Mesophilic Anaerobic Digestion of Waste Activated Sludge: Influence of the Solid Retention Time in the Wastewater Treatment Process. *Process Biochem.* **2005**, *40*, 1453–1460. [CrossRef]
90. Bres, P.; Beily, M.E.; Young, B.J.; Gasulla, J.; Butti, M.; Crespo, D.; Candal, R.; Komilis, D. Performance of Semi-Continuous Anaerobic Co-Digestion of Poultry Manure with Fruit and Vegetable Waste and Analysis of Digestate Quality: A Bench Scale Study. *Waste Manag.* **2018**, *82*, 276–284. [CrossRef]
91. Moretto, G.; Ardolino, F.; Piasentin, A.; Girotto, L.; Cecchi, F. Integrated Anaerobic Codigestion System for the Organic Fraction of Municipal Solid Waste and Sewage Sludge Treatment: An Italian Case Study. *J. Chem. Technol. Biotechnol.* **2019**, *95*, 418–426. [CrossRef]
92. Islam, M.S.; Ranade, V. V Enhancing BMP and Digestibility of DAF Sludge via Hydrodynamic Cavitation. *Chem. Eng. Process. Process Intensif.* **2024**, *198*, 109733. [CrossRef]
93. Kafle, G.K.; Chen, L. Comparison on Batch Anaerobic Digestion of Five Different Livestock Manures and Prediction of Biochemical Methane Potential (BMP) Using Different Statistical Models. *Waste Manag.* **2016**, *48*, 492–502. [CrossRef] [PubMed]
94. Parvez, K.; Mansoor Ahammed, M. Development of an Anaerobic Digestibility Index for Organic Solid Wastes. *Biomass Bioenergy* **2024**, *187*, 107298. [CrossRef]
95. Ozsefil, I.C.; Miraloglu, I.H.; Ozbayram, E.G.; Ince, B.; Ince, O. Bioaugmentation of Anaerobic Digesters with the Enriched Lignin-Degrading Microbial Consortia through a Metagenomic Approach. *Chemosphere* **2024**, *355*, 141831. [CrossRef] [PubMed]
96. Üveges, Z.; Damak, M.; Klátyik, S.; Ramay, M.W.; Fekete, G.; Varga, Z.; Csaba, G.; Szekacs, A.; Aleksza, L. Biomethane Potential in Anaerobic Biodegradation of Commercial Bioplastic Materials. *Fermentation* **2023**, *9*, 261. [CrossRef]
97. Sindall, R.; Dapelo, D.; Leadbeater, T.; Bridgeman, J. Positron Emission Particle Tracking (PEPT): A Novel Approach to Flow Visualisation in Lab-Scale Anaerobic Digesters. *Flow Meas. Instrum.* **2017**, *54*, 250–264. [CrossRef]

Disclaimer/Publisher's Note: The statements, opinions and data contained in all publications are solely those of the individual author(s) and contributor(s) and not of MDPI and/or the editor(s). MDPI and/or the editor(s) disclaim responsibility for any injury to people or property resulting from any ideas, methods, instructions or products referred to in the content.

Review

Unraveling the Potential of Microbial Flocculants: Preparation, Performance, and Applications in Wastewater Treatment

Yang Yang [1,2], Cancan Jiang [2,3,*], Xu Wang [2,3], Lijing Fan [1], Yawen Xie [2,3], Danhua Wang [4], Tiancheng Yang [2,3], Jiang Peng [5], Xinyuan Zhang [4] and Xuliang Zhuang [2,3,6,*]

1. Henan Institute of Advanced Technology, Zhengzhou University, Zhengzhou 450000, China; yangy0604@126.com (Y.Y.); lj6116@163.com (L.F.)
2. Research Center for Eco-Environmental Sciences, Chinese Academy of Sciences, Beijing 100085, China; xuwang_st@rcees.ac.cn (X.W.); ywxie_st@rcees.ac.cn (Y.X.); yangtiancheng23@mails.ucas.ac.cn (T.Y.)
3. College of Resources and Environment, University of Chinese Academy of Sciences, Beijing 100049, China
4. School of Life Sciences, Division of Life Sciences and Medicine, University of Science and Technology of China, Hefei 230027, China; dhw@mail.ustc.edu.cn (D.W.); zhangxinyuar@mail.ustc.edu.cn (X.Z.)
5. The Institute of International Rivers and Eco-Security, Yunnan University, Kunming 650500, China; 18819784314@163.com
6. State Key Laboratory of Tibetan Plateau Earth System, Environment and Resources (TPESER), Institute of Tibetan Plateau Research, Chinese Academy of Sciences, Beijing 100101, China
* Correspondence: ccjiang@rcees.ac.cn (C.J.); xlzhuang@rcees.ac.cn (X.Z.)

Citation: Yang, Y.; Jiang, C.; Wang, X.; Fan, L.; Xie, Y.; Wang, D.; Yang, T.; Peng, J.; Zhang, X.; Zhuang, X. Unraveling the Potential of Microbial Flocculants: Preparation, Performance, and Applications in Wastewater Treatment. *Water* **2024**, *16*, 1995. https://doi.org/10.3390/w16141995

Academic Editor: Helvi Heinonen-Tanski

Received: 14 June 2024
Revised: 10 July 2024
Accepted: 12 July 2024
Published: 14 July 2024

Copyright: © 2024 by the authors. Licensee MDPI, Basel, Switzerland. This article is an open access article distributed under the terms and conditions of the Creative Commons Attribution (CC BY) license (https://creativecommons.org/licenses/by/4.0/).

Abstract: Microbial flocculants (MBFs), a class of eco-friendly and biodegradable biopolymers produced by various microorganisms, have gained increasing attention as promising alternatives to conventional chemical flocculants in wastewater treatment and pollutant removal. This review presents a comprehensive overview of the current state of MBF research, encompassing their diverse sources (bacteria, fungi, and algae), major categories (polysaccharides, proteins, and glycoproteins), production processes, and flocculation performance and mechanisms. The wide-ranging applications of MBFs in removing suspended solids, heavy metals, dyes, and other pollutants from industrial and municipal wastewater are critically examined, highlighting their superior efficiency, selectivity, and environmental compatibility compared to traditional flocculants. Nonetheless, bioflocculants face significant challenges including high substrate costs, low production yields, and intricate purification methodologies, factors that impede their industrial scalability. Moreover, the risk of microbial contamination and the attendant health implications associated with the use of microbial flocculants (MBFs) necessitate thorough evaluation. To address the challenges of high production costs and variable product quality, strategies such as waste valorization, strain improvement, process optimization, and biosafety evaluation are discussed. Moreover, the development of multifunctional MBF-based flocculants and their synergistic use with other treatment technologies are identified as emerging trends for enhanced wastewater treatment and resource recovery. Future research directions are outlined, emphasizing the need for in-depth mechanistic studies, advanced characterization techniques, pilot-scale demonstrations to accelerate the industrial adoption of MBF, and moreover, integration with novel wastewater treatment processes, such as partial nitrification and the anammox process. This review is intended to inspire and guide further research and development efforts aimed at unlocking the full potential of MBFs as sustainable, high-performance, and cost-effective bioflocculants for addressing the escalating challenges in wastewater management and environmental conservation.

Keywords: microbial flocculant; wastewater treatment; pollutant removal; flocculation mechanism; resource recovery

1. Introduction

Rapid industrialization and population growth have enabled the rapid increase in the generation of wastewater containing a variety of pollutants, including suspended solids, heavy metals, dyes, and both organic and inorganic contaminants. The inadequate treatment of such discharge poses serious dangers to aquatic environments, human health, and ecological safety [1]. Consequently, there is an urgent requirement for the advancement of effective and sustainable techniques for wastewater treatment and pollutant removal. Flocculation has been broadly employed as a primary wastewater treatment technique due to its simplicity, cost-effectiveness, and notable efficiency in the removal of suspended and colloidal particles [2,3]. Traditional flocculants utilized in the wastewater treatment sector can be categorized into inorganic flocculants (such as aluminum sulfate, ferric chloride, and polyaluminum chloride) and organic synthetic polymers (such as polyacrylamide and polyethylene imine) [4,5]. Although these chemical flocculants have been widely adopted, they are associated with drawbacks, including high cost, potential toxicity, low biodegradability, and secondary pollution resulting from the residual flocculants and their degradation products [6].

In recent years, there has been a notable surge in the development of eco-friendly and sustainable flocculants as alternatives to conventional chemical counterparts [7]. MBFs, in particular, have attracted significant attention due to their unique benefits, including high efficiency, selective action, minimal toxicity, and superior biodegradability [8,9]. These eco-friendly agents are essentially extracellular polymeric substances (EPSs) produced by microorganisms during their growth and metabolic processes [10]. Composed of polysaccharides, proteins, nucleic acids, and lipids, these EPSs are characterized by diverse functional groups (such as hydroxyl, carboxyl, amino, and phosphate) facilitating interactions with pollutants through mechanisms including charge neutralization, adsorption, and flocculation [11]. The utilization of MBFs in wastewater treatment and pollution remediation has gained considerable traction in recent times. Their application in treating diverse types of wastewaters, including municipal sewage and industrial effluents from sectors such as textiles, dyeing, pulp and paper, tanneries, agricultural runoff, and landfill leachate, has demonstrated their versatility and capability [12,13]. MBFs have proven effective in removing suspended solids, turbidity, chemical oxygen demand (COD), heavy metals, dyes, and other contaminants [14–16], often achieving removal efficiencies on par with or surpassing those of traditional chemical flocculants [17–19]. Moreover, the biodegradable and nontoxic nature of MBFs reduces environmental liabilities associated with the disposal of residuals and effluents, contributing to sustainable waste management practices [20].

Although MBFs offer numerous attractive advantages and hold significant promise for application, they still face several obstacles that impede their large-scale production and use. Major hurdles include high manufacturing costs, low yield, complex compositions, and the uncertain safety of MBF products [7]. To address these challenges and facilitate the industrial application of MBFs, considerable research has been directed towards identifying high-producing strains, optimizing fermentation processes, sourcing low-cost substrates, and enhancing downstream processing methods [21]. With the advancement of wastewater treatment technology, the benefits and potential of biological flocculants in wastewater treatment are increasingly evident. Furthermore, as science and technology progress, we have access to more advanced technologies for further developing the principles and applications of biological flocculants.

The objective of this review is to offer a comprehensive summary of the current research landscape on MBFs and their prospective applications in wastewater treatment and pollution control. This article starts by outlining the origins, classifications, and production process of MBFs. It then reviews the effectiveness and mechanisms of MBFs in flocculation. Following this, the application of MBFs in treating various types of wastewaters and removing different pollutants is critically reviewed and discussed. This review also highlights the existing challenges and future directions in MBF research and application. This review is anticipated to provide insightful knowledge and guidance for professionals involved in wastewater management and environmental restoration.

2. Sources and Categories of MBFs

MBFs are a diverse group of extracellular polymeric substances (EPSs) produced by various microorganisms, including bacteria, fungi, and algae [22]. These EPSs serve important biological functions, such as cell adhesion, protection, and nutrient uptake, and can also facilitate the aggregation and sedimentation of suspended particles in aqueous systems [23]. The composition, structure, and properties of MBFs vary depending on the microbial species, growth conditions, and extraction methods.

2.1. MBF-Producing Microorganisms

A wide range of microorganisms have been reported to produce MBFs with different flocculating activities and characteristics. Table 1 presents some typical MBF-producing microorganisms and their MBF yield under experimental conditions. A major disadvantage of an MBF is its low yield, and the yield of some strains is less than 1 g/L. However, currently, through screening and genetic engineering methods, MBF-producing strains with high yields have been obtained, such as *Lipomyces starkeyi* in Table 1. The yield can reach about 62.1 ± 1.2 g/L. Of all the MBF-producing microorganisms, bacteria are the most common and extensively studied MBF producers, especially those belonging to the genera *Bacillus*, *Pseudomonas*, *Klebsiella*, *Rhodococcus*, and *Paenibacillus* [24]. These bacteria can be isolated from various sources, such as activated sludge, soil, wastewater, and marine environments, and are known for their high growth rate, easy cultivation, and adaptability to different substrates [25].

Table 1. Typical MBF-producing microorganisms.

Microorganism	Source	MBF Yield	Reference
Bacillus mucilaginosus	Farmland soil	1.58–2.19 g/L	[26]
Bacillus licheniformis	Soil	2.84 g/L	[27]
Bacillus velezensis	Activated sludge	7.6 g/L	[28]
Bacillus mojavensis	Agricultural soil	1.33 g/L	[29]
Aspergillus parasiticus	Activated sludge	0.54 g/L	[30]
Enterobacter cloacae	Recycled sludge	2.27 g/L	[31]
Klebsiella variicola	Soil	6.96 g/L	[32]
Serratia ficaria	Soil	2.41 g/L	[33]
Aspergillus flavus	—	0.4 g/L	[34]
Penicillium purpurogenum	Laboratory	6.4 g/L	[35]
Phanerochaete chrysosporium	Wastewater sludge	2.2 g/L	[36]
Paenibacillus mucilaginosus	Soil	7.8 g/L	[37]
Lipomyces starkeyi	Mangrove ecosystem	62.1 ± 1.2 g/L	[38]
Alcaligenes faecalis	Sediment sample	4 g/L	[39]

Fungi and yeasts are also significant producers of MBFs, with species from genera such as Aspergillus, Penicillium, Phanerochaete, Streptomyces, and Lipomyces being notable [40,41]. Fungal MBFs typically exhibit higher molecular weights and enhanced flocculating activity compared to their bacterial counterparts. However, their production cost can be relatively higher, due to the slower growth rate and more complex fermentation requirements characteristic of fungi.

Algae, including species such as *Scenedesmus obliquus* AS-6-1 and *Chlorella vulgaris* JSC-7, have also been investigated as potential sources for MBF production [42,43]. Algal MBFs are primarily composed of polysaccharides and can be produced through photosynthetic processes that require minimal energy and nutrients. However, typically, the yield of MBFs from algae is lower compared to bacteria and fungi.

In addition to the natural MBF producers, genetically engineered microorganisms have been developed to enhance MBF production or confer specific functionalities. For example, the overexpression of *bcsB* can stimulate an increase in EPS production and enhance the flocculation effect of *Escherichia coli* [44]. Genetic engineering approaches offer new opportunities to design and optimize MBFs with desired properties for specific applications [45].

2.2. Categories of MBFs

Based on their chemical composition and structure, MBFs can be broadly classified into three major categories: polysaccharides, proteins, and glycoproteins [46], as shown in Figure 1. Polysaccharide MBFs are the most common and extensively studied type of MBFs [47]. They are composed of repeating monosaccharide units (e.g., glucose, galactose, mannose) connected by glycosidic bonds, forming linear or branched polymers with high molecular weights (10^4–10^7 Da) [48]. Polysaccharide MBFs often contain functional groups (e.g., hydroxyl, carboxyl, amino, sulfate) that can interact with pollutants through hydrogen bonding, electrostatic attraction, and complexation mechanisms. Some examples of polysaccharide MBFs include xanthan, dextran, pullulan, and alginate-like exopolysaccharides. Protein MBFs are another important category of MBFs, which are composed of polypeptide chains with a high content of acidic (e.g., aspartic acid, glutamic acid) and hydrophobic (e.g., alanine, valine) amino acids. Protein MBFs usually have lower molecular weights (10^3–10^5 Da) than polysaccharide MBFs but exhibit higher charge density and flocculating activity. The flocculation mechanism of protein MBFs involves charge neutralization and bridging effects between the positively charged amino groups and negatively charged particles. Glycoprotein MBFs are a special type of MBF that contain both polysaccharide and protein moieties covalently linked together [49]. The polysaccharide component provides the backbone structure and hydrogen bonding sites, while the protein component confers the charge and hydrophobic interactions. Glycoprotein MBFs often have a complex and heterogeneous structure, with a molecular weight ranging from 10^5 to 10^7 Da. The synergistic effect of polysaccharide and protein components can enhance the flocculation performance and stability of glycoprotein MBFs.

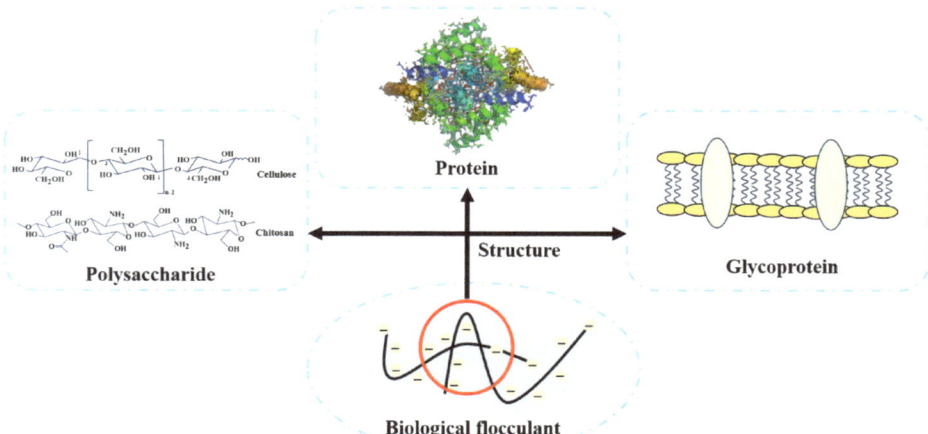

Figure 1. Chemical structures of different categories of MBFs.

In addition to these major categories, some MBFs may also contain other components, such as nucleic acids (e.g., DNA, RNA), lipids (e.g., fatty acids, phospholipids), and inorganic substances (e.g., metals, minerals) [50]. These minor components can modulate the physicochemical properties and flocculating activity of MBFs, depending on their content and interaction with the main components. The diversity and complexity of MBF composition and structure reflect the adaptation and optimization of microorganisms to different environmental conditions and substrates. Understanding the structure–function relationship of MBFs is crucial for selecting and designing MBFs with desirable properties for specific applications in wastewater treatment and pollutant removal. A wide variety of microorganisms have been discovered as MBF-producing strains, including bacteria, fungi, and algae. Table 1 summarizes some typical MBF-producing microorganisms reported in the literature. Among them, *Bacillus*, *Pseudomonas*, *Klebsiella*, *Aspergillus*, and *Penicillium* are the most common genera that can produce MBFs with high flocculating activity [51,52]. Most of the MBF-producing strains are isolated from activated sludge, soils, and wastewater, probably due to the natural selection of microorganisms under stress conditions. Some strains are also obtained through mutation or genetic engineering to enhance MBF production.

3. MBF Production, Flocculation Performance, and Mechanisms

3.1. MBF Production

MBFs are usually produced by microbial fermentation under optimized culture conditions. The growth medium for MBF production generally contains a carbon source, nitrogen source, mineral salts, and trace elements. Carbon sources are crucial for MBF biosynthesis, which are usually supplied as glucose, sucrose, starch, or agricultural wastes like rice straw, wheat bran, and molasses [53]. Nitrogen sources and the C/N ratio also play important roles in microbial growth and MBF accumulation. Inorganic nitrogen sources like ammonium salts are commonly used, while organic nitrogen sources like peptone, yeast extract, and corn steep liquor are sometimes added to improve MBF production [54]. Mineral salts (e.g., magnesium, calcium, potassium salts) are essential for maintaining microbial metabolism and enzyme activity. The optimal medium composition depends on the microbial species and should be optimized through experimental design. The cultivation conditions such as temperature, pH, aeration, and agitation speed also need to be optimized for efficient MBF production [55]. Most MBF-producing microorganisms favor temperatures between 25 and 37 °C and neutral pH (6.0–8.0). The aeration rate and agitation speed should provide sufficient oxygen supply while avoiding shear stress on microbial cells. Fed-batch fermentation is often adopted to achieve high cell density and MBF yield by preventing substrate inhibition [56]. Some inducers like ethanol, methanol, and organic acids can be added to stimulate MBF biosynthesis [57]. A two-stage fermentation strategy, where MBF production is separated from the cell growth stage, is also employed to enhance yield.

After fermentation, MBFs are present in the fermentation broth as soluble metabolites or in association with the microbial cell surface. Centrifugation or filtration is used to remove the microbial cells and obtain the cell-free supernatant. Alcohol (ethanol or acetone) precipitation is the most common method to extract MBFs from the supernatant, followed by centrifugation or filtration to collect the precipitates [7]. The crude MBF can be further purified by dialysis, ion exchange chromatography, size exclusion chromatography, or other purification techniques to obtain MBFs with higher purity [58]. Alternatively, MBFs can be recovered by directly treating the fermentation broth with an alkaline solution (e.g., NaOH) to dissolve and extract the MBF from the cell surface. Acid precipitation (e.g., HCl) is then used to precipitate MBFs from the alkaline extract [59]. Membrane filtration technology is also explored to simultaneously concentrate and purify MBFs from the fermentation broth [60,61]. The extraction efficiency and product purity may be affected by the extraction techniques and operating parameters, which should be optimized for each specific MBF. The extracted MBFs are usually dried by lyophilization or spray-drying to obtain a stable product for application.

3.2. Flocculation Performance

The flocculation performance of MBFs is usually evaluated by a jar test, where a certain dosage of the MBF is added into the wastewater under stirring, followed by a slow mixing and sedimentation process. The supernatant is then analyzed for the residual contaminant concentration or turbidity to calculate the removal efficiency [62]. Flocculation capacity can also be assessed by measuring the turbidity, particle size distribution, and settling velocity of the flocs formed [63]. The dosage effect, pH tolerance, temperature stability, and selectivity of MBFs are important parameters to evaluate their flocculation performance under different conditions. Compared with conventional chemical flocculants, many MBFs exhibit higher or comparable flocculation efficiencies towards various wastewater contaminants. For example, the MBF produced by *Bacillus agaradhaerens* C9 could achieve over 99% removal of chemical oxygen demand (COD) and turbidity from textile wastewater, outperforming polyaluminum chloride [64]. The MBF from *Klebsiella pneumoniae* could remove 97% of Cu(II) and 94% Cr(VI) from electroplating wastewater, showing superior performance over polyacrylamide [65]. The MBFs isolated from *Bacillus velezensis* (40B), *Bacillus mojavensis* (32A), and *Pseudomonas* (38A) had excellent C.I 28 basic yellow dye removal capability, and their maximum decolorization efficiencies were 91%, 89%, and 88% [66]. MBFs can maintain high flocculation activity over a wide range of pH (3–11) and temperature (20–80 °C), probably owing to the stability of hydroxyl and carboxyl groups on MBFs [67,68]. The dosage of an MBF is typically lower than chemical flocculants, with the optimal dose being several mg/L compared to tens or hundreds of mg/L for the latter [69]. This is attributed to the higher charge density and molecular weight of MBFs that can facilitate bridge formation and charge neutralization [70]. MBFs sometimes show selectivity towards some contaminants, which may be related to the specific binding affinity between the functional groups on MBFs and the target pollutants [71]. Due to the significant impact of the environment on microorganisms, biological flocculants' flocculation performance is easily influenced by factors such as temperature and pH, leading to poor environmental stability. Prior to application, flocculant-producing bacteria might require a period of acclimatization. Nonetheless, after a screening process, strains that can adapt to various conditions have been successfully identified. For instance, a strain of Klebsiella pneumoniae, which produces a flocculant, was isolated from H acid wastewater. This flocculant boasts a high molecular mass, thermal stability, and pH responsiveness, and exhibits notably high flocculation activity [72]. B. mojavensis strain 32A has high flocculation efficiency, high yield, and thermal stability in the temperature range of 5~60 °C, and is suitable for use in neutral, weakly acidic, and weakly alkaline environments [73].

3.3. Flocculation Mechanisms

The flocculation mechanisms of MBFs have been extensively studied, and several hypotheses are proposed, as shown in Figure 2. Charge neutralization is regarded as a key mechanism, where the negatively charged MBF can neutralize the positively charged particles or vice versa, resulting in the destabilization and aggregation of colloidal particles [74]. Adsorption bridging is another important mechanism, in which the long-chain MBF can adsorb onto the surface of multiple particles and form a three-dimensional network, thus bridging the particles into large flocs [75]. The hydroxyl and carboxylic groups in biological flocculants also enhance the flocculation effect because they bind strongly to particles and other chemical contaminants [76]. Sweeping flocculation may also occur, where the MBF can capture and enmesh the fine particles into the precipitates as the MBF settles down [77]. Other mechanisms such as hydrophobic interactions and hydrogen bonding are also proposed to play a role in MBF flocculation. The hydrophobic regions on MBFs (e.g., lipid fraction) can facilitate the aggregation of hydrophobic pollutants through hydrophobic interactions. The hydrogen bonds formed between hydroxyl groups on MBFs and water molecules or particles may aid the adsorption and stability of flocs [78]. In some cases, Ca^{2+} or other divalent cations are involved in the flocculation process by crosslinking the negatively charged MBF and particles, thus acting as bridging agents [79].

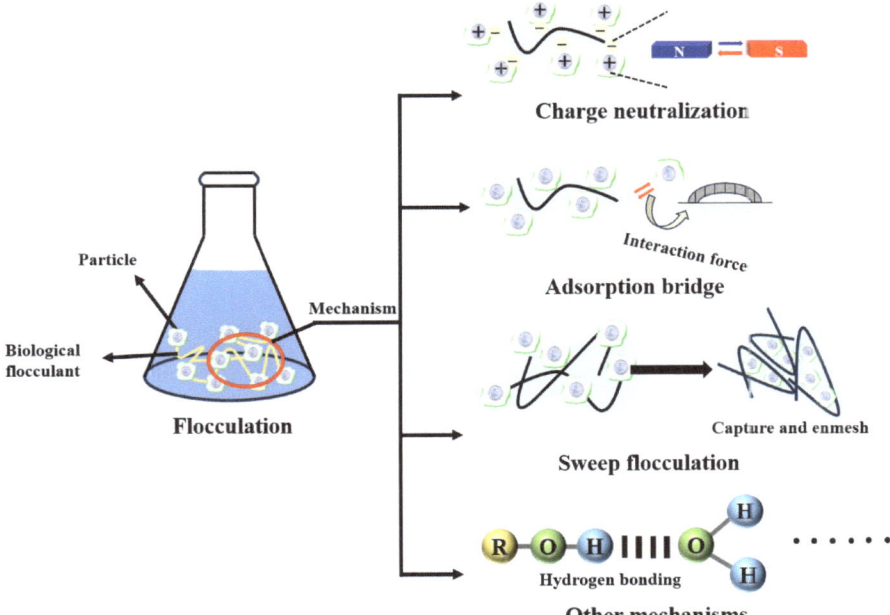

Figure 2. Proposed flocculation mechanisms of MBFs. (At present, there are three main hypotheses about the mechanism of bioflocculants: charge neutralization, adsorption bridging, and precipitation net trapping. In addition, there are other hypotheses about the mechanism, such as the formation of hydrogen bonds between bioflocculants and hydroxyl groups on MBFs and water molecules or particles.)

At present, the flocculation mechanism of MBFs has not been thoroughly studied, but the advancement of science and technology provides us with more opportunities to explore this mechanism. It is important to note that the actual flocculation mechanism of MBFs may involve a combination of multiple mechanisms, which can vary depending on the characteristics of the MBF and the wastewater substrate. Additionally, the dominant mechanism can differ under varying pH, concentration, and other conditions [80,81]. The types of biological flocculants can also influence the flocculation mechanism. For instance, bioflocculants can initiate the destabilization of kaolin suspension through charge neutralization, followed by enhancing the aggregation of suspended particles through adsorption and bridging [82]. The bioflocculant B4-PS, prepared by *Arthrobacter* B4, may primarily rely on ionization and charge neutralization as its main flocculation mechanism [83]. Therefore, when studying the mechanism of biological flocculants, it is crucial to consider the specific circumstances. The investigation of the flocculation mechanism of MBFs remains a key focus for future research. A deeper understanding of the underlying mechanism is essential for designing and optimizing MBF for specific wastewater treatment applications.

4. Applications in Wastewater Treatment

4.1. Removal of Suspended Solids

MBFs play a crucial role in the treatment of wastewater, particularly in the removal of suspended solids (SSs). An SS refers to small solid particles that remain suspended in water, such as clay particles, inorganic sediments, organic sediments, organic scale, corrosion products, and other similar substances. These suspended solids can cause turbidity and give rise to aesthetic and safety concerns [84]. Due to their large surface area and charge density, MBFs can efficiently adsorb and flocculate SSs, leading to the rapid sedimentation of particles. For instance, the *Aspergillus flavus*-produced MBF achieved 92% SS removal from paper-making wastewater under ideal conditions [30]. Similarly, the bioflocculant IC-1 from *Isaria cicadae* GZU6722 was highly effective in treating coal-washing wastewater, achieving a maximum SS removal rate of 91.81% [85]. Furthermore, a protein-based MBF from *Rhizopus* sp. was capable of removing over 90% of SSs from domestic wastewater [86]. Similarly, a bioflocculant prepared via *Serratia marcescens* was observed to eliminate 83.95% of turbidity and 78.82% of suspended solids [4]. These findings underscore the promising potential of MBFs as bioflocculants for the control of SSs and turbidity in wastewater management.

4.2. Removal of Heavy Metals

The removal of toxic heavy metals from industrial effluents is another promising application of MBFs. Heavy metals like lead, chromium, cadmium, and mercury are highly toxic and can accumulate in the food chain, posing severe health risks [87]. Conventional treatment methods like chemical precipitation and ion exchange are costly and may cause secondary pollution [88]. MBFs offer an eco-friendly and cost-effective alternative for heavy metal removal owing to their strong affinity and selectivity towards metal ions. The hydroxyl, carboxyl, phosphate, and amino groups on MBFs can serve as binding sites for metal ions through electrostatic interaction or complexation. Sathiyanarayanan et al. reported that the MBF generated by *Bacillus subtilis* could adsorb 97% of Cr(VI) from tannery effluents [89]. Gomaa investigated the potential of MBFs from *Pseudomonas aeruginosa* for heavy metal removal, achieving an 85%, 80%, and 79% removal efficiency for Pb(II), Cu(II), and Cd(II) respectively [90]. By using MBFs as flocculation aids together with traditional alkaline precipitation, over 99% removal of Zn(II), Pb(II), Cu(II) and Cd(II) was obtained from electroplating wastewater [91]. The removal rates of Zn(II), Cd(II), Cu(II), and Hg by biological flocculants can reach 82.63 ± 1.20, 72.076 ± 0.42, 57.36 ± 1.05, and $44.7 \pm 1.053\%$ [92]. The metal-laden MBF flocs could be easily separated from water, and the heavy metals could be further recovered from the flocs by desorption or an incineration process. At present, some cation-dependent biological flocculants require the addition of Fe(II), Ca(II), and Al(III) ions to enhance their flocculation effect, and although this practice can reduce the dosage of inorganic flocculants and has certain advantages, it will introduce certain metal ions in the water body, weakening the green non-toxic advantage of biological flocculants. Therefore, the screening of biological flocculants with strong flocculation effects and the ability to remove metal ions such as Al(III) and Fe(II) is also a place worthy of attention in future research.

4.3. Removal of Dyes

Dyes are a major class of pollutants in the textile, leather, printing, and dyeing industries. The presence of dyes in water bodies can reduce light penetration, inhibit photosynthesis, and may cause carcinogenic and mutagenic effects [93]. Due to the complex structure and poor biodegradability of synthetic dyes, they are difficult to remove by conventional biological treatment methods. MBFs have shown excellent performance in decolorizing various synthetic dyes from wastewater. The mechanisms of dye removal by MBFs include adsorption, charge neutralization, and bridging effect. The functional groups (e.g., amino, hydroxyl, carboxyl) on MBFs can bind with dye molecules through hydrogen bonding, electrostatic attraction, and π-π stacking interaction [5]. Moreover, some MBFs exhibit enzyme-like activities (e.g., peroxidase, laccase) that can degrade the dye molecules into smaller fragments [94]. A moderately basophilic endophytic bacterium, *Bacillus ferribacterium* (Kx898362), was isolated from Asiatica sinensis. After optimization, the biodegradation rate of this strain could reach 92.76% after 72 h under the conditions of a DB-14 dye concentration of 68.78 ppm and the addition of 1 g of sucrose and 2.5% (v/v) inoculants [95]. The halo-alkaliphilic *Nesterenkonia lacusekhoensis* EMLA3 strain, isolated from textile effluents, degraded 94% of the dye in just 1 h [96]. Pu et al. found that polysaccharide B2, produced by *Bacillus gigantium* strain PL8, can effectively remove

Congo red dye (88.14%) and Pb(II) ions (82.64%) [50]. The MBF showed superior dye removal performance compared to conventional flocculants like polyaluminum chloride and polyacrylamide. The dye-containing MBF flocs could be easily separated by settling or filtration, thus facilitating the removal and recovery of dye pollutants [97].

4.4. Removal and Recovery of Sulfur Compounds

MBFs play a vital role in the treatment of sulfur-laden wastewater, facilitating the separation and recovery of sulfur compounds. While polyaluminum chloride (PAC) demonstrates higher efficiency, MBFs are still effective in flocculating biogenic elemental sulfur (S^0), especially when combined with other flocculants like polymer iron sulfate, achieving high turbidity and color removal [98]. Sulfide-laden water can be treated microbially, with *Thiobacillus* denitrificans oxidizing sulfides to sulfate and a co-culture with floc-forming heterotrophs enabling stable flocculation over extended periods [99]. The sulfate removal efficiency of CYBF biological flocculant reached 54.4%, significantly surpassing that of $FeCl_3$ and alum at the optimal dosage of 8 mg/L and pH 6 [100]. Microbial fuel cells (MFCs) utilizing sulfate-reducing bacteria (SRB) and sulfide-oxidizing bacteria (SOB) biofilms efficiently convert sulfate to sulfide and then to elemental sulfur, enhancing MFC performance [101]. The sulfur cycling mediated by microorganisms has significant environmental implications, particularly in wastewater treatment and pollution bioremediation [102]. Introducing elemental sulfur in denitrification systems can enhance nitrogen removal efficiency in organic-limited nitrate wastewater, with key microbial species contributing to sulfur and nitrogen metabolism [103]. Optimization techniques such as response surface methodology (RSM) can improve the efficiency of MBFa, and solid composite microbial inoculants (SCMIs) offer better storage stability and enhanced removal performance for volatile organic sulfide compounds (VOSCs) compared to microbial suspensions [98,104]. In conclusion, MBFs are effective and environmentally significant in treating sulfur-laden wastewater, with the potential for optimization and improved stability through solid composite inoculants.

4.5. Other Applications

As shown in Figure 3, apart from the above-mentioned contaminants, MBFs have also shown potential in the removal of other pollutants from wastewater, such as nutrients (e.g., nitrogen, phosphorus), oils, microplastics, and organic matter. For instance, the MBF was found to remove 99.2% of arsenite from an aqueous solution [105]. The marine *B. cereus*-derived MBF exhibited robust flocculation performance, effectively catalyzed the synthesis of antibacterial silver nanoparticles, and facilitated the removal of heavy metals [106]. The MBF from *Klebsiella oxytoca* GS-4-08 can degrade nitriles in a continuous flow reactor [107]. Microplastics have emerged as a significant environmental and health concern due to their widespread presence in water bodies, soil, and even air [108]. A Promising Approach Biological method for removing microplastics from water and soil offers a sustainable and eco-friendly solution, such as the MBF from *Bacillus enclensis* being able to biodegrade polyethylene, polypropylene, and polystyrene microplastics [109]. *Bacillus gottheilii* also appeared as a better potential microplastic degrader [110]. It was proposed that the removal mechanisms involve the complexation and charge neutralization between the pollutants and the functional groups on MBFs. MBFs are also used as eco-friendly flocculants for microalgae harvesting and sludge dewatering in wastewater treatment processes. The MBF can facilitate the flocculation and settling of microalgal cells, thus improving the harvesting efficiency while avoiding the use of harmful chemical flocculants [111]. Similarly, MBFs can enhance the dewaterability of waste sludge by promoting the formation of larger and stronger flocs, thus reducing the moisture content and volume of the sludge [112]. The application of MBFs in microalgae harvesting and sludge dewatering not only reduces the economic and environmental costs but also improves the quality and safety of the harvested biomass. Moreover, bioflocculants can function as an efficient eco-friendly corrosion inhibitor, significantly reducing negative environmental impacts and offering a sustainable alternative to synthetic polymers [113].

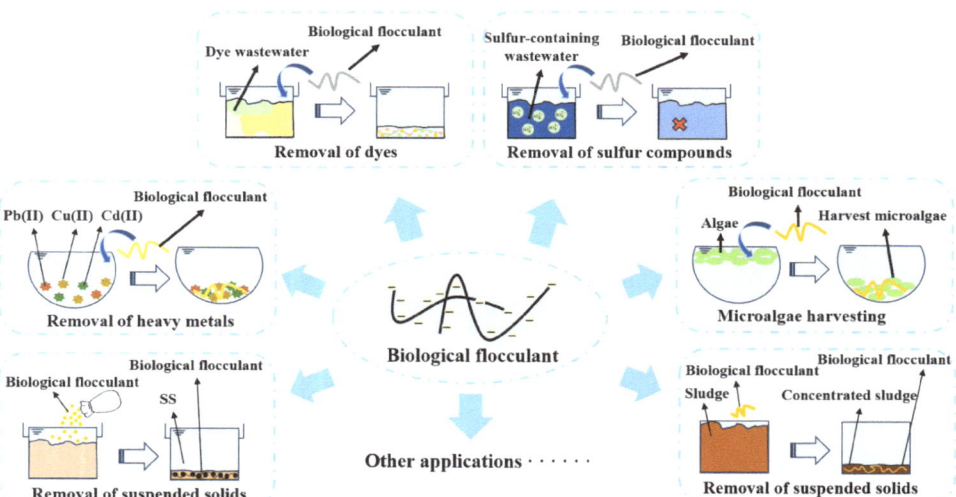

Figure 3. Applications in wastewater treatment. (Biological flocculants hold significant promise for sewage treatment, as they can efficiently eliminate suspended solids, heavy metals, dyes, and sulfur compounds from water. They also have numerous other applications, such as microalgae harvesting and enhancing sludge dewatering capacity.)

Meanwhile, concerning the safety evaluation of bioflocculants, some researchers have uncovered their notable bactericidal and antibacterial properties, for example, in a series of experiments using model water, different raw water sources, and mixed water containing a mixture of these two waters, *M. oleifera* was found to remove 88%, 82%, and 66% of *E. coli* [114]. Biological flocculants kill mosquito larvae. With the increase in the dosage of biological flocculant, the mortality rate of larvae increased in *Ae. aegypti* and *C. quinquefasciatus* mosquitoes [115]. Tsilo et al. found that bioflocculants showed significant antibacterial properties against both Gram-positive and Gram-negative bacteria [116]. These studies indicate their potential utility as antibacterial agents.

5. Challenges and Future Perspectives

As shown in Figure 4, a microbial flocculant (MBF) exhibits environmentally friendly characteristics, biodegradability, superior efficiency, and selectivity, and shows promise in various applications such as the removal of suspended solids (SSs) and the treatment of heavy metal wastewater, dye wastewater, and sulfur-containing wastewater. Additionally, it can be used for microalgae harvesting, microplastic removal, and bacteria inhibition, making it highly attractive.

However, despite these advantages, MBFs have not yet become a mainstream treatment process due to several challenges and limitations. The main obstacles to MBF production are high substrate cost, low yield, and a complex purification process, leading to increased production costs. Furthermore, the poor stability of biological flocculants, caused by the susceptibility of microorganisms to environmental influences, hinders the large-scale use of these flocculants. Nevertheless, significant progress has been made in screening strains with high yield, high flocculation rates, and improved stability, enhancing the stability of biological flocculants.

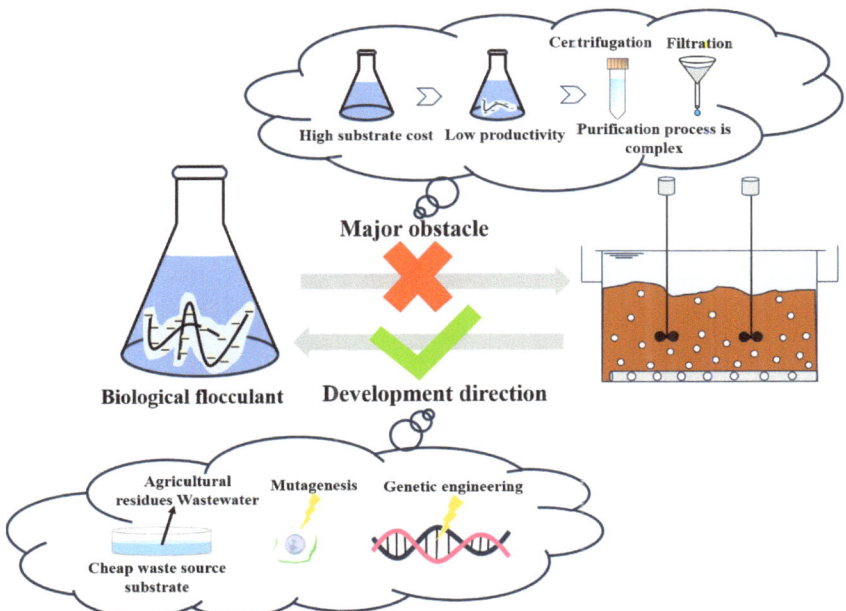

Figure 4. Challenges and future perspectives.

Although many hypotheses exist regarding the flocculation mechanism of biological flocculants, further in-depth studies are still required. The advancement of science and technology can aid in exploring the mechanism of flocculation. Before bioflocculants can be produced and used, safety assessments need to be conducted. While some scholars have researched this area, there is currently no unified industry standard in place.

To reduce the production cost, cheap and waste-derived substrates such as agricultural residues, industrial effluents, and wastewater can be used as low-cost carbon and nitrogen sources for MBF production [117]. Strain improvement by mutagenesis and genetic engineering is also a promising approach to enhance the MBF yield and reduce the substrate cost [118]. The variable composition and impurity of MBF products is another challenge that affects their flocculation performance and stability. The composition and properties of MBFs depend on the microbial species, cultivation conditions, and extraction methods, which may lead to batch-to-batch variation and inconsistent product quality [1,119]. Therefore, quality control and standardization of the MBF production process are crucial to ensure the reliability and reproducibility of MBF applications. An in-depth characterization of MBF composition and structure by advanced analytical techniques like FTIR, NMR, GC-MS, and MALDI-TOF is helpful for establishing the quality criteria and optimizing the production process [24,120,121]. The potential microbial contamination and health risks associated with MBF application should also be considered and evaluated. Although MBFs are generally regarded as safe and non-toxic, some MBF-producing strains are opportunistic pathogens that may cause infections in immunocompromised individuals. The presence of residual microbial cells, endotoxins, or pyrogens in MBF products may also pose health hazards [122]. Therefore, strict biosafety assessment and quality control are required to ensure the safety and hygiene of MBF products [123]. The use of non-pathogenic strains, screening biological flocculants with antibacterial activity, cell removal by ultrafiltration, and sterilization by UV or heat treatment can minimize the microbial contamination risks [115]. The potential of bioflocculants as bacteriostatic agents should also be considered in the safety assessment of bioflocculants. Bioflocculants with antibacterial activity serve as effective natural disinfectants for wastewater treatment. Screening biological flocculants with high bacteriostatic performance can make the MBF have both flocculating and disinfecting effects, offering a cost-effective solution for sewage treatment [124].

For future research, more efforts are needed to elucidate the flocculation mechanisms of MBFs and establish the structure–activity relationship [125]. Advanced characterization techniques like SEM, TEM, AFM, and QCM-D can be used to visualize the floc structure and probe the molecular interactions during the MBF flocculation process [126]. Molecular dynamics simulations and quantum chemical calculations are also powerful tools for studying the binding mechanism and energy involved in the MBF–contaminant interaction. A better mechanistic understanding will guide the rational screening and design of MBF for specific wastewater treatment. The combination of MBFs with other physical, chemical, or biological treatment methods is another research direction to enhance the overall treatment efficiency and economy. For example, the sequential use of MBFs and membrane filtration can improve the removal efficiency and flux of the membrane process while reducing the membrane fouling [127]. The modified bioflocculant carrier doped with polylactic acid (PLA) can obviously promote the colonization rate and number of microorganisms on the carrier, and improve the purification efficiency of COD and ammonia nitrogen in wastewater [128]. The development of multifunctional MBFs and their synergistic application with other technologies are expected to bring new opportunities for wastewater treatment and reuse.

Moreover, the emergence of novel nitrogen removal processes, such as partial nitrification, anaerobic ammonium oxidation (anammox), and sulfide-driven autotrophic denitrification, has opened up new avenues for the application of MBFs in wastewater treatment. These innovative biological nitrogen removal methods offer distinct advantages over conventional nitrification–denitrification processes, including reduced energy consumption, lower carbon source requirements, and decreased sludge production [129–131]. As these cutting-edge nitrogen removal technologies continue to advance and gain traction, the potential of MBFs in enhancing wastewater treatment efficiency and sustainability becomes increasingly apparent. MBFs can play a pivotal role in wastewater treatment through multiple mechanisms. Firstly, the synergistic use of MBFs with other flocculants can significantly improve pollutant removal efficiency, creating optimal conditions for subsequent nitrogen removal processes [132–134]. Secondly, MBFs can promote sulfide-driven autotrophic denitrification, enabling the use of sulfide as an electron donor and the coupling of this process with anammox to further optimize nitrogen removal performance [135,136]. The application of solid composite microbial inoculants in lieu of microbial suspensions can further enhance the storage stability and pollutant removal capabilities of MBFs, contributing to the development of more robust and efficient wastewater treatment systems. As research continues to delve into the integration of MBFs with novel nitrogen removal processes, the potential of MBFs in revolutionizing wastewater treatment and enabling resource recovery becomes increasingly evident. By leveraging the synergies between these technologies, it is possible to develop sustainable, efficient, and cost-effective solutions for wastewater treatment, fostering the protection of aquatic environments and the advancement of a circular economy.

6. Conclusions

This review summarizes the research progress and application potential of MBFs in wastewater treatment and resource recovery. As eco-friendly and sustainable alternatives to synthetic chemical flocculants, MBFs have demonstrated excellent flocculation performance and versatile functions in removing various contaminants (e.g., suspended solids, dyes, heavy metals) from wastewater. The biodegradability, low toxicity, and high selectivity of MBFs make them promising tools for wastewater purification and sludge dewatering processes.

To address the challenges related to high production costs and quality inconsistencies, certain strategies can be adopted, such as waste valorization, strain improvement, and process optimization. Additionally, the biosafety evaluation and standardization of MBF production are essential to ensure reliability and safety in practical applications. Future research should focus on elucidating the flocculation mechanisms, designing novel composite MBF flocculants, and integrating MBFs with other technologies to achieve synergistic wastewater treatment.

This paper reviews the research progress and application potential of MBFs in wastewater treatment and resource recovery from different sources, main categories, production processes, flocculation properties, and mechanisms, and combines new scientific research technologies. Advanced analytical techniques were used to characterize the composition and structure of MBFs to study the mechanism of flocculation, to develop the collaborative application of MBFs with other technologies, to evaluate the safety of biological flocculants, to study MBFs with both flocculation and bactericidal functions, and to develop the application potential in novel nitrogen removal processes.

As the demand for sustainable and cost-effective water treatment solutions continues to grow, the development and application of MBFs in wastewater purification and resource recovery are expected to receive increased attention and momentum in the future.

Author Contributions: Conceptualization, Y.Y. and C.J.; resources, X.W., Y.X., D.W., L.F., J.P., X.Z. (Xinyuan Zhang) and T.Y.; writing—original draft preparation, Y.Y.; writing—review and editing, C.J. and X.Z. (Xuliang Zhuang); supervision, X.Z. (Xuliang Zhuang); funding acquisition, X.Z. (Xuliang Zhuang). All authors have read and agreed to the published version of the manuscript.

Funding: This research was funded by the National Natural Science Foundation of China (nos. 42177099 and 42230411), the Provincial Science and Technology Innovative Program for Carbon Peak and Carbon Neutrality of Jiangsu of China, the "Leading Goose" R&D Program of Zhejiang (no. 2023C03131 and no. 2023C03132), and the Key Technology and Equipment System for Intelligent Control of the Water Supply Network (no. 2022YFC3203800).

Data Availability Statement: The data presented in this study are available upon request from the corresponding author.

Acknowledgments: We are grateful for the meticulous guidance provided by the teachers, and we extend our thanks to all the students who contributed to this article.

Conflicts of Interest: The authors declare no conflicts of interest. The funders had no role in the design of the study; in the collection, analyses, or interpretation of data; in the writing of the manuscript; or in the decision to publish the results.

References

1. Vimala, R.T.V.; Lija, E.; Chik, C.E.N.C.E.; Owodunni, A.A.; Ahmad, A.; Alnawajha, M.M.; Rahim, N.F.M.; Said, N.S.M.; Abdullah, S.R.S.; Kasan, N.A.; et al. What Compound inside Biocoagulants/Bioflocculants Is Contributing the Most to the Coagulation and Flocculation Processes? *Sci. Total Environ.* **2022**, *806*, 150902. [CrossRef]
2. Abu Bakar, S.N.H.; Abu Hasan, H.; Abdullah, S.R.S.; Kasan, N.A.; Muhamad, M.H.; Kurniawan, S.B. A Review of the Production Process of Bacteria-Based Polymeric Flocculants. *J. Water Process Eng.* **2021**, *40*, 101915. [CrossRef]
3. Wei, H.; Gao, B.; Ren, J.; Li, A.; Yang, H. Coagulation/Flocculation in Dewatering of Sludge: A Review. *Water Res.* **2018**, *143*, 608–631. [CrossRef]
4. Kurniawan, S.B.; Imron, M.F.; Abdullah, S.R.S.; Othman, A.R.; Purwanti, I.F.; Hasan, H.A. Treatment of Real Aquaculture Effluent Using Bacteria-Based Bioflocculant Produced by *Serratia marcescens*. *J. Water Process Eng.* **2022**, *47*, 102708. [CrossRef]
5. Artifon, W.; Cesca, K.; de Andrade, C.J.; Ulson de Souza, A.A.; de Oliveira, D. Dyestuffs from Textile Industry Wastewaters: Trends and Gaps in the Use of Bioflocculants. *Process Biochem.* **2021**, *111*, 181–190. [CrossRef]
6. Lee, C.S.; Robinson, J.; Chong, M.F. A Review on Application of Flocculants in Wastewater Treatment. *Process Saf. Environ. Protection* **2014**, *92*, 489–508. [CrossRef]
7. Nwodo, U.U.; Green, E.; Mabinya, L.V.; Okaiyeto, K.; Rumbold, K.; Obi, L.C.; Okoh, A.I. Bioflocculant Production by a Consortium of *Streptomyces* and *Cellulomonas* Species and Media Optimization via Surface Response Model. *Colloids Surf. B Biointerfaces* **2014**, *116*, 257–264. [CrossRef]
8. Yin, Y.-J.; Tian, Z.-M.; Tang, W.; Li, L.; Song, L.-Y.; McElmurry, S.P. Production and Characterization of High Efficiency Bioflocculant Isolated from *Klebsiella* sp. ZZ-3. *Bioresour. Technol.* **2014**, *171*, 336–342. [CrossRef] [PubMed]
9. Siddharth, T.; Sridhar, P.; Vinila, V.; Tyagi, R.D. Environmental Applications of Microbial Extracellular Polymeric Substance (EPS): A Review. *J. Environ. Manag.* **2021**, *287*, 112307. [CrossRef]

10. Okaiyeto, K.; Nwodo, U.U.; Okoli, S.A.; Mabinya, L.V.; Okoh, A.I. Implications for Public Health Demands Alternatives to Inorganic and Synthetic Flocculants: Bioflocculants as Important Candidates. *Microbiologyopen* **2016**, *5*, 177–211. [CrossRef]
11. Okaiyeto, K.; Nwodo, U.U.; Mabinya, L.V.; Okoli, A.S.; Okoh, A.I. Evaluation of Flocculating Performance of a Thermostable Bioflocculant Produced by Marine *Bacillus* sp. *Environ. Technol.* **2016**, *37*, 1829–1842. [CrossRef] [PubMed]
12. Pu, S.; Qin, L.; Che, J.; Zhang, B.; Xu, M. Preparation and Application of a Novel Bioflocculant by Two Strains of *Rhizopus* sp. Using Potato Starch Wastewater as Nutrilite. *Bioresour. Technol.* **2014**, *162*, 184–191. [CrossRef]
13. Huang, J.; Huang, Z.-L.; Zhou, J.-X.; Li, C.-Z.; Yang, Z.-H.; Ruan, M.; Li, H.; Zhang, X.; Wu, Z.-J.; Qin, X.-L.; et al. Enhancement of Heavy Metals Removal by Microbial Flocculant Produced by *Paenibacillus polymyxa* Combined with an Insufficient Hydroxide Precipitation. *Chem. Eng. J.* **2019**, *374*, 880–894. [CrossRef]
14. Feng, J.; Xu, Y.; Ding, J.; He, J.; Shen, Y.; Lu, G.; Qin, W.; Guo, H. Optimal Production of Bioflocculant from *Pseudomonas* sp. GO2 and Its Removal Characteristics of Heavy Metals. *J. Biotechnol.* **2022**, *344*, 50–56. [CrossRef]
15. Fan, H.; Yu, J.; Chen, R.; Yu, L. Preparation of a Bioflocculant by Using Acetonitrile as Sole Nitrogen Source and Its Application in Heavy Metals Removal. *J. Hazard. Mater.* **2019**, *363*, 242–247. [CrossRef]
16. Mohamed Hatta, N.S.; Lau, S.W.; Takeo, M.; Chua, H.B.; Baranwal, P.; Mubarak, N.M.; Khalid, M. Novel Cationic Chitosan-like Bioflocculant from *Citrobacter youngae* GTC 01314 for the Treatment of Kaolin Suspension and Activated Sludge. *J. Environ. Chem. Eng.* **2021**, *9*, 105297. [CrossRef]
17. Molaei, N.; Chehreh Chelgani, S.; Bobicki, E.R. A Comparison Study between Bioflocculants and PAM for Dewatering of Ultrafine Phyllosilicate Clay Minerals. *Appl. Clay Sci.* **2022**, *218*, 106409. [CrossRef]
18. Guo, J.; Chen, C. Sludge Conditioning Using the Composite of a Bioflocculant and PAC for Enhancement in Dewaterability. *Chemosphere* **2017**, *185*, 277–283. [CrossRef]
19. Cao, G.; Zhang, Y.; Chen, L.; Liu, J.; Mao, K.; Li, K.; Zhou, J. Production of a Bioflocculant from Methanol Wastewater and Its Application in Arsenite Removal. *Chemosphere* **2015**, *141*, 274–281. [CrossRef]
20. Li, H.; Wu, S.; Du, C.; Zhong, Y.; Yang, C. Preparation, Performances, and Mechanisms of Microbial Flocculants for Wastewater Treatment. *Int. J. Environ. Res. Public Health* **2020**, *17*, 1360. [CrossRef]
21. Gan, L.; Huang, X.; He, Z.; He, T. Exopolysaccharide Production by Salt-Tolerant Bacteria: Recent Advances, Current Challenges, and Future Prospects. *Int. J. Biol. Macromol.* **2024**, *264*, 130731. [CrossRef] [PubMed]
22. Alias, J.; Abu Hasan, H.; Sheikh Abdullah, S.R.; Othman, A.R. Properties of Bioflocculant-Producing Bacteria for High Flocculating Activity Efficiency. *Environ. Technol. Innov.* **2022**, *27*, 102529. [CrossRef]
23. de Jesus, C.S.; de Jesus Assis, D.; Rodriguez, M.B.; Menezes Filho, J.A.; Costa, J.A.V.; de Souza Ferreira, E.; Druzian, J.I. Pilot-Scale Isolation and Characterization of Extracellular Polymeric Substances (EPS) from Cell-Free Medium of *Spirulina* sp. LEB-18 Cultures under Outdoor Conditions. *Int. J. Biol. Macromol.* **2019**, *124*, 1106–1114. [CrossRef] [PubMed]
24. Nie, Y.; Wang, Z.; Zhang, R.; Ma, J.; Zhang, H.; Li, S.; Li, J. *Aspergillus oryzae*, a Novel Eco-Friendly Fungal Bioflocculant for Turbid Drinking Water Treatment. *Sep. Purif. Technol.* **2021**, *279*, 119669. [CrossRef]
25. Oyewole, O.A.; Jagaba, A.; Abdulhammed, A.A.; Yakubu, J.G.; Maude, A.M.; Abioye, O.P.; Adeniyi, O.D.; Egwim, E.C. Production and Characterization of a Bioflocculant Produced by Microorganisms Isolated from Earthen Pond Sludge. *Bioresour. Technol. Rep.* **2023**, *22*, 101492. [CrossRef]
26. Lian, B.; Chen, Y.; Zhao, J.; Teng, H.H.; Zhu, L.; Yuan, S. Microbial Flocculation by *Bacillus mucilaginosus*: Applications and Mechanisms. *Bioresour. Technol.* **2008**, *99*, 4825–4831. [CrossRef] [PubMed]
27. Zhao, H.; Liu, H.; Zhou, J. Characterization of a Bioflocculant MBF-5 by *Klebsiella pneumoniae* and Its Application in *Acanthamoeba* Cysts Removal. *Bioresour. Technol.* **2013**, *137*, 226–232. [CrossRef] [PubMed]
28. Moghannem, S.A.M.; Farag, M.M.S.; Shehab, A.M.; Azab, M.S. Exopolysaccharide Production from *Bacillus velezensis* KY471306 Using Statistical Experimental Design. *Braz. J. Microbiol.* **2018**, *49*, 452–462. [CrossRef]
29. Xia, S.; Zhang, Z.; Wang, X.; Yang, A.; Chen, L.; Zhao, J.; Leonard, D.; Jaffrezic-Renault, N. Production and Characterization of a Bioflocculant by *Proteus mirabilis* TJ-1. *Bioresour. Technol.* **2008**, *99*, 6520–6527. [CrossRef]
30. Deng, S.; Yu, G.; Ting, Y.P. Production of a Bioflocculant by *Aspergillus parasiticus* and Its Application in Dye Removal. *Colloids Surf. B Biointerfaces* **2005**, *44*, 179–186. [CrossRef]
31. Prasertsan, P.; Dermlim, W.; Doelle, H.; Kennedy, J.F. Screening, Characterization and Flocculating Property of Carbohydrate Polymer from Newly Isolated *Enterobacter cloacae* WD7. *Carbohydr. Polym.* **2006**, *66*, 289–297. [CrossRef]
32. Xia, X.; Lan, S.; Li, X.; Xie, Y.; Liang, Y.; Yan, P.; Chen, Z.; Xing, Y. Characterization and Coagulation-Flocculation Performance of a Composite Flocculant in High-Turbidity Drinking Water Treatment. *Chemosphere* **2018**, *206*, 701–708. [CrossRef] [PubMed]
33. Gong, W.-X.; Wang, S.-G.; Sun, X.-F.; Liu, X.-W.; Yue, Q.-Y.; Gao, B.-Y. Bioflocculant Production by Culture of *Serratia ficaria* and Its Application in Wastewater Treatment. *Bioresour. Technol.* **2008**, *99*, 4668–4674. [CrossRef] [PubMed]
34. Aljuboori, A.H.R.; Idris, A.; Abdullah, N.; Mohamad, R. Production and Characterization of a Bioflocculant Produced by *Aspergillus flavus*. *Bioresour. Technol.* **2013**, *127*, 489–493. [CrossRef] [PubMed]
35. Liu, L.-F.; Cheng, W. Characteristics and Culture Conditions of a Bioflocculant Produced by *Penicillium* sp. *Biomed. Environ. Sci.* **2010**, *23*, 213–218. [CrossRef] [PubMed]
36. Shih, I.L.; Van, Y.T.; Yeh, L.C.; Lin, H.G.; Chang, Y.N. Production of a Biopolymer Flocculant from *Bacillus licheniformis* and Its Flocculation Properties. *Bioresour. Technol.* **2001**, *78*, 267–272. [CrossRef] [PubMed]

37. Chen, S.; Cheng, R.; Xu, X.; Kong, C.; Wang, L.; Fu, R.; Li, J.; Wang, S.; Zhang, J. The Structure and Flocculation Characteristics of a Novel Exopolysaccharide from a *Paenibacillus* Isolate. *Carbohydr. Polym.* **2022**, *291*, 119561. [CrossRef] [PubMed]
38. Yu, X.; Wei, X.; Chi, Z.; Liu, G.-L.; Hu, Z.; Chi, Z.-M. Improved Production of an Acidic Exopolysaccharide, the Efficient Flocculant, by *Lipomyces starkeyi* U9 Overexpressing UDP-Glucose Dehydrogenase Gene. *Int. J. Biol. Macromol.* **2020**, *165*, 1656–1663. [CrossRef] [PubMed]
39. Maliehe, T.S.; Basson, A.K.; Dlamini, N.G. Removal of Pollutants in Mine Wastewater by a Non-Cytotoxic Polymeric Bioflocculant from *Alcaligenes faecalis* HCB2. *Int. J. Environ. Res. Public Health* **2019**, *16*, 4001. [CrossRef]
40. Sun, R.; Sun, P.; Zhang, J.; Esquivel-Elizondo, S.; Wu, Y. Microorganisms-Based Methods for Harmful Algal Blooms Control: A Review. *Bioresour. Technol.* **2018**, *248*, 12–20. [CrossRef]
41. Pei, X.-Y.; Ren, H.-Y.; Liu, B.-F. Flocculation Performance and Mechanism of Fungal Pellets on Harvesting of Microalgal Biomass. *Bioresour. Technol.* **2021**, *321*, 124463. [CrossRef] [PubMed]
42. Characterization of the Flocculating Agent from the Spontaneously Flocculating Microalga *Chlorella vulgaris* JSC-7. *J. Biosci. Bioeng.* **2014**, *118*, 29–33. [CrossRef] [PubMed]
43. Guo, S.-L.; Zhao, X.-Q.; Wan, C.; Huang, Z.-Y.; Yang, Y.-L.; Asraful Alam, M.; Ho, S.-H.; Bai, F.-W.; Chang, J.-S. Characterization of Flocculating Agent from the Self-Flocculating Microalga *Scenedesmus obliquus* AS-6-1 for Efficient Biomass Harvest. *Bioresour. Technol.* **2013**, *145*, 285–289. [CrossRef] [PubMed]
44. Nguyen, M.H.; Ojima, Y.; Sakka, M.; Sakka, K.; Taya, M. Probing of Exopolysaccharides with Green Fluorescence Protein-Labeled Carbohydrate-Binding Module in *Escherichia coli* Biofilms and Flocs Induced by bcsB Overexpression. *J. Biosci. Bioeng.* **2014**, *118*, 400–405. [CrossRef] [PubMed]
45. Ummalyma, S.B.; Gnansounou, E.; Sukumaran, R.K.; Sindhu, R.; Pandey, A.; Sahoo, D. Bioflocculation: An Alternative Strategy for Harvesting of Microalgae—An Overview. *Bioresour. Technol.* **2017**, *242*, 227–235. [CrossRef] [PubMed]
46. Pathak, M.; Sarma, H.K.; Bhattacharyya, K.G.; Subudhi, S.; Bisht, V.; Lal, B.; Devi, A. Characterization of a Novel Polymeric Bioflocculant Produced from Bacterial Utilization of N-Hexadecane and Its Application in Removal of Heavy Metals. *Front. Microbiol.* **2017**, *8*, 170. [CrossRef] [PubMed]
47. Salehizadeh, H.; Yan, N.; Farnood, R. Recent Advances in Polysaccharide Bio-Based Flocculants. *Biotechnol. Adv.* **2018**, *36*, 92–119. [CrossRef] [PubMed]
48. Tang, W.; Song, L.; Li, D.; Qiao, J.; Zhao, T.; Zhao, H. Production, Characterization, and Flocculation Mechanism of Cation Independent, pH Tolerant, and Thermally Stable Bioflocculant from *Enterobacter* sp. ETH-2. *PLoS ONE* **2014**, *9*, e114591. [CrossRef]
49. Li, Z.; Zhong, S.; Lei, H.; Chen, R.; Yu, Q.; Li, H.-L. Production of a Novel Bioflocculant by *Bacillus licheniformis* X14 and Its Application to Low Temperature Drinking Water Treatment. *Bioresour. Technol.* **2009**, *100*, 3650–3656. [CrossRef]
50. Pu, L.; Zeng, Y.-J.; Xu, P.; Li, F.-Z.; Zong, M.-H.; Yang, J.-G.; Lou, W.-Y. Using a Novel Polysaccharide BM2 Produced by *Bacillus megaterium* Strain PL8 as an Efficient Bioflocculant for Wastewater Treatment. *Int. J. Biol. Macromol.* **2020**, *162*, 374–384. [CrossRef]
51. Nie, Y.; Wang, Z.; Wang, W.; Zhou, Z.; Kong, Y.; Ma, J. Bio-Flocculation of *Microcystis aeruginosa* by Using Fungal Pellets of *Aspergillus oryzae*: Performance and Mechanism. *J. Hazard. Mater.* **2022**, *439*, 129606. [CrossRef] [PubMed]
52. Shahadat, M.; Teng, T.T.; Rafatullah, M.; Shaikh, Z.A.; Sreekrishnan, T.R.; Ali, S.W. Bacterial Bioflocculants: A Review of Recent Advances and Perspectives. *Chem. Eng. J.* **2017**, *328*, 1139–1152. [CrossRef]
53. Guo, J.; Yu, J.; Xin, X.; Zou, C.; Cheng, Q.; Yang, H.; Nengzi, L. Characterization and Flocculation Mechanism of a Bioflocculant from Hydrolyzate of Rice Stover. *Bioresour. Technol.* **2015**, *177*, 393–397. [CrossRef] [PubMed]
54. Nontembiso, P.; Sekelwa, C.; Leonard, M.V.; Anthony, O.I. Assessment of Bioflocculant Production by *Bacillus* sp. Gilbert, a Marine Bacterium Isolated from the Bottom Sediment of Algoa Bay. *Mar. Drugs* **2011**, *9*, 1232–1242. [CrossRef] [PubMed]
55. Gao, Q.; Zhu, X.-H.; Mu, J.; Zhang, Y.; Dong, X.-W. Using *Ruditapes philippinarum* Conglutination Mud to Produce Bioflocculant and Its Applications in Wastewater Treatment. *Bioresour. Technol.* **2009**, *100*, 4996–5001. [CrossRef] [PubMed]
56. Salehizadeh, H.; Shojaosadati, S.A. Removal of Metal Ions from Aqueous Solution by Polysaccharide Produced from *Bacillus firmus*. *Water Res.* **2003**, *37*, 4231–4235. [CrossRef] [PubMed]
57. Toeda, K.; Kurane, R. Microbial Flocculant from *Alcaligenes cupidus* KT201. *Agric. Biol. Chem.* **1991**, *55*, 2793–2799. [CrossRef]
58. Kumar, C.G.; Joo, H.; Kavali, R.; Choi, J.; Chang, C. Characterization of an Extracellular Biopolymer Flocculant from a Haloalkalophilic *Bacillus* Isolate. *World J. Microbiol. Biotechnol.* **2004**, *20*, 837–843. [CrossRef]
59. Cosa, S.; Mabinya, L.V.; Olaniran, A.O.; Okoh, O.O.; Bernard, K.; Deyzel, S.; Okoh, A.I. Bioflocculant Production by *Virgibacillus* sp. Rob Isolated from the Bottom Sediment of Algoa Bay in the Eastern Cape, South Africa. *Molecules* **2011**, *16*, 2431–2442. [CrossRef]
60. Li, Q.; Liu, H.; Qi, Q.; Wang, F.-S.; Zhang, Y. Isolation and Characterization of Temperature and Alkaline Stable Bioflocculant from *Agrobacterium* sp. M-503. *New Biotechnol.* **2010**, *27*, 789–794. [CrossRef]
61. Zhou, Y.; Li, X.; Zhou, Z.; Feng, J.; Sun, Y.; Ren, J.; Lu, Z. New Insights into Biopolymers: In Situ Collection and Reuse for Coagulation Aiding in Drinking Water Treatment Plants and Microbial Mechanism. *Sep. Purif. Technol.* **2024**, *337*, 126448. [CrossRef]
62. Flaten, T.P. Aluminium as a Risk Factor in Alzheimer's Disease, with Emphasis on Drinking Water. *Brain Res. Bull.* **2001**, *55*, 187–196. [CrossRef] [PubMed]
63. Brostow, W.; Pal, S.; Singh, R.P. A Model of Flocculation. *Mater. Lett.* **2007**, *61*, 4381–4384. [CrossRef]
64. Mabinya, L.V.; Cosa, S.; Nwodo, U.; Okoh, A.I. Studies on Bioflocculant Production by *Arthrobacter* sp. Raats, a Freshwater Bacteria Isolated from Tyume River, South Africa. *Int. J. Mol. Sci.* **2012**, *13*, 1054–1065. [CrossRef] [PubMed]

65. Wu, J.-Y.; Ye, H.-F. Characterization and Flocculating Properties of an Extracellular Biopolymer Produced from a *Bacillus subtilis* DYU1 Isolate. *Process Biochem.* **2007**, *42*, 1114–1123. [CrossRef]
66. Elkady, M. Bioflocculation of Basic Dye onto Isolated Microbial Biopolymers. *Chem. Biochem. Eng. Q.* **2017**, *31*, 209–224. [CrossRef]
67. Gao, J.; Bao, H.; Xin, M.; Liu, Y.; Li, Q.; Zhang, Y. Characterization of a Bioflocculant from a Newly Isolated *Vagococcus* sp. W31. *J. Zhejiang Univ. Sci. B* **2006**, *7*, 186–192. [CrossRef] [PubMed]
68. Yang, Q.; Luo, K.; Liao, D.; Li, X.; Wang, D.; Liu, X.; Zeng, G.; Li, X. A Novel Bioflocculant Produced by *Klebsiella* sp. and Its Application to Sludge Dewatering. *Water Environ. J.* **2012**, *26*, 560–566. [CrossRef]
69. Lu, W.-Y.; Zhang, T.; Zhang, D.-Y.; Li, C.-H.; Wen, J.-P.; Du, L.-X. A Novel Bioflocculant Produced by *Enterobacter aerogenes* and Its Use in Defecating the Trona Suspension. *Biochem. Eng. J.* **2005**, *27*, 1–7. [CrossRef]
70. Yokoi, H.; Natsuda, O.; Hirose, J.; Hayashi, S.; Takasaki, Y. Characteristics of a Biopolymer Flocculant Produced by *Bacillus* sp. PY-90. *J. Ferment. Bioeng.* **1995**, *79*, 378–380. [CrossRef]
71. Zhang, T.; Lin, Z.; Zhu, H. Microbial Flocculant and Its Application in Environmental Protection. *J. Environ. Sci.* **1999**, *11*, 1–12.
72. Zhong, C.; Xu, A.; Wang, B.; Yang, X.; Hong, W.; Yang, B.; Chen, C.; Liu, H.; Zhou, J. Production of a Value Added Compound from the H-Acid Waste Water—Bioflocculants by *Klebsiella pneumoniae*. *Colloids Surf. B Biointerfaces* **2014**, *122*, 583–590. [CrossRef] [PubMed]
73. Elkady, M.F.; Farag, S.; Zaki, S.; Abu-Elreesh, G.; Abd-El-Haleem, D. *Bacillus mojavensis* Strain 32A, a Bioflocculant-Producing Bacterium Isolated from an Egyptian Salt Production Pond. *Bioresour. Technol.* **2011**, *102*, 8143–8151. [CrossRef]
74. Levy, N.; Magdassi, S.; Bar-Or, Y. Physico-Chemical Aspects in Flocculation of Bentonite Suspensions by a Cyanobacterial Bioflocculant. *Water Res.* **1992**, *26*, 249–254. [CrossRef]
75. Deng, S.; Bai, R.; Hu, X.; Luo, Q. Characteristics of a Bioflocculant Produced by *Bacillus mucilaginosus* and Its Use in Starch Wastewater Treatment. *Appl. Microbiol. Biotechnol.* **2003**, *60*, 588–593. [CrossRef] [PubMed]
76. Hassimi, A.H.; Ezril Hafiz, R.; Muhamad, M.H.; Sheikh Abdullah, S.R. Bioflocculant Production Using Palm Oil Mill and Sago Mill Effluent as a Fermentation Feedstock: Characterization and Mechanism of Flocculation. *J. Environ. Manag.* **2020**, *260*, 110046. [CrossRef] [PubMed]
77. Yim, J.H.; Kim, S.J.; Ahn, S.H.; Lee, H.K. Characterization of a Novel Bioflocculant, p-KG03, from a Marine Dinoflagellate, *Gyrodinium impudicum* KG03. *Bioresour. Technol.* **2007**, *98*, 361–367. [CrossRef] [PubMed]
78. Zheng, Y.; Ye, Z.-L.; Fang, X.-L.; Li, Y.-H.; Cai, W.-M. Production and Characteristics of a Bioflocculant Produced by *Bacillus* sp. F19. *Bioresour. Technol.* **2008**, *99*, 7686–7691. [CrossRef] [PubMed]
79. Liu, Z.; Hu, Z.; Wang, T.; Chen, Y.; Zhang, J.; Yu, J.; Zhang, T.; Zhang, Y.; Li, Y. Production of Novel Microbial Flocculants by *Klebsiella* sp. TG-1 Using Waste Residue from the Food Industry and Its Use in Defecating the Trona Suspension. *Bioresour. Technol.* **2013**, *139*, 265–271. [CrossRef] [PubMed]
80. Kurniawan, S.B.; Ahmad, A.; Imron, M.F.; Abdullah, S.R.S.; Othman, A.R.; Hasan, H.A. Potential of Microalgae Cultivation Using Nutrient-Rich Wastewater and Harvesting Performance by Biocoagulants/Bioflocculants: Mechanism, Multi-Conversion of Biomass into Valuable Products, and Future Challenges. *J. Clean. Prod.* **2022**, *365*, 132806. [CrossRef]
81. Li, R.; Gao, B.; Huang, X.; Dong, H.; Li, X.; Yue, Q.; Wang, Y.; Li, Q. Compound Bioflocculant and Polyaluminum Chloride in Kaolin-Humic Acid Coagulation: Factors Influencing Coagulation Performance and Floc Characteristics. *Bioresour. Technol.* **2014**, *172*, 8–15. [CrossRef] [PubMed]
82. Zhang, X.; Sun, J.; Liu, X.; Zhou, J. Production and Flocculating Performance of Sludge Bioflocculant from Biological Sludge. *Bioresour. Technol.* **2013**, *146*, 51–56. [CrossRef] [PubMed]
83. Li, Y.; Li, Q.; Hao, D.; Hu, Z.; Song, D.; Yang, M. Characterization and Flocculation Mechanism of an Alkali-Activated Polysaccharide Flocculant from *Arthrobacter* sp. B4. *Bioresour. Technol.* **2014**, *170*, 574–577. [CrossRef] [PubMed]
84. Salim, S.; Bosma, R.; Vermuë, M.H.; Wijffels, R.H. Harvesting of Microalgae by Bio-Flocculation. *J. Appl. Phycol.* **2011**, *23*, 849–855. Available online: https://link.springer.com/article/10.1007/s10811-010-9591-x (accessed on 6 June 2024). [CrossRef] [PubMed]
85. Zou, X.; Sun, J.; Li, J.; Jia, Y.; Xiao, T.; Meng, F.; Wang, M.; Ning, Z. High Flocculation of Coal Washing Wastewater Using a Novel Bioflocculant from *Isaria cicadae* GZU6722. *Pol. J. Microbiol.* **2020**, *69*, 55–64. [CrossRef] [PubMed]
86. Buthelezi, S.P.; Olaniran, A.O.; Pillay, B. Production and Characterization of Bioflocculants from Bacteria Isolated from Wastewater Treatment Plant in South Africa. *Biotechnol. Bioprocess Eng.* **2010**, *15*, 874–881. [CrossRef]
87. Fu, F.; Wang, Q. Removal of Heavy Metal Ions from Wastewaters: A Review. *J. Environ. Manag.* **2011**, *92*, 407–418. [CrossRef] [PubMed]
88. Sathiyanarayanan, G.; Dineshkumar, K.; Yang, Y.-H. Microbial Exopolysaccharide-Mediated Synthesis and Stabilization of Metal Nanoparticles. *Crit. Rev. Microbiol.* **2017**, *43*, 731–752. [CrossRef]
89. Sathiyanarayanan, G.; Seghal Kiran, G.; Selvin, J. Synthesis of Silver Nanoparticles by Polysaccharide Bioflocculant Produced from Marine *Bacillus subtilis* MSBN17. *Colloids Surf. B Biointerfaces* **2013**, *102*, 13–20. [CrossRef]
90. Eman Zakaria, G. Production and Characteristics of a Heavy Metals Removing Bioflocculant Produced by *Pseudomonas aeruginosa*. *Pol. J. Microbiol.* **2012**, *61*, 281–289. [CrossRef]
91. Liu, J.; Ma, J.; Liu, Y.; Yang, Y.; Yue, D.; Wang, H. Optimized Production of a Novel Bioflocculant M-C11 by *Klebsiella* sp. and Its Application in Sludge Dewatering. *J. Environ. Sci.* **2014**, *26*, 2076–2083. [CrossRef]
92. Vimala, R.T.V. Role of Bacterial Bioflocculant on Antibiofilm Activity and Metal Removal Efficiency. *J. Pure Appl. Microbiol.* **2019**, *13*, 1823–1830. [CrossRef]

93. Ma, X.; Duan, D.; Chen, X.; Feng, X.; Ma, Y. A Polysaccharide-Based Bioflocculant BP50-2 from Banana Peel Waste: Purification, Structure and Flocculation Performance. *Int. J. Biol. Macromol.* **2022**, *205*, 604–614. [CrossRef]
94. Ayed, L.; Khelifi, E.; Jannet, H.B.; Miladi, H.; Cheref, A.; Achour, S.; Bakhrouf, A. Response Surface Methodology for Decolorization of Azo Dye Methyl Orange by Bacterial Consortium: Produced Enzymes and Metabolites Characterization. *Chem. Eng. J.* **2010**, *165*, 200–208. [CrossRef]
95. Neetha, N.J.; Sandesh, K.; Girish Kumar, K.; Chidananda, B.; Ujwal, P. Optimization of Direct Blue-14 Dye Degradation by *Bacillus fermus* (Kx898362) an Alkaliphilic Plant Endophyte and Assessment of Degraded Metabolite Toxicity. *J. Hazard. Mater.* **2019**, *364*, 742–751. [CrossRef]
96. Prabhakar, Y.; Gupta, A.; Kaushik, A. Microbial Degradation of Reactive Red-35 Dye: Upgraded Progression through Box–Behnken Design Modeling and Cyclic Acclimatization. *J. Water Process Eng.* **2021**, *40*, 101782. [CrossRef]
97. Artifon, W.; Mazur, L.P.; de Souza, A.A.U.; de Oliveira, D. Production of Bioflocculants from Spent Brewer's Yeast and Its Application in the Treatment of Effluents with Textile Dyes. *J. Water Process Eng.* **2022**, *49*, 102997. [CrossRef]
98. Chen, F.; Yuan, Y.; Chen, C.; Zhao, Y.; Tan, W.; Huang, C.; Xu, X.; Wang, A. Investigation of Colloidal Biogenic Sulfur Flocculation: Optimization Using Response Surface Analysis. *J. Environ. Sci.* **2016**, *42*, 227–235. [CrossRef]
99. Lee, C.-M.; Sublette, K.L. Microbial Treatment of Sulfide-Laden Water. *Water Res.* **1993**, *27*, 839–846. [CrossRef]
100. Shende, A.P.; Chidambaram, R. Cocoyam Powder Extracted from *Colocasia antiquorum* as a Novel Plant-Based Bioflocculant for Industrial Wastewater Treatment: Flocculation Performance and Mechanism. *Heliyon* **2023**, *9*, e15228. [CrossRef]
101. Lee, D.-J.; Liu, X.; Weng, H.-L. Sulfate and Organic Carbon Removal by Microbial Fuel Cell with Sulfate-Reducing Bacteria and Sulfide-Oxidising Bacteria Anodic Biofilm. *Bioresour. Technol.* **2014**, *156*, 14–19. [CrossRef]
102. Wu, B.; Liu, F.; Fang, W.; Yang, T.; Chen, G.-H.; He, Z.; Wang, S. Microbial Sulfur Metabolism and Environmental Implications. *Sci. Total Environ.* **2021**, *778*, 146085. [CrossRef]
103. Han, F.; Zhang, M.; Shang, H.; Liu, Z.; Zhou, W. Microbial Community Succession, Species Interactions and Metabolic Pathways of Sulfur-Based Autotrophic Denitrification System in Organic-Limited Nitrate Wastewater. *Bioresour. Technol.* **2020**, *315*, 123826. [CrossRef] [PubMed]
104. Chen, D.-Z.; Zhao, X.-Y.; Miao, X.-P.; Chen, J.; Ye, J.-X.; Cheng, Z.-W.; Zhang, S.-H.; Chen, J.-M. A Solid Composite Microbial Inoculant for the Simultaneous Removal of Volatile Organic Sulfide Compounds: Preparation, Characterization, and Its Bioaugmentation of a Biotrickling Filter. *J. Hazard. Mater.* **2018**, *342*, 589–596. [CrossRef] [PubMed]
105. Guo, J.; Chen, C. Removal of Arsenite by a Microbial Bioflocculant Produced from Swine Wastewater. *Chemosphere* **2017**, *181*, 759–766. [CrossRef] [PubMed]
106. Sajayan, A.; Seghal Kiran, G.; Priyadharshini, S.; Poulose, N.; Selvin, J. Revealing the Ability of a Novel Polysaccharide Bioflocculant in Bioremediation of Heavy Metals Sensed in a *Vibrio* Bioluminescence Reporter Assay. *Environ. Pollut.* **2017**, *228*, 118–127. [CrossRef]
107. Yu, L.; Hua, J.; Fan, H.; George, O.; Lu, Y. Simultaneous Nitriles Degradation and Bioflocculant Production by Immobilized *K. oxytoca* Strain in a Continuous Flow Reactor. *J. Hazard. Mater.* **2020**, *387*, 121697. [CrossRef]
108. Sharma, V.K.; Ma, X.; Lichtfouse, E.; Robert, D. Nanoplastics Are Potentially More Dangerous than Microplastics. *Environ. Chem. Lett.* **2023**, *21*, 1933–1936. [CrossRef]
109. Ahmad Shukri, Z.N.; Che Engku Chik, C.E.N.; Hossain, S.; Othman, R.; Endut, A.; Lananan, F.; Terkula, I.B.; Kamaruzzan, A.S.; Abdul Rahim, A.I.; Draman, A.S.; et al. A Novel Study on the Effectiveness of Bioflocculant-Producing Bacteria *Bacillus enclensis*, Isolated from Biofloc-Based System as a Biodegrader in Microplastic Pollution. *Chemosphere* **2022**, *308*, 136410. [CrossRef]
110. Padervand, M.; Lichtfouse, E.; Robert, D.; Wang, C. Removal of Microplastics from the Environment. A Review. *Environ. Chem. Lett.* **2020**, *18*, 807–828. [CrossRef]
111. Shurair, M.; Almomani, F.; Bhosale, R.; Khraisheh, M.; Qiblawey, H. Harvesting of Intact Microalgae in Single and Sequential Conditioning Steps by Chemical and Biological Based–Flocculants: Effect on Harvesting Efficiency, Water Recovery and Algal Cell Morphology. *Bioresour. Technol.* **2019**, *281*, 250–259. [CrossRef] [PubMed]
112. Kurade, M.B.; Murugesan, K.; Selvam, A.; Yu, S.-M.; Wong, J.W.C. Ferric Biogenic Flocculant Produced by *Acidithiobacillus ferrooxidans* Enable Rapid Dewaterability of Municipal Sewage Sludge: A Comparison with Commercial Cationic Polymer. *Int. Biodeterior. Biodegrad.* **2014**, *96*, 105–111. [CrossRef]
113. Muthulakshmi, L.; Seghal Kiran, G.; Ramakrishna, S.; Cheng, K.Y.; Ampadi Ramachandran, R.; Mathew, M.T.; Pruncu, C.I. Towards Improving the Corrosion Resistance Using a Novel Eco-Friendly Bioflocculant Polymer Produced from *Bacillus* sp. *Mater. Today Commun.* **2023**, *35*, 105438. [CrossRef]
114. Pritchard, M.; Craven, T.; Mkandawire, T.; Edmondson, A.S.; O'Neill, J.G. A Comparison between *Moringa oleifera* and Chemical Coagulants in the Purification of Drinking Water—An Alternative Sustainable Solution for Developing Countries. *Phys. Chem. Earth Parts A/B/C* **2010**, *35*, 798–805. [CrossRef]
115. Vimala, R.T.V.; Lija Escaline, J.; Sivaramakrishnan, S. Characterization of Self-Assembled Bioflocculant from the Microbial Consortium and Its Applications. *J. Environ. Manag.* **2020**, *258*, 110000. [CrossRef]
116. Tsilo, P.H.; Basson, A.K.; Ntombela, Z.G.; Maliehe, T.S.; Pullabhotla, V.S.R.R. Production and Characterization of a Bioflocculant from *Pichia kudriavzevii* MH545928.1 and Its Application in Wastewater Treatment. *Int. J. Environ. Res. Public Health* **2022**, *19*, 3148. [CrossRef] [PubMed]

117. Liu, Z.; Hao, N.; Hou, Y.; Wang, Q.; Liu, Q.; Yan, S.; Chen, F.; Zhao, L. Technologies for Harvesting the Microalgae for Industrial Applications: Current Trends and Perspectives. *Bioresour. Technol.* **2023**, *387*, 129631. [CrossRef] [PubMed]
118. Liu, C.; Sun, D.; Liu, J.; Zhu, J.; Liu, W. Recent Advances and Perspectives in Efforts to Reduce the Production and Application Cost of Microbial Flocculants. *Bioresour. Bioprocess.* **2021**, *8*, 51. [CrossRef]
119. Zha, X.; Li, C.; Li, X.; Huang, Y. Hydrothermal Liquid Fraction of Concentrated Organic Matter in Sewage for Coarse Flocculant Preparation: The Role and Regulation Mechanism. *Sep. Purif. Technol.* **2024**, *339*, 126654. [CrossRef]
120. Kurniawan, S.B.; Abdullah, S.R.S.; Othman, A.R.; Purwanti, I.F.; Imron, M.F.; Ismail, N.; Izzati; Ahmad, A.; Hasan, H.A. Isolation and Characterisation of Bioflocculant-Producing Bacteria from Aquaculture Effluent and Its Performance in Treating High Turbid Water. *J. Water Process Eng.* **2021**, *42*, 102194. [CrossRef]
121. Li, N.-J.; Lan, Q.; Wu, J.-H.; Liu, J.; Zhang, X.-H.; Zhang, F.; Yu, H.-Q. Soluble Microbial Products from the White-Rot Fungus *Phanerochaete chrysosporium* as the Bioflocculant for Municipal Wastewater Treatment. *Sci. Total Environ.* **2021**, *780*, 146662. [CrossRef]
122. Kurniawan, S.B.; Imron, M.F.; Sługocki, Ł.; Nowakowski, K.; Ahmad, A.; Najiya, D.; Abdullah, S.R.S.; Othman, A.R.; Purwanti, I.F.; Hasan, H.A. Assessing the Effect of Multiple Variables on the Production of Bioflocculant by *Serratia marcescens*: Flocculating Activity, Kinetics, Toxicity, and Flocculation Mechanism. *Sci. Total Environ.* **2022**, *836*, 155564. [CrossRef] [PubMed]
123. Rajivgandhi, G.; Vimala, R.T.V.; Maruthupandy, M.; Alharbi, N.S.; Kadaikunnan, S.; Khaled, J.M.; Manoharan, N.; Li, W.-J. Enlightening the Characteristics of Bioflocculant of Endophytic Actinomycetes from Marine Algae and Its Biosorption of Heavy Metal Removal. *Environ. Res.* **2021**, *200*, 111708. [CrossRef] [PubMed]
124. Giri, S.S.; Ryu, E.; Park, S.C. Characterization of the Antioxidant and Anti-Inflammatory Properties of a Polysaccharide-Based Bioflocculant from *Bacillus subtilis* F9. *Microb. Pathog.* **2019**, *136*, 103642. [CrossRef] [PubMed]
125. Nouha, K.; Kumar, R.S.; Balasubramanian, S.; Tyagi, R.D. Critical Review of EPS Production, Synthesis and Composition for Sludge Flocculation. *J. Environ. Sci.* **2018**, *66*, 225–245. [CrossRef] [PubMed]
126. Chen, L.; Zhao, B.; An, Q.; Qiu Guo, Z.; Huang, C. The Characteristics and Flocculation Mechanisms of SMP and B-EPS from a Bioflocculant-Producing Bacterium *Pseudomonas* sp. XD-3 and the Application for Sludge Dewatering. *Chem. Eng. J.* **2024**, *479*, 147584. [CrossRef]
127. Yang, Y.; Guo, W.; Ngo, H.H.; Zhang, X.; Liang, S.; Deng, L.; Cheng, D.; Zhang, H. Bioflocculants in Anaerobic Membrane Bioreactors: A Review on Membrane Fouling Mitigation Strategies. *Chem. Eng. J.* **2024**, *486*, 150260. [CrossRef]
128. Li, L.; He, Z.; Song, Z.; Sheng, T.; Dong, Z.; Zhang, F.; Ma, F. A Novel Strategy for Rapid Formation of Biofilm: Polylactic Acid Mixed with Bioflocculant Modified Carriers. *J. Clean. Prod.* **2022**, *374*, 134023. [CrossRef]
129. Hou, J.; Xia, L.; Ma, T.; Zhang, Y.; Zhou, Y.; He, X. Achieving Short-Cut Nitrification and Denitrification in Modified Intermittently Aerated Constructed Wetland. *Bioresour. Technol.* **2017**, *232*, 10–17. [CrossRef]
130. Yao, S.; Chen, L.; Guan, D.; Zhang, Z.; Tian, X.; Wang, A.; Wang, G.; Yao, Q.; Peng, D.; Li, J. On-Site Nutrient Recovery and Removal from Source-Separated Urine by Phosphorus Precipitation and Short-Cut Nitrification-Denitrification. *Chemosphere* **2017**, *175*, 210–218. [CrossRef]
131. Zheng, X.; Zhou, W.; Wan, R.; Luo, J.; Su, Y.; Huang, H.; Chen, Y. Increasing Municipal Wastewater BNR by Using the Preferred Carbon Source Derived from Kitchen Wastewater to Enhance Phosphorus Uptake and Short-Cut Nitrification-Denitrification. *Chem. Eng. J.* **2018**, *344*, 556–564. [CrossRef]
132. Liu, Y.; Zeng, Y.; Yang, J.; Chen, P.; Sun, Y.; Wang, M.; Ma, Y. A Bioflocculant from *Corynebacterium glutamicum* and Its Application in Acid Mine Wastewater Treatment. *Front. Bioeng. Biotechnol.* **2023**, *11*, 1136473. [CrossRef] [PubMed]
133. Zeng, F.; Xu, L.; Sun, C.; Liu, H.; Chen, L. A Novel Bioflocculant from Raoultella Planticola Enhances Removal of Copper Ions from Water. *J. Sens.* **2020**, *2020*, 1–10. [CrossRef]
134. Ni, F.; Peng, X.; He, J.; Yu, L.; Zhao, J.; Luan, Z. Preparation and Characterization of Composite Bioflocculants in Comparison with Dual-Coagulants for the Treatment of Kaolin Suspension. *Chem. Eng. J.* **2012**, *213*, 195–202. [CrossRef]
135. Polizzi, C.; Gabriel, D.; Munz, G. Successful Sulphide-Driven Partial Denitrification: Efficiency, Stability and Resilience in SRT-Controlled Conditions. *Chemosphere* **2022**, *295*, 133936. [CrossRef]
136. Liu, H.; Liu, D.; Huang, Z.; Chen, Y. Bioaugmentation Reconstructed Nitrogen Metabolism in Full-Scale Simultaneous Partial Nitrification-Denitrification, Anammox and Sulfur-Dependent Nitrite/Nitrate Reduction (SPAS). *Bioresour. Technol.* **2023**, *367*, 128233. [CrossRef]

Disclaimer/Publisher's Note: The statements, opinions and data contained in all publications are solely those of the individual author(s) and contributor(s) and not of MDPI and/or the editor(s). MDPI and/or the editor(s) disclaim responsibility for any injury to people or property resulting from any ideas, methods, instructions or products referred to in the content.

Review

A Systematic Literature Review for Addressing Microplastic Fibre Pollution: Urgency and Opportunities

Carmen Ka-Man Chan [1], Chris Kwan-Yu Lo [2,3] and Chi-Wai Kan [1,3,*]

1. School of Fashion and Textiles, The Hong Kong Polytechnic University, Hung Hom, Kowloon, Hong Kong; carmenkm-itc.chan@connect.polyu.hk
2. Department of Logistics and Maritime Studies, The Hong Kong Polytechnic University, Hung Hom, Kowloon, Hong Kong; tcclo@polyu.edu.hk
3. Research Centre for Resources Engineering towards Carbon Neutrality, The Hong Kong Polytechnic University, Hung Hom, Kowloon, Hong Kong
* Correspondence: tccwk@polyu.edu.hk

Abstract: Microplastic fibre (MPF) pollution is a pressing concern that demands urgent attention. These tiny synthetic textile fibres can be found in various ecosystems, including water and air, and pose significant environmental risks. Despite their size (less than 5 mm), they can harm aquatic and terrestrial organisms and human health. Studies have demonstrated that these imperceptible pollutants can contaminate marine environments, thereby putting marine life at risk through ingestion and entanglement. Additionally, microplastic fibres can absorb toxins from the surrounding water, heightening their danger when consumed by aquatic organisms. Traces of MPFs have been identified in human food chains and organs. To effectively combat MPF pollution, it is crucial to understand how these fibres enter ecosystems and their sources. Primary sources include domestic laundry, where synthetic textile fibres are released into wastewater during washing. Other significant sources include industrial effluents, breakdown of plastic materials, and atmospheric deposition. Additionally, MPFs can be directly released into the environment by improperly disposing of consumer products containing these fibres, such as non-woven hygienic products. A comprehensive approach is necessary to address this pressing issue, including understanding the sources, pathways, and potential risks of MPFs. Immediate action is required to manage contamination and mitigate MPF pollution. This review paper provides a systematic literature analysis to help stakeholders prioritise efforts towards reducing MPFs. The key knowledge gaps identified include a lack of information regarding non-standardised test methodology and reporting units, and a lack of information on manufacturing processes and products, to increase understanding of life cycle impacts and real hotspots. Stakeholders urgently need collaborative efforts to address the systematic changes required to tackle this issue and address the proposed opportunities, including targeted government interventions and viable strategies for the industry sector to lead action.

Keywords: microplastic fibres; microfibres; fibre fragmentation; shedding; microplastic pollution; domestic laundry; wastewater; wastewater treatment

Citation: Chan, C.K.-M.; Lo, C.K.-Y.; Kan, C.-W. A Systematic Literature Review for Addressing Microplastic Fibre Pollution: Urgency and Opportunities. *Water* 2024, 16, 1988. https://doi.org/10.3390/w16141988

Academic Editor: Alexandre T. Paulino

Received: 31 May 2024
Revised: 9 July 2024
Accepted: 10 July 2024
Published: 13 July 2024

Copyright: © 2024 by the authors. Licensee MDPI, Basel, Switzerland. This article is an open access article distributed under the terms and conditions of the Creative Commons Attribution (CC BY) license (https://creativecommons.org/licenses/by/4.0/).

1. Introduction

The issue of microplastic pollution, mainly caused by microplastic fibres (MPFs), has been widely recognised for some time. Despite being first identified in 2011 by Browne et al. [1], progress in addressing this problem has been slow compared to other forms of microplastic pollution, as specific research on MPFs remains relatively limited. Researchers have consistently recommended taking measures to standardise methodologies and reporting units and adopt a more systematic approach to evaluating textile parameters. They have also suggested taking a balanced approach to regulations and public education [2–5]. This systematic literature review aims to identify existing knowledge gaps and priority steps that the textile sector can take to mitigate this issue promptly and where it can help to speed up progress.

According to the Textile Exchange [6], synthetic textiles will have the largest global textile fibre production share in 2021, at 64%. The global fibre production demand has significantly grown from 8.4 kg per person in 1975 to 14.3 kg per person in 2021, with less than 0.6% recycled from pre-consumer or post-consumer textiles. If the business persists with the status quo, the definite growth trajectory will be 34% from 2020 to 2030. The reason for the growth and shift in the use of fossil fuel-based plastic fibres is because they cost less and provide better functional performance than natural fibres such as cotton [7,8].

MPFs/MFs are widely distributed and are found in diverse environments such as oceans, freshwater, and air. Browne et al. [1] first estimated that the accumulation of MPs is associated with shoreline population density worldwide, indicating that 85% of MPs are MPFs. According to Boucher and Friot [9], synthetic textiles account for 35% of the global release of MPs into the ocean, with 25% emerging from the wastewater pathway. Based on research conducted by the Ellen MacArthur Foundation [10], the current rate of domestic washing is projected to release approximately 22 million tons of MPFs into the ocean by 2050. Like MPs, MPFs are non-degradable, accumulate, and take hundreds of years to decompose [11,12], thus inevitably building up in the environment. This alarming statistic highlights the pressing need to address MPF pollution before it becomes increasingly challenging.

Increasing evidence indicates that MPFs are the most common type of secondary MP present in marine environments [12]. The most prominent and broadly reported sources of MPFs are shedding during domestic laundry and manufacturing processes [13]. For an average wash load of 6 kg, over 700,000 fibres could be released per wash and discharged into the aquatic environment via wastewater treatment plants (WWTPs). Although WWTPs are reported to be 95–99% effective, they are not explicitly designed for MP/MPF retention so they can bypass WWTPs. Owing to their enormous discharge volumes, there is compelling evidence that WWTPs are significant sinks for MP pollution [14]. Apart from wastewater from WWTPs, other release sources may be solid waste or sludge disposal. Sludge, a by-product of WWTPs commonly applied to agricultural land as fertilisers, has been found to contain MPFs/MFs. Tao et al. [15] estimated that during a 15 min drying process, over 90,000 microfibres could be released from 1 kg of polyester and cotton textiles. The release of MPFs from drying clothes can represent a source of airborne MP pollution [15–17]. These by-products, which appear as airborne textile fibres, can also cause persistent terrestrial contamination [18–21]. These are all pathways for the release of MPFs into the environment.

The ecosystem impacts of MPs/MPFs have the potential to occur through physical, chemical, and biological pathways. Woodall et al. [12] initially reported the ingestion of MPFs by deep sea organisms in a natural setting. The concern regarding MF buildup in the environment is due to their fibrous nature. MPFs tend to entangle and block the digestive tract, leading to the starvation and impaired growth of microorganisms [21]. The longer-term impacts include the capacity to absorb harmful chemical substances from fibres associated with dyes or additives used in textile manufacturing. Several studies have reported that MPFs can transfer contaminants. The ingestion of MPFs by various aquatic species, including turtles, seabirds, fish, and lobsters, has been associated with reduced feeding and reproductive abilities.

Furthermore, owing to the large surface areas of MPs/MPFs, toxic compounds can be introduced into humans by them eating higher trophic level species because the compounds are transferred along the food chain [22]. A recent study by Ragusa et al. [23] identified MPs in human placentas. Increasing evidence for the presence of MPFs in the atmosphere has recently been reported. The flying MPFs inhaled by humans can be deposited in lung tissue [24] and may lead to tumours. According to Cole [25], nanoscopic and microscopic fibrous materials can also be carcinogenic and fibrotic, whereas particles of the same content are comparatively benign. The fact that MPFs are increasing and accumulating toxins dangerous to marine life and potentially humans will only magnify and become more challenging to resolve if further action is delayed.

Microplastics (MPs) are synthetic polymers often defined as plastic particles smaller than 5 mm [26], which include particles in the nano-size range (1 nm) [27]. Figure 1 shows a microscopic view of different types of MPs. Microplastic fibres (MPFs) are petrochemicals derived from synthetic fibre-based textiles and are considered a subset of MPs [28]. MPFs are also commonly described as fibrous or thread-like pieces of plastic with a length between 100 µm and 5 mm and a width of at least 1.5 orders of magnitude shorter [7]. MPFs are extensively distributed across various environments, including air, marine, landfill, and terrestrial. Like MPs, once in the environment, they take hundreds of years to degrade and are challenging to remove. In this study, our primary focus was on MPFs.

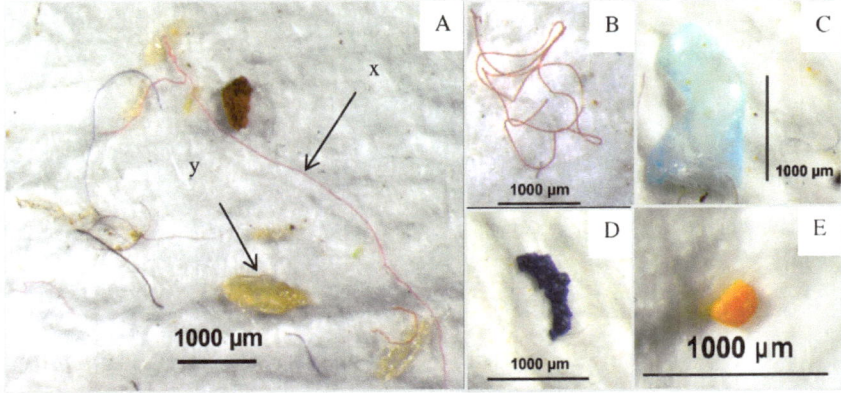

Figure 1. Microscopic views of the different types of microplastics. Adapted from [29]. (**A**) Photograph of a typical filter paper under a stereomicroscope, where x = microplastic fibre and y = biogenic material paper; (**B**) synthetic microplastic fibre, (**C,D**) synthetic fragments; and (**E**) synthetic bead.

Microfibres (MFs) were first produced in Japan in the 1970s and have exceptionally fine diameters for the textile industry. According to the industry definition, MFs generally have a linear density smaller than one decitex (as shown in Figure 2), a linear density unit of one gram per 10,000 m, or a diameter < 10 µm [30]. They are typically made of synthetic fibres. Nonetheless, growing concerns about fibre fragmentation have led to expanding the definition of fragmented fibres. This is commonly generalised to encompass both natural and man-made cellulosic fibres that exhibit characteristics of lengths less than 5 mm [31] and length-to-diameter ratios greater than three [32]. The most remarkable difference between MPFs and MFs is the origin of the fibres. MPF refers to fibres of synthetic origin only, and MF refers to all types of fibres. Therefore, when MFs are referred to in this study, this refers to the generalised definition of fragmented fibres of all material types that are shorter than 5 mm.

Sanchez-Vidal et al. [33] noted that while MPFs are the most prevalent form of MP pollution, research has focused primarily on MPs. Despite the thousands of research papers published on MPs, a mere two hundred studies specifically address the issue of MPF pollution, according to a search of Web of Science using the keywords "microplastic", "microfibres or microfibers or fibres or fibers" and "textiles or apparel or clothing" as of November 2022. Studies on MPFs have primarily focused on synthetic textiles and their shed rates during domestic laundering. Research on the loss of MPFs during the rest of the manufacturing life cycle still needs to be expanded. Our knowledge is incomplete and biased, which may lead to a conclusion that favours washing discharges directed to municipal WWTPs as the primary sources of fibres found in the sea and surface waters [1].

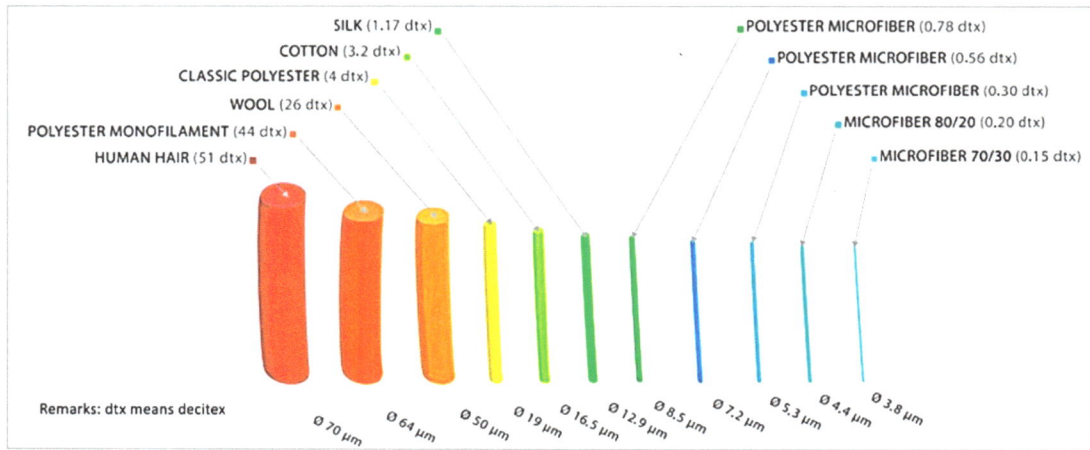

Figure 2. The illustration shows how microfibres are defined in terms of fineness (less than 1 decitex) compared to common textile fibres and human hair. Modified from [8].

Dealing with MPFs can be a challenging task, especially when it comes to managing and measuring irregular shapes that are easily tangled. Using a mesh size of 80 μm for sampling, researchers have captured fibres up to 250 times more effectively than with a mesh size of 300 μm, which has been commonly used in previous studies [34]. However, owing to inconsistent sampling and quantification methods at various locations, it is difficult to understand the magnitude of the MPF problem entirely, and MPFs are mostly underestimated [7,35].

Studies on MPFs should be connected to a broad range of industries. Instead, research has been mostly limited to laundry in the consumer phase without covering the entire production process. Therefore, current solutions focus only on a small part of the problem and cannot address MPF release across the complete product lifecycle.

Since there is yet to be a standardised method for analysing MPs and MPFs [36], the sampling and extraction techniques used to quantify MPs/MFs are inconsistent. As a result, the reported MPF concentrations among the studies were conflicting, varying, and divided, without a standardised concentration reference. Many studies have reported in oversimplified tones that are missing critical details. Concentrations of MPs/MPFs are relatively complex to compare because of the different units used [37]. Although efforts have been made to convert these units [38], the results can be biased or misleading because assumptions must be made.

As MPFs are currently the most prevalent MP source, targeted research is essential to understand this issue. This study used a systematic bibliometric approach to objectively review this subject and understand the knowledge gaps and how proliferation has led to the current state. It then focuses on insights that can help to prioritise mitigation actions for researchers, practitioners, and regulators to address this urgent need.

2. Materials and Methods

2.1. Data Source

A structured bibliometric approach or citation network analysis (CNA) is an objective way to capture research that shows knowledge proliferation over time. Establishing a research scope based on keyword selection is a more accurate evaluation than a traditional literature review [39].

Bibliometric analysis requires obtaining and analysing data on a specific topic or a range of issues. The data were captured through the Web of Science (WOS) in November 2022. The following search criteria were used. TS refers to the topic fields, which include titles, abstracts, keywords, and indexing. "TS = (microplastic*) AND TS = (microfibre* or microfiber* or fibre* or fiber*) AND TS = (textile* or clothing* or apparel* or fashion*) and Environmental Sciences or Engineering Environmental or Marine Freshwater Biology or Water Resources or Materials Science Textiles or Multidisciplinary Sciences or Polymer Science or Green Sustainable Science Technology or Engineering Chemical or Toxicology (Web of Science Categories) and Review Article or Proceeding Paper or Correction or Early Access or Editorial Material or Meeting Abstract (Exclude–Document Types) and 7.70 Thermodynamics or 3.87 Paper and Wood Materials Science or 3.32 Entomology or 6.73 Social Psychology or 2.74 Photocatalysts or 7.139 Energy and Fuels (Exclude–Citation Topics Meso) and Energy Fuels or Food Science Technology or Education Educational Research or Fisheries or Public Environmental Occupational Health or Meteorology Atmospheric Sciences or Oceanography or Toxicology or Microbiology (Exclude–Research Areas) and 3.60.1998 Pfos (Exclude–Citation Topics Micro) and Chinese (Exclude–Languages)". In total, 219 articles were identified. Figure 3 shows a flowchart summarising the identification, screening and bibliometric analysis used.

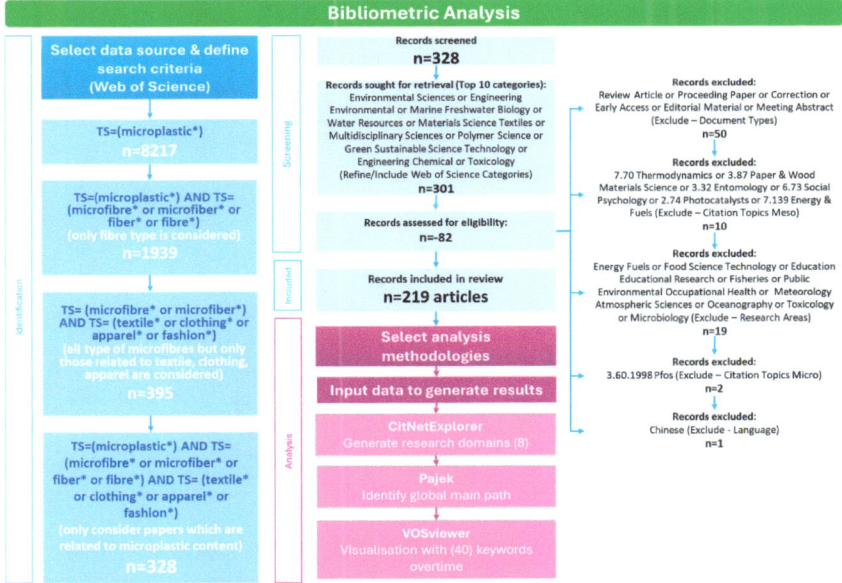

Figure 3. Flowchart of the bibliometric analysis showing the identification, screening and bibliometric analysis used. (* stands for wildcard which means include variants).

As microfibres have been developed in the industry since the 1970s, there were over 6246 papers related to this agenda, too many to be considered for a literature review. If the focus on microplastics and fibres was not specific to textiles and clothing, 1939 papers were available. We refined the parameters to be more specific and aligned them with the objectives of identifying opportunities from a sector perspective. In the early stage of knowledge development, this topic proliferated with research based on MPs without focusing on fibres, which were subsequently found to be the most substantial form. Therefore, instead of searching for keywords in the abstract, it was kept to the whole content, which resulted in 69 more papers, as these were critical papers. The results are summarised in Table 1.

Table 1. List of keyword research results from Web of Science in November 2022. (* stands for wildcard which means include variants).

Keywords	No of Articles
TS = (microplastic*)	8217
TS = (microplastic*) AND TS = (microfibre* or microfiber* or fibre* or fiber*)	1939
TS = (microfibre* or microfiber*) AND TS = (textile* or clothing* or apparel* or fashion*)	395
TS = (microplastic*) AND TS= (microfibre* or microfiber* or fibre* or fiber*) AND TS = (textile* or clothing* or apparel* or fashion*)	219

From the WOS, the graph in Figure 4 illustrates that there have been few published studies in the research domain since 2011, and a significant growth in numbers and citations since 2019. Data were exported and analysed using WOS, Microsoft Excel 365, CitNetExplorer 1.0.0, Pajek 5.16 and VOSviewer 1.6.18.

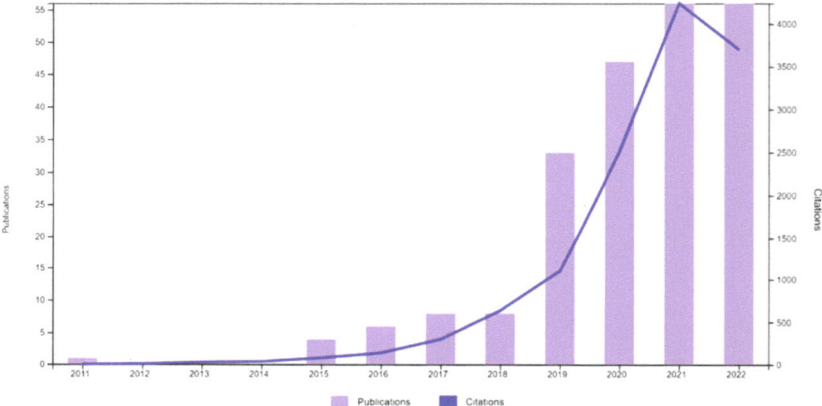

Figure 4. Times cited and the distribution of publications over time (2011–2022) were generated from the Web of Science using the search keywords listed in Table 1. (The Citation Report graphic is derived from the Clarivate Web of Science, Copyright: Clarivate 2022. All rights reserved. Date of search: November 2022).

2.2. Methodology

CNA aims to identify research domains, evaluate research traditions, and map changing paradigms [40]. The concept of CNA is that citation networks constitute a theory of connections between articles and systematic channels, transforming scientific knowledge, particularly when developing the main path [41]. Researchers in the same field tend to cite each other and add new knowledge to their latest research. Thus, the advancing knowledge in a particular research field will continue to grow [40].

CitNetExplorer is software that maps the topmost cited publications in the database. All non-matching cited references were excluded to ensure that every citation in the database in CitNetExplorer could be matched with a publication. A resolution for clustering of 2.00 with a minimum of ten citations and merging of small clusters was selected, and the analysis generated eight clusters with six publications that did not belong to a cluster. Clustering is used to identify the number of research domains [42].

According to Van Eck and Waltman [43], circles indicate publications, curved lines indicate citation relations, and the vertical axis represents time in visualisation. The proximity of publications in the citation network determines publication locations in the horizontal direction. The closer the two publications are to each other, the closer they are positioned in the horizontal direction. The citation score is the number of citations of a publication received from other physical publications in the WOS Database. The core publication has citation relations with at least ten different publications. The longest path was the longest knowledge network developed during the study period. Therefore, the core publication and longest path were also generated for evaluation, in addition to a complete network cluster analysis.

Pajek is software for analysing and visualising large networks (networks containing up to one billion vertices) [44]. This was applied to determine the global main path in the acyclic citation network using a traversal weight scheme to examine the research domain's most significant knowledge dissemination.

Publications assigned to the same cluster tend to be closely related to the citation networks. Each cluster was reviewed and assigned to a research domain. The longest path analysis from CitiNetExplorer and the main path analysis from Pajek were used to identify the key contributors to the leading articles.

VOSviewer is software that analyses publication data to create network maps. It creates distance-based visualisation, in which more closely related terms are closer to the visual display [42]. This software was used to map the co-occurrence of all keywords in publications. Full counting was performed to ensure that each keyword had the same weight as all the other keywords, regardless of the number of keywords per document. The minimum occurrence requirement for a keyword to be included in the map was selected as 40. Only the top 40 keywords on the map are included.

3. Results

3.1. Publication Profile

Table 2 shows that the environmental science communities continue to drive major studies. Five of the top ten journals comprised 53% (116) of the total environmental science or pollution research publications, and 13% (29) were from chemical and polymer disciplines. Again, such small numbers affirm that MPFs/MFs are less focused subjects.

Table 2. Number of publications by top 10 journals.

Name of Journal	No of Publications	%
Science of the Total Environment	34	16%
Environmental Pollution	27	12%
Marine Pollution Bulletin	22	10%
Environmental Science Technology	15	7%
Environmental Science and Pollution Research	14	6%
Chemosphere	9	4%
Journal of Hazardous Materials	8	4%
PLoS ONE	6	3%
Frontiers in Marine Science	4	2%
Polymers	4	2%

Table 3 illustrates that over 70% (155) of the articles from the top ten countries were from developed countries in the USA and Europe. China is the largest producer of textiles and clothing in the world. Although it had the highest number of publications, it accounted for only 15% (n = 34) of the total publications. This may explain why the research was highly skewed toward the usage phase of textiles in garment laundry. From Table 4, the largest cluster generated by CitNetExplorer was assigned to the research domain of domestic laundry and drying, with 85 publications, constituting 39% of the total.

Table 3. Distribution of articles by top 10 countries.

Publication Countries	No of Publications	%
People's Republic of China	36	16%
USA	34	16%
England	27	12%
Italy	26	12%
Canada	15	7%
Germany	12	5%
Spain	12	5%
Switzerland	11	5%
Australia	9	4%
Finland	9	4%

Table 4. CitNetExplorer-generated clusters and the number of publications in each cluster.

Group No	Colour	No of Publications	Research Domains
0	NA	6	Scattered Samples
1	Blue	85	Domestic laundry and drying
2	Green	28	Test methodology
3	Purple	22	Aquatic ecosystem
4	Orange	21	Atmosphere environment
5	Yellow	19	Wastewater source
6	Brown	17	Abundance and distribution
7	Pink	11	Terrestrial ecosystem
8	Light Blue	10	Hazardous nature

3.2. Citation Network

Figure 5 shows the cluster analysis of all 219 publications imported from WOS. Eight clusters were found with 1483 citation links, and six papers did not belong to any cluster. The colours and breakdown of each cluster are summarised in Table 4. The research domains were allocated according to their content by evaluating each cluster relationship. The details of each cluster are discussed in Section 3.4.

As shown in Figure 6, there were 153 core publications with 1363 citation links. From the longest path analysis between Browne et al. [1] and Volgare et al. [45], there were 17 publications with 92 citation links, all of which were built on MPFs/MFs released from domestic laundry with different textile and laundry parameters, and the uptake of MPFs from aquatic organisms. In addition to MPFs from polyester, Zambrano et al. [46] raised concerns regarding MFs, including from cotton and rayon. Kärkkäinen et al. [16] expanded the MPF release to include tumble drying. Two articles proliferated and were not found in the main path. De Falco et al. [47] shared the potential of using pectin finishes to reduce MPF release, and Herweyers et al. [48] attempted to evaluate MPF/MF pollution from a consumer perspective.

In Figure 7 of the global main path, Rochman et al. [49] diverted from Browne [50] to assess MPs/MPFs ingested by fishes and bivalves. This study highlights the potential of MPs/MPFs for human consumption, such as from seafood. Instead of studying the release of MPFs from textiles, Talvitie et al. [51] continued to dive deeper and found direct evidence that WWTPs are a point source of MPs/MPFs released into aquatic environments. Pirc et al. [52] focused on MPF emissions from different textile and laundry parameters.

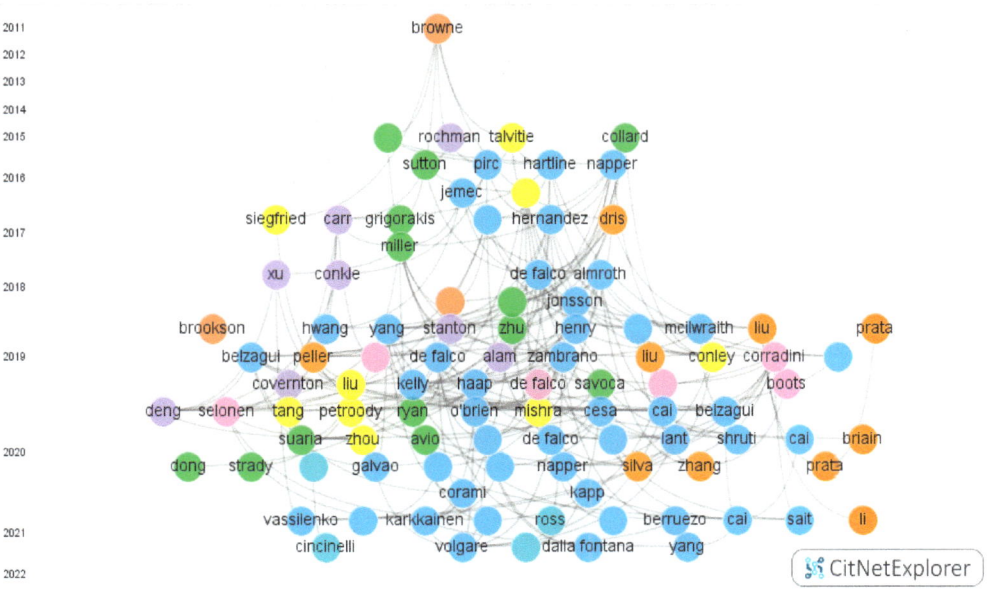

Figure 5. Full citation network generated by cluster analysis. (Research domain clusters are coloured according to Table 4).

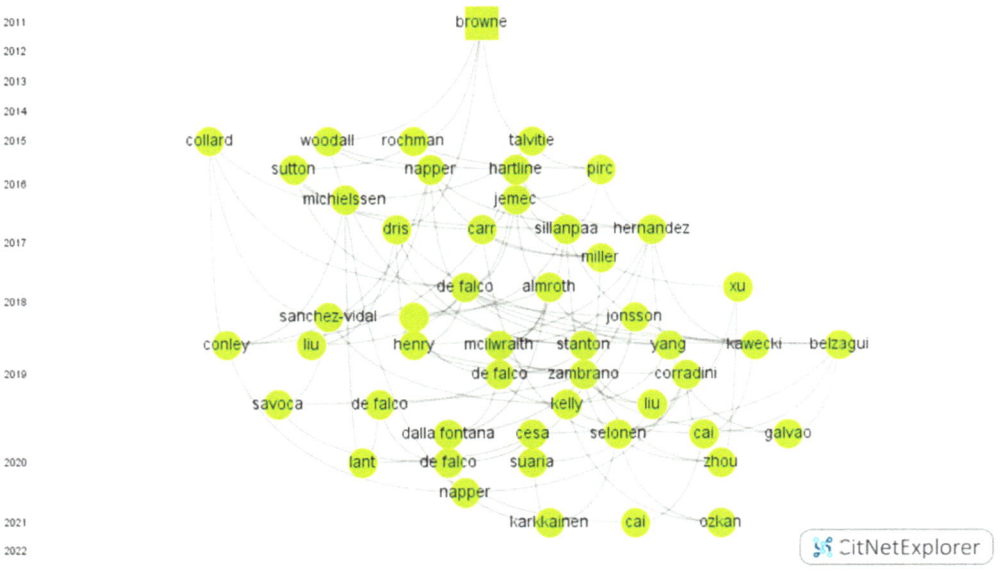

Figure 6. Citation network by core publication analysis.

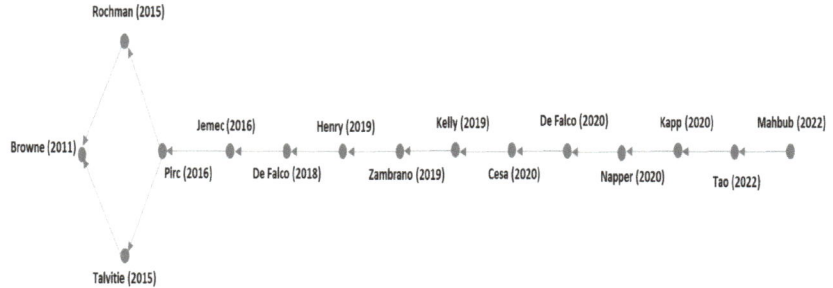

Figure 7. The global main path was generated using Pajek.

3.3. Connection between Keywords

Figure 8 shows that MPFs/MFs have originated as a branch of MP research with a long history. However, the scope has broadened to cover more comprehensive fibres, including natural and semi-synthetic fibres [46]. Since then, more papers have reported the release or shedding from these non-synthetic sources using the fibre type as a function of the test parameters [53]. The dynamics of the problem have shifted from MP pollution to the more significant issue of fibre fragmentation from the most commonly used textile materials (polyester and cotton) and their applications in the last few years.

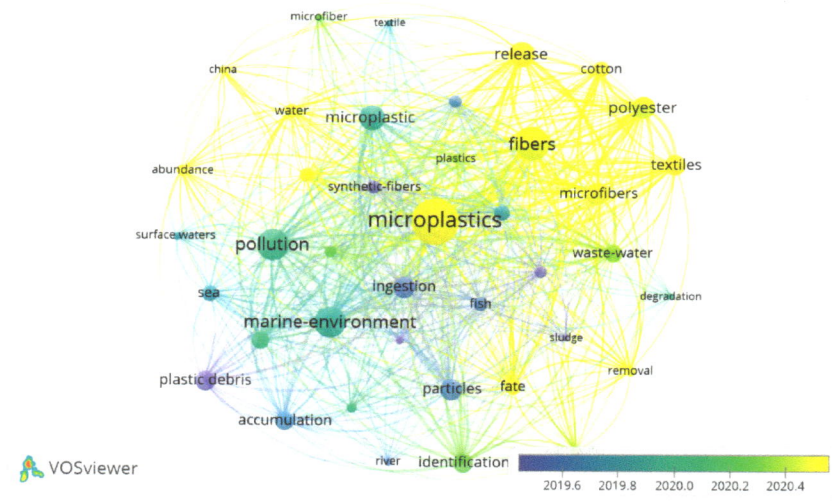

Figure 8. Overlay visualisation result by 40 keywords co-occurrence generated from VOSviewer.

Although studies have shown that textile finishing agents have prolonged their biodegradability from 85% to 75% [54], merging non-synthetic, natural, and semi-synthetic fibres in one piece has increased the challenges and made the problem more difficult to unravel. This requires a more complex system approach across the entire textile value chain as it becomes challenging to differentiate between accountabilities and responsibilities in the complex chain. Nonetheless, this may hinder using more sustainable bio-based synthetic materials as they are exposed to a similar MF fragmentation issue. Therefore, it is crucial to develop methodologies to differentiate common textile materials and their blends to find appropriate solutions for different fibre compositions and reduce their release into the environment. Although natural or semi-synthetic materials can behave differently depending on their degradability and the chemicals added, generalising natural MFs with MPFs is not recommended as it may hamper the resolution of MPF pollution.

De Falco et al. [47] raised concerns about MPF release from washing, and evaluated other laundry and textile parameters that impact the release quantity. Domestic washing was concluded to be the source of the MFs released into the ocean via wastewater treatment plants. De Falco et al. [55] demonstrated for the first time that the immediate release of MFs from clothing to air during wearing is equally as important as when released into the waterway from washing. Attention has also shifted from the aquatic environment to the terrestrial, as a proliferation of studies has raised similar concerns about airborne MFs in indoor and outdoor environments with additional releases from electronic dryers [56]. This can potentially outweigh aquatic emissions, notably if the retention rate of WWTP is improved. In addition, the concentration of MFs in sludge will increase, polluting terrestrial ecosystems.

It is also not difficult to recognise that China has become one of the most researched countries, as most polyester fibres are produced in this region [57]. The ability to initiate remedial action at an early stage of the production process should be more effective, as some studies have found that the emission quantity from industrial WWTPs can be up to a thousand times greater than that from municipal WWTPs [36,58]. Therefore, the ability to mitigate at the production stage is a more impactful mitigation approach than to do so during its use and disperses in diversified pathways.

3.4. Summary of Each Research Domain

3.4.1. Domestic Laundry and Drying

This research domain has the largest group of publications and citation links in CNA. The global main path was first introduced by Browne et al. [1], who reported that more than 1900 fibres were released per garment per wash. This was followed by Pirc et al. [52], who first stated that the release of MFs was 3.5 times higher from tumble drying.

Several studies have reviewed the shedding of MFs during the usage phase. Fabrics and garments are washed under different conditions, such as with different detergents, temperatures, and machines. Variations in the sampling methods, testing methodologies, and reporting units employed in numerous studies make it difficult to compare the results. The impacts are summarised in Table 5.

Front-load washing machines reduce MF release because their mechanical actions are less severe. This reduction can be by up to seven times [59]. Most studies have found a decreasing trend in the release of fibres by sequential washing [16,53]. Washing load is considered to be one of the most influential factors in fibre release. On the other hand, textile parameters are expected to play a critical role. Limited studies are available for reference in development, although it is known that more compact structures, higher twists, and long filaments can reduce MF release. However, this information is more complex to translate to textile design without reporting more details and comparing meaningful parameters in different production processes. Rathinamoorthy and Balasaraswathi [60] also identified that, among all laundry parameters, water volume was the most influential on MF release, followed by washing duration, mechanical agitation, and temperature. Therefore, a full load can maintain the water-to-fabric ratio at a minimum level, which is the most effective setting to reduce fibre shedding during washing, which was also shown by the results of Kelly et al. [61].

Table 5. Summary of effect on MPF/MF release from domestic laundry and drying from selected articles.

Parameters	Articles	Effect on MPF/MF Release
Textile Parameters		
Structure	[62]	Increase with loose construction
	[55]	Reduce with compact to loose structure
	[63]	Reduce as interlacing coefficient and weft density increase
Composition	[64]	Recycled polyester > virgin polyester
	[65]	Acrylic > polyester > nylon
Spinning method	[66]	Ring > rotor or air-jet
Yarn twist	[62]	Reduce with higher twist
	[67]	Spun > non-twist filament > hard-twist filament
	[55]	Reduce with a higher twist
Fibre length	[55]	Reduce with continuous filament over short staples
	[53]	Increased release with shorter irregular fibres
	[68]	Reduce from staple to textured filament.
Finishing	[47]	Reduce with a pectin-based finish
	[66]	The processed surface can produce five times more
Cutting	[66]	Scissor-cut 3–31 times higher than laser-cut
Washing and Drying Parameters		
Machine type	[59]	The top load releases seven times more than the front load
Subsequent washes	[52]	Successive washes decrease emissions
	[69]	Reduce and typically stabilise from the 4th and 5th cycle
	[61]	Reduce after 4 cycles
	[53]	Reduce after the peak at 3rd cycle
	[17]	Reduce and stabilise from 5th cycle
	[67]	Reduce
	[68]	Reduce significantly from 5th cycle
	[70]	Reduce and stabilise at the 7th cycle
Water volume-to-fabric ratio/washing load decrease	[61]	Increase as the most influential factor
	[45]	Increase by five times
	[60]	Increase as the most influential factor
Washing temperature	[46]	Increase with temperature
	[61]	No significant effect between 15 and 30 °C and increase at 60 °C
	[71]	Increase with temperature
	[70]	1.8 times more if the temperature is increased from 20 to 40 °C
Washing and drying time	[61]	No impact if the increase is from 15 to 60 min
	[71]	Increase with duration and spin speed
	[70]	Increase if duration increases from 30 to 60 min

Table 5. Cont.

Parameters	Articles	Effect on MPF/MF Release
Using detergent and softener	[52]	Reduce (both detergent and softener)
	[72]	Reduce (softener only)
	[46]	Increase (detergent only)
	[61]	No effect (detergent only)
	[53]	Reduce (detergent only)
	[73]	No significant impact (both detergent and softener)
	[71]	Increase (detergent and conditioner)
	[74]	Reduce (softener only)
	[70]	Increase (detergent only)

Both McIlwraith et al. [75] and Napper et al. [76] proved that mitigation devices had some impact on reducing fibre release in the washing process, and the Lint LUV-R filter was the best performer, with a retention rate of approximately 80%. The remaining devices had significantly lower rates (approximately 30%). Herweyers et al. confirmed that further education is required if these devices are introduced to consumers. They must be user-friendly and easy for consumers to understand how they can make an impact, so as to secure long-term behavioural changes [48].

Owing to the vast variations in textile materials, there is an inevitably considerable variation between the sample specimens used, which makes comparisons inconclusive. Researchers from non-textiles backgrounds may collaborate with industry or academic disciplines to support the selection of fabrics that have an enormous impact and for which detailed textile parameters can be provided.

3.4.2. Test Methodology

Measuring and identifying microplastics is difficult because of inconsistent sampling and quantification methods, resulting in challenges in comparing data and estimating their prevalence [35,77]. For many years, the characterisation of polymers has relied on optical spectroscopy to provide information on polymeric materials' identity and chemical composition. One major finding is that relying solely on microscopic visual identification of MPs or MFs is inaccurate; each particle must be confirmed as plastic using techniques capable of tackling small particle sizes, such as micro-Fourier-transform infrared microscopy (µFTIR), micro-Raman(µRaman), Pyrolysis Gas Chromatography/Mass Spectrometry(Py-GC/MS), and High-Performance Liquid Chromatography(HLPC) [78]. However, µFTIR and µRaman are the most commonly used techniques and are highly recommended for characterising MPs, with similar performance, especially in wastewater samples [78,79]. Raman spectroscopy is a laser-based method that provides better resolution than infrared spectroscopy. This is well suited when the process requires focusing on small regions of a sample. It can also address the identification of MPs as small as 1 µm. The Raman spectrum yielded similar but complementary information to that obtained by FTIR. To establish a standardised method for the qualitative analysis and characterisation of MPs, both instruments can be effectively combined with optical microscopy [80]. However, its high cost and time requirements render it unscalable.

However, a few commercial standards have been published for quantifying MFs from domestic washing. They are published by the Microfibre Consortium [81], AATCC [30], and ISO [31] (which are under development). They are remarkably similar and consistent; therefore, the results are likely comparable. The minor difference is that the TMC method requires eight specimens but no blank test. In comparison, the AATCC provides two options for pretreatment and drying: detergent and wash temperature, as per the label.

The AATCC and ISO standards are less sophisticated versions of the TMC standards, and require only four sample specimens.

Adopting new test methods will require industry collaboration, which can be accelerated by stakeholder pressure and regulatory interventions to introduce aligned measures. Currently, most identification methods are based on MPs and laboratory scales, with MPFs being a subgroup of MPs. This requires scalable investment and the ability to include non-synthetic fibres. However, current commercialised standards do not consider the influence of other particle contaminants, differentiation of fibre types, and length distribution. The accuracy and reliability of the control parameters are yet to be improved. This relies on specific polymer characteristics not possessed by natural fibres. Some test protocols entail using solvents to remove contaminants that can dissolve or remove organic matter, including natural fibres. Therefore, in developing a methodology, it is essential to include fibre types commonly used by the textile industry.

3.4.3. Aquatic Ecosystem

Pollution in aquatic ecosystems is the first concern regarding the release of MPs/MPFs/MFs into the environment. Indeed, there is compelling evidence that their impact was previously underestimated because 300 µm sieves were used in the sampling process, and most later studies confirmed the most prevalent length to be <100 µm. According to Conkle et al. [82], approximately 80% of previously published papers did not account for MP size <300 µm. The discrepancies can be as high as one to four orders of magnitude when a sieve <100 µm is used [83]. Therefore, earlier studies using coarse sieves for sampling underestimated MP prevalence.

MPFs have been found in almost every marine habitat's seawater and sediments. In a review by Gago et al. [84], the most abundant fibre length of surface water was established as 500–1000 µm and up to 4750 µm. In sub-surface waters, this was found to be 1–5 mm, with the most abundant fibres in sediment ranging in size between 800 and 1000 µm, consistent with those found in WWTPs and domestic laundry studies. MPs and MPFs can be viewed as unavoidable byproducts of contemporary lifestyles which can be transmitted directly to oceans, rivers, and lakes without assistance from WWTPs [85]. Nevertheless, a Great Lakes study revealed that the estimate for MPFs in aquatic environments was disproportionately higher than the concentrations found in wastewater effluents [86]. As a result, research into aquatic environments may consider a more balanced approach of using a smaller sample volume, like Barrows et al. [87], instead of coarse sieves and a large volume sample size, which may result in underestimation.

3.4.4. Atmospheric Environment

MPFs are ubiquitous, but their long-term effects on human health and the environment are poorly understood. Dris et al. [88] first reported the concentration of 1.0 and 60.0 fibres/m^3 and 0.3 and 1.5 fibres/m^3 in indoor and outdoor air, respectively. Most were natural or cellulosic, which is unsurprising because they are short, staple fibres and are more vulnerable to breakage and fallout during use and wear. Studies of long-term exposure to UV radiation from sunlight and fluctuations in ambient temperature have proven that textiles break down and release MFs into the atmosphere [89].

It is generally believed that the most consequential shedding of clothing fibres occurs during laundering. However, this assumption may need to be revised as it only accounts for some products. Non-laundered fabrics such as flags and sails also undergo predictable disintegration, with associated fibre losses over time [85]. Fibres were also readily noticeable on clothing closets, floors, display monitors, and other undisturbed indoor surfaces. In later studies, there have been emerging discussions on their fate in the air we breathe, the dust we inhale at home, the water we drink, and other elements of our being and environment. Liu et al. [89] and Zhang et al. [90] suggested that textile clothes and soft furnishings are likely to be the major sources of airborne MPs in indoor environments. Over 60% of MPs were MPFs, which could be an essential source of MP pollution.

However, the concentration of MPFs or MFs largely varies between indoor and outdoor environments in different regions, as reported by [19,34,89,91]. Clothing and textile furniture are the dominant sources [92]. However, differences in lifestyle and the use of textile-derived products are expected to be the main reasons for the variations in indoor environments. At the same time, the distribution in the outdoor environment is vastly influenced by wind, airflow, and other factors, such as consumption habits, socioeconomic status, traffic, and urbanisation [93].

Most airborne fibres in various studies [19,91,94] were below 1000 μm, posing an inhalation risk to humans. More recently, it was demonstrated that MPFs in the atmosphere may cause issues related to inhalation and their presence in human lungs [24].

Nevertheless, the airborne MF phenomenon requires strict contamination control from sampling to the point of the experiment. This is particularly important because the potential presence of airborne MFs, such as from researchers' clothing or other textile materials, can significantly impact the experiment's outcome. Therefore, it is essential to implement effective control procedures to minimise contamination.

3.4.5. Wastewater Source

Doubtlessly, wastewater is a well-known source of microplastic fibres released into the aquatic environment. It is accepted that laundering fabrics leads to abrasion and wear of textiles, leading to fibre-shedding and their subsequent release within effluent from washing machines through sewage effluent. Browne et al. [1] reported that the proportions of polyester and acrylic fibres in sewage effluent resemble those of MPs found worldwide contaminating sediments. This was interpreted as an indication that at least some of the MFs in the marine environment originated from textile washing, with WWTPs acting as pathways.

WWTPs as pathways for MP release have drawn attention recently, with an exponentially growing number of related publications in the last few years [95]. Mintenig et al. [96] suggested that WWTPs could be a sink and a source of MPs and thus play an essential role in MP pollution. There is strong evidence that MPs/MPFs can easily bypass WWTP filtration and other solid separation processes, which are not designed for such a purpose [97–99]. Magnusson and Norén [100] and Talvitie et al. [101] demonstrated that the supply of MPs from WWTP effluents to the aquatic environment may be substantial because of the enormous volume discharged daily.

In the study by Conley et al. [96], MPF removal efficiency was reported at 80.2–97.2%, which is relatively high but still significantly less than the total MPs. Talvitie et al. [102] suggested that advanced wastewater treatment (e.g., a membrane bioreactor) can improve the removal efficiency of small-sized MPs (<100 μm). This was further supported by Ziajahromi et al. [98], who found that fibres were the dominant MPs detected in most effluent samples and were not completely removed even after advanced treatment processes.

WWTPs in developed countries are believed to be more efficient. For example, Mintenig et al. [96] showed a removal efficiency of 98% for MFs after advanced filtration in a German WWTP. Whereas, in developing economies, WWTPs usually have lower standards because of inadequate sewage infrastructure. In 2014, China alone accounted for 69% of all polyester fibre production globally, with the combined output of China, India, and Southeast Asia representing over 80% of the global total [13]. Nevertheless, developing countries produce and consume a higher proportion of synthetic textile materials, at 62.7% compared to 48.2% in developed countries [9]. Developing countries tend not to have commonly available tertiary treatment standards, which is of greater concern.

Xu et al. [28] measured MPFs directly discharged from textile mills to an industrial WWTP that collects effluents from mills in the region for treatment before their discharge to the aquatic environment. This is a typical textile industrial CETP with a daily treatment capacity of 30,000 tons in the same region where the present study was conducted. This plant receives production wastewater from 33 printing and dyeing mills in a textile industrial park, accounting for approximately 95% of the influent. The average abundance of

MFs, both natural and synthetic, was reported as 334.1 (±24.3) items/L in the influent before being reduced to 16.3 (±1.2) items/L in the final effluent, with a retention efficiency of 95.1%.

When comparing industrial textile effluents from Xu et al. [28] and municipal effluents from Yang et al. [103] and Lv et al. [104] in China, there were 28–1310 times more MFs discharged directly in concentrated effluent to aquatic environments from industrial sources than from municipal WWTPs. The scale of textile effluents is vast, as it is one of the highest water-use industries globally [105] and is considered the second largest consumer and polluter of clean water [10]. Once discharged, no practical solution exists to remove it from freshwater and the ocean. It is essential to avoid discharging MPFs/MFs to aquatic environments, as there is no alternative way to remediate them. Therefore, it is imperative to take immediate action to control the release of wastewater. The magnitude of this pathway suggests that it is of primary importance and high priority. Regulators should be informed and educated without delay.

3.4.6. Abundance and Distribution

Estimating MPF pollution globally with many unknown factors and uncertainties is challenging, which explains the discrepancies between studies. These uncertainties can be either structural (related to the understanding of the mechanisms and pathways of leakage) or data-related (associated with the availability of reliable datasets, which are particularly difficult to obtain in certain countries) [106]. The predominant global estimates of MPF leakage are listed in Table 6.

Table 6. Global microfibres release yearly estimates from listed sources.

Estimates (Weight)	Source of References
190,000 tonnes/year	[107]
525,000 tonnes/year	[9]
260,000 tonnes/year	[108]
280,000 tonnes/year	[109]

Boucher and Billard [98] defined the leakage of primary MPs as a function of loss and release rates. Therefore, the loss rate measures the quantity of MPFs lost from a specific activity such as domestic washing, and the data are usually more accessible. The release rate estimates the fraction of this loss reaching the ocean, which is not captured in waste treatment plants or other infrastructure. This leads to significant uncertainties owing to the high complexity of release pathways. The estimation often requires validation from field studies, which require further improvement. Perhaps the latest estimate by Belzagui et al. [109] was closer to the true scenario for release based on domestic laundry, as the estimation model considered extended parameters such as the volume of laundry effluents, percentage of municipal water that has been treated, type of water treatment applied, and proportion of front- versus top-loading washing machines.

Owing to their high density, MPFs are the most prevalent MPs in the natural environment, particularly in sediment and surface water. Miller et al. [110] and Dris et al. [111] identified both synthetic and non-synthetic fibres in freshwater and Lusher et al. [112] identified them in the gastrointestinal tracts of fish. Surprisingly, despite synthetic fibre production surpassing natural fibre production for over a decade, a few studies [73,113] have found that natural MFs are more abundant than synthetic ones. Apart from this, other sources show that MFs are more diverse during the use phase through the air system, with some findings suggesting that MFs from natural sources are even higher density [111].

The most commonly detected polymers are polyesters (PES), polyamides (PA), and polyethylene (PE). The fibre types examined could have been more consistent in different studies. Ziajahromi et al.'s [98] findings resemble the global fibre production market, with polyester as the most significant contributor. Fishing nets, ropes, and gears are potential sources of MPF pollution. The main concern regarding the use or abandonment of fishing gear is that it is a source of secondary MP pollution. Generally, it is perceived that ropes discarded following their use in fishing activities and marine transportation are the primary source of MPFs following weathering and breakdown of material over time [114–116].

The MPFs shed during washing vary from 11.9 to 17.7 µm in diameter and 5.0 to 7.8 mm in length across polyester, polyester/cotton blends, and acrylic. They typically range in diameter from 6 to 175 µm [117]. The most common fibre lengths found in marine environments are 100–1500 µm [87] and in wastewater treatment plants (WWTPs) are 100–1000 µm [118]. The size range of <500 µm was dominated by surface waste, whereas 500–5000 µm was dominated by sediments [119].

Assessing the worldwide impact of MPF contamination is complex due to various unknown factors and uncertainties. Thus, there may be divergent findings from different studies. As textile materials are diversified, more accurate estimates require a deeper understanding of the relationship between textile parameters and their fragmentation performance over the entire product life cycle. For example, industrial wet processing factories which discharge through industrial WWTPs can potentially release higher amounts of MPFs than municipal WWTPs with input from domestic washing [58,120]. This can contribute to significantly more releases than in the current estimation. In addition, more recent studies claim that terrestrial pathways from the use of electronic dryers can contribute more MFs than the aquatic pathway [15,111]. Considering these additional sources and pathways is crucial for comprehensively assessing the global abundance of MFs.

3.4.7. Terrestrial Ecosystem

Synthetic fibres have been reported in municipal sewage sludge in soil. They can be dated back to the studies of Habib et al. [121] and Zubris and Richards [122], which indicated fibre presence in wastewater and its diverging pathways. Fibres within the sludge material cannot biodegrade quickly in nature and can persist for more than 15 years in the soil. Such pollutants in the soil can be retained in terrestrial environments or eventually enter marine environments via runoff.

Because of its nutritional and organic content, sludge produced from WWTPs is still used as a fertiliser in the agricultural field. Most of the MPFs found in WWTPs were retained in sludge [101]. WWTPs are not designed to filter out MPs/MPFs [123]. According to Corradini et al. [124], MPs were found at 18–49 particles/g, and the majority were fibres, accounting for 90% of sludge and 97% of soil.

Evidence of MPs'/MPFs' impact on soil ecosystem functioning and soil stability is varied. Owing to the presence of MPs and the size distribution of water-stable soil aggregates, evidence showed that fewer seeds germinated. The experiment using MPs and MPFs to grow the Lolium Perenne plant showed reduced shoot height [125]. Prendergast-miller et al. [126] studied polyester MFs in the litter-feeding earthworm Lumbricus Terrestris. They showed that they were not fatal but caused transcriptional responses related to general stress, and there was evidence of a change in casting behaviour. On the other hand, textiles treated with silver nanoparticles released into wastewater and sludge are further transported to the soil and may cause reproductive impairment in earthworms [127]. In addition, shorter microplastic fibres can enter the terrestrial food chain via ingestion by soil vertebrates [128].

A more recent estimate by Gavigan et al. [129] suggested that after emissions from apparel washing between 1950 and 2016, waterbodies would have received 2.9 Mt of MPFs. In contrast, terrestrial environments and landfills receive almost 2.5 Mt. The quantity was nearly as large (and increasing) in terrestrial areas and landfills. Improved wastewater treatment can shift the MPF emissions from water bodies to terrestrial environments. Further examination is necessary to fully understand the transfer of MPFs and the potential consequences of such transfers, particularly when implementing WWTP retention improvements to mitigate this issue. It is essential to ensure that this strategy does not simply shift the problem from aquatic to terrestrial environments.

3.4.8. Hazardous Risk

The concerns regarding MPFs compared to cellulosic fibres, typically those from cotton and considered a pollution issue, are mainly because polymers are not easily degraded or decomposed in the natural environment [130]. Once created, they can persist for hundreds of years and are transferred between different media, resulting in a bioaccumulation effect. Another concern is that there are thousands of types of chemicals added to the textile production process to provide specific functionalities and durability that depend on the intended product applications and are frequently identified in marine environments, which leach out and transfer to aquatic organisms [131]. Several studies have reported that MPFs transfer contaminants, such as plasticisers [132], dyes [133], polycyclic aromatic hydrocarbons, and polychlorinated biphenyls [134]. MPFs can contaminate deep sea organisms via various physical, chemical, and biological pathways [12].

Several studies have documented that marine MPs were covered by biofilm communities [135,136]. These organic layers likely acted as reservoirs for hazardous substances such as persistent organic pollutants (POPs). In addition, this prolongs degradation, as biofilms can form a protective layer against UV radiation [137]. Furthermore, MPs/MPFs can function as vehicles to carry harmful pathogenic microorganisms and parasites from stable biofilms on their surfaces when exposed to wastewater. Their surfaces provide habitats for microbial colonisation and biofilm formation, allowing the migration of opportunistic pathogens and invasive species [135,138].

Fibrous MPFs may be more harmful than their spherical counterparts, causing cancer, scarring, and harm to marine life [139]. Many aquatic species ingested MPFs/MFs, leading to entanglement, slowed growth, and diminished feeding and reproduction [140,141]. The vast presence of MPFs/MFs in the marine environment has supported evidence that biota has ingested them directly or potentially along the food chain. They were present in mussels [142], oysters [143], sea urchins [144], sea cucumbers [145], fish, shrimp, shellfish [49,146], and even in deep sea organisms [138]. More recent studies have reported oxidative stress responses in clams [147] and others, and cellular disturbances and thermal stress responses in mussels [148]. Moreover, the large surface area of MPFs means that environmental pollutants may be absorbed onto the surface of the particles, with the potential to be transferred into body tissues once ingested [149,150]. These contaminated residues can adsorb and concentrate organic pollutants that may be ingested by marine fauna, which can be subsequently transferred to the food chain and potentially reach humans [14,22,151,152].

The release of MPFs from drying clothes could represent a source of airborne MPFs. Recently, increasing evidence of the presence of MPFs in the atmosphere has been reported [153,154]. The flying MPFs inhaled by humans are deposited in the lung tissue and may lead to tumours. Several studies have shown that respiratory inflammation, pulmonary fibrosis, and cancer can also be caused by regular and prolonged exposures [85].

There have been growing concerns that natural fibre sources also contributed to the issue, not to a lesser extent [68]. This may be because of the faster degradation rate and metabolites available in the marine environment [46,155]. In addition, the higher biodegradability of natural fibres may also result in a higher chance of releasing chemical additives [156]. Environmental concerns regarding natural and synthetic fibres also differ owing to their chemical pollutant sorption behaviours. They possess different fibre surface properties and chemical bonds which control sorption behaviours, such as electronegativity, making the adsorption of natural fibres on cationic surfactants relatively higher than on synthetic fibres [46,62]. The difference in surface attractions showed the potential for different threads to play distinct roles in hazardous chemical sorption and fate. Therefore, shifting from synthetic fibres to natural fibres from this perspective does not appear to solve the pollution issue. The use of less and more sustainable, non-toxic, and nonpersistent chemicals is believed to be a more holistic approach. Perhaps further research should evaluate this in greater detail to find a non-regrettable option for different hazardous impacts from various types of fibres.

4. Opportunities

Research on MPFs/MFs associated with textile materials has been a prominent area of focus since 2011. This subject gained more attention in 2019, reaching 61 studies in 2022. Despite some advancements, this field is still in its nascent stages, necessitating additional research, particularly a systematic approach to sampling, methodology, and reporting standards that can be applied. Below is a list of priorities with actions that are recommended to leverage and accelerate progress.

4.1. Interdisciplinary Collaboration

Plastics are used in a vast number of products. Further studies and resources are being employed to address MP pollution. Since MPs and MPFs have similar chemical structures, the knowledge gained from MP studies is expected to be referenced, especially regarding ecological impact and toxicity. However, there is an expectation to report MPFs/MFs in more detail regarding textile parameters [4], so that the information can be comparable and reproducible. Nonetheless, it should only partially depart from the MP mainstream because they have larger communities and stakeholders in the textile sector to leverage their knowledge, resources, and solutions. MPF studies can also tap into MP studies through collaborations between academic disciplines. Textile and polymer science researchers should be desperate to collaborate with ecologists and toxicologists to determine the life cycle impact of MPFs/MFs [53].

Solutions for improving the effectiveness of washing practices require a multifaceted approach involving collaborations with washing machine manufacturers, wastewater treatment facility builders, and promotion of responsible consumer washing habits. This involves engagement beyond primary research communities in the current subject, including exploring new technologies, implementing stricter regulations, and educating the public.

4.2. Textile Parameters

Textile production uses materials with unique compositions, structures, and properties to cater to specific aesthetic and functional demands. Therefore, without considering critical specification parameters, such as fibre count, yarn twist, fabric density, structure, and finishes, selecting fabrics at random can lead to inconclusive outcomes and hinder a comprehensive understanding of fibre fragmentation [157].

A better understanding of textile manufacturing processes that result in the increased release of MPFs/MFs could lead to the development of a textile process that reduces the release of MF fragments during domestic washing. For example, Pangaia partnered with MTIX using multiplexed laser surface enhancement (MLSE®) technology to modify fibre surfaces within a fabric to prevent microfibre shedding [158].

Regardless of the small size and shedding performance of MFs, they may not wholly resemble the same version, as they are much finer and more vulnerable. Fibre shedding is a well-known phenomenon in the textile industry, and there is a long history and set of control and effective reduction measures available for its use [148]. Therefore, remedial measures can be built on the existing knowledge of shedding.

A good practice guideline was published by [159], and reducing the MF release in the synthetic textile supply chain could be a starting point. However, it is more critical to begin referencing it and promote its adoption in the textile industry without further delay in waiting for the perfect version. Since it was prepared a few years ago, it can be updated with more refined details and parameters from the latest research.

4.3. Laundry Parameters

The laundry parameter studies concluded that the laundry setting plays a critical role in reducing MF release. Consumers are advised to wash in cold water and with fewer cycles and launder clothes when necessary and in full loads [73,160], which indirectly extends the lifetime of a garment irrespective of conflicting parameters when using detergents and softeners. This requires changing consumer behaviour when purchasing and caring for clothes. Switching to milder washing and drying conditions can be the most cost-effective solution because it merely involves changing the existing machine settings, consequently reducing energy.

Most consumers may find it more challenging to adapt to options that incur additional costs of purchasing a front-loading washing machines that generate fewer MFs or purchasing less high-quality clothing [13]. Indeed, manufacturers of washing machines, tumble dryers, detergents, and softeners should also contribute by improving their performance in releasing fewer MPFs/MFs and communicating their impact to consumers.

4.4. Sustainable Chemicals

Although some studies have suggested that fibres from biomass can mitigate MF release, some people are still sceptical as they may persist after chemical treatment. It is widely understood that MPs/MPFs/MFs are considered a threat and risk to the ecosystem and human health due to contaminants [161]. Therefore, it is of higher importance to ban these hazardous chemicals and move to the use of natural or non-toxic substances in the production process, which can be more effective in reducing their toxicity and hazard risk than avoiding synthetic textiles.

However, it is necessary to distinguish the non-degradability of synthetic textiles, which can become vectors or carriers of other pathogens or pose a physical hazard to microorganisms, as this will take time to be resolved through sustainable chemistry. The outcome needs to be differentiated from that of natural fibres [101].

4.5. Renewable Materials and Circularity

Existing solutions are available to shift the use of renewable materials. A commercial example is Tandem Repeat Technologies, which uses genetic sequencing and synthetic biology to produce a new fibres based on a unique protein structure originally found in squid tentacles. In contrast, Natural Fibre Welding provides a sustainable and circular material that uses plants as a renewable resource.

Since the Earth's resources are finite, reducing, reusing, and repurposing with an extended product lifespan is still preferable. Although improving shedding performance can reduce MF emission quantity, reducing the use, production, and enhanced circularity of textile materials within the system can effectively reduce MF release, which addresses the root cause of pollution [153–155].

Currently, the lint filter is the most effective device. Yousef et al. [162] have provided new modelling to collect MFs and use them as a renewable energy source. This innovative development transforms MF waste into a new valuable energy resource capable of reducing the carbon footprint and has the potential to be scaled up.

4.6. Wastewater Treatment

As WWTPs are the major sinks and sources of MPs/MPFs/MFs, a higher retention rate using advanced treatment technology should considerably influence release reduction. However, proper sludge handling must be performed simultaneously to ensure that the problem is not transferred from aquatic to terrestrial environments. Although, practically, other advanced treatment technologies such as Membrane Bioreactor (MBR) and Zero Liquid Discharge (ZLD) are available in the industry to reduce water stress and pollution, they all involve significant investment and time for implementation [7].

Increasing recycled water within WWTPs and reducing wastewater discharge can be a quick fix. Innovations in dry finishing technologies, such as using carbon dioxide as a dyeing carrier, is cleaner than wet processing. This is also a new trend that can help resolve MFs and water pollution in one solution. Combining these benefits has multiple effects on accelerating the investment decisions. Lack of knowledge and test methodology, inability to include MPF requirements in sourcing policies from retailers, and limited financial incentives to invest are the most significant barriers to industrial transformation [156]. Apart from standardising the testing methodology, incentives such as green funds supporting the transformation to innovative technologies are essential to resolve MF discharge through the aquatic pathway.

4.7. Mitigation Devices

Several studies have evaluated mitigation devices, and non-profit organisations (NGOs) have been educating and promoting the use of mitigation devices and proper wash care at the consumer level [163,164]. However, these devices have gained little popularity [48], as some of them were less effective than they had thought, and the users found them difficult to recognise owing to their impact level.

Although these external devices are convenient, improper waste handling and disposal potentially mean that the disposal route of MFs is transferred from one end to another. Thus, they are still incapable of mitigating this issue. Washing machine manufacturers should have a role to play by building similar performance features into new models, making them more accessible and practical. It is essential to recognise that relying solely on these devices may not adequately reduce MF release from this source. Embracing sustainable washing habits, such as decreasing the frequency of washing and altering purchasing behaviours to acquire fewer but longer lasting garments, can result in a more impactful outcome.

4.8. Standardised Test Method

Developing methodologies that effectively measure MPF/MF release is necessary to evaluate solutions and enable science-based decisions. Effective and standardised methods can help regulatory bodies by holding all stakeholders accountable and defining the maximum thresholds for releasing MPFs/MFs into the environment, thus realistically stopping them. A more unified approach using AATCC TM212 [30], ISO/DIS 4484-1 [31], and the TMC method-2019 [81] is now available for the industry to evaluate MFs from laundry. There is continued demand to expand its scope to cover drying, air, wastewater, and sludge. Since synthetic and natural fibre impacts and toxicity mechanisms intrinsically differ, methods that can enumerate size and quantify separately when the materials are blended are also important.

The different reporting parameters and insufficient transparency of the methodologies used have hampered the progress of MP studies. A group of researchers has developed harmonised reporting guidelines [165]. A similar approach should also be followed in MPF/MF studies, alongside developing a standardised method to guide researchers and commercial practitioners and ensure that the results are comparable, reproducible, and reliable for estimating its impact and developing mitigation solutions. Furthermore, it can facilitate the implementation of regulatory measures.

4.9. Government Interventions

Global legislation to regulate MPFs is imperative; however, only a few regions have taken regulatory action. Currently, France is the front runner. In 2020, it was the first country to pass a law that required all new washing machines to have MF filters by 2025. Apart from taking responsibility for the producer scheme through legislative intervention, governments also need to improve their waste management and WWTP infrastructures, which requires large-scale investment [58,159]. On the other hand, funding research to accelerate knowledge and engage stakeholders is an inevitable action to support policy development, making them practical and pragmatic for implementation.

Moreover, providing education and information to the public is an indispensable part of government intervention. Educating consumers to take responsibility for reducing release in the usage phase must be targeted to reach decision makers who can make purchasing choices. Consumer behaviours are typically challenging to change, as they require a shift in values and beliefs, to understand their social responsibilities, and become part of the solution. Initiatives driven by civil societies that engage communities more effectively remain vital to continuing the shifting momentum [151,154].

Existing mitigation solutions are voluntary and market driven. Innovative technologies require financial investments to scale and make them affordable. Government intervention and investors in scaling are inevitable and often play a critical role in reaching the tipping point. Socioeconomic factors should not be excluded from resolving sustainability issues involving systematic change. In reality, some of the solutions are interdependent. For example, reducing synthetic fibres can increase the demand for natural fibres with a larger carbon footprint and competing land use and biomaterials in other systems. Therefore, wider stakeholder groups from cross-disciplines must collaborate to address this cross-disciplinary agenda and evaluate its impact on the complete lifecycle of related ecosystems to ensure that it does not negatively impact other industries or the socioeconomic system.

5. Conclusions

5.1. Limitations

Bibliometric analysis is generally viewed as a reliable method, but some degree of subjectivity remains when choosing articles based on specific keywords. While a set of 219 articles is adequate for conducting main path analysis, expanding the sample size can yield greater insights into emerging trends, particularly concerning less common keywords like "fibrous microplastics", "airborne microplastics", and "microfibre fragmentation" found in journal databases. This review used the WOS database, and it is recommended that other databases and non-traditional publications be included.

WOS consistently updates its database; data downloaded for a specific period maybe added to after a particular time. According to Pilkington and Meredith [166], CNA treats all citations as equally important irrespective of the significance of their citation or utility type. Consequently, the research trends were more robust than the specific numbers per group. Negative citation is unavoidable, but it is rare and insignificant.

5.2. Outlook

The success of the industry relies on practising sustainable solutions throughout a product's life cycle [167]; however, these solutions cannot be isolated because there has been a systematic change from low technology to an industry that depends on innovation and sophisticated knowledge of the impact of MPFs/MFs on the environment, human health, and socioeconomic aspects. Furthermore, this depends on the capability of stakeholders to work together in a collaborative approach.

The bibliometric analysis used in this review has enabled the systematic utilisation of publications from all science areas, likely to influence multidisciplinary research topics and reach a broader audience. Moreover, it is remarkably applicable in the study of MPFs/MF, as it is more of an interdisciplinary subject of interest requiring a collaborative approach to accelerate progress.

Author Contributions: C.K.-M.C. identified the objective of this study, developed the content, conducted the research and drafted the manuscript. C.K.-Y.L. contributed to the methodology, design of the study, and revision of the manuscript. C.-W.K. identified the objective of this study, the design of the study, and the manuscript revision. All authors have read and agreed to the published version of the manuscript.

Funding: This research was funded by The Hong Kong Polytechnic University (grant number: R-ZDE1 and 1-BBC6) And The APC was funded by R-ZDE1 and 1-BBC6.

Acknowledgments: Authors would like to thank the financial support from The Hong Kong Polyetchnic University for this work (Account Code: R-ZDE1 and 1-BBC6). Certain data included herein are derived from Clarivate™ (Web of Science™). © Clarivate 2022. All rights reserved.

Conflicts of Interest: The authors declare no conflicts of interest.

References

1. Browne, M.A.; Crump, P.; Niven, S.J.; Teuten, E.; Tonkin, A.; Galloway, T.; Thompson, R. Accumulation of Microplastic on Shorelines Woldwide: Sources and Sinks. *Environ. Sci. Technol.* **2011**, *45*, 9175–9179. [CrossRef] [PubMed]
2. Cesa, F.S.; Turra, A.; Baruque-Ramos, J. Synthetic fibers as microplastics in the marine environment: A review from textile perspective with a focus on domestic washings. *Sci. Total Environ.* **2017**, *598*, 1116–1129. [CrossRef] [PubMed]
3. Palacios-Marín, A.V.; Tausif, M. Fragmented fibre (including microplastic) pollution from textiles. *Text. Prog.* **2021**, *53*, 123–182. [CrossRef]
4. Periyasamy, A.P.; Tehrani-Bagha, A. A review on microplastic emission from textile materials and its reduction techniques. *Polym. Degrad. Stab.* **2022**, *199*, 109901. [CrossRef]
5. Rathinamoorthy, R.; Balasaraswathi, S.R. Microfiber Pollution Prevention—Mitigation Strategies and Challenges. In *Sustainable Textiles*; Springer Nature: Singapore, 2022; pp. 205–243. [CrossRef]
6. Textile Exchange. Preferred Fiber & Materials: Market Report 2022. Volume 60. Available online: https://textileexchange.org/knowledge-center/reports/materials-market-report-2022/ (accessed on 22 August 2019).
7. Barrows, A.P.W.; Neumann, C.A.; Berger, M.L.; Shaw, S.D. Grab vs. neuston tow net: A microplastic sampling performance comparison and possible advances in the field. *Anal. Methods* **2017**, *9*, 1446–1453. [CrossRef]
8. Decitex Smart Textiles. Microfiber. Available online: https://www.decitex.com/en/microfiber# (accessed on 16 October 2023).
9. Boucher, J.; Friot, D. *Primary Microplastics in the Oceans: A Global Evaluation of Sources*; IUCN: Gland, Switzerland, 2017; Volume 43. [CrossRef]
10. Iowe, D. *A New Textiles Economy: Redesigning Fashion's Future*; Ellen MacArthur Foundation: Isle of Wight, UK, 2017; pp. 1–150.
11. Lv, M.; Jiang, B.; Xing, Y.; Ya, H.; Zhang, T.; Wang, X. Recent advances in the breakdown of microplastics: Strategies and future prospectives. *Environ. Sci. Pollut. Res. Int.* **2022**, *29*, 65887–65903. [CrossRef] [PubMed]
12. Woodall, L.C.; Sanchez-Vidal, A.; Canals, M.; Paterson, G.L.J.; Coppock, R.; Sleight, V.; Calafat, A.; Rogers, A.D.; Narayanaswamy, B.E.; Thompson, R.C. The Deep Sea Is a Major Sink for Microplastic Debris. *R. Soc. Open Sci.* **2014**, *1*, 140317. [CrossRef] [PubMed]
13. Henry, B.; Laitala, K.; Klepp, I.G. Microplastic Pollution from Textiles: A Literature Review. 2018. Available online: https://oda.oslomet.no/oda-xmlui/bitstream/handle/20.500.12199/5360/OR1%20-%20Microplastic%20pollution%20from%20textiles%20-%20A%20literature%20review.pdf (accessed on 10 February 2023).
14. Carr, S.A.; Liu, J.; Tesoro, A.G. Transport and Fate of Microplastic Particles in Wastewater Treatment Plants. *Water Res.* **2016**, *91*, 174–182. [CrossRef]
15. Tao, D.; Zhang, K.; Xu, S.; Lin, H.; Liu, Y.; Kang, J.; Yim, T.; Giesy, J.P.; Leung, K.M.Y.Y. Microfibers Released into the Air from a Household Tumble Dryer. *Environ. Sci. Technol. Lett.* **2022**, *9*, 120–126. [CrossRef]
16. O'Brien, S.; Okoffo, E.D.; O'Brien, J.W.; Ribeiro, F.; Wang, X.; Wright, S.L.; Samanipour, S.; Rauert, C.; Toapanta, T.Y.A.; Albarracin, R.; et al. Airborne Emissions of Microplastic Fibres from Domestic Laundry Dryers. *Sci. Total Environ.* **2020**, *747*, 141175. [CrossRef]
17. Kärkkäinen, N.; Sillanpää, M.; Karkkainen, N.; Sillanpaa, M. Quantification of Different Microplastic Fibres Discharged from Textiles in Machine Wash and Tumble Drying. *Environ. Sci. Pollut. Res.* **2021**, *28*, 16253–16263. [CrossRef] [PubMed]
18. Wright, S.L.; Ulke, J.; Font, A.; Chan, K.L.A.; Kelly, F.J. Atmospheric Microplastic Deposition in an Urban Environment and an Evaluation of Transport. *Environ. Int.* **2020**, *136*, 105411. [CrossRef] [PubMed]
19. Prata, J.C.; Castro, J.L.; da Costa, J.P.; Duarte, A.C.; Rocha-Santos, T.; Cerqueira, M. The Importance of Contamination Control in Airborne Fibers and Microplastic Sampling: Experiences from Indoor and Outdoor Air Sampling in Aveiro, Portugal. *Mar. Pollut. Bull.* **2020**, *159*, 111522. [CrossRef] [PubMed]
20. Huang, Y.; He, T.; Yan, M.; Yang, L.; Gong, H.; Wang, W.; Qing, X.; Wang, J. Atmospheric transport and deposition of microplastics in a subtropical urban environment. *J. Hazard. Mater.* **2021**, *416*, 126168. [CrossRef] [PubMed]
21. Desforges, J.-P.W.; Galbraith, M.; Ross, P.S. Ingestion of microplastics by zooplankton in the Northeast Pacific Ocean. *Arch. Environ. Contam. Toxicol.* **2015**, *69*, 320–330. [CrossRef] [PubMed]
22. Mizraji, R.; Ahrendt, C.; Perez-Venegas, D.; Vargas, J.; Pulgar, J.; Aldana, M.; Ojeda, F.P.; Duarte, C.; Galbán-Malagón, C.; Patricio Ojeda, F.; et al. Is the Feeding Type Related with the Content of Microplastics in Intertidal Fish Gut? *Mar. Pollut. Bull.* **2017**, *116*, 498–500. [CrossRef] [PubMed]
23. Ragusa, A.; Svelato, A.; Santacroce, C.; Catalano, P.; Notarstefano, V.; Carnevali, O.; Papa, F.; Rongioletti, M.C.A.; Baiocco, F.; Draghi, S.; et al. Plasticenta: First Evidence of Microplastics in Human Placenta. *Environ. Int.* **2021**, *146*, 1–2. [CrossRef] [PubMed]
24. Amato-Lourenço, L.F.; Carvalho-Oliveira, R.; Júnior, G.R.; dos Santos Galvão, L.; Ando, R.A.; Mauad, T. Presence of Airborne Microplastics in Human Lung Tissue. *J. Hazard. Mater.* **2021**, *416*, 126124. [CrossRef] [PubMed]
25. Cole, M. A novel method for preparing microplastic fibers. *Sci. Rep.* **2016**, *6*, 34519. [CrossRef]
26. Costa, M.F.; Ivar Do Sul, J.A.; Silva-Cavalcanti, J.S.; Christina, M.; Araújo, B.; Spengler, Â.; Tourinho, P.S. On the Importance of Size of Plastic Fragments and Pellets on the Strandline: A Snapshot of a Brazilian Beach. *Environ. Monit. Assess.* **2010**, *168*, 299–304. [CrossRef]
27. GESAMP. Sources, Fate and Effects of Microplastics in the Marine Environment (Part 1). Available online: http://www.gesamp.org/publications/reports-and-studies-no-90 (accessed on 20 February 2023).
28. Xu, X.; Hou, Q.; Xue, Y.; Jian, Y.; Wang, L.P. Pollution Characteristics and Fate of Microfibers in the Wastewater from Textile Dyeing Wastewater Treatment Plant. *Water Sci. Technol.* **2018**, *78*, 2046–2054. [CrossRef] [PubMed]

29. Lusher, A.L.; Burke, A.; O'Connor, I.; Officer, R. Microplastic Pollution in the Northeast Atlantic Ocean: Validated and Opportunistic Sampling. *Mar. Pollut. Bull.* **2014**, *88*, 325–333. [CrossRef] [PubMed]
30. *AATCC TM212-2021*; Test Method for Fiber Fragment Release during Home Laundering. AATCC: Research Triangle Park, NC, USA, 2021.
31. *ISO/DIS 4484-1*; Textiles and Textile Products—Microplastics from Textile Sources—Part 1: Determination of Material Loss from Fabrics during Washing. ISO: Geneva, Switzerland, 2023. Available online: https://www.iso.org/obp/ui/en/#iso:std:iso:4484:-1:ed-1:v1:en (accessed on 21 April 2024).
32. *ISO/DIS 4484-2*; Textiles and Textile Products—Microplastics from Textile Sources—Part 2: Qualitative and Quantitative Evaluation of Microplastics. ISO: Geneva, Switzerland, 2023. Available online: https://www.iso.org/obp/ui/en/#iso:std:iso:4484:-2:ed-1:v1:en (accessed on 21 April 2024).
33. Sanchez-Vidal, A.; Thompson, R.C.; Canals, M.; De Haan, W.P. The Imprint of Microfibres in Southern European Deep Seas. *PLoS ONE* **2018**, *13*, e0207033. [CrossRef] [PubMed]
34. Dris, R.; Gasperi, J.; Tassin, B. Sources and Fate of Microplastics in Urban Areas: A Focus on Paris Megacity. In *Handbook of Environmental Chemistry*; Springer: Cham, Switzerland, 2018; Volume 58, pp. 69–83. [CrossRef]
35. Mishra, S.; Rath, C.C.; Das, A.P. Marine Microfiber Pollution: A Review on Present Status and Future Challenges. *Mar. Pollut. Bull.* **2019**, *140*, 188–197. [CrossRef] [PubMed]
36. Brodin, M.; Norin, H.; Hanning, A.-C.; Persson, C.; Okcabol, S. Microplastics from Industrial Laundries—A Laboratory Study of Laundry Effluents. 2018. Available online: http://www.diva-portal.org/smash/get/diva2:1633776/FULLTEXT01.pdf (accessed on 8 July 2019).
37. Hidalgo-Ruz, V.; Gutow, L.; Thompson, R.C.; Thiel, M. Microplastics in the Marine Environment: A Review of the Methods Used for Identification and Quantification. *Environ. Sci. Technol.* **2012**, *46*, 3060–3075. [CrossRef] [PubMed]
38. Ballent, A.; Corcoran, P.L.; Madden, O.; Helm, P.A.; Longstaffe, F.J. Sources and Sinks of Microplastics in Canadian Lake Ontario Nearshore, Tributary and Beach Sediments. *Mar. Pollut. Bull.* **2016**, *110*, 383–395. [CrossRef] [PubMed]
39. Vejvar, M.; Lai, K.H.; Lo, C.K.Y. A Citation Network Analysis of Sustainability Development in Liner Shipping Management: A Review of the Literature and Policy Implications. *Marit. Policy Manag.* **2020**, *47*, 1–26. [CrossRef]
40. Colicchia, C.; Strozzi, F. Supply Chain Risk Management: A New Methodology for a Systematic Literature Review. *Supply Chain Manag.* **2012**, *17*, 403–418. [CrossRef]
41. Hummon, N.P.; Dereian, P. Connectivity in a Citation Network: The Development of DNA Theory. *Soc. Netw.* **1989**, *11*, 39–63. [CrossRef]
42. Van Eck, N.J.; Waltman, L. Visualizing Bibliometric Networks. In *Measuring Scholarly Impact*; Springer: Cham, Switzerland, 2014; pp. 285–320. [CrossRef]
43. Van Eck, N.J.; Waltman, L. CitNetExplorer: A New Software Tool for Analyzing and Visualizing Citation Networks. *J. Informetr.* **2014**, *8*, 802–823. [CrossRef]
44. Mrvar, A.; Batagelj, V. Analysis and Visualization of Large Networks with Program Package Pajek. *Complex Adapt. Syst. Model.* **2016**, *4*, 1–8. [CrossRef]
45. Volgare, M.; De Falco, F.; Avolio, R.; Castaldo, R.; Errico, M.E.; Gentile, G.; Ambrogi, V.; Cocca, M. Washing Load Influences the Microplastic Release from Polyester Fabrics by Affecting Wettability and Mechanical Stress. *Sci. Rep.* **2021**, *11*, 19479. [CrossRef]
46. Zambrano, M.C.; Pawlak, J.J.; Daystar, J.; Ankeny, M.; Cheng, J.J.; Venditti, R.A. Microfibers Generated from the Laundering of Cotton, Rayon and Polyester Based Fabrics and Their Aquatic Biodegradation. *Mar. Pollut. Bull.* **2019**, *142*, 394–407. [CrossRef]
47. De Falco, F.; Gentile, G.; Avolio, R.; Errico, M.E.; Di Pace, E.; Ambrogi, V.; Avella, M.; Cocca, M. Pectin Based Finishing to Mitigate the Impact of Microplastics Released by Polyamide Fabrics. *Carbohydr. Polym.* **2018**, *198*, 175–180. [CrossRef]
48. Herweyers, L.; Carteny, C.C.; Scheelen, L.; Watts, R.; Du Bois, E. Consumers' Perceptions and Attitudes toward Products Preventing Microfiber Pollution in Aquatic Environments as a Result of the Domestic Washing of Synthetic Clothes. *Sustainability* **2020**, *12*, 2244. [CrossRef]
49. Rochman, C.M.; Tahir, A.; Williams, S.L.; Baxa, D.V.; Lam, R.; Miller, J.T.; Teh, F.-C.; Werorilangi, S.; Teh, S.J. Anthropogenic debris in seafood: Plastic debris and fibers from textiles in fish and bivalves sold for human consumption. *Sci. Rep.* **2015**, *5*, 14340. [CrossRef]
50. Browne, M.A. Sources and Pathways of Microplastics to Habitats. In *Marine Anthropogenic Litter*; Springer International Publishing: Cham, Switzerland, 2015; Volume 334, pp. 229–244. [CrossRef]
51. Talvitie, J.; Heinonen, M.; Pääkkönen, J.-P.; Vahtera, E.; Mikola, A.; Setälä, O.; Vahala, R. Do Wastewater Treatment Plants Act as a Potential Point Source of Microplastics? Preliminary Study in the Coastal Gulf of Finland, Baltic Sea. *Water Sci. Technol.* **2015**, *72*, 1495–1504. [CrossRef]
52. Pirc, U.; Vidmar, M.; Mozer, A.; Kržan, A.; Krzan, A.; Kržan, A.; Kr, A. Emissions of Microplastic Fibers from Microfiber Fleece during Domestic Washing. *Environ. Sci. Pollut. Res.* **2016**, *23*, 22206–22211. [CrossRef] [PubMed]
53. Cesa, F.S.; Turra, A.; Checon, H.H.; Leonardi, B.; Baruque-Ramos, J. Laundering and Textile Parameters Influence Fibers Release in Household Washings. *Environ. Pollut.* **2020**, *257*, 113553. [CrossRef]
54. Lykaki, M.; Zhang, Y.Q.; Markiewicz, M.; Brandt, S.; Kolbe, S.; Schrick, J.; Rabe, M.; Stolte, S. The Influence of Textile Finishing Agents on the Biodegradability of Shed Fibres. *Green Chem.* **2021**, *23*, 5212–5221. [CrossRef]

55. De Falco, F.; Cocca, M.; Avella, M.; Thompson, R.C.; De Falco, F.; Cocca, M.; Avella, M.; Thompson, R.C. Microfiber Release to Water, Via Laundering, and to Air, via Everyday Use: A Comparison between Polyester Clothing with Differing Textile Parameters. *Environ. Sci. Technol.* **2020**, *54*, 3288–3296. [CrossRef] [PubMed]
56. Kapp, K.J.; Miller, R.Z. Electric Clothes Dryers: An Underestimated Source of Microfiber Pollution. *PLoS ONE* **2020**, *15*, e0239165. [CrossRef] [PubMed]
57. YarnsandFibers.com. Where Is Polyester Produced in the World? Available online: https://www.yarnsandfibers.com/textile-resources/synthetic-fibers/polyester/polyester-production-raw-materials/where-is-polyester-produced-in-the-world/ (accessed on 12 December 2022).
58. Chan, C.K.M.; Park, C.; Chan, K.M.; Mak, D.C.W.; Fang, J.K.H.; Mitrano, D.M. Microplastic Fibre Releases from Industrial Wastewater Effluent: A Textile Wet-Processing Mill in China. *Environ. Chem.* **2021**, *18*, 93–100. [CrossRef]
59. Hartline, N.L.; Bruce, N.J.; Karba, S.N.; Ruff, E.O.; Sonar, S.U.; Holden, P.A. Microfiber Masses Recovered from Conventional Machine Washing of New or Aged Garments. *Environ. Sci. Technol.* **2016**, *50*, 11532–11538. [CrossRef] [PubMed]
60. Rathinamoorthy, R.; Balasaraswathi, S.R. Investigations on the Interactive Effect of Laundry Parameters on Microfiber Release from Polyester Knitted Fabric. *Fibers Polym.* **2022**, *23*, 2052–2061. [CrossRef]
61. Kelly, M.R.; Lant, N.J.; Kurr, M.; Burgess, J.G. Importance of Water-Volume on the Release of Microplastic Fibers from Laundry. *Environ. Sci. Technol.* **2019**, *53*, 11735–11744. [CrossRef] [PubMed]
62. Carney Almroth, B.M.; Åström, L.; Roslund, S.; Petersson, H.; Johansson, M.; Persson, N.-K. Quantifying Shedding of Synthetic Fibers from Textiles; A Source of Microplastics Released into the Environment. *Environ. Sci. Pollut. Res.* **2017**, *25*, 1191–1199. [CrossRef] [PubMed]
63. Berruezo, M.; Bonet-Aracil, M.; Montava, I.; Bou-Belda, E.; Díaz-García, P.; Gisbert-Payá, J.; Gisbert-Paya, J.; Bou-Belda, E.; Díaz-García, P. Preliminary Study of Weave Pattern Influence on Microplastics from Fabric Laundering. *Text. Res. J.* **2021**, *91*, 1037–1045. [CrossRef]
64. Özkan, İ.; Gündoğdu, S. Investigation on the Microfiber Release under Controlled Washings from the Knitted Fabrics Produced by Recycled and Virgin Polyester Yarns. *J. Text. Inst.* **2021**, *112*, 264–272. [CrossRef]
65. Choi, S.; Kim, J.; Kwon, M. The Effect of the Physical and Chemical Properties of Synthetic Fabrics on the Release of Microplastics during Washing and Drying. *Polymers* **2022**, *14*(16), 3384. [CrossRef]
66. Cai, Y.; Mitrano, D.M.; Heuberger, M.; Hufenus, R.; Nowack, B. The Origin of Microplastic Fiber in Polyester Textiles: The Textile Production Process Matters. *J. Clean. Prod.* **2020**, *267*, 121970. [CrossRef]
67. Choi, S.; Kwon, M.; Park, M.-J.; Kim, J. Analysis of Microplastics Released from Plain Woven Classified by Yarn Types during Washing and Drying. *Polymers* **2021**, *13*, 2988. [CrossRef] [PubMed]
68. Palacios-Marín, A.V.; Jabbar, A.; Tausif, M. Fragmented Fiber Pollution from Common Textile Materials and Structures during Laundry. *Text. Res. J.* **2022**, *92*, 2265–2275. [CrossRef]
69. Sillanpää, M.; Sainio, P. Release of Polyester and Cotton Fibers from Textiles in Machine Washings. *Environ. Sci. Pollut. Res.* **2017**, *24*, 19313–19321. [CrossRef] [PubMed]
70. Mahbub, M.S.; Shams, M. Acrylic Fabrics as a Source of Microplastics from Portable Washer and Dryer: Impact of Washing and Drying Parameters. *Sci. Total Environ.* **2022**, *834*, 155429. [CrossRef] [PubMed]
71. Periyasamy, A.P. Evaluation of Microfiber Release from Jeans: The Impact of Different Washing Conditions. *Environ. Sci. Pollut. Res.* **2021**, *28*, 58570–58582. [CrossRef] [PubMed]
72. De Falco, F.; Gullo, M.P.; Gentile, G.; Di Pace, E.; Cocca, M.; Gelabert, L.; Brouta-Agnésa, M.; Rovira, A.; Escudero, R.; Villalba, R.; et al. Evaluation of microplastic release caused by textile washing processes of synthetic fabrics. *Environ. Pollut.* **2018**, *236*, 916–925. [CrossRef] [PubMed]
73. Lant, N.J.; Hayward, A.S.; Peththawadu, M.M.D.D.; Sheridan, K.J.; Dean, J.R. Microfiber Release from Real Soiled Consumer Laundry and the Impact of Fabric Care Products and Washing Conditions. *PLoS ONE* **2020**, *15*, e0233332. [CrossRef] [PubMed]
74. Rathinamoorthy, R.; Raja Balasaraswathi, S. Investigations on the Impact of Handwash and Laundry Softener on Microfiber Shedding from Polyester Textiles. *J. Text. Inst.* **2022**, *113*, 1428–1437. [CrossRef]
75. McIlwraith, H.K.; Lin, J.; Erdle, L.M.; Mallos, N.; Diamond, M.L.; Rochman, C.M. Capturing Microfibers—Marketed Technologies Reduce Microfiber Emissions from Washing Machines. *Mar. Pollut. Bull.* **2019**, *139*, 40–45. [CrossRef]
76. Napper, I.E.; Barrett, A.C.; Thompson, R.C. The Efficiency of Devices Intended to Reduce Microfibre Release during Clothes Washing. *Sci. Total Environ.* **2020**, *738*, 140412. [CrossRef] [PubMed]
77. Ryan, P.G.; Suaria, G.; Perold, V.; Pierucci, A.; Bornman, T.G.; Aliani, S. Sampling Microfibres at the Sea Surface: The Effects of Mesh Size, Sample Volume and Water Depth. *Environ. Pollut.* **2020**, *258*, 113413. [CrossRef] [PubMed]
78. Dyachenko, A.; Mitchell, J.; Arsem, N. Extraction and Identification of Microplastic Particles from Secondary Wastewater Treatment Plant (WWTP) Effluent. *Anal. Methods* **2017**, *9*, 1412–1418. [CrossRef]
79. Prata, J.C.; da Costa, J.P.; Duarte, A.C.; Rocha-Santos, T. Methods for Sampling and Detection of Microplastics in Water and Sediment: A Critical Review. *TrAC-Trends Anal. Chem.* **2019**, *110*, 150–159. [CrossRef]
80. Lares, M.; Ncibi, M.C.; Sillanpää, M.; Sillanpää, M. Intercomparison Study on Commonly Used Methods to Determine Microplastics in Wastewater and Sludge Samples. *Environ. Sci. Pollut. Res.* **2019**, *26*, 12109–12122. [CrossRef] [PubMed]
81. The Microfibre Consortium. *FAQs on the TMC Test Method*. 2021. Available online: https://www.microfibreconsortium.com/s/TMC-Test-Method_FAQs-bzch.pdf (accessed on 20 February 2023).

82. Conkle, J.L.; Báez Del Valle, C.D.; Turner, J.W. Are We Underestimating Microplastic Contamination in Aquatic Environments? *Environ. Manag.* **2018**, *61*, 1–8. [CrossRef] [PubMed]
83. Covernton, G.A.; Pearce, C.M.; Gurney-Smith, H.J.; Chastain, S.G.; Ross, P.S.; Dower, J.F.; Dudas, S.E. Size and Shape Matter: A Preliminary Analysis of Microplastic Sampling Technique in Seawater Studies with Implications for Ecological Risk Assessment. *Sci. Total Environ.* **2019**, *667*, 124–132. [CrossRef] [PubMed]
84. Gago, J.; Carretero, O.; Filgueiras, A.V.; Viñas, L. Synthetic Microfibers in the Marine Environment: A Review on Their Occurrence in Seawater and Sediments. *Mar. Pollut. Bull.* **2018**, *127*, 365–376. [CrossRef]
85. Carr, S.A. Sources and Dispersive Modes of Micro-Fibers in the Environment. *Integr. Environ. Assess. Manag.* **2017**, *13*, 466–469. [CrossRef]
86. Baldwin, A.K.; Corsi, S.R.; Mason, S.A. Plastic Debris in 29 Great Lakes Tributaries: Relations to Watershed Attributes and Hydrology. *Environ. Sci. Technol.* **2016**, *50*, 10377–10385. [CrossRef]
87. Barrows, A.P.W.; Cathey, S.E.; Petersen, C.W. Marine Environment Microfiber Contamination: Global Patterns and the Diversity of Microparticle Origins. *Environ. Pollut.* **2018**, *237*, 275–284. [CrossRef]
88. Dris, R.; Gasperi, J.; Mirande, C.; Mandin, C.; Guerrouache, M.; Langlois, V.; Tassin, B. A First Overview of Textile Fibers, including Microplastics, in Indoor and Outdoor Environments. *Environ. Pollut.* **2017**, *221*, 453–458. [CrossRef] [PubMed]
89. Liu, K.; Wang, X.H.; Fang, T.; Xu, P.; Zhu, L.X.; Li, D.J. Source and Potential Risk Assessment of Suspended Atmospheric Microplastics in Shanghai. *Sci. Total Environ.* **2019**, *675*, 462–471. [CrossRef] [PubMed]
90. Zhang, Q.; Zhao, Y.; Du, F.; Cai, H.; Wang, G.; Shi, H. Microplastic Fallout in Different Indoor Environments. *Environ. Sci. Technol.* **2020**, *54*, 6530–6539. [CrossRef] [PubMed]
91. Torres-Agullo, A.; Karanasiou, A.; Moreno, T.; Lacorte, S. Airborne Microplastic Particle Concentrations and Characterization in Indoor Urban Microenvironments. *Environ. Pollut.* **2022**, *308*, 119707. [CrossRef] [PubMed]
92. Gaston, E.; Woo, M.; Steele, C.; Sukumaran, S.; Anderson, S. Microplastics Differ between Indoor and Outdoor Air Masses: Insights from Multiple Microscopy Methodologies. *Appl. Spectrosc.* **2020**, *74*, 1079–1098. [CrossRef] [PubMed]
93. Tunahan Kaya, A.; Yurtsever, M.; Çiftçi Bayraktar, S. Ubiquitous Exposure to Microfiber Pollution in the Air. *Eur. Phys. J. Plus* **2018**, *133*, 488. [CrossRef]
94. Abbasi, S.; Turner, A. Dry and Wet Deposition of Microplastics in a Semi-Arid Region (Shiraz, Iran). *Sci. Total Environ.* **2021**, *786*, 147358. [CrossRef] [PubMed]
95. Sun, J.; Dai, X.; Wang, Q.; van Loosdrecht, M.C.M.; Ni, B.J. Microplastics in Wastewater Treatment Plants: Detection, Occurrence and Removal. *Water Res.* **2019**, *152*, 21–37. [CrossRef]
96. Mintenig, S.M.; Int-Veen, I.; Löder, M.G.J.; Primpke, S.; Gerdts, G. Identification of Microplastic in Effluents of Wastewater Treatment Plants Using Focal Plane Array-Based Micro-Fourier-Transform Infrared Imaging. *Water Res.* **2017**, *108*, 365–372. [CrossRef]
97. Murphy, F.; Ewins, C.; Carbonnier, F.; Quinn, B. Wastewater Treatment Works (WwTW) as a Source of Microplastics in the Aquatic Environment. *Environ. Sci. Technol.* **2016**, *50*, 5800–5808. [CrossRef]
98. Ziajahromi, S.; Neale, P.A.; Rintoul, L.; Leusch, F.D.L.L. Wastewater Treatment Plants as a Pathway for Microplastics: Development of a New Approach to Sample Wastewater-Based Microplastics. *Water Res.* **2017**, *112*, 93–99. [CrossRef]
99. Conley, K.; Clum, A.; Deepe, J.; Lane, H.; Beckingham, B. Wastewater Treatment Plants as a Source of Microplastics to an Urban Estuary: Removal Efficiencies and Loading per Capita over One Year. *Water Res. X* **2019**, *3*, 100030. [CrossRef] [PubMed]
100. Magnusson, K.; Norén, F. *Screening of Microplastic Particles in and Down-Stream a Wastewater Treatment Plant*; Report C55; IVL Swedish Environmental Research Institute: Stockholm, Sweden, 2014; Volume 22.
101. Talvitie, J.; Mikola, A.; Setälä, O.; Heinonen, M.; Koistinen, A. How Well Is Microlitter Purified from Wastewater?—A Detailed Study on the Stepwise Removal of Microlitter in a Tertiary Level Wastewater Treatment Plant. *Water Res.* **2017**, *109*, 164–172. [CrossRef] [PubMed]
102. Talvitie, J.; Mikola, A.; Koistinen, A.; Setälä, O. Solutions to Microplastic Pollution—Removal of Microplastics from Wastewater Effluent with Advanced Wastewater Treatment Technologies. *Water Res.* **2017**, *123*, 401–407. [CrossRef] [PubMed]
103. Yang, L.; Li, K.; Cui, S.; Kang, Y.; An, L.; Lei, K. Removal of Microplastics in Municipal Sewage from China's Largest Water Reclamation Plant. *Water Res.* **2019**, *155*, 175–181. [CrossRef] [PubMed]
104. Lv, X.; Dong, Q.; Zuo, Z.; Liu, Y.; Huang, X.; Wu, W.M. Microplastics in a Municipal Wastewater Treatment Plant: Fate, Dynamic Distribution, Removal Efficiencies, and Control Strategies. *J. Clean. Prod.* **2019**, *225*, 579–586. [CrossRef]
105. Rather, L.J.; Jameel, S.; Dar, O.A.; Ganie, S.A.; Bhat, K.A.; Mohammad, F. Advances in the Sustainable Technologies for Water Conservation in Textile Industries. In *Water in Textiles and Fashion*; Woodhead Publishing: Sawston, UK, 2019; pp. 175–194. [CrossRef]
106. Boucher, J.; Billard, G. *The Challenges of Measuring Plastic Pollution*; Institut Veolia: Aubervilliers, France, 2019. [CrossRef]
107. Sherrington, C. Plastics in the Marine Environment. Eunomia. Available online: https://www.eunomia.co.uk/reports-tools/plastics-in-the-marine-environment/ (accessed on 10 February 2023).
108. Ryberg, M.W.; Laurent, A.; Hauschild, M. *Mapping of Global Plastics Value Chain*; United Nations Environment Programme: Nairobi, Kenya, 2018; p. 96.

109. Belzagui, F.; Gutiérrez-Bouzán, C.; Álvarez-Sánchez, A.; Vilaseca, M.; Gutierrez-Bouzan, C.; Alvarez-Sanchez, A.; Vilaseca, M. Textile Microfibers Reaching Aquatic Environments: A New Estimation Approach. *Environ. Pollut.* **2020**, *265*, 114889. [CrossRef] [PubMed]
110. Miller, R.Z.; Watts, A.J.R.; Winslow, B.O.; Galloway, T.S.; Barrows, A.P.W. Mountains to the Sea: River Study of Plastic and Non-Plastic Microfiber Pollution in the Northeast USA. *Mar. Pollut. Bull.* **2017**, *124*, 245–251. [CrossRef]
111. Dris, R.; Gasperi, J.; Rocher, V.; Tassin, B. Synthetic and Non-Synthetic Anthropogenic Fibers in a River under the Impact of Paris Megacity: Sampling Methodological Aspects and Flux Estimations. *Sci. Total Environ.* **2018**, *618*, 157–164. [CrossRef]
112. Lusher, A.; Hollman, P.; Mendozal, J. *Microplastics in Fisheries and Aquaculture: Status of Knowledge on Their Occurrence and Implications for Aquatic Organisms and Food Safety*; FAO: Rome, Italy, 2017.
113. Stanton, T.; Johnson, M.; Nathanail, P.; MacNaughtan, W.; Gomes, R.L. Freshwater and airborne textile fibre populations are dominated by 'natural', not microplastic, fibres. *Sci. Total Environ.* **2019**, *666*, 377–389. [CrossRef]
114. Sundt, P.; Schulze, P.-E.; Syversen, F. Sources of Microplastic-pollution to the Marine Environment; Project Report M-321 I 2015; Norwegian Environment Agency. 2014. Available online: https://www.miljodirektoratet.no/globalassets/publikasjoner/M321/M321.pdf (accessed on 10 February 2023).
115. Zhao, S.; Zhu, L.; Wang, T.; Li, D. Suspended Microplastics in the Surface Water of the Yangtze Estuary System, China: First Observations on Occurrence, Distribution. *Mar. Pollut. Bull.* **2014**, *86*, 562–568. [CrossRef]
116. Bergmann, M.; Gutow, L.; Klages, M. *Marine Anthoropogenic Litter*; Springer: Berlin/Heidelberg, Germany, 2015. [CrossRef]
117. Napper, I.E.; Thompson, R.C. Release of Synthetic Microplastic Plastic Fibres from Domestic Washing Machines: Effects of Fabric Type and Washing Conditions. *Mar. Pollut. Bull.* **2016**, *112*, 39–45. [CrossRef] [PubMed]
118. Wolff, S.; Kerpen, J.; Prediger, J.; Barkmann, L.; Müller, L. Determination of the Microplastics Emission in the Effluent of a Municipal Wastewater Treatment Plant Using Raman Microspectroscopy. *Water Res. X* **2019**, *2*, 100014. [CrossRef] [PubMed]
119. Yahaya, T.; Abdulazeez, A.; Oladele, E.; Funmilayo, W.E.; Dikko, O.C.; Ja'afar, U.; Salisu, N. Microplastics Abundance, Characteristics, and Risk in Badagry Lagoon in Lagos State, Nigeria. *Pollution* **2022**, *8*, 1325–1337. [CrossRef]
120. Zhou, H.; Zhou, L.; Ma, K. Microfiber from Textile Dyeing and Printing Wastewater of a Typical Industrial Park in China: Occurrence, Removal and Release. *Sci. Total Environ.* **2020**, *739*, 140329. [CrossRef]
121. Habib, D.; Locke, D.C.; Cannone, L.J. Synthetic Fibers as Indicators of Municipal Sewage Sludge, Sludge Products, and Sewage Treatment Plant Effluents. *Water Air Soil Pollut.* **1998**, *103*, 1–8. [CrossRef]
122. Zubris, K.A.V.; Richards, B.K. Synthetic Fibers as an Indicator of Land Application of Sludge. *Environ. Pollut.* **2005**, *138*, 201–211. [CrossRef]
123. Li, X.; Chen, L.; Mei, Q.; Dong, B.; Dai, X.; Ding, G.; Zeng, E.Y. Microplastics in Sewage Sludge from the Wastewater Treatment Plants in China. *Water Res.* **2018**, *142*, 75–85. [CrossRef]
124. Corradini, F.; Meza, P.; Eguiluz, R.; Casado, F.; Huerta-Lwanga, E.; Geissen, V. Evidence of Microplastic Accumulation in Agricultural Soils from Sewage Sludge Disposal. *Sci. Total Environ.* **2019**, *671*, 411–420. [CrossRef]
125. Boots, B.; Russell, C.W.; Green, D.S. Effects of Microplastics in Soil Ecosystems: Above and below Ground. *Environ. Sci. Technol.* **2019**, *53*, 11496–11506. [CrossRef] [PubMed]
126. Prendergast-miller, M.T.; Katsiamides, A.; Abbass, M.; Sturzenbaum, S.R.; Thorpe, K.L.; Hodson, M.E. Polyester-Derived Micro Fi Bre Impacts on the Soil-Dwelling Earthworm. *Environ. Pollut.* **2019**, *251*, 453–459. [CrossRef] [PubMed]
127. Tourinho, P.S.; Loureiro, S.; Talluri, V.S.S.L.P.; Dolar, A.; Verweij, R.; Chvojka, J.; Michalcová, A.; Kočí, V.; van Gestel, C.A.M. Microplastic Fibers Influence Ag Toxicity and Bioaccumulation in Eisenia Andrei but Not in Enchytraeus Crypticus. *Ecotoxicology* **2021**, *30*, 1216–1226. [CrossRef] [PubMed]
128. Selonen, S.; Dolar, A.; Jemec Kokalj, A.; Skalar, T.; Parramon Dolcet, L.; Hurley, R.; van Gestel, C.A.M. Exploring the Impacts of Plastics in Soil—The Effects of Polyester Textile Fibers on Soil Invertebrates. *Sci. Total Environ.* **2020**, *700*, 134451. [CrossRef] [PubMed]
129. Gavigan, J.; Kefela, T.; Macadam-Somer, I.; Suh, S.; Geyer, R. Synthetic Microfiber Emissions to Land Rival Those to Waterbodies and Are Growing. *PLoS ONE* **2020**, *15*, e0237839. [CrossRef] [PubMed]
130. Royer, S.J.; Wiggin, K.; Kogler, M.; Deheyn, D.D. Degradation of Synthetic and Wood-Based Cellulose Fabrics in the Marine Environment: Comparative Assessment of Field, Aquarium, and Bioreactor Experiments. *Sci. Total Environ.* **2021**, *791*, 148060. [CrossRef] [PubMed]
131. Hermabessiere, L.; Dehaut, A.; Paul-Pont, I.; Lacroix, C.; Jezequel, R.; Soudant, P.; Duflos, G. Occurrence and Effects of Plastic Additives on Marine Environments and Organisms: A Review. *Chemosphere* **2017**, *182*, 781–793. [CrossRef] [PubMed]
132. Mathalon, A.; Hill, P. Microplastic Fibers in the Intertidal Ecosystem Surrounding Halifax Harbor, Nova Scotia. *Mar. Pollut. Bull.* **2014**, *81*, 69–79. [CrossRef]
133. Collard, F.; Gilbert, B.; Eppe, G.; Parmentier, E.; Das, K. Detection of Anthropogenic Particles in Fish Stomachs: An Isolation Method Adapted to Identification by Raman Spectroscopy. *Arch. Environ. Contam. Toxicol.* **2015**, *69*, 331–339. [CrossRef]
134. Keshavarzifard, M.; Zakaria, M.P.; Sharifi, R. Ecotoxicological and Health Risk Assessment of Polycyclic Aromatic Hydrocarbons (PAHs) in Short-Neck Clam (Paphia Undulata) and Contaminated Sediments in Malacca Strait, Malaysia. *Arch. Environ. Contam. Toxicol.* **2017**, *73*, 474–487. [CrossRef]
135. Michels, J.; Stippkugel, A.; Lenz, M.; Wirtz, K.; Engel, A. Rapid Aggregation of Biofilm-Covered Microplastics with Marine Biogenic Particles. *Proc. R. Soc. B Biol. Sci.* **2018**, *285*, 20181203. [CrossRef] [PubMed]

136. Liu, P.; Zhan, X.; Wu, X.; Li, J.; Wang, H.; Gao, S. Effect of Weathering on Environmental Behavior of Microplastics: Properties, Sorption and Potential Risks. *Chemosphere* **2020**, *242*, 125193. [CrossRef] [PubMed]
137. Lassen, C.; Hansen, S.F.; Magnusson, K.; Norén, F.; Hartmann, N.I.B.; Jensen, P.R.; Nielsen, T.G.; Brinch, A. *Microplastics—Occurrence, Effects and Sources of Releases to the Environment in Denmark*; The Danish Environmental Protection Agency: Copenhagen, Denmark, 2015. Available online: https://mst.dk/publikationer/2015/november/microplastics-occurrence-effects-and-sources-of-releases-to-the-environment-in-denmark (accessed on 10 February 2023).
138. Girard, E.B.; Kaliwoda, M.; Schmahl, W.W.; Wörheide, G.; Orsi, W.D. Biodegradation of Textile Waste by Marine Bacterial Communities Enhanced by Light. *Environ. Microbiol. Rep.* **2020**, *12*, 406–418. [CrossRef] [PubMed]
139. Cole, M.; Lindeque, P.; Halsband, C.; Galloway, T.S. Microplastics as Contaminants in the Marine Environment: A Review. *Mar. Pollut. Bull.* **2011**, *62*, 2588–2597. [CrossRef] [PubMed]
140. Watts, A.J.R.; Urbina, M.A.; Corr, S.; Lewis, C.; Galloway, T.S. Ingestion of Plastic Microfibers by the Crab Carcinus Maenas and Its Effect on Food Consumption and Energy Balance. *Environ. Sci. Technol.* **2015**, *49*, 14597–14604. [CrossRef]
141. Taylor, M.L.; Gwinnett, C.; Robinson, L.F.; Woodall, L.C. Plastic Microfibre Ingestion by Deep-Sea Organisms. *Sci. Rep.* **2016**, *6*, 33997. [CrossRef] [PubMed]
142. Duflos, G.; Dehaut, A.; Cassone, A.-L.; Frère, L.; Hermabessiere, L.; Himber, C.; Rinnert, E.; Rivière, G.; Lambert, C.; Soudant, P.; et al. Microplastics in Seafood: Identifying a Protocol for Their Extraction and Characterization. *Fate Impact Microplastics Mar. Ecosyst.* **2017**, *215*, 74. [CrossRef]
143. Weinstein, J.E.; Ertel, B.M.; Gray, A.D. Accumulation and Depuration of Microplastic Fibers, Fragments, and Tire Particles in the Eastern Oyster, Crassostrea Virginica: A Toxicokinetic Approach. *Environ. Pollut.* **2022**, *308*, 119681. [CrossRef]
144. Murano, C.; Vaccari, L.; Casotti, R.; Corsi, I.; Palumbo, A. Occurrence of Microfibres in Wild Specimens of Adult Sea Urchin Paracentrotus Lividus (Lamarck, 1816) from a Coastal Area of the Central Mediterranean Sea. *Mar. Pollut. Bull.* **2022**, *176*, 113448. [CrossRef]
145. Mohsen, M.; Sun, L.; Lin, C.; Huo, D.; Yang, H. Mechanism Underlying the Toxicity of the Microplastic Fibre Transfer in the Sea Cucumber Apostichopus Japonicus. *J. Hazard. Mater.* **2021**, *416*, 125858. [CrossRef]
146. Pradit, S.; Noppradit, P.; Goh, B.P.; Sornplang, K.; Ong, M.C.; Towatana, P. Occurrence of Microplastics and Trace Metals in Fish and Shrimp from Songkhla Lake, Thailand during the COVID-19 Pandemic. *Appl. Ecol. Environ. Res.* **2021**, *19*, 1085–1106. [CrossRef]
147. Esterhuizen, M.; Buchenhorst, L.; Kim, Y.J.; Pflugmacher, S. In Vivo Oxidative Stress Responses of the Freshwater Basket Clam Corbicula Javanicus to Microplastic Fibres and Particles. *Chemosphere* **2022**, *296*, 134037. [CrossRef] [PubMed]
148. Pittura, L.; Nardi, A.; Cocca, M.; De Falco, F.; D'Errico, G.; Mazzoli, C.; Mongera, F.; Benedetti, M.; Gorbi, S.; Avella, M.; et al. Cellular Disturbance and Thermal Stress Response in Mussels Exposed to Synthetic and Natural Microfibers. *Front. Mar. Sci.* **2022**, *9*, 1–15. [CrossRef]
149. Wright, S.L.; Thompson, R.C.; Galloway, T.S. The Physical Impacts of Microplastics on Marine Organisms: A Review. *Environ. Pollut.* **2013**, *178*, 483–492. [CrossRef] [PubMed]
150. Erni-Cassola, G.; Gibson, M.I.; Thompson, R.C.; Christie-Oleza, J.A. Lost, but Found with Nile Red: A Novel Method for Detecting and Quantifying Small Microplastics (1 Mm to 20 Mm) in Environmental Samples. *Environ. Sci. Technol.* **2017**, *51*, 13641–13648. [CrossRef]
151. Bakir, A.; Rowland, S.J.; Thompson, R.C. Enhanced Desorption of Persistent Organic Pollutants from Microplastics under Simulated Physiological Conditions. *Environ. Pollut.* **2014**, *185*, 16–23. [CrossRef] [PubMed]
152. Yang, D.; Shi, H.; Li, L.; Li, J.; Jabeen, K.; Kolandhasamy, P. Microplastic Pollution in Table Salts from China. *Environ. Sci. Technol.* **2015**, *49*, 13622–13627. [CrossRef] [PubMed]
153. Dris, R.; Gasperi, J.; Saad, M.; Mirande, C.; Tassin, B. Synthetic Fibers in Atmospheric Fallout: A Source of Microplastics in the Environment? *Mar. Pollut. Bull.* **2016**, *104*, 290–293. [CrossRef] [PubMed]
154. Gasperi, J.; Wright, S.L.; Dris, R.; Collard, F.; Mandin, C.; Guerrouache, M.; Langlois, V.; Kelly, F.J.; Tassin, B. Microplastics in air: Are we breathing it in? *Curr. Opin. Environ. Sci. Health* **2018**, *1*, 1–5. [CrossRef]
155. Ladewig, S.M.; Bao, S.; Chow, A.T. Natural Fibers: A Missing Link to Chemical Pollution Dispersion in Aquatic Environments. *Environ. Sci. Technol.* **2015**, *49*, 12609–12610. [CrossRef]
156. Acharya, S.; Rumi, S.S.; Hu, Y.; Abidi, N. Microfibers from Synthetic Textiles as a Major Source of Microplastics in the Environment: A Review. *Text. Res. J.* **2021**, *91*, 2136–2156. [CrossRef]
157. Athey, S.N.; Erdle, L.M. Are We Underestimating Anthropogenic Microfiber Pollution? A Critical Review of Occurrence, Methods, and Reporting. *Environ. Toxicol. Chem.* **2022**, *41*, 822–837. [CrossRef] [PubMed]
158. Conservation X labs. PANGAIA x MTIX Microfiber Mitigation. Available online: https://www.microfiberinnovation.org/innovation/pangaia-x-mtix (accessed on 2 April 2022).
159. Mermaids. Ocean Clean Wash. Handbook for Zero Microplastics from Textiles and Laundry. 2018. Available online: http://life-mermaids.eu/en/deliverables-mermaids-life-2/ (accessed on 21 January 2022).
160. Cotton, L.; Hayward, A.S.; Lant, N.J.; Blackburn, R.S. Improved garment longevity and reduced microfibre release are important sustainability benefits of laundering in colder and quicker washing machine cycles. *Dye. Pigment.* **2020**, *177*, 108120. [CrossRef]
161. Li, J.; Lemstra, P.J.; Ma, P. Chapter 7: Can High-Performance Fibers Be(Come) Bio-Based and Also Biocompostable? *Adv. Ind. Eng. Polym. Res.* **2022**, *5*, 117–132. [CrossRef]

162. Yousef, S.; Eimontas, J.; Zakarauskas, K.; Striūgas, N.; Mohamed, A. A New Strategy for Using Lint-Microfibers Generated from Clothes Dryer as a Sustainable Source of Renewable Energy. *Sci. Total Environ.* **2021**, *762*, 143107. [CrossRef] [PubMed]
163. Plastic Pollution Coalition. 15 Ways to Stop Microfiber Pollution Now. Plastic Pollution Coalition. Available online: https://www.plasticpollutioncoalition.org/blog/2017/3/2/15-ways-to-stop-microfiber-pollution-now (accessed on 13 December 2022).
164. Plastic Soup Foundation. Microfiber Pollution: What Solutions for the Oceans? Available online: https://www.oceancleanwash.org/solutions/solutions-for-consumers/ (accessed on 13 December 2022).
165. Cowger, W.; Booth, A.M.; Hamilton, B.M.; Thaysen, C.; Primpke, S.; Munno, K.; Lusher, A.L.; Dehaut, A.; Vaz, V.P.; Liboiron, M.; et al. Reporting Guidelines to Increase the Reproducibility and Comparability of Research on Microplastics. *Appl. Spectrosc.* **2020**, *74*, 1066–1077. [CrossRef] [PubMed]
166. Pilkington, A.; Meredith, J. The Evolution of the Intellectual Structure of Operations Management—1980–2006: A Citation/Co-Citation Analysis. *J. Oper. Manag.* **2009**, *27*, 185–202. [CrossRef]
167. Daukantienė, V. Analysis of the Sustainability Aspects of Fashion: A Literature Review. *Text. Res. J.* **2022**, *93*, 991–1002. [CrossRef]

Disclaimer/Publisher's Note: The statements, opinions and data contained in all publications are solely those of the individual author(s) and contributor(s) and not of MDPI and/or the editor(s). MDPI and/or the editor(s) disclaim responsibility for any injury to people or property resulting from any ideas, methods, instructions or products referred to in the content.

Article

Mercury Concentrations in Dust from Dry Gas Cleaning of Sinter Plant and Technical Removal Options

Claudia Hledik [1,*], Yilan Zeng [2,3], Tobias Plattner [4] and Maria Fuerhacker [1]

1. Department of Water, Atmosphere and Environment, University of Natural Resources and Life Sciences Vienna, Muthgasse 18, 1190 Vienna, Austria; maria.fuerhacker@boku.ac.at
2. Department of Inorganic Chemistry, Faculty of Natural Sciences, Comenius University Bratislava, Ilkovicova 6, 842 15 Bratislava, Slovakia; zeng2@uniba.sk
3. Department of Environmental Ecology and Landscape Management, Faculty of Natural Sciences, Comenius University Bratislava, Ilkovicova 6, 842 15 Bratislava, Slovakia
4. Primetals Technologies Austria GmbH, 4031 Linz, Austria; tobias.plattner@primetals.com
* Correspondence: claudia.hledik@boku.ac.at

Citation: Hledik, C.; Zeng, Y.; Plattner, T.; Fuerhacker, M. Mercury Concentrations in Dust from Dry Gas Cleaning of Sinter Plant and Technical Removal Options. *Water* 2024, 16, 1948. https://doi.org/10.3390/w16141948

Academic Editors: Hamidi Abdul Aziz, Issam A. Al-Khatib, Rehab O. Abdel Rahman, Tsuyoshi Imai and Yung-Tse Hung

Received: 30 May 2024
Revised: 27 June 2024
Accepted: 6 July 2024
Published: 10 July 2024

Copyright: © 2024 by the authors. Licensee MDPI, Basel, Switzerland. This article is an open access article distributed under the terms and conditions of the Creative Commons Attribution (CC BY) license (https://creativecommons.org/licenses/by/4.0/).

Abstract: Mercury (Hg) is a naturally occurring element and has been released through human activities over an extended period. The major source is the steel industry, especially sinter plants. During a sintering process, high amounts of dust and gaseous emission are produced. These gases contain high loads of SO_X and NO_X as well as toxic pollutants, such as heavy metals like Hg. These toxic pollutants are removed by adsorbing to solids, collected as by-products and deposited as hazardous waste. The by-products contain a high amount of salt, resulting in a high water solubility. In this study, to ultimately reduce the waste amount in landfills, leachates of the by-products have been produced. The dissolved Hg concentration and its distribution across different charges were determined. Hg concentrations between 3793 and 12,566 µg L^{-1} were measured in the leachates. The objective was to lower the Hg concentration in leachates by chemical precipitation with sodium sulfide (Na_2S) or an organic sulfide followed by filtration. Both reagents precipitate Hg with removal rates of up to 99.6% for the organic sulfide and 99.9% for Na_2S, respectively. The dose of the precipitator as well as the initial Hg concentration affected the removal rate. In addition to Hg, other relevant heavy metals have to be included in the calculation of the amount of precipitator as well. Between relevant heavy metals including Hg and sulfide, the ratio should be more than 1.5. The novelty of this study is the measurement and treatment of Hg in wastewater with a high ionic strength. The high salt concentrations did not influence the efficiency of the removal methods. An adjustment of the precipitator dose for each sample is necessary, because an overdose potentially leads to the re-dissolving of Hg. It could be shown that the emission limit of 0.005 mg L^{-1} could be reached especially by precipitation with Na_2S.

Keywords: sinter plant; mercury; dry gas cleaning; chemical precipitation; leaching

1. Introduction

Mercury (Hg) is a global pollutant with a high toxicity for humans and ecosystem health. Hg is a naturally occurring element but also is emitted into the ecosystem by humans over a long time. Generally, Hg is present in the environment in three different chemical fractions: elemental Hg, inorganic Hg and organic Hg [1]. Predominant in the atmosphere is the gaseous elemental Hg, which can undergo a long-range transport and atmospheric deposition into the aquatic systems [2]. Under reducing conditions, bacteria can form the highly toxic methylmercury, which biomagnifies in aquatic and terrestrial food chains, resulting in exposure to humans and wildlife throughout the world [3–5]. In EU freshwaters, mainly Hg, besides a few other ubiquitous priority pollutants, is responsible for the failure of the good chemical status of the water bodies in many member states [6]. Thermal industrial processes including steel production and especially sinter plants are significant emitters of elemental Hg [2,7]. Sintering means the agglomeration of iron ore in combination with other elements, which is then used in blast furnaces. This leads to the emission of high amounts of flue gas, accounting for approximately 40% of the total waste gas volume in the iron and steel industry. Due to the high temperatures, some volatile heavy metals, like Hg, are released from the feed material to the exhaust gas stream. The flue gases also contain dust, CO_X, NO_X, polychlorinated biphenyls (PCBs), polycyclic aromatic hydrocarbons (PAHs), polychlorinated (PCDDs) as well as polybrominated dibenzodioxins (PBDDs) and furans (PCDFs), acid gases (SO_X, HF, HCl, etc.), alkali metals, organic carbon and other pollutants. For the removal of the above-mentioned components from the flue gas of sinter plants, a very efficient dry gas cleaning system, such as the MEROS (Maximized Emission Reduction of Sintering) process, can be used. It is a dry-type gas cleaning process, using carbon-based adsorbents and either hydrated lime or sodium bicarbonate as absorbent [8]. The absorbent materials used are hazardous waste in secure landfills. Due to a high concentration of soluble salts, the material is not stable and has to be stabilized to prevent the release of Hg and other elements into the environment [9]. By leaching the by-product with water as a pretreatment before landfill disposal, the soluble Hg is transferred to the water phase and is removed from the waste [10]. Afterwards, the leached water phase has to be treated to remove Hg and other dissolved heavy metals to fulfill the emission limits of surface water [8,11,12]. In Austria, the limitation value for Hg in surface water in accordance with the General Wastewater Emission Regulation is 10 $\mu g\ L^{-1}$ (unfiltered) [13]. The Austrian Quality Target Ordinances for Chemistry of surface water prescribes a maximal allowed Hg concentration of 0.07 $\mu g\ L^{-1}$ (filtered with a pore size of 0.45 μm) [14].

Conventional techniques for the removal of heavy metals from industrial wastewater include chemical precipitation, adsorption, floatation, ion exchange, coagulation/flocculation and electrochemical processes [15]. Chemical precipitation is one of the most mature and cost-effective methods, including hydroxide precipitation, carbonate precipitation and sulfide precipitation. Metal sulfide precipitation is a crucial treatment in industrial wastewater as well as extractive metallurgy [16,17]. Sulfide sources such as solid FeS, CaS and Na_2S, aqueous NaHS, NH_4S and gaseous H_2S are the most commonly used reagents [18]. To control the sulfide concentration in the solution and to prevent an overdosage on an industrial scale, thioacetamide (CH_3CSNH_2) is mainly applied as an addition to the precipitator [19,20]. Many studies have proven the feasibility of sulfide precipitation for heavy metals in synthetic solutions [15–26]. Most of the studies do not address Hg or their application in high salt-containing wastewater with a high ionic strength and other precipitable heavy metals at high pH values of 8–8.5.

This study investigates the uptake of Hg from by-products of dry gas cleaning plants by the water phase (eluates) using leaching and how the Hg concentrations were distributed in the wastewater produced by different charges of dry gas cleaning dust. Additionally, two different precipitation agents, Na_2S and an organic sulfide ($NaS_2CN(C_2H_5)_2$), for the removal of Hg from the leaching eluates were compared using calculated removal rates and discharge limits. Due to the high salt amounts and the presence of other precipitable heavy metals in the eluates, side reactions of the precipitators can occur, and the optimal dosing of the precipitator has to be found to reach an effective Hg removal and to avoid the re-dissolution of HgS. Two different calculation strategies of the precipitator dosing were applied. To investigate the efficiency of the Na_2S treatment and to provide different handling options for the industry, Na_2S was applied as a solid and as a prepared solution and compared to the Hg removal efficiency of an organic sulfide as solution. The mechanism of Hg removal in this study is the precipitation with S. Compared to the literature, Hg was always treated in synthetic solutions or wastewater without high ionic strength. In this study, high salt-containing wastewater samples were treated, and the results are compared with the literature.

2. Materials and Methods

2.1. Description of the Sinter Plant

Solid residues of different charges were collected from the dry gas cleaning system of a sinter plant. The following design parameters give an overview of the plant size. The sinter strand is 250 m^2, the sinter capacity is 7920 t d^{-1} and the waste gas stream is approximately 600,000 Nm^3 h^{-1}. The designated desulfurization rate and by-product formation is between 60% and 70% and approx. 620 kg h^{-1}, respectively.

2.2. Sampling

The samples were taken as grab samples from different charges in the period from 2019 to 2021. From sample A-D1, an amount of 120 kg was directly delivered from the production site to the university and stored in 4 different barrels in the technical lab of the SIG Institute of the University of Natural Resources and Life Sciences Vienna. The samples A-D2 to A-D4 were submitted in sizes of about 2 kg in plastic bottles.

2.3. Production of the Eluates and Their Quality

For the water-soluble phase of the samples, 1:10 (mass per volume) eluates were prepared. First, 100 g of dust was weighted in a volumetric flask and filled up to 1000 mL with deionized water. The mixtures were stirred for 120 min by magnetic stirrers at room temperature (20 °C). After sedimentation of the undissolved particles, the samples were decanted and filtered by membrane pressure filtration (cellulose nitrate, 47 mm diameter, pore size of 0.45 μm; 6 bar). These filtered 1:10 solutions in water are called eluates (e.g., A stands for the plant, Dn for the number of dust and E for the eluate: A-Dn-E). These eluates have an initial pH of 8 to 8.5 (20 °C); the quality of the eluates (ions and heavy metals) is shown in Table 1. In a screening, the ions were measured by ion chromatography and the heavy metals by ICP-MS, as described in Section 2.7. The Hg concentration in Table 2 is a calculated mean value of six Hg measurements (see Figure 1).

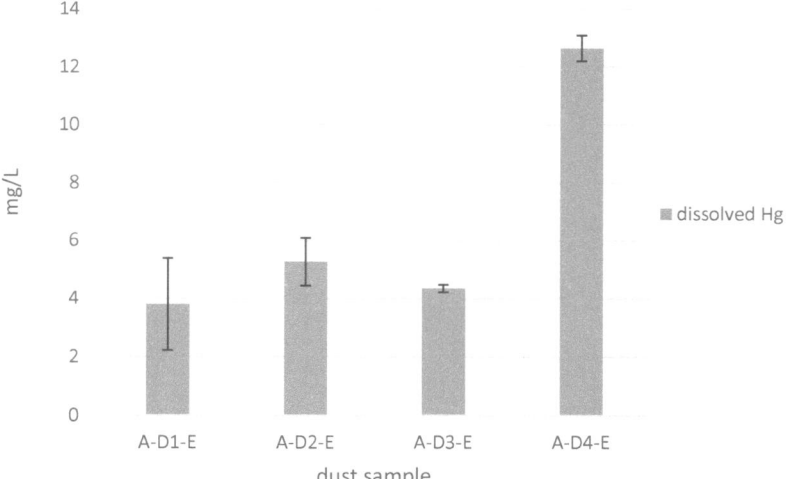

Figure 1. Distribution of dissolved Hg in the eluates of the dry gas cleaning dusts of the sinter plant. Values given as mean values (6 samples) and standard deviation.

Table 1. Anion and cation concentrations in the eluates of the dust samples used.

Ions (g L^{-1})	A-D1-E	A-D2-E	A-D3-E	A-D4-E
Cl	13	6.7	10	13
K	10	11	9.5	10
SO$_4$	44	45	45	44
Na	24	23	21	24
F	0.09	0.056	0.060	0.11
Br	0.18	0.21	0.22	0.18
NO$_3$-N	0.034	n. a.	0.046	0.044
Mg	0.018	0.48	<0.0020	0.025
NH$_4$-N	<0.0050	0.22	0.091	<0.0050
Ca	0.10	0.23	n. a.	0.060

Note: n. a.: not analyzed.

Table 2. Heavy metals concentrations in the eluates of the dust samples used.

Heavy Metals (mg L^{-1})	A-D1-E	A-D2-E	A-D3-E	A-D4-E
Pb	0.31	0.026	0.80	0.38
Cd	1.2	0.2	1.0	0.00024
Cu	1.1	0.12	0.069	0.42
Fe	0.17	0.014	0.0058	0.0058
Zn	7.1	0.21	1.4	2.9
Hg	3.8 ± 1.5	5.6 ± 0.8	4.3 ± 0.1	12.5 ± 0.4
Mn	2.2	2.0	2.0	2.1
Cr	0.0018	0.0049	0.038	1.8
As	0.28	0.16	0.016	0.21
V	0.096	0.0017	0.0028	0.042
Tl	3.0	0.24	0.59	1.3
Mo	1.5	1.1	1.1	1.2

We found that 85% to 88% of dusts A-D1 to A-D3 and 92% of dust A-D4 are water soluble. K and Na made up the highest number of cations and Cl and SO$_4$ made up the highest number of anions in the solutions. Although there are differences within the composition of the charges, the salts are in the same high-concentration ranges.

The heavy metal concentrations show a high variability between the samples of different charges with potentially high dissolved concentrations (>1000 µg L^{-1}) of Pb, Cd, Cu, Zn, Mn, Cr, Tl and Mo and especially Hg.

2.4. Hg Removal Methods

Two different chemical agents were used for the removal of dissolved Hg. The dose of the precipitator was calculated either based on the molecular concentration of Hg including different equivalent concentrations between Hg and the precipitators or on the molar concentration of the heavy metals with a low water solubility in the sulfide form. As precipitator, Na$_2$S as solid or in solution (1 g L^{-1}) as well as an organic sulfide solution (30% NaS$_2$CN(C$_2$H$_5$)$_2$) were used. The theoretical dosage of each precipitator was calculated according to two strategies presented below.

2.5. Calculation Strategies of the Theoretical Precipitator Dosage and the Removal Rate

The theoretical dosage of precipitators was calculated according to two strategies.

Strategy 1: The calculation was based on the initial dissolved Hg concentration in the eluates. Theoretically, 1 mol of Hg^{2+} consumes 1 mol of S^{2-} and produces 1 mol of HgS. The mass of the precipitator required by the theory was obtained according to the following Equation (1):

$$m(precipitator) = \frac{m\left(Hg^{2+}\right)}{M\left(Hg^{2+}\right)} \times V \times M(precipitator) \times n \quad (1)$$

Strategy 2: The calculation was based on the initial concentration of precipitable heavy metals (Pb, Cd, Cu, Fe, Zn, Hg, Mn, Ni, Cr, Tl and Mo) in the eluates. For the calculation of the amount of precipitator, the following Equation (2) was used:

$$m(precipitator) = \sum \frac{m_{heavy\ metals}}{M_{heavy\ metals}} \times V \times M(precipitator) \times n \quad (2)$$

The dosage of reagents is indicated, e.g., n4, which means 4 times the theoretical mass of Na_2S required to precipitate all Hg as well as other heavy metals in the eluate.

The removal rate (R) was calculated according to the following Equation (3):

$$R = 100\% \times \left(\frac{C_0 - C_A}{C_0}\right) \quad (3)$$

In all equations, "m" represents the mass of substance with the unit mg; "c" represents the concentration in mg L^{-1}; "M" represents the molar mass with the unit g mol^{-1}; "V" represents the volume in mL; "n" is the ratio between the mole Hg and the mole precipitator; "C_0" represents the initial concentration; and "C_A" represents the final dissolved Hg concentration in μg L^{-1} after sulfide precipitation and filtration.

2.6. Treatments for the Removal of Hg

All removal experiments were carried out in laboratory scale under the same operational conditions: 20 °C, initial pH of the eluates 8 to 8.5, 120 mL sample volume, 20 min mixing time and 40 min of sedimentation. The solids were removed by pressure filtration (pore size 0.45 μm). Table 3 shows the calculated molar ratios for the experiments carried out during this research. For all experiments, the Hg concentrations in the untreated eluates were determined, too.

Table 3. Overview of all treatments to reduce the Hg concentration in the eluates.

Experimental Run	Sample	Treatment	Ratio 1	Ratio 2	Calculation Strategy
Run 1		Na_2S	n0.7	n4	
Run 2	A-D1-E	Na_2S	n1.1	n6	Strategy 1
Run 3		Na_2S	n1.8	n10	
Run 4	A-D1-E	solid Na_2S	n3.3	n19	
Run 5		Na_2S sol.	n3.3	n19	
Run 6	A-D2-E	solid Na_2S	n1.6	n5	
Run 7		Na_2S sol.	n1.6	n5	
Run 8	A-D3-E	solid Na_2S	n3.1	n12	
Run 9		Na_2S sol.	n3.1	n12	
Run 10	A-D4-E	solid Na_2S	n2.1	n8	
Run 11		Na_2S sol.	n2.1	n8	Strategy 2
Run 12	A-D1-E	solid Na_2S	n3.3	n19	
Run 13		organic sul.	n3.3	n19	
Run 14	A-D2-E	solid Na_2S	n1.1	n6	
Run 15		organic sul	n1.1	n6	
Run 16	A-D3-E	solid Na_2S	n1.9	n11	
Run 17		organic sul	n1.9	n11	
Run 18	A-D4-E	solid Na_2S	n1.4	n8	
Run 19		organic sul	n1.4	n8	

Notes: organic sul: organic sulfide solution. Ratio 1: ratio referring to all heavy metals. Ratio 2: ratio referring to Hg.

Run 1 to Run 3 tested the effect of the addition of different equivalents of Hg and solid sodium sulfide (Na$_2$S water-free): 1:4, 1:6 and 1:10. The solid Na$_2$S treatment applied in this research was based on the studies of AHM Veeken (2003) [27], Li et al. (2019) [22] and Prokkola (2020) [23]. The studies confirmed the feasibility of using sulfide precipitation to remove heavy metals in a multi-metals system from different wastewaters, especially Hg [28]. Run 4 to Run 11 were designed to study the Hg removal efficiency of sulfide precipitates by using solid Na$_2$S and a Na$_2$S solution (1 g L^{-1}).

In Run 12 to 19, the efficiency of Hg removal by the treatment with solid Na$_2$S compared to organic sulfide (NaS$_2$CN(C$_2$H$_5$)$_2$) solution was examined.

2.7. Analytical Methods

The dissolved fractions (eluate A-Dn-E) were analyzed for selected ions and dissolved heavy metals. The heavy metal components were determined by means of ionized plasma and detection by mass spectroscopy (ICP-MS, ELAN DRC-e, Perkin Elmer, Waltham, MA, USA) and then quantified according to DIN EN ISO 17294 [29]. The samples were acidified with 2% HNO$_3$ suprapure, and for the Hg measurement, the samples were stabilized with 5 ppb Au solution additionally. Several dilutions of the samples were analyzed to cover the different concentration ranges and stay within the working range of the various ions and heavy metals, and the limits of quantification (LOQ) are listed in Table 4.

Table 4. Limits of quantification for selected heavy metals.

Heavy Metals (µg L^{-1})	LOQ
Pb	0.50
Cd	0.05
Cu	0.50
Fe	1.0
Zn	5.0
Hg	0.10
Mn	0.50
Cr	0.50
As	0.50
V	1.0
Tl	1.0
Mo	1.0

Ions were separated by a liquid chromatograph (DIONEX ICS 3000, DIONEX Softron, Germering, Germany) equipped with an autosampler, suppressor and conductivity detector. For cation separation, a CS12A 250 mm × 2 mm + CG12A 50 mm × 2 mm column was used, and for anion separation, an AS15 250 mm × 2 mm + AG15 50 mm × 2 mm column was used. The eluent (mobile phase) for the cation determination was methane sulfonic acid solution, and for the anion determination, it was potassium hydroxide solution. All standards were prepared using stock standards (Merck, Darmstadt, Germany), and dilutions and solutions were prepared with deionized water. Several dilutions of the samples were analyzed to cover the different concentration ranges and stay within the working range of the various ions. The dilutions resulted in different limits of quantification listed in Table 5.

Table 5. Limits of quantification of selected ions.

Ions (mg L^{-1})	LOQ
Cl	0.25–3.0
NO$_3$-N1	0.22–2.5
SO$_4$	1.0–10
PO$_4$-P	0.65–4.9
F	0.1–3.0
Na	0.25–6.5
Ca	0.25–6.5
Mg	0.25–6.5
NH$_4$-N	0.78–14.8

3. Results and Discussion

3.1. Dissolved Hg Distribution within the Eluates

Over a time period of one year, six eluates per dust were produced, and the Hg concentrations were measured directly after the production. Figure 1 shows the distribution of dissolved Hg in the eluates of samples A-D1-E to A-D4-E.

The eluates contain Hg concentrations between 3.7 and 12.6 mg L^{-1}. Differences can be determined between and within the charges of the plant (for A-D1-E, between 1.4 and 4.7 mg L^{-1} (Table 6) and the standard deviation in Figure 1) that were taken over a time period of one year. The mean value and standard deviation for all samples were calculated with six values. Especially the sample A-D4-E contains a high concentration of dissolved Hg (12.6 mg L^{-1}). The Hg concentrations in all eluates exceed the emission limit value of 0.005 mg L^{-1} and have to be reduced. Although the dusts were sampled in small amounts (sample A-D1, A-D2 and A-D4), the results show inhomogeneities and high standard deviations within the dust samples.

Table 6. List of all experimental runs with related sample used, precipitation agent and applied ratio, the initial Hg concentration of the sample as well as the reached Hg concentration after the treatment and the calculated removal rate.

Run	Sample	Treatment	Ratio 1	Ratio 2	Initial Hg (mg/L)	Final Hg (mg/L)	Removal Rate (%)
Run 1		Na$_2$S	n0.7	n4		0.0078	99.7
Run 2	A-D1-E	Na$_2$S	n1.1	n6	2.4	0.0056	99.8
Run 3		Na$_2$S	n1.8	n10		0.0056	99.8
Run 4	A-D1-E	solid Na$_2$S	n3.3	n19	4.7	0.021	99.6
Run 5		Na$_2$S sol.	n3.3	n19		0.015	99.7
Run 6	A-D2-E	solid Na$_2$S	n1.6	n5	5.6	0.0060	99.9
Run 7		Na$_2$S sol.	n1.6	n5		0.0083	99.9
Run 8	A-D3-E	solid Na$_2$S	n3.1	n12	4.3	0.0018	99.9
Run 9		Na$_2$S sol.	n3.1	n12		0.013	99.7
Run 10	A-D4-E	solid Na$_2$S	n2.1	n8	12.6	0.014	99.9
Run 11		Na$_2$S sol.	n2.1	n8		0.031	99.7
Run 12	A-D1-E	Solid Na$_2$S	n3.3	n19	1.4	0.010	99.3
Run 13		organic sul.	n3.3	n19		0.018	98.7
Run 14	A-D2-E	Solid Na$_2$S	n1.1	n6	3.8	0.070	98.1
Run 15		organic sul	n1.1	n6		0.29	92.2
Run 16	A-D3-E	Solid Na$_2$S	n1.9	n11	4.5	0.00098	99.9
Run 17		organic sul	n1.9	n11		0.016	99.6
Run 18	A-D4-E	Solid Na$_2$S	n1.4	n8	12.9	0.054	99.6
Run 19		organic sul	n1.4	n8		1.6	87.5

Notes: organic sul: organic sulfide solution. Ratio 1: ratio referring to all heavy metals. Ratio 2: ratio referring to Hg.

The variation in the Hg concentrations can be explained by differences in the raw materials, especially the lignite coal and the inhomogeneity in the dusts caused during formation [10]. The Hg amount in the lignite coal depends on the origin; many authors have presented data between 0.01 and 1.5 mg kg^{-1} lignite coal depending on the country of origin [30–32]. The literature presents studies where leaching is used to remove Hg from contaminated soil. But there is a lack of studies where leaching is used as a Hg removal method of the specific by-products of dry gas cleaning plants. Leaching as pretreatment before depositing avoids the necessity of removing the Hg from the contaminated soil [33–35].

3.2. Treatments for Removal of Hg

Table 6 shows the results of the treatment experiments for the Hg removal with Na_2S and organic sulfide solution. The results of Run 1 to Run 3 indicate the effect of different Na_2S concentrations on the Hg removal. The dosage of the two precipitators in Run 1 to 3 was calculated according to Strategy 1 and in Run 4 to 19 according to Strategy 2; for a better comparison, both calculated ratios are given in Table 6.

Generally, higher ratios resulted in better removal rates of Hg with Na_2S. The addition of solid Na_2S resulted in mostly lower final concentrations than the addition of a prepared Na_2S solution. By comparing the final Hg concentrations after Na_2S treatment to that after organic sulfide addition, the values are always lower after the Na_2S treatment. This indicates that treatments with solid Na_2S are more effective. In practical applications, treating with a Na_2S solution increases the total amount of water volume by approximately 0.7% to 2%. Given that a substantial amount of water is already required to dissolve the dust, this would result in a waste of water resources. The use of solid Na_2S gave slightly better results, but the technical feasibility for the application and dosage for treatment needs to be evaluated. Unfortunately, a high removal rate does not result in fulfilling the different national emission standards. Even if the removal rate of the same dust sample reached >99%, the methods have to be adjusted and optimized to reach the limited emission values of 0.005 or 0.01 mg L^{-1} for the discharge into a receiving water body. Therefore, the removal rate is not sufficient to assess the appropriateness of the removal methods. Although a higher Hg removal with Na_2S could be reached, for the application at an industrial scale, the use of organic sulfide solution is recommended, because under acidic conditions, toxic H_2S gas is released from Na_2S. As the treatment with organic sulfide obtained removal rates over 87%, with an optimization and adjustment, the limitation values could probably be reached, too. Although the focus was on the precipitation of Hg, other water-insoluble sulfides of heavy metals in the eluates need to be considered. Therefore, the resulting final concentrations of Hg in relation to the applied sulfide ratios according to the two calculation strategies are shown in Figure 2.

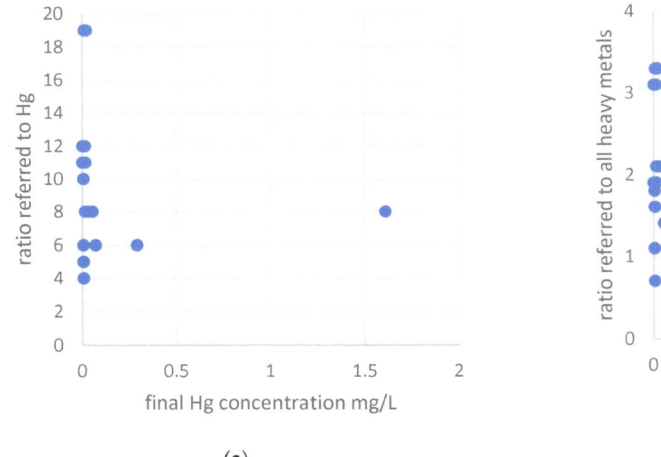
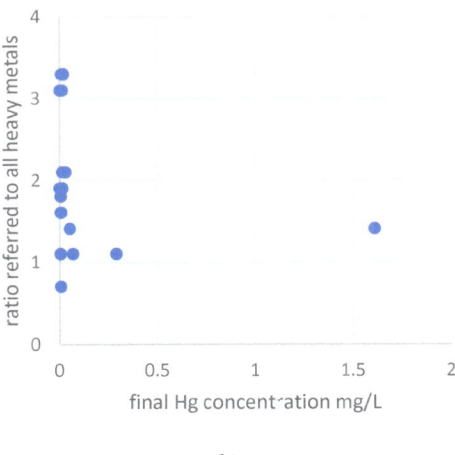

Figure 2. (a) Relation between the final dissolved Hg concentrations and the ratio referred to the initial Hg concentration; (b) relation between the final dissolved Hg concentrations and the ratio referred to the initial concentrations of precipitable heavy metals. Y-axis has different scales.

Figure 2 demonstrates that the consideration of the concentrations of relevant heavy metals including Hg gives more consistent data, as they are also relevant because they react with sulfide, too. Figure 2 shows that ratios in which all heavy metals are below 1.5 do not result in sufficient low Hg concentrations. Therefore, the optimized dosage has to be calculated according to Strategy 2, referring to all heavy metals in ratio exceeding 1.5. In the study of Han et al. (2014), Hg was removed from synthetic wastewater by FeS particles [36]. The molar ratios used of Hg and FeS were n 1, n 2 and n 2.5. With the ratio n 1, more than 99% of Hg could be removed. The ratios n 2 and 2.5 showed removal rates of 96% to 97%. Summarizing, Han et al. found that a ratio of 1 resulted in a better Hg removal compared to the higher doses. Our study showed that the ratios referring to Hg are not consistent for an effective Hg removal. Our results are not in line with the study of Han et al., but the important difference is the composition of the wastewater used. For the experiments of Han et al. (2014) [36], synthetic solutions were used without any other substances that could react with sulfide. These results confirm the hypothesis that in real wastewater with a high ionic strength and other heavy metals, sulfide also precipitated with the heavy metals, and less of the total added sulfide is available for Hg [17,18,37]. The study of Chai et al. (2010) showed a Hg removal with Na_2S from 48 to 0.12 mg L^1 in a synthetic solution without any other sulfide reactants contained and a molar Hg to Na_2S ratio of 1:16 at pH 9 [38]. These results are in line with the removal effort of our study. Hg can also form complexes with salts like Cl and the soluble $HgCl_2$ complex, and due to that, not all Hg ions are available for the precipitation with sulfide [39,40]. Pb, Cd, Cu Fe and Zn react with sulfide too, and according to their low chemical solubility, they form insoluble sulfide precipitates and thereby can be removed from the wastewater by filtration or sedimentation afterwards [41]. It is an effective technique for the Hg removal but sensitive to overdosage due to the formation of soluble Hg–polysulfide complexes [37,39,42]. Sulfide precipitates are not amphoteric, so a high degree of metal removal in a shorter time over a wide pH range can be achieved [17,18].

A number of studies showed that Na_2S is also a method used to remove other heavy metals (for example, Zn, Cu, and Pb). Table 7 provides measurements for different HMs for Run 1 to 3. For these experiments, relevant heavy metal concentrations were measured in the untreated eluates as well as after the treatment with solid Na_2S in different ratios (Table 7 and Figure 3).

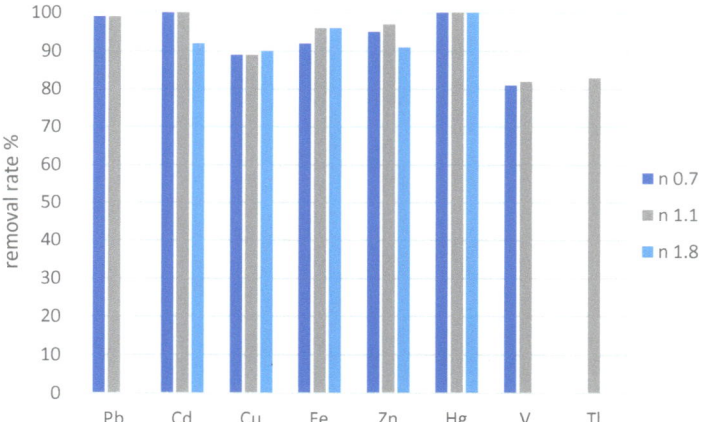

Figure 3. Removal rates of selected heavy metals after treatment with solid Na_2S in different concentrations in sample A-D1-E.

Table 7. Dissolved concentrations of heavy metals in eluate A-D1-E before and after treatment with different concentrations of solid Na_2S, and the ratios refer to all heavy metals.

Heavy Metal (mg L^{-1})	Initial Conc.	Final Concentrations		
	A-D1-E	n0.7	n1.1	n1.8
Pb	0.31	0.0027	0.0016	n.a.
Cd	1.2	0.0048	0.0051	0.093
Cu	1.1	0.012	0.11	0.11
Fe	0.17	0.013	0.0062	0.0068
Zn	7.1	0.37	0.018	0.63
Hg	2.4	0.0078	0.0056	0.0056
V	0.096	0.018	0.017	n.a.
Tl	3.0	n.a.	0.51	n.a.

Notes: n.a: not analyzed.

Table 7 and Figure 3 show that a higher ratio of solid Na_2S does not necessarily lead to a higher removal of HM. A Na_2S ratio of 1.1 resulted in the removal rates of Hg, Pb, Cd, and Zn being higher than 95%, while those of Fe and Cu were higher than 90% and V and that for Tl was over 80%. For Hg, the ratio (Strategy 2) should be more than 1.5, but the application of an excessive amount of Na_2S would cause material waste and a secondary pollution due to the excess of sulfide [37]. Due to the high salt concentrations and the other present metals, it is expected that the precipitants do not only form insoluble complexes with Hg; the other metals are precipitated, too. For an efficient removal, the adjustment of the precipitator in the initial concentration of all precipitable heavy metals will be necessary. Due to the formation of the soluble Hg–polysulfide complexes, the Hg concentration can increase after the precipitation again, because this complex is not stable. The results show that also other heavy metals like Cu, Zn, Ni and Sn form soluble complexes with S, which supports the reported results in the literature [24,25,43]. The work of Fukuta et al. (2004) showed removal rates of 94.5% for Cu (pH 1.5), 75.9% for Zn (pH 2.5) and 65.9% for Ni (pH 5.5–6.0) [25]. A further study of Mahdi et al. (2012) showed that 90% of Cu, Ni and Zn could be reduced from a synthetic solution with Na_2S in 30–60 min with the additional control of the sulfide concentration in the solution with thioacetamide (CH_3CSNH_2) as an addition to the precipitator to prevent the overdosage [19]. The work of Silvia et al. (2017) presents Cu, Zn and Ni removal over 90% with H_2S gas [20]. The results of both studies are comparable with our study and show that the use of H_2S is as efficient as Na_2S for the removal of Cu, Zn and Ni. In another study, heavy oil fly ash with over 9.0% (weight per weight) of sulfur content has removed 99.99% of Cu from a synthetic solution under optimal operational conditions [26]. The removal treatments in all of the mentioned studies were applied to synthetic solutions, and different sulfide sources were used. The comparison of the results of the experiments of this study show that the removal rates are in the same range, although the wastewater used in our study had a high ionic strength. The concentrations of salts in the treated solution did not have an impact on the efficiency of heavy metal removal by sulfide precipitation from different sulfide sources, but as the comparison with the study of Han et al. showed, other heavy metals react with sulfide too and are removed as precipitates [36].

3.3. Cost Analysis of the Hg Removal Process

The amount of precipitator used in all of the experimental Runs is projected to the necessary amount for an industrial-scale plant, and the costs are calculated for each precipitator. The gas cleaning plant produces 620 kg by-product per hour. By dissolving this amount in a 1:10 dilution in water, approximately 6200 L eluate is produced per hour. The average price of Na_2S with a purity +90% is 7440€ kg^{-1} (Merck), and for organic sulfide ($NaS_2CN(C_2H_5)_2$) with a purity +90%, it is 396 € kg^{-1} (Merck). Table 8 shows the costs for each Run.

Table 8. Statement of costs for the Hg removal for 1 h.

Na2S		Organic Sulfide	
Run 1	369 € h^{-1}	Run 13	16 € h^{-1}
Run 2	600 € h^{-1}	Run 15	5 € h^{-1}
Run 3	876 € h^{-1}	Run 17	9 € h^{-1}
Run 4	1661 € h^{-1}	Run 19	7 € h^{-1}
Run 5	1661 € h^{-1}		
Run 6	461 € h^{-1}		
Run 7	461 € h^{-1}		
Run 8	1015 € h^{-1}		
Run 9	1015 € h^{-1}		
Run 10	692 € h^{-1}		
Run 11	692 € h^{-1}		
Run 12	1661 € h^{-1}		
Run 14	554 € h^{-1}		
Run 16	876 € h^{-1}		
Run 18	692 € h^{-1}		

The cost analysis shows that the use of Na$_2$S is more expensive than the organic sulfide. In Run s 13 and 17, low final Hg concentrations of 18.2 µg L^{-1} and 15.9 µg L^{-1} Hg were reached. These results conform the use of organic sulfide for the industrial application.

4. Conclusions

The eluate produced from dry gas cleaning dusts in this study contains 85% to 92% of salts, and therefore the reaction might not be easily predicted from chemical reaction constants. Measurement of different eluates, produced separately from the same dust samples at different times, resulted in a high standard deviation of the dissolved Hg concentrations, and this was also true within single samples. High differences were also observed between different sample charges of the considered plant. Therefore, intensive investigations of the relevant heavy metals in the eluates are required for the optimized dosage of sulfide precipitators.

All dissolved Hg concentrations in the eluates exceeded the Austrian emission standard for surface water of 0.010 µg L^{-1}. Due to this fact, the Hg concentration has to be reduced before discharging the wastewater into surface water.

The treatment with Na$_2$S and organic sulfide solution showed high Hg removal rates between 87.5% and 99.99%, depending on the type of precipitator and its dosage as well as the initial Hg concentrations. Sulfide does not only react with Hg; there are also other relevant heavy metals that have to be considered in the calculation of the optimal amount of precipitator; but an overdosing of the precipitator can re-dissolve Hg.

The testing of different ratios between heavy metals and the precipitator showed that a ratio of more than 1.5 (according to Strategy 2) is necessary to reduce Hg appropriately. With Na$_2$S, a better Hg removal was reached compared to the organic sulfide solution, but for the application at an industrial scale, the use of organic sulfide solution is recommended because under acidic conditions, Na$_2$S reacts to toxic H$_2$S gas.

Na$_2$S can be applied as a solid but also as prepared solution. The solution would increase the amount of water up to 2%, but as a solution, the adding of the precipitator in the wastewater and the homogenization would be easier at a technical scale. For the evaluation of the removal performance, the removal rate is not an appropriate parameter, but the final dissolved concentrations must be used. The precipitation experiments in this study showed that with Na$_2$S treatment, it is possible to reach the effluent emission limit for Hg for different countries.

In addition to Hg, some other heavy metals precipitate with sulfide too and can be removed with Na$_2$S from the wastewater produced.

The cost analysis confirms the use of org. sulfide, because it is less expensive, and with optimization, it can remove Hg efficiently as well.

Author Contributions: Conceptualization, M.F., C.H. and Y.Z.; data curation, C.H. and Y.Z.; funding acquisition, T.P.; writing—review and editing all authors, project administration, M.F. All authors have read and agreed to the published version of the manuscript.

Funding: This research was funded by FFG (Österreichische Forschungsfördergesellschaft) within the COMET Competence Centers for Excellent Technologies—COMET Centre (K1-Met), FFG Project No.: 869295. Open access funding provided by University of Natural Resources and Life Sciences Vienna (BOKU).

Data Availability Statement: The data presented in this study are available on request from the corresponding author. The data are not publicly available.

Acknowledgments: This study was elaborated within the COMET Competence Centers for Excellent Technologies—COMET Centre (K1-Met) and financially supported by the FFG (Österreichische Forschungsfördergesellschaft).

Conflicts of Interest: Author Tobias Plattner was employed by the company Primetals Technologies Austria GmbH. The remaining authors declare that the research was conducted in the absence of any commercial or financial relationships that could be construed as a potential conflict of interest.

References

1. Tuzen, M.; Sarı, A.; Mogaddam, M.R.A.; Kaya, S.; Katin, K.P.; Altunay, N. Synthesis of carbon modified with polymer of diethylenetriamine and trimesoyl chloride for the dual removal of Hg (II) and methyl mercury ([CH_3Hg]$^+$) from wastewater: Theoretical and experimental analyses. *Mater. Chem. Phys.* **2022**, *277*, 125501. [CrossRef]
2. Ariya, P.A.; Peterson, K.A. Chemical transformation of gaseous elemental Hg in the atmosphere. In *Dynamics of Mercury Pollution on Regional and Global Scales*; Springer: Boston, MA, USA, 2005; pp. 261–294.
3. Chen, C.Y.; Driscoll, C.T.; Eagles-Smith, C.A.; Eckley, C.S.; Gay, D.A.; Hsu-Kim, H.; Keane, S.E.; Kirk, J.L.; Mason, R.P.; Obrist, D.; et al. A critical time for mercury science to inform global policy. *Environ. Sci. Technol.* **2018**, *52*, 9556–9561. [CrossRef] [PubMed]
4. Sakamoto, M.; Nakamura, M.; Murata, K. Mercury as a global pollutant and mercury exposure assessment and health effects. *Nihon Eiseigaku Zasshi. Jpn. J. Hyg.* **2018**, *73*, 258–264. [CrossRef] [PubMed]
5. Feng, X.; Li, P.; Fu, X.; Wang, X.; Zhang, H.; Lin, C.-J. Mercury pollution in China: Implications on the implementation of the Minamata Convention. *Environ. Sci. Process. Impacts* **2022**, *24*, 634–648. [CrossRef] [PubMed]
6. European Environment Agency. *Impact of Mercury on European Water Quality*; European Environment Agency: Copenhagen, Denmark, 2018.
7. Carpi, A. Mercury from combustion sources: A review of the chemical species emitted and their transport in the atmosphere. *Water Air Soil Pollut.* **1997**, *98*, 241–254. [CrossRef]
8. Fleischanderl, A.; Steinparzer, T.; Plattner, T.; Neuhold, R.; Goetz, M. Green Solutions for Iron Ore Agglomeration Off-gas Treatment and By-Product Utilization. In Proceedings of the ESTAD 2023, Düsseldorf, Germany, 12–16 June 2023.
9. Song, M.; Liu, J.; Xu, S. Characterization and solidification/stabilization of iron-ore sintering gas cleaning residue. *J. Mater. Cycles Waste Manag.* **2015**, *17*, 790–797. [CrossRef]
10. Mukherjee, A.B.; Zevenhoven, R.; Bhattacharya, P.; Sajwan, K.S.; Kikuchi, R. Mercury flow via coal and coal utilization by-products: A global perspective. *Resour. Conserv. Recycl.* **2008**, *52*, 571–591. [CrossRef]
11. Hledik, C.; Goetz, M.; Ottner, F.; Fürhacker, M. MEROS Dust Quality of Different Plants and Its Potential Further Uses. *Metals* **2021**, *11*, 840. [CrossRef]
12. Schroeder, W.H.; Munthe, J. Atmospheric mercury—An overview. *Atmos. Environ.* **1998**, *32*, 809–822. [CrossRef]
13. Bundesministerium für Landwirtschaft, Regionen und Tourismus. *Verordnung des Bundesministers für Land- und Forstwirtschaft über die Allgemeine Begrenzung von Abwasseremissionen in Fließgewässer und öffentliche Kanalisationen (Allgemeine Abwasseremissionsverordnung–AAEV)*; Bundesministerium für Landwirtschaft, Regionen und Tourismus: Vienna, Austria, 1996.
14. Bundesgesetzblatt für die Republik Österreich. *Verordnung des Bundesministers für Land- und Forstwirtschaft, Umwelt und Wasserwirtschaft über die Festlegung des Zielzustandes für Oberflächengewässer (Qualitätszielverordnung Chemie Oberflächengewässer–QZV Chemie OG)*; Bundesgesetzblatt für die Republik Österreich: Vienna, Austria, 2006.
15. Shrestha, R.; Ban, S.; Devkota, S.; Sharma, S.; Joshi, R.; Tiwari, A.P.; Kim, H.Y.; Joshi, M.K. Technological trends in heavy metals removal from industrial wastewater: A review. *J. Environ. Chem. Eng.* **2021**, *9*, 105688. [CrossRef]
16. Estay, H.; Barros, L.; Troncoso, E. Metal sulfide precipitation: Recent breakthroughs and future outlooks. *Minerals* **2021**, *11*, 1385. [CrossRef]
17. Pohl, A. Removal of heavy metal ions from water and wastewaters by sulfur-containing precipitation agents. *Water Air Soil Pollut.* **2020**, *231*, 503. [CrossRef]

18. Lewis, A.E. Review of metal sulphide precipitation. *Hydrometallurgy* **2010**, *104*, 222–234. [CrossRef]
19. Gharabaghi, M.; Irannajad, M.; Azadmehr, A.R. Selective Sulphide Precipitation of Heavy Metals from Acidic Polymetallic Aqueous Solution by Thioacetamide. *Ind. Eng. Chem. Res.* **2012**, *51*, 954–963. [CrossRef]
20. Silva, P.M.; Raulino, G.S.; Vidal, C.B.; do Nascimento, R.F. Selective precipitation of Cu^{2+}, Zn^{2+} and Ni^{2+} ions using H_2S (g) produced by hydrolysis of thioacetamide as the precipitating agent. *Desalination Water Treat.* **2017**, *95*, 220–226. [CrossRef]
21. Grau, J.; Akinc, M. Synthesis of nickel sulfide by homogeneous precipitation from acidic solutions of thioacetamide. *J. Am. Ceram. Soc.* **1996**, *79*, 1073–1082. [CrossRef]
22. Li, H.; Zhang, H.; Long, J.; Zhang, P.; Chen, Y. Combined Fenton process and sulfide precipitation for removal of heavy metals from industrial wastewater: Bench and pilot scale studies focusing on in-depth thallium removal. *Front. Environ. Sci. Eng.* **2019**, *13*, 49. [CrossRef]
23. Prokkola, H.; Nurmesniemi, E.-T.; Lassi, U. Removal of metals by sulphide precipitation using Na_2S and HS^--solution. *ChemEngineering* **2020**, *4*, 51. [CrossRef]
24. Tokuda, H.; Kuchar, D.; Mihara, N.; Kubota, M.; Matsuda, H.; Fukuta, T. Study on reaction kinetics and selective precipitation of Cu, Zn, Ni and Sn with H2S in single-metal and multi-metal systems. *Chemosphere* **2008**, *73*, 1448–1452. [CrossRef] [PubMed]
25. Fukuta, T.; Ito, T.; Sawada, K.; Kojima, Y.; Matsuda, H.; Seto, F. Separation of Cu, Zn and Ni from plating solution by precipitation of metal sulfides. *Kagaku Kogaku Ronbunshu* **2004**, *30*, 227–232. [CrossRef]
26. Rostamnezhad, N.; Kahforoushan, D.; Sahraei, E.; Ghanbarian, S.; Shabani, M. A method for the removal of Cu (II) from aqueous solutions by sulfide precipitation employing heavy oil fly ash. *Desalination Water Treat.* **2016**, *57*, 17593–17602. [CrossRef]
27. Veeken, A.; De Vries, S.; Van der Mark, A.; Rulkens, W. Selective precipitation of heavy metals as controlled by a sulfide-selective electrode. *Sep. Sci. Technol.* **2003**, *38*, 1–19. [CrossRef]
28. Kaksonen, A.H.; Riekkola-Vanhanen, M.-L.; Puhakka, J. Optimization of metal sulphide precipitation in fluidized-bed treatment of acidic wastewater. *Water Res.* **2003**, *37*, 255–266. [CrossRef] [PubMed]
29. ISO 17294-2:2023; Water Quality—Application of Inductively Coupled Plasma Mass Spectrometry (ICP-MS)—Part 2: Determination of Selected Elements Including Uranium Isotopes. Austrian Standards; ISO: Geneva, Switzerland, 2023. Available online: https://www.iso.org/standard/82245.html (accessed on 26 June 2024).
30. Diehl, S.; Goldhaber, M.; Hatch, J. Modes of occurrence of mercury and other trace elements in coals from the warrior field, Black Warrior Basin, Northwestern Alabama. *Int. J. Coal Geol.* **2004**, *59*, 193–208. [CrossRef]
31. Kolker, A.; Senior, C.L.; Quick, J.C. Mercury in coal and the impact of coal quality on mercury emissions from combustion systems. *Appl. Geochem.* **2006**, *21*, 1821–1836. [CrossRef]
32. Park, J.Y.; Won, J.H.; Lee, T.G. Mercury analysis of various types of coal using acid extraction and pyrolysis methods. *Energy Fuels* **2006**, *20*, 2413–2416. [CrossRef]
33. Reis, A.T.; Lopes, C.B.; Davidson, C.M.; Duarte, A.C.; Pereira, E. Extraction of mercury water-soluble fraction from soils: An optimization study. *Geoderma* **2014**, *213*, 255–260. [CrossRef]
34. Xu, J.; Kleja, D.B.; Biester, H.; Lagerkvist, A.; Kumpiene, J. Influence of particle size distribution, organic carbon, pH and chlorides on washing of mercury contaminated soil. *Chemosphere* **2014**, *109*, 99–105. [CrossRef] [PubMed]
35. Xie, F.; Dong, K.; Wang, W.; Asselin, E. Leaching of mercury from contaminated solid waste: A mini-review. *Miner. Process. Extr. Metall. Rev.* **2020**, *41*, 187–197. [CrossRef]
36. Han, D.S.; Orillano, M.; Khodary, A.; Duan, Y.; Batchelor, B.; Abdel-Wahab, A. Reactive iron sulfide (FeS)-supported ultrafiltration for removal of mercury (Hg (II)) from water. *Water Res.* **2014**, *53*, 310–321. [CrossRef] [PubMed]
37. Hsu-Kim, H.; Sedlak, D.L. Similarities between inorganic sulfide and the strong Hg (II)-complexing ligands in municipal wastewater effluent. *Environ. Sci. Technol.* **2005**, *39*, 4035–4041. [CrossRef] [PubMed]
38. Chai, L.-y.; Wang, Q.-w.; Wang, Y.-y.; Li, Q.-z.; Yang, Z.-h.; Shu, Y.-d. Thermodynamic study on reaction path of Hg (II) with S (II) in solution. *J. Cent. South Univ. Technol.* **2010**, *17*, 289–294. [CrossRef]
39. Jay, J.A.; Morel, F.M.; Hemond, H.F. Mercury speciation in the presence of polysulfides. *Environ. Sci. Technol.* **2000**, *34*, 2196–2200. [CrossRef]
40. Mason, R.P.; Reinfelder, J.R.; Morel, F.M. Uptake, toxicity, and trophic transfer of mercury in a coastal diatom. *Environ. Sci. Technol.* **1996**, *30*, 1835–1845. [CrossRef]
41. Weast, R.C. *Handbook of Chemistry and Physics*; CRC Press: Boca Raton, FL, USA, 1979–1980.
42. Paquette, K.E.; Helz, G.R. Inorganic speciation of mercury in sulfidic waters: The importance of zero-valent sulfur. *Environ. Sci. Technol.* **1997**, *31*, 2148–2153. [CrossRef]
43. Kondo, H.; Fujita, T.; Kuchar, D.; Fukuta, T.; Matsuda, H.; Kubota, M.; Yagishita, K. Separation of metal sulfides from plating wastewater containing Cu, Zn and Ni by selective sulfuration with hydrogen sulfide. *J. Surf. Finish. Soc. Jpn.* **2006**, *57*, 901–906. [CrossRef]

Disclaimer/Publisher's Note: The statements, opinions and data contained in all publications are solely those of the individual author(s) and contributor(s) and not of MDPI and/or the editor(s). MDPI and/or the editor(s) disclaim responsibility for any injury to people or property resulting from any ideas, methods, instructions or products referred to in the content.

Review

Permeable Concrete Barriers to Control Water Pollution: A Review

Rehab O. Abdel Rahman [1,*], Ahmed M. El-Kamash [1] and Yung-Tse Hung [2]

1 Hot Laboratory Center, Atomic Energy Authority of Egypt, Cairo P.O. Box 13759, Egypt
2 Department of Civil and Environmental Engineering, Cleveland State University, Cleveland, OH 44115, USA
* Correspondence: alaarehab@yahoo.com; Tel.: +20-010-614-044-62

Citation: Abdel Rahman, R.O.; El-Kamash, A.M.; Hung, Y.-T. Permeable Concrete Barriers to Control Water Pollution: A Review. *Water* **2023**, *15*, 3867. https://doi.org/10.3390/w15213867

Academic Editor: Enedir Ghisi

Received: 19 September 2023
Revised: 28 October 2023
Accepted: 1 November 2023
Published: 6 November 2023

Copyright: © 2023 by the authors. Licensee MDPI, Basel, Switzerland. This article is an open access article distributed under the terms and conditions of the Creative Commons Attribution (CC BY) license (https://creativecommons.org/licenses/by/4.0/).

Abstract: Permeable concrete is a class of materials that has long been tested and implemented to control water pollution. Its application in low-impact development practices has proved its efficiency in mitigating some of the impacts of urbanization on the environment, including urban heat islands, attenuation of flashfloods, and reduction of transportation-related noise. Additionally, several research efforts have been directed at the dissemination of these materials for controlling pollution via their use as permeable reactive barriers, as well as their use in the treatment of waste water and water purification. This work is focused on the potential use of these materials as permeable reactive barriers to remediate ground water and treat acid mine drainage. In this respect, advances in material selection and their proportions in the mix design of conventional and innovative permeable concrete are presented. An overview of the available characterization techniques to evaluate the rheology of the paste, hydraulic, mechanical, durability, and pollutant removal performances of the hardened material are presented and their features are summarized. An overview of permeable reactive barrier technology is provided, recent research on the application of permeable concrete technology is analyzed, and gaps and recommendations for future research directions in this field are identified. The optimization of the mix design of permeable reactive concrete barriers is recommended to be directed in a way that balances the performance measures and the durability of the barrier over its service life. As these materials are proposed to control water pollution, there is a need to ensure that this practice has minimal environmental impacts on the affected environment. This can be achieved by considering the analysis of the alkaline plume attenuation in the downstream environment.

Keywords: permeable concrete; mix design; preamble reactive barriers; acid mine drainage; remediation; pollutant removal

1. Introduction

All life forms on our planet need water to function properly; this need is attributed to the chemical composition of this compound, in which water is a polar covalent molecule, which supports its use as a solvent for many nutrients and, subsequently, their transport throughout living organisms. In addition, water is the main ingredient used to form different fluids that are needed to protect and lubricate biological tissues; it helps in controlling the body's temperature and acts as a medium for the chemical reactions of enzymes [1]. Nearly 97.5% of this natural compound is saline, and only 0.3% is surface fresh liquid water that is distributed in lakes, rivers, swamps, soil moisture, and the atmosphere. The limited amount of easily accessible fresh water and its necessity for life's continuation were the main drivers of the identification of the provision of clean water and sanitation as one of the Sustainable Development Goals (Goal 6). In addition, the availability of freshwater of acceptable quality is the driving force to achieve Goals 2, 3, 8, and 9 (i.e., zero hunger; good health and well-being; decent work and economic growth; and industry, innovation, and infrastructure) [2]. Moreover, preventing and controlling pollution that can spread because of the improper management of solid wastes and wastewater is a key aspect of achieving Goals 11, 14, 15, and 17 (i.e., sustainable cities and communities; life below water; life on Earth; and partnership for goals) [2]. Hence, several efforts are being carried out worldwide that aim to ensure the sustainable provision of water with acceptable quality by preventing and controlling the pollution of surface and ground waters. These efforts focus on the investigation of new materials and/or systems to evaluate their potential implementation in preventing and controlling water pollution, where passive engineering barriers play an important role in this field via the application of permeable barriers that can remove pollutants from different types of water.

Conventional cement-based materials is a class of materials that have been used for decades to support human civilization, and they depend on the use of a hydraulic binder to bond fragmented particles. Upon the hydration of the binder, new hardened materials with enhanced physical and mechanical properties are formed. The properties of the fresh and hardened materials can be tolerated by changing the mix design, i.e., changing the additives and admixtures, water-to-cement ratio, finesse, and type of cement used [3]. The hydration reaction products include major hydration phases (i.e., calcium silicate hydrate (C-S-H), portlandite (CH), and ettringite (AFt)); minor hydration phases (e.g., monosulfate and hydrogarnet); and the heat of hydration. The proportion of these phases is dependent on the mix design and the curing conditions [3–6]. These hydrated phases determine the properties of the hardened cement-based materials and the evolution of these properties overtime [6,7]. Conventional cement-based materials are heavily used in the construction sector and in environmental protection and restoration. In particular, the application of these materials in environmental protection encompass many fields, including rock repair and enforcement, the design of disposal facilities for hazardous and radioactive wastes, the stabilization and solidification of hazardous and radioactive contaminants, and water and wastewater treatment [8–21]. This wide range of applications is supported by the low cost of these materials and their availability, the accumulated knowledge and experience from operating these materials, and the ease of engineering the hardened materials that ensures the attainment of the required performances.

Permeable concrete, also known as pervious concrete, is a subclass of cement-based materials that is tailored to have a characteristic interconnected and tortuous macroporous structure by eliminating the use of fine aggregates [22–25]. This subclass of cement-based materials is receiving increased interest in research to enhance its application in the prevention and control of water pollution due to its following characteristics [5,8,16–20,22,25–31]:

- Input materials that are available and produced through standardized production methods. Additionally, they are low cost, which reduces the cost of pollution prevention and control practices;
- Preparation/construction requiring the use of simple devices (e.g., mixers at room temperature) that have a record of long use and experience;
- Characteristic porous structure allowing for its use as a filter and the passage of water without the need to enforce this passage, in addition to having a high specific surface area that enhances the sorption of containments;
- The presence of the amorphous and crystalline hydration phases providing sites for the chemical and physical entrapment of different anions and cations;
- The hydration of the cement creates highly alkaline conditions leading to the precipitation of most of the metallic contaminants;
- Ecofriendly permeable concrete has the potential to reduce the material and energy footprints of these water pollution prevention and control practices, as well as to reduce greenhouse gas emissions.

The interest of the scientific community in studying permeable concrete, its durability, and its use as barriers can be recognized by analyzing the number of publications indexed in the Scopus database (Figure 1). This analysis was conducted by constraining the search in the database using the Boolean operator "AND", whereby the total number of documents that mentioned the words "permeable" and "concrete", starting from 1944 to 15 September 2023, was 2056 documents. The restriction of the search to the durability of permeable concrete and permeable concrete barriers reduced the total number of documents to 411 and 121, respectively. Figure 1a,b visualize the variation of the annual published works over the past twenty years and their distribution based on the type of document. It is clear that the scientific interest in this research field increased over time, declined in 2014, then increased again, only to decline again in 2022. It should be noted that the records in 2023 are not final, as the search was conducted in the fourth quarter of the year. The average ratios between the number of publications that addressed permeable concrete durability and permeable concrete barriers relative to those that addressed permeable concrete are 0.21 and 0.09, respectively. This indicates that durability studies on these materials are addressed in one-fifth of the publications, and their use as barriers represent approximately 10%. Most of the published works are research articles published in journals, as indicated in Figure 1b, in which the average ratios of the publications, independent of the type of document, are fairly constant. The geographical distributions of the published works are illustrated in Figure 1c–e; for this part of the analysis, the national contributions that had less than 1% of the total publications were summed with the undefined category under the name "Other". The United States and China represent the major contributors in the "permeable AND concrete" (Figure 1c) and "permeable AND concrete AND barrier" (Figure 1d) research topics, whereas India and China are the main contributors in the topic "permeable AND concrete AND durability" (Figure 1e). Through applications of permeable concrete materials in low-impact development practices, their efficiency in mitigating some urbanization impacts on the environment has been proven, e.g., urban heat islands, attenuation of flashfloods, and reduction of transportation-related noise [22,32–35]. Some recent review papers have been published that summarize different aspects of these applications, including the state-of-the-art development of these materials and their characteristics, performance, application, and sustainability [32–35]. In addition, research efforts have been directed at disseminating the use of permeable concrete not only for preventing water pollution through their use in low-impact development practices but also for controlling pollution via their use as permeable reactive barriers and their use in the treatment of waste water and water purification [18,21,24,25,30,31,36,37]. These applications for controlling water pollution need to be assessed in light of the acquired knowledge from the application of these materials in low-impact development practices over more than three decades and recently published research in these areas. The aim of this work is to assess the application of permeable concrete materials in the control of water pollution through their application as permeable reactive barriers. In this respect, an overview of both conventional and innovative permeable concrete mix design and their characterization is presented. Permeable reactive barrier technology is reviewed, recent studies on the application of permeable concrete with this technology are analyzed, and gaps in this field are identified.

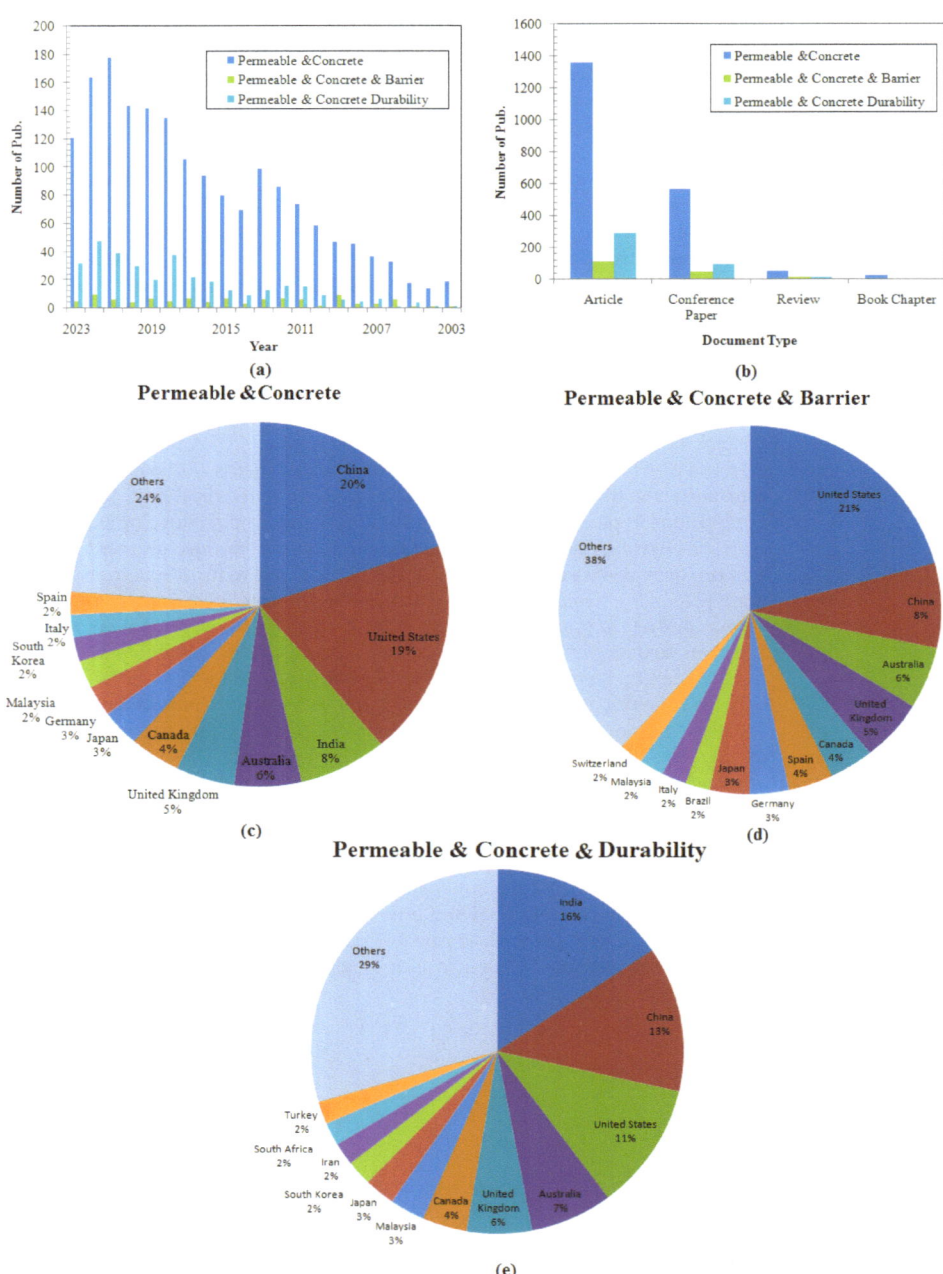

Figure 1. Bibliometric data analysis: (**a**) annual variation of the published works; (**b**) type of the published documents; (**c**–**e**) geographical distribution of the published works.

2. Permeable Concrete Mix Design

In general, the design of any cement-based material mix is dependent on the identified function of the hardened material. Depending on the application of these materials, processing and performance requirements should be defined [38]. Most of the review articles that addressed the preparation of permeable concrete only focused on the performance requirements of the produced material (i.e., mechanical, hydraulic, and durability requirements) with little focus on their processing requirements [26,32–34,39–43]. Currently, there is only one available valid standard test for this class of cement-based materials that address the measurements of the infiltration rate through the in-place barrier: ASTMC 1701/C1707M-17a. This test is a performance test and is related to the application of the permeable concrete as pavement. Both conventional and innovative cement-based materials were tested for their application in the production of permeable concrete, in which alkali-activated materials were recently investigated for this purpose [26,30,39–43]. In this section, advances in material selection and their proportion in the mix design of conventional and innovative permeable concrete are presented.

2.1. Conventional Permeable Concrete Mix Design

Conventional permeable concrete is manufactured using ordinary Portland cement (OPC), where type 1 is the most used in research investigations. with coarse additives, little or no fine aggregate, admixtures, and water. The hardened materials are designed to have interconnected pores (2–8 mm in size), and dead-end and capillary pores [32]. They should contain 15–25% voids with an acceptable compressive strength in the range of 2.8–28 MPa [44]. In this respect, OPC is used to provide a coating layer on the coarse aggregate, which is required to bond the aggregates, sustain certain mechanical and hydraulic properties, and ensure its durability over its service life. During the optimization of the mix design, the following general requirements maybe considered:

- A water-to-cement (w/c) ratio optimized in the range 0.25–0 45.Increasing the water ratio is an advantage for the formation of porous materials, but it will reduce the thickness of the final cement coat on the aggregates, as the paste will have a high flowability [23,32,44,45];
- A coarse-aggregate-to-cement ratio in the range of 4:1 to 6:1, in which the volume of aggregate in the hardened materials occupy 50–65% [23]. The grading of the aggregates should be optimized to control the void ratio of the hardened product, whereby single-sized coarse aggregate or narrow-grading coarse aggregate (e.g., between 9.5 and 19 mm) can be used [44,46];
- The use of admixtures in the permeable concrete to control the workability of the paste without greatly increasing the water content, retarding or accelerating the hydration process, and improving the freeze–thaw durability [44,46].

Alternative materials have been investigated to replace cement and/or aggregates to reduce the environmental impacts associated with their use and to improve the performance of hardened permeable concrete [33]. In this respect, natural materials and industrial and agricultural wastes have been investigated for their applications as supplementary cementitious materials (SCMs) and/or replacement of aggregates. Figure 2 represents the mix design of permeable concrete with no replacement (Figure 2a) and the partial replacement of cement (Figure 2b) and aggregates (Figure 2c) [47]. Additionally, several materials have been tested for their application as admixture, fibers, or functional materials to enhance certain properties of the paste/fresh and hardened permeable concrete. Table 1 lists some examples of different materials that have been investigated for their application as alternative SCMs, aggregates, and fibers [11,20,25,31,37,48–95]. Recent review articles were devoted to addressing the effect of solid waste reuse [47] and the use of functional materials on the properties of hardened permeable concrete [48,49]. Several research papers investigated the use of different fibers to reinforce permeable concrete, in which natural fibers, industrial wastes, and organic and inorganic chemicals were investigated [51,54,57,67,75,78,81,85,86,90]. Additionally, some research papers employed several admixtures, i.e., superplasticizer and water-reducing and air-entrapping admixtures to control the fresh and hardened permeable concrete properties [25,52,61,94,96]. Finally, functional materials (i.e., chemicals) have been investigated for their effects on enhancing the mechanical properties and pollutant-removal performance of permeable concrete, e.g., reduced graphene [36], nano-iron oxide [52,59,62], and nano-titanium oxide [63]. Limited research has proposed the coating of aggregates to improve the fresh paste concentration at the aggregates' joints [73] or to enhance their mechanical, sorption, and leaching performance [93]. It should be noted that coating the aggregates was reported to be a nonstandard practice for concrete producers and can increase the overall price of the final produced materials [97].

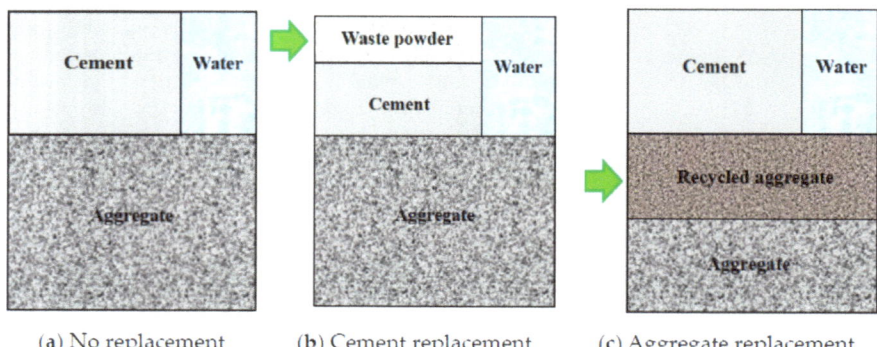

Figure 2. Schematic presentation of the use of alternative materials to replace the cement and aggregated in conventional cement mix design (copyrighted [47]).

Table 1. Alternative SCMs, aggregates, and fibers investigated to prepare permeable concrete.

	SCMs		Aggregates		Fibers	
	Material	Refs.	Material	Refs.	Material	Refs.
Natural materials	Metakaolin (MK)	[53,61]	Basalt	[51,56,63,79]	Basalt fibers	[78,86]
	Nano-clay	[81,90]	Lignite	[50]	Jute fibers	[54]
			Limestone	[20,51,52,59,66,71,80]		
			Granite	[31,84,88,92]		
			Pumice	[51,90]		
			Zeolite	[20]		
Industrial wastes	Fly ash	[11,25,31,52,53,55–60,89,94]	Recycled concrete aggregate	[60,64,66,70–73,81,90]	Plastic fibers	[81,90]
	Blast furnace slag	[25,54,64,66–69]	Recycled brick aggregate	[74–77]		
	Volcanic ash	[65]	Iron slag	[20]		
	Copper slag	[64]	Steel slag	[79]	Fine saw dust	[57]
	Silica fume	[25,56,61,80–82,89]	Crumb rubber	[82]		
	Calcium carbide	[84]				
	Sugarcane bagasse ash	[95]				
Agro-waste	Biochar	[37,83]	Petioles from Sterculia foetida plant	[91]	N-A *	
	Rice husk	[84,85]	Oil palm kernel shell	[92]		

Table 1. Cont.

	SCMs		Aggregates		Fibers	
	Material	Refs.	Material	Refs.	Material	Refs.
Chemicals	Nano silica	[59,87,88]	N-A *		Steel and steel wool fibers	[51,75,81,85,86,90]
					Polypropylene	[51,67,75]
					Polyphenylene sulfide	[85]
					Glass	[85]

Note: * N-A not available

2.2. Innovative Permeable Concrete Mix Design

Permeable concretes based on the use of innovative cements has also been prepared and investigated, and this includes geopolymers, magnesium phosphate cement, and calcium sulfoaluminate cement [26,30,39–43,98–102]. Most of these research efforts have focused on the use of geopolymers with limited investigation of the use of magnesium phosphate cements, and only one paper, to the knowledge of the authors, proposed a combination of ordinary Portland cement and calcium sulfoaluminate cement [102].

2.2.1. Permeable Geopolymers Concrete

Geopolymer is a relatively new class of alkali-activated cement-based materials that is prepared using aluminosilicate source(s) and alkali-activating solution(s). Natural minerals, industrial wastes, and chemicals can be used to prepare geopolymers [41,103,104]. Metakaoline (MK) is the most used natural aluminosilicate source, whereas fly ash (FA), slag, red mud, and biomass fly ash are the most used waste materials for the same purpose [39,41,105–108]. Other natural materials have been used in the preparation of geopolymers, including bentonite and feldspar, but to the knowledge of the authors these natural aluminosilicate sources have not been investigated to prepare permeable concrete. Different alkali-activating solutions are used for the preparation of geopolymers, including NaOH, KOH, water glass, or a combination of them [39,41,103]. The main binding phase in hardened geopolymers is aluminosilicate gel, and it is classified based on its Si/Al ratio into poly(si-alate) (Si/Al = 1) and poly(sialatesiloxo) (Si/Al = 2), and poly(sialate-disiloxo) (Si/Al = 3) [103,104]. As in any cement-based materials, the final the properties of the hardened geopolymers are strongly dependent on their mix design and curing conditions, in which the value of the silicon-to-alumina ratio effectively changes the composition of the hardened geopolymer as follow [6,103,105,106]:

- Crystalline zeolite is formed in geopolymers at Si/Al < 1;
- Geopolymers of reduced porosity are formed at 1 < Si/Al < 2;
- The porosity of geopolymers are dependent on the solubility of the Si source at 2 < Si/Al.

Permeable geopolymers can be prepared using different methods to produce a wide variety of materials with distinctive porous characteristics, and these methods can be categorized into the self-forming method (SFM; Figure 3A), direct foaming method (DFM; Figure 3B), adding filler method (AFM; Figure 3C), and particle stacking method (PSM; Figure 3D) [41]. Table 2 lists the features of these methods, the general pore structure characteristics of the hardened geopolymers, and their applications.

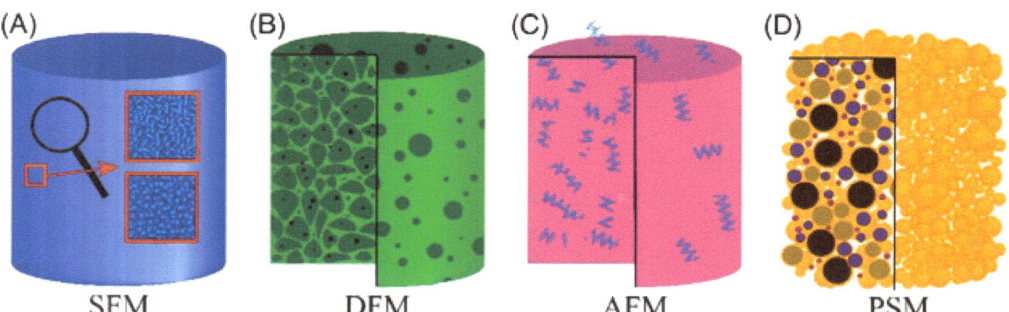

Figure 3. Schematic diagram and photos of the preparation method of porous geopolymer (copy righted from [41]).

Table 2. Features of the porous geopolymer preparation methods [41].

Method	Preparation	Pore Characteristics	Applications
SFM	The porous structure is self-formed without the addition of any material	Porous structure cannot be observed directly Pores are small	Sorption Membrane filtration
DFM	Foaming agents, surfactants, or both are used	Large pore diameter that can directly be observed Noted circular pores on the surface and irregular internal bubbles pores	Building insulation Building lightweight
AFM	Porous filler or materials are added	Reflects the filler's pore structure rather than that between the filler and geopolymer	Various applications including Adsorption Ultralight weight Building insulation
PSM	Bonding of the aggregates	The pore structure is formed in geopolymers or between the aggregates and geopolymers The pore diameter is related to the aggregate size Pores are observed directly	Porous pavement Permeable concrete

2.2.2. Permeable Magnesium Phosphate Concrete

Magnesium phosphate cements are formed through reactions between MgO and phosphates to form a magnesium phosphate salt with cementitious properties [5,6,98,100,101]. Magnesium phosphate cement is characterized by its fast hardening, near-neutral pH, low water demand, high adhesive strength to metals and concrete and high bending and compressive strength [5,6]. Acid phosphate anions (e.g., mono-potassium di-hydrogen phosphate, mono-sodium di-hydrogen phosphate, mono-ammonium di-hydrogen phosphate, and di-ammonium hydrogen phosphate) are used as phosphate sources or aqueous phosphoric acid [5,6,98,100,101,109,110]. The main final phase in hardened magnesium phosphate cement is struvite, where the ratio between the Mg and PO_4 largely affects the produced hardened cement's properties. At a low Mg/PO_4 ratio (<4), the crystallization of the struvite is enhanced and denser microstructure is attained. Because of its fast setting, a retarder is usually used to control the rate of the reaction, e.g., sodium tripolyphosphate, glacial acetic acid, and boric acid [5,98,100,101]. The limited research on the preparation of permeable magnesium phosphate concrete has tailored the mix design of this paste to include an aluminosilicate source (e.g., FA, MK, granulated blast furnace slag [98,101], steel slag [98], and crushed stone [101]) as aggregates and borax as retarder.

2.3. Permeable Concrete Mix Design: Future Prospects

As mentioned in the previous subsections, several research efforts have been directed toward the incorporation of natural materials as SCMs, aggregates, and fibers in conventional permeable concrete. Other research efforts have been directed toward the incorporation of industrial or agricultural wastes for the same purposes or for their use as aluminosilicate sources in the preparation of innovative permeable concrete. These studies have been motivated by the need to reduce the environmental burdens of the conventional cement industry. Within this quest, special attention should be paid to the characteristics of the used solid wastes and their compliance with national regulations. In this respect, it should be noted that some solid wastes (e.g., coal ash, copper slag, rice husk) may contain considerable amounts of heavy metals, sulfur, and chlorine [47,52,111,112]. Subsequently, the extent of the presence of these contaminants and their dissolution and mobility should be assessed to determine their compliance with national regulations. In the case of noncompliance, a pretreatment process should be designed to reduce the risk of the presence of these contaminants and to mitigate their release. These topics have not been fully investigated in the literature and need to be addressed in depth to ensure the sustainable use of permeable concrete materials in preventing and controlling the pollution of water.

Extensive scientific efforts in the preparation and testing of permeable concrete have generated an acceptable range of values for the water-to-cement ratio and the coarse-aggregate-to-cement ratio, as indicated in Section 2.1. Yet, a standardized method that can be followed to produce universal permeable concrete products for specified applications is still missing [23,32,44–46,97]. In this respect, a standard practice to proportionate the used materials, identify the mixing procedure, and determine the optimum curing conditions is missing not only for innovative cement-based materials but also for conventional permeable concrete [97]. This quest to establish a procedure for the optimization of the mix design is also scarce in the literature. To the knowledge of the authors, only one paper elaborated on the development of a mix design procedure that was recommended for the optimization of the permeability and compressive strength of the permeable concrete [113]. The optimization procedure comprises three steps that benefit from a set of constitutive relationships, and the performance of a film-forming ability test to allow for the determination of the optimum mix design, as presented in Figure 4 [113].

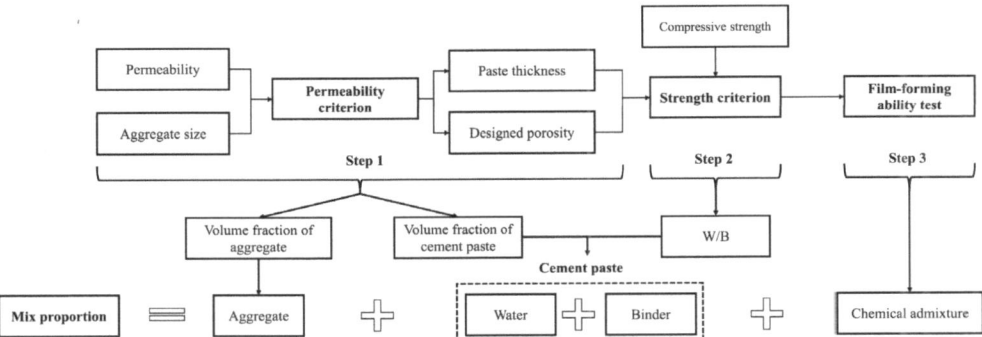

Figure 4. Proposed procedure to optimize the mix design of permeable concrete based on the permeability and compressive strength (copyrighted from [113]).

3. Characterization and Functional and Durability Performances of Permeable Concrete Materials

Different testing and evaluation techniques are used to characterize and evaluate the important properties of cement-based materials. These techniques are used to ensure that the requirements of both of the processing properties of the paste/fresh concrete and the desired performance and degradation resistance of the hardened materials are met [114,115]. On the one hand, standardized and nonstandardized characterization techniques have been developed to qualify the raw materials (e.g., grading), the paste (e.g., rheological properties), and the hardened materials (e.g., pore structure and permeability).

The details of these tests are found elsewhere [114,116,117]. On the other hand, functional performance evaluation and durability tests have been developed to ensure that the hardened cement-based materials meet the functional and durability requirements, respectively. For instance, permeable concrete used in permeable pavement applications is required to exhibit adequate hydraulic, mechanical, thermal, and sorption performances that enable the final hardened materials to effectively allow for the management of rainfall during their service life, meet the requirements on their strength (i.e., compressive, tensile, flexural, and abrasion resistance), and have improved sound absorption and temperature mitigation [32]. For permeable reactive barriers, their hydraulic performance requires that the flow is maintained under a natural hydraulic gradient without considerable retention within the barrier and to maintain good pollutant removal over their service life. In this section, the available characterization techniques and functional and durability tests applied to the permeable concrete materials are presented, and their features are summarized with some highlights provided regarding the effect of variations of the mix design on the measured properties.

3.1. Rheological Properties of the Paste

The composition of the mix design will affect the rheological properties of the paste and, subsequently, its adherence on the aggregates and the overall quality of the hardened material [113,118]. As the rheology of the paste will affect the thickness of the formed cement-based materials on the aggregates, it will affect its mechanical and hydraulic performances. In the case of permeable concrete, as the minimum amount of binder is applied, the importance of adjusting the rheological properties of the pastes becomes a necessity to avoid the formation of inhomogeneous paste and its segregation and the formation of hardened nonporous material or materials with low mechanical strength. As flowability is an important paste property, the adequacy of this mix design to produce a paste with adequate flowability can be checked via the application of the following tests [113,118–131]:

- Hand compacting method: This is a qualitative, easy method for testing the adequacy of the water in the paste and, hence, provides an indication of the paste's flowability (Figure 5a–c). Scarce water will yield a crumbling of the ball (Figure 5a), and excess water will yield an accumulation of paste on the glove, leaving the aggregates with a minimum coat of cement (Figure 5c). Adequate water will lead to the formation of a ball without excess paste accumulated on the glove (Figure 5c);
- Slump flow test: This test is used to examine the horizontal flow of the paste (ASTM 143/C143M-12) [126]. In this respect, the slump is recommended to be adjusted to near zero and less than 5 cm [122,123,126,128];
- The flow table test: this is used to measure the flowability of the paste (ASTM C230/C230M-14), where acceptable values are in the range 15–23 cm [87,129–131].

Additionally, the viscosity of the paste, yield stress, and adhesive force of permeable concrete pastes are measured using a rheometer or viscosimeter, and some researchers have studied the effect of these parameters on the compressive strength, thickness of the cement-based material on the aggregates, and on the pore structure of the hardened material [118,124,125,129,131]. There is an available standardized test to characterize the rheological properties of the paste using a rotational rheometer: ASTM C1749. Furthermore, there are available methods to measure the thickness of the formed paste layer on the aggregate: ideal paste thickness or actual paste thickness methods [113]. A modification of the actual paste thickness method was proposed using a flow table, and Figure 6 illustrates the procedure for conducting this method [124]. Another modification of this procedure was proposed by applying vibration or limiting the number of drops to 10 [113]. Additionally, image processing can be used to calculate the thickness.

Figure 5. Hand compacting method: (**a**) scarce water content; (**b**) adequate water content; (**c**) excess water content (copyrighted from [124]).

Figure 6. Procedure to evaluate the paste coating thickness on the surface of the aggregates (copyrighted from [124]).

Figure 7a–d illustrate the effect of the variation of the conventional cement mix design on the flowability, i.e., slump spread, yield stress, viscosity, and the maximum paste coating thickness on the aggregates [124]. Figure 7a,c show the effect of the variation of the w/c ratio in a simple permeable concrete samples composed of 100% cement as binder on these properties. The general behavioral trend can easily be deduced in these simple systems, where the flowability (i.e., slump spread) increases by increasing the w/c ratio and the rest of the properties display decreasing behaviors [124]. The behavior of these properties can be described using polynomial equations with correlation coefficients larger than 0.99. The effect of the variation of the SCM incorporation percentage in the permeable concrete (i.e., granulated blast furnace slag (GBFS)) on the same properties has a limited effect compared to that of the w/c ratio, shown in Figure 7a–c, as reported by Xie et. al. [124]. These studied samples were prepared at a lower w/c ratio (i.e., 0.28) and contained superplasticizer at a dosage equal to 0.4%. For those samples, the flowability slightly increased with the increased incorporation of the GBFS up to 50% and with the rest of the properties being reduced. The increasing correlation between the thickness of the cement-based material on the aggregates and the yield stress and the viscosity of the paste were confirmed for all of the studied mix designs (Figure 7a–d insert) [113,124,125].

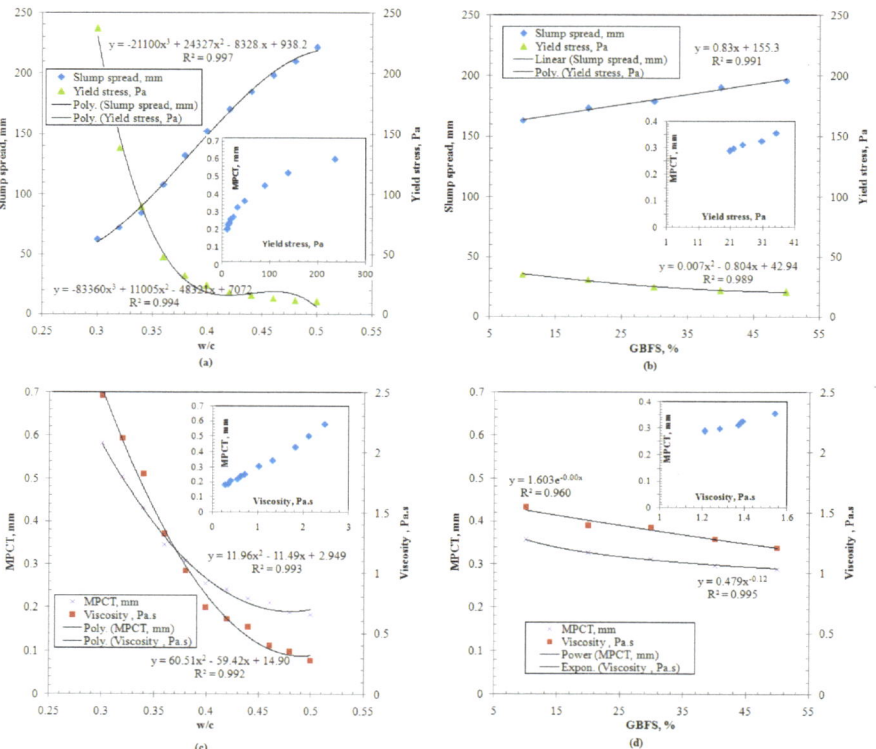

Figure 7. Effect of the mix design variation on the paste properties in conventional permeable concrete: (**a,c**) effect of the w/c ratio variation; (**b,d**) effect of the GBFS in cooperation (source data were extracted from [124]).

The effect of the variation of the sodium content in the alkali activator of the mix design of the permeable geopolymer concrete on the flowability of the paste, viscosity, shear stress, and the thickness of the hardened geopolymer coat was studied by Geng et al. [131]. The studied samples were BFS-based geopolymers prepared with water glass and sodium hydroxide in the presence of a retarder and limestone as coarse aggregate. Figure 8a–c illustrate the results in which the flowability, viscosity, and coat thickness increased with the increase in the Na_2O content up to 8%; then, the values of these properties decline. The increased behavior of the flowability and reduced behavior of the viscosity and shear stress of the paste is attributed to the formation on an electric double layer with repulsive forces on the precursor particles (i.e., BFS) due to the adsorption of activator silicate anions (Figure 8a,b). A further increase in the amount of Na_2O is claimed to lead to the enhanced formation of C-S-H gel that reduces the flowability and increases the viscosity and shear stress [131]. The variation of the geopolymeric coat's thickness on the aggregates (Figure 8c) shows a similar trend to that of the flowability [131]. The effect of the admixture on the rheological properties of the geopolymeric paste is an active area of study that aims to identify the effect of various admixtures on the workability of the paste. A study was devoted to investigating the effect of the water-reducing admixture and the molar strength of the alkali activator on the rheological properties of the fly-ash-based geopolymer paste [132]. The studied systems included both fine and coarse aggregates, as well as the alkali activator composed of a mixture of sodium hydroxide and sodium silicate. The results revealed that there was a critical molar strength (i.e., 4 M) beyond which the plasticizer and the superplasticizer had contradicting effects on the plastic viscosity, slump spread, and yield stress of the studied samples. In this respect, above this threshold value (i.e., <4 and =10 M), a clear decrease in the slump spread and increase in the yield stress and viscosity were recorded [132]. This study indicates that the employment of a lignin-based plasticizer leads to a better performance in terms of the paste's workability over that of a polycarboxylic–ether-based high-range water reducer.

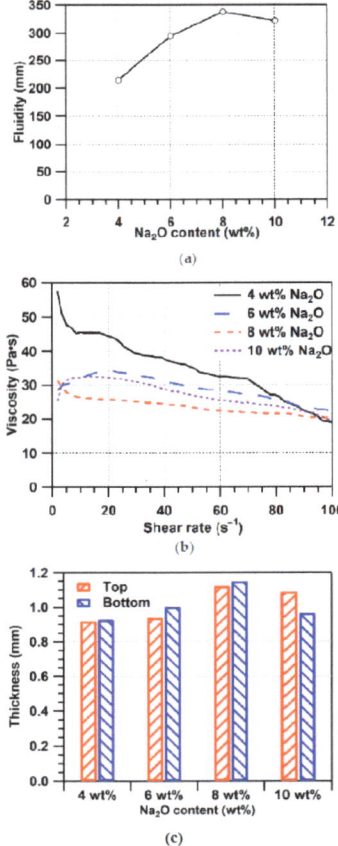

Figure 8. Effect of the variation of the mix design on the paste's properties in innovative permeable concrete: (**a**) fluidity; (**b**) viscosity and shear stress; (**c**) coat thickness (copyrighted from [131]).

3.2. Hydraulic Properties of Hardened Permeable Concrete

There are different standardized tests that have been issued to measure the pore structure and permeability of hardened cement-based materials, and there are other non-standardized tests that have been developed and used. These tests are used to measure the porosity, specific surface area, total connected porosity, pore volume, and air and water permeability [114]. These tests include:

- Gravimetric techniques: These are employed to measure the porosity using general standardized ASTM tests: ATSM C457/C457 M-16. It should be noted that a standardized test to measure the porosity of permeable concrete (ASTM C1754) was recently withdrawn;
- Absorption tests: These include the BET and MIP, and standardized ASTM tests are available for fragmented materials: ASTM D5604-96 and ASTM D4404-18, respectively. In addition, auto-clam and Figg tests are used to measure in situ water and air permeability [114];
- Ultrasonic techniques: These can be used to determine both the permeability and compressive strength of materials [116,133];
- Imaging techniques: These are used to construct 3D models for a sample, either using X-ray computed tomography or 2D scanning images and suitable image processing software. These models are used to drive empirical relationships for calculating the pore size and distribution and to model the mechanical and hydrological behavior of the material [97,114–138].

The permeability reflects the ability of the material to allow water to flow through it; it is dependent on the pore characteristics of the material, i.e., pore size, shape, connectivity, and tortuosity [22]. The permeability (k, cm/s) can be measured using the constant head method (Equation (1)) or the falling head method (Equation (2)) [22,97,139–141]:

$$k = \frac{QL}{Ah} \tag{1}$$

$$k = \left(\frac{aL}{At}\right) ln\left(\frac{h_1}{h_2}\right) \tag{2}$$

where Q is the flow rate, h is the head in the constant head method, A and L are the surface area and length of the sample. In the falling head method, a and t are the cross-sectional area of the pipe encasing the sample and the time required for the water pressure head to drop within predetermined levels (h_1 and h_2), respectively. Research papers that compare the validity and accuracy of both methods indicate that the permeability measured using the falling head method is lower than that of the constant head method, in which the latter is reported as viable and provides economic benefits [139,140]. It should be noted that the flow rate measured in permeable concrete is claimed to be in the transient flow regime between laminar and turbulent flow, which necessitates the use of a valid equation to describe the flow, i.e., Darcy–Forchheimer [124,140]. Several researchers have fitted the experimental data to obtain a relationship between the porosity and the permeability of the permeable concrete, in which linear, power, and exponential equations are derived [111]. It should be noted that these equations should be treated as mix-design dependent, and their validity should be tested before their application.

The effect of the variation of conventional cement mix design on the porosity and permeability of hardened permeable concrete is illustrated in Figure 9a–c [123]. The studied samples were prepared at a constant aggregate-to-cement ratio of 4 and different w/c ratios (0.24–0.32) and admixture dosages (0.4–1.1%) to control the final porosity. The investigations of the effect of the admixture on the values and relationships of the porosity and permeability at a fixed w/c ratio revealed that the use of a 0.4% admixture yielded a lower porosity and permeability of the hardened sample. The authors attributed this behavior to the paste's drainage into the interconnected voids [123]. The known effect of increasing the w/c ratio on the porosity and permeability was not clear in this study (Figure 9b). The visualization of the statistical distribution of the permeability of these samples, as shown in Figure 9c, indicates that the permeability decreases with an increase in the w/c. This is attributed by the authors in that work to the increase in the degree of lubrication, which results in a better densification of the mixture, consequently reducing the permeability [123]. It should be noted that the sample with the highest water content was prepared without any admixture, and the samples with the lowest water content were prepared with the highest admixture ratio; this clearly shows the importance of considering the rheological properties of the paste and their critical role in tailoring the hydraulic properties of the hardened permeable concrete.

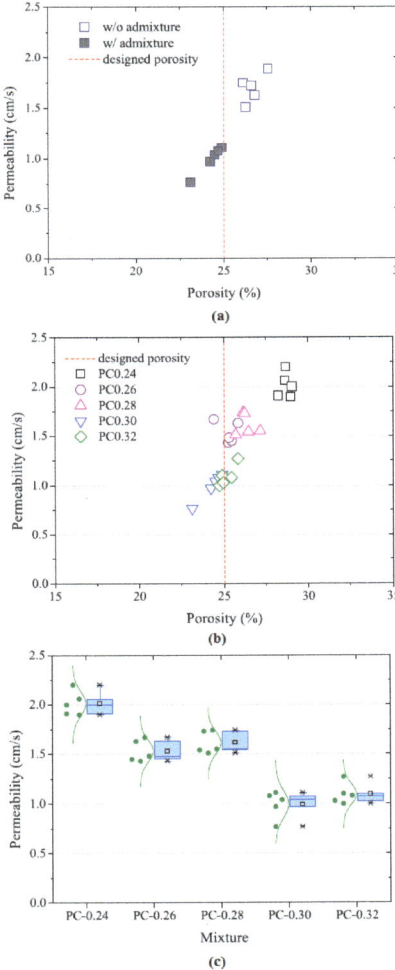

Figure 9. Effects of the mix design on the hydraulic properties of the hardened permeable concrete of different water to cement ration PC-X (X = 0.24; 0.26; 0.28; 0.3; 0.32): (**a**) effect of the incorporation of the admixture; (**b**) effect of the w/c ratio; (**c**) statistical distribution of the permeability at different w/c ratios (copyrighted from [123]).

The porosity and water permeability in the innovative permeable concrete is affected by the mix design of the innovative cement. In this respect, increasing the Na_2O content led to a reduction in the total and connected porosity and the permeability of BFS-based geopolymer prepared with water glass and NaOH in the presence of a retarder using limestone aggregates [133]. Another study was dedicated to investigating the effect of the substitution of GBFS with red mud to produce permeable geopolymer concrete using a combination of NaOH, Na_2SiO_3, and water glass as an alkali activator at an aggregate-to-binder ratio of 5 [40]. This study indicates that a 30%substitution led to an increase in the total void ratio and permeability by 7.69 and 6.35%, respectively [40]. Further, an increase in the red mud incorporation did not affect the void ratio and led to a reduction in the increase permeability to half of its value [40]. This behavior was not further discussed in that work. The substitution of the GBFS with red mud was investigated in another study, which confirms the increase in the porosity and permeability with an increase in the substitution by up to 50% [31].

3.3. Mechanical Properties

Mechanical properties are among the most important performance measures that need to be optimized during the design of any cement-based material. The hardened material strength should be preserved during its service life to allow for the sustainable functionality of the barrier. ASTM C109/C109M-20b [142]; ASTM C39 [143]; ASTM C496 [144]; and ASTM C293 [145] were developed to measure the compressive strength for cubic and cylindrical samples, splitting tensile strength; flexural strength, respectively. Several authors have used measured experimental data to deduce empirical models that can describe the relation between mix design components (e.g., water-to-cement ratio, cement content, aggregate size, and their porosity) and the compressive strength at a specified age (e.g., 28 days), flexural strength, tensile strength, elastic modulus, and fatigue [97].

As permeable concrete is designed to attain a specified mechanical performance and because of the limited amount of cementitious material in these composites, several authors have investigated the enhancement of this performance using different SCMs, fine aggregates, and fibers [25,51–61,79–89,94,95]. In this respect, it should be noted that the mechanical performance does not show a linear relationship over a wide range of these materials' incorporation. Therefore, an extensive number of review papers have addressed these effects in a comprehensive way [22,33,35,39,41,47,97,104]. For conventional permeable concrete, using fine aggregates was reported to improve the mechanical properties of the concrete. Thus, the reactivity of the used material plays an important role in determining their contributions to the build-up of the mechanical properties. In particular, the use of pozzolanic materials (e.g., volcanic ash, BFS, and FA) contributes to the long-term build-up of the strength of the hardened materials [47]. The inclusion of the fibers was proposed to improve the mechanical properties and durability of the permeable concrete, and plastic fibers were reported to have a limited positive effect on the mechanical performance of the hardened material [22,51]. Similarly, the use of permeable geopolymer concrete can be affected negatively with the use of low reactive materials in the mix design, and the use of red mud in GBFS geopolymers led to a reduction of the compressive strength [40]. This reduction increased with an increase in the GBFS substitution [30,40].

3.4. Durability of Hardened Permeable Concrete

In general, cement-based materials, as any other material, are affected by their presence in the environment, and the porous nature of these materials facilitates the penetration of water, gases, and aggressive materials leading to the activation of different reactions with the hydrated phases in the hardened materials affecting their durability. Both physical and chemical reactions occur between cement-based materials and ambient environmental components, leading to a reduction in their permeability [141,146], e.g., carbonation [79];sulfate attacks [80];induction of cracks through, for example, freeze–thaw cycles [61,73,80,88,128]; aggregate reactions [69]; and loss of materials via, for instance, leaching [24,36,61], abrasion [64,79,84,86,89], and erosion. In addition, the performance of these materials might be affected by the applied loads on them that can, in conjunction with environmental conditions, lead to serious failures. Subsequently, it could be concluded that both the hydraulic and mechanical performances of the cement-based barriers are affected by these conditions. Methods to solely assess the effects of these conditions are standardized, e.g., sulfate resistance (ASTM C452-21 [147]), aggregate reaction (ASTM C1778 [148]), freeze and thaw (ASTMC666 [149]), and aggregate soundness (ASTM C88 [150]). It should be noted that the standard test ASTM C1747 has been withdrawn.

Another important durability aspect of permeable concrete material is the clogging of the pores that affects the drainage performance of these barriers [20,22,35,49,124,151]. An evaluation of permeable concrete clogging is not only important for permeable pavement but for any permeable barrier, as it can seriously affect one of its main characteristic properties, i.e., permeability. Clogging is very similar to the well-known fouling phenomena in membranes, in which it can be traced to [22]:

- Physical clogging: an accumulation of suspended particles within the porous structure; this phenomenon does not include a chemical reaction;
- Chemical clogging: which occurs because of the penetration of chemical components into the flow of water through the barrier, leading to scale formation that clogs the porous structure;
- Biological clogging: which occurs because of the reproduction of algae and bacteria within the porous structure of the material.

As the clogging is mainly affected the hydraulic performance of the barrier, it is usually evaluated by measuring the permeability of the sample, as mentioned above, or by relying on the imaging techniques.

The effects of the variation of the mix design on the durability of the permeable concrete have been investigated [59,61,73,79,80,84,86,89,128]. The effect of using a combination of SCMs (i.e., nano silica (22.8%), spent fluid catalytic cracking catalyst (11.4%), and paper sludge waste (2.86%)) in conventional permeable concrete prepared using three gravel sizes at an aggregate-to-binder ratio equals to 5.6:1 on the leaching and freeze and thaw resistance of the hardened materials in the presence of melamine-based superplasticizer was addressed [128]. The study revealed that the use these materials enabled a reduction in the portlandite leaching and increased the resistance to freezing–thawing even after 50 cycles [128]. Another study addressed the effect of using nanomaterials on the physical durability of the hardened permeable concrete [59]. In this context, different samples were prepared using varying nano silica (0–4%) and nano iron (6%) contents in the IP and GU Portland cement in the presence of water-reducing admixture (0–1%) and cement substitution with FA in the range of 10–50% [59]. A multivariant method was followed to optimize the mix design to achieve a specified mechanical and hydraulic performance based on the maximum achievable compressive strength and target permeability equal to 8.8mm/s. The recommended optimum mix design comprised 24% fly ash, 1.9% nano silica/fly ash, and 0.35%admixture. The study concluded that the use of the nanomaterials improved the physical durability of the hardened materials but increased their costs [59]. The effect of modifying the surface of the recycled aggregates to improve the durability of the hardened permeable conventional concrete waste was investigated [73]. The study utilized a simple mix design composed of cement and recycled aggregate. The study indicated that the surface modification of the aggregates using a hydrophobic silicone membrane improved the durability of the hardened materials due to the following [73]:

- Its role in preventing the water accumulation on the aggregate surface, which allowed for the increased formation of dense C-S-H in the interface transitional zone;
- Its ability to mitigate the water absorption and migration into the aggregate that suppresses the ice pressure.

Well-optimized mix designs for innovative cement-based materials have been reported to provide beneficial durability performances under varying conditions including different chemical and freeze and thaw resistances [39,41,103,104,146]. In particular, the effect of the type of permeable geopolymer concrete (i.e., GBFS, FA, and MK) on the durability of the hardened materials was considered [26]. These three investigated systems at a target porosity of 20% were prepared and tested, including a control conventional permeable concrete and two permeable geopolymer concretes prepared from MK-GBFS and FA-GBFS. The MK geopolymer was found to have the best mechanical performance in terms of compressive and tensile strength and durability in terms of freeze and thaw resistance. The study indicated that the FA geopolymer provided remarkable benefits in terms of reducing energy consumption and greenhouse gas emissions.

3.5. Pollutants Removal Performance

As mentioned in Section 1, the presence of amorphous and crystalline hydration phases within the hardened permeable concrete provides sites for the chemical and physical entrapment of different anions and cations. This structural feature is employed in permeable concrete to remove different organic and inorganic pollutants from the water and entrap them within the material structure [6,8,12,15–19,21,24,25,27–31,36,38,40,41,48,49,62,63,93,103,151–153]. In this respect, monolith samples of permeable concrete are prepared either in the form of disks, cylinders, or cubes, characterized, and then tested for their potential application in removing pollutants (Figure 10). Both static [24,25,40,63] and dynamic [30,31,36,40,62] tests can be applied to test the ability of these materials to remove the pollutants of interests, including heavy metals, CODs, BODs, DODs, dyes, etc. In both types of experiments, the performance of the monolith material is expressed by recording the concentration measured in the aqueous phase or calculating the removal percentage (P, %) or the reduction ratio (Equation (3)) [40],

$$P,\% = \left(\frac{C_o - C_m}{C_o}\right) \times 100 \qquad (3)$$

where C_x is the concentration of the studied pollutant before the first sorption cycle ($x=0$) and after m number of cycles ($x = m$). Some of these investigations were directed toward assessing the effect of adding materials of a known sorption capacity [48], e.g., zeolite [20,154,155], iron oxide [62], graphene oxide [36], or the addition of materials of known photocatalytic effects, such as iron oxide or titanium oxide to enhance the sorption ability. Finally different nanomaterials were investigated for the same purpose [59,62,63,151,156].

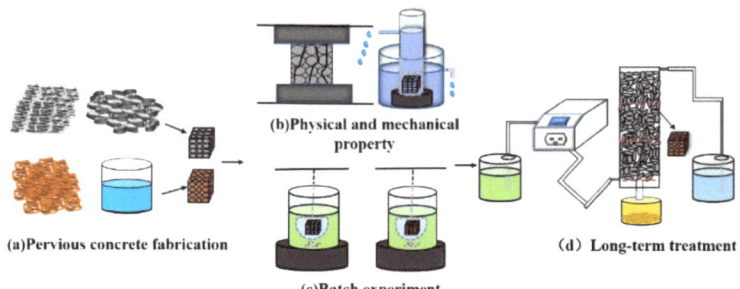

Figure 10. Experimental testing scheme for potential application of permeable materials in pollutant removal: (**a**) preparation of the monolith sample; (**b**) physical and mechanical characterizations; (**c**) static removal test; (**d**) dynamic removal test (copyrighted from [30]).

The performance of conventional permeable concrete prepared at a specified w/c ratio (0.3) and varying coarse aggregate-to-cement ratios in the presence of silica fume (SF; 10%) and FA (20%) aimed at the removal of total phosphorus (TP) and total nitrogen (TN) was investigated [25]. The study indicated that higher removal performances were obtained for samples that contained smaller sizes of aggregates and higher void contents. In this respect, 1.7 and 2.8 times more TP and TN concentrations were reported to be removed using permeable concrete containing 5–10 mm aggregates. The amount of attached microorganisms on the permeable concrete was concluded to control the amount of removed TP and TN [25]. In the context of the use of permeable concrete for water purification, the use of low pH concrete was suggested to reduce the leaching of the alkaline components that can affect an aquatic ecosystem [25]. The utilization of conventional permeable concrete prepared with OPC, basalt, FA, and SF for water purification was investigated [56]. In their work, Wang et al. investigated the performance of different mix designs of permeable concrete in the removal of suspended solids (SS), ammonia–nitrogen (AN), and total phosphorus (TP). The study concluded that SS removal was the highest followed by AN then TP, and the increased incorporation of FA led to a slight reduction in the removal performance. The study indicated that there is a need to improve the removal performance of AN and TP using auxiliary purification materials [56]. The performances of different permeable conventional concrete in the removal of some heavy metals were studied using batch static procedures by some investigators [24,40,155]. The studied systems were OPC-based permeable concrete with different types of aggregates, such as gravel, limestone, soda lime glass beads, and pumice, in the presence and absence of additive materials, e.g., FA and silica fumes. Table 3 lists the features of the investigated mix designs and the removal experiment conditions: initial contaminant concentration (C_o, ppm), time (t, h), and percentage removal (P. %). Holmes et al. confirm that there are several mechanisms that contribute to the removal of the heavy metals from the aqueous solutions onto the conventional permeable concrete [24]. These mechanisms include both chemical and physical sorption, precipitation, co-precipitation, and internal diffusion. Their study indicates that the use of calcareous aggregates improved the removal of heavy metals and reduced their leachability [24]. Finally, the dynamic removal of nitrate was studied using conventional permeable concrete prepared at varying coarse-pumice-to-cement (3–5), w/c (0.26–0.35), fine aggregates (0–20%), and nano-silica (0–6%) ratios [157]. The results indicate that the removal performance increased from 18.5–29% to 53.5–64.2% with an increase in the incorporation of nano silica [157].

Table 3. Contaminant removal performances for conventional permeable concrete.

Contaminant	Permeable Concrete Mix Design				Removal Conditions		P, %	Refs.
	Aggregates	A/C	w/c	Additives	C_o, ppm	t, h		
Pb	Gravel	5.5	0.4	-	2-207.2	72	84–91	[24]
	Limestone	5.5	0.4	-	2-207.2		87–88	
	GB *	5.5	0.4	-	2-207.2		88.5–92	
	Gravel	5.5	0.4	FA, 33.5%	2-207.2		31.5–92	
	Limestone	5.5	0.4	FA, 33.5%	2-207.2		87–88	
	GB *	5.5	0.4	FA, 33.5%	2-207.2		68.5–95.5	
	Na	5	0.37	-	50	0.5	38	[40]
Cd	Gravel	5.5	0.4	-	0.11-112.4	72	95–97	[24]
	Limestone	5.5	0.4	-	0.11-112.4		56–80	
	GB *	5.5	0.4	-	0.11-112.4		16–99	
	Gravel	5.5	0.4	FA, 33.5%	0.11-112.4		48–97	
	Limestone	5.5	0.4	FA, 33.5%	0.11-112.4		54–64	
	GB *	5.5	0.4	FA, 33.5%	0.11-112.4		39.5–78	
Cu	Pumice	3	0.35	Pumice, 10% SF **, 5%	Na	Na	97	[155]

Table 3. Cont.

Contaminant	Permeable Concrete Mix Design				Removal Conditions		P, %	Refs.
	Aggregates	A/C	w/c	Additives	C_o, ppm	t, h		
Ni	Pumice	3	0.35	Pumice, 10% SF, 5% **	Na	Na	71	[155]
Zn	Gravel	5.5	0.4	-	0.65-65.38	72	96	[24]
	Limestone	5.5	0.4	-	0.65-65.38		72–80	
	GB *	5.5	0.4	-	0.65-65.38		67–100	
	Gravel	5.5	0.4	FA, 33.5%	0.65-65.38		56–96.5	
	Limestone	5.5	0.4	FA, 33.5%	0.65-65.38		76	
	GB *	5.5	0.4	FA, 33.5%	0.65-65.38		32–88	

Notes: * GB soda lime glass beads. ** SF silica fume.

An investigation of the potential use of innovative permeable concrete in the removal of different heavy metals was conducted [40]. In comparison with conventional permeable concrete, the innovative permeable concrete was found to have higher removal performance, and this increase is contaminant specific, e.g., Cd (18.75%), Pb (17.91%), Cu (25.07%), and Cr (39.18%) [40]. This performance was further increased by increasing the substitution of GBFS with red mud due its fine particle and high surface reactivity [40].

4. Permeable Reactive Concrete Barriers

The remediation of contaminated ground water is an essential activity to ensure the sustainability of this source of water, and both active and passive ground water remediation technologies have been implemented. Specifically, passive remediation via permeable reactive barrier have been applied since 1995. The basic idea behind this technology is to install a porous reactive material below the ground surface in the down gradient direction of the plume of the contaminated ground water, which allows for its passage under a natural hydraulic gradient [156–163]. Figure 11 illustrates the configuration of the funnel and gates (F&GPRB) and the continuous trench (CPRB) installation of the permeable reactive barrier [161]. This technology provides several economic and technical benefits compared to active ex situ groundwater remediation technology, i.e., pump and treat. These benefits include the following [158,161–163]:

- Low energy consumption, which reduces the carbon footprint of the process and the operating costs;
- Requires monitoring its activity with only minimum scheduled maintenance after a specified period of operation, if needed;
- Easily installation and removal procedures;
- Provides efficient and targeted remediation, where these barriers convert specified pollutants to fewer toxic species and/or retains them.

The major limitations of this technology include the depletion of the reactive chemical compounds over time due to their reaction with the contaminant and the clogging of the barrier pores. These limitations can be addressed by relying on the efficient design of the barrier that addresses the plume and site characteristics.

These barriers are designed to activate several mechanisms to decontaminate the groundwater plume, e.g., precipitation, sorption, and degradation [162]. Conventional reactive materials have been investigated and implemented for this purpose including zero-valent iron (ZVI), carbonaceous materials, sulfate-reducing bacteria, metal oxide/sulfides, mineral materials, and industrial wastes [159,161,162,164–166]. Innovative materials include the use of single materials, e.g., meso-zero-valent iron, permeable concrete, basic oxygen furnace slag, or modified/composite materials [164,166]. The factors that affect the selection of materials for permeable reactive barriers include [159]:

- Reactivity: the capacity of the material should be high enough to allow for the precipitation and/or sorption and/or degradation during its service life;
- Hydraulic conductivity: the barrier should have adequate permeability to allow for the passage of the contaminated ground water without considerable retardation in its velocity;
- Environmental compatibility: the used material should not have the potential to release toxic species into the host environment;
- Long-term physical and chemical stability: the material should have adequate long-term stability to eliminate the need for maintenance of the barrier during its service life.

Table 4 lists illustrative examples of the large-scale implementation of permeable reactive barriers in remediating different contaminants [164,167–172]. In this respect, it is clear that this technology has been successfully implemented to remediate different types of contaminants including both organic and inorganic. Only one of these examples, based in Willisau, Switzerland, claimed to require additional remediation action, as it did not achieve its target performance, and this was attributed to the complicated hydrogeological conditions at the site and the geometry of the installed barrier [164,170].

Permeable concrete has been investigated for its potential use in the construction of permeable reactive barriers for ground water remediation [30,31,154,156,173–183]. In particular, these research efforts were directed toward the assessment of the potential use of these materials for the remediation of acidic contaminated groundwater, in particular contamination from acid mine drainage, in which the high buffering capacity of these materials will sustain its performance over long periods of time. Published studies in this field are very limited compared to those related to the application of these materials in the low-development impact practices. These studies addressed the pollutant removal performance of these materials under acidic conditions in the mine acid drainage plume, where the effect of the incorporation of the aggregates and SCMs was studied, the permeable concrete performance in removing the pollutants was compared to that of the mature zero-valent iron (ZVI) materials, and the use of these materials in combination with the active method was proposed. Most of the published works focused on conventional concrete, where the use of innovative permeable concrete was addressed in a limited research. In this section, the published research in this area is summarized.

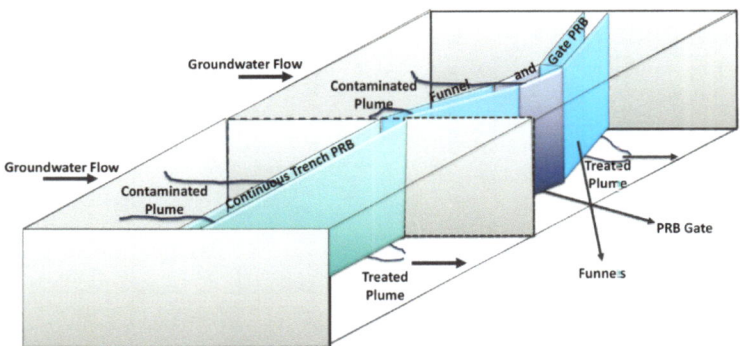

Figure 11. Configuration of the PRB designs: (copyrighted from [161]).

Table 4. Large-scale application of permeable reactive barriers in the remediation of different contaminants.

Location	Nature of the Contamination			PRB Specification			Refs.
	Target Contaminant	Site Characteristics	Plume Characteristic	Dimension	Barrier Materials and Specification	Performance	
Ontario, Canada	Perchloroethene (PCE) and trichloroethene (TCE)	Medium-fine sand underlain by a clayey silt deposit at 9 m below ground level	1 m wide and 1 m thick, C_o = 270 ppm (TCE) C_o = 50 ppm (PCE)	L = 5.5 m, W = 1.6 m, D = 2.2 m	CPRB mixture of ZVI and commercial coarse sand (22:78)	After 299 days, the TCE and PCE were reduced by 90% and 85% with an uncertainty of less than ±5%.	[167]
Northern Ireland, UK	TCE	Sand and gravel over Sherwood sandstone	C_o = 390 ppm	Full scale D = 8 m	F&GPRB with ZVI as the active material	Calcite precipitation observed in the upstream improved the ZVI reactivity. No biological fouling.	[164,168]
Willisau, Switzerland	Chromate	Clayey and silty sand underlined by sand and gravel	C_o < 10 ppm	Full scale	Double array of vertical piles containing iron shavings and gravel	Did not achieve its removal target due to its location in nearly oxygen and calcium carbonate saturated aquifer in a regime of high groundwater velocities.	[169,170]
Florida, USA	Iron and manganese	Clayey sand	C_o = 30 ppm (Fe) C_o = 1.62 ppm (Mn)	L = 6 m, W = 0.9 m, D = 4.6 m	Limestone and crushed concrete	In the first year, Fe removal efficiency 91–95% Reduced performance after three years showed due to clogging.	[171]

4.1. Effect of the Aggregates

The effect of the aggregates type on the pollutant removal performance of permeable concrete was investigated by evaluating the nitrate sorption onto permeable concrete samples prepared with different aggregates [173]. In this respect, type 2 OPC, coarse aggregates (3.8 in no. 4), and fine aggregate (no. 6 and no. 8) were used to prepare the studied samples, and pumice, zeolite, and perlite were tested. The water absorption, alkali–silica reactivity, and permeability were measured for all samples, and it was found that pumice showed the lowest water absorption (7.84%), maximum permeability (1.64 cm/s), and lowest expansion during the alkali–silica reactivity tests (0.02%). In batch experiments, the reaction between the different samples and nitrate solution (C_o = 70 ppm) was studied, it was reported that the reactions reached equilibrium in 30 min for samples containing perlite and pumice aggregates and required longer times for the samples containing zeolite aggregates. In addition, the activation of the aggregates using HCl and H_2SO_4 was concluded to improve the nitrate removal performance, but it did not affect the time required to reach equilibrium. In this regard, the activation of pumice aggregates using HCl was found to improve the nitrate removal from 39% to 50%.

The removal performance of permeable concrete containing granite (Gr) or dolomite (D) coarse aggregates was compared for their potential use in the management of acid mine drainage [182]. The study indicated that granite aggregates resulted in better treatment of manganese (+32%) compared to that of the dolomite aggregates. Both types of aggregates had a fairly equal removal performance for sulfates (Gr = 30% and D =29.3%); calcium (Gr = 84.7% and D = 85%); and iron (Gr = 99.5% and D = 99.6%). Another study assessed the hydraulic performance of permeable concrete of different aggregate types that was used in the treatment of acid mine drainage [180]. In that work, the effect of using 30% fly ash as an SCM was addressed at different water-to-cement ratios (0.25–0.27), and different coarse aggregates were used, including dolomite (67 and 95), granite (67–132), shale (67), and andesite (67) [180]. The studied samples were reported to have effective removal performances after 90 days for Al (98.1%), Mg (86.5%), Mn (99.8%), Zn (97.4%), and Fe (99.4%), with an increased chromium concentration due to chromium leaching (−112%) from the cement and fly ash. The samples showed high pore connectivity (95–99.7%), which reflects the maintenance of the hydraulic performance. In addition, it was reported that the isolated porosity increased from 0.1% to 0.86%, while the pore connectivity was reduced from 99.7% to 95.3% for the samples that contained fly ash. A study was dedicated to addressing the effect of the aggregates on the removal performance and costs of the remediation project [24]. In that study, the permeable reactive barrier mix design relied on the use of gravel, limestone, or soda lime glass beads in the presence or absence of fly ash as supplementary material to remediate mine acid drainage. A fixed water-to-cement ratio of 0.4 was used and a fly ash substitution of 25% was considered to prepare permeable concrete with a 25% void content. The results revealed that with high cement content, the hydration products buffered the water leading to the precipitation of lead, zinc, and cadmium. Barriers containing limestone were proved to provide a better removal performance for the studied heavy metals (see Table 3). The study analyzed the costs of the needed materials for the permeable concrete barriers in comparison with the aggregate alone and activated carbon to remediate a plume consisting of a single heavy metal, in which the lime stone aggregate alone had the least costs (USD 35,000,000 for zinc removal assuming a total mass of 1650×10^6 kg). It should be noted that these preliminary cost estimates did not consider the breakthrough characteristics of the barrier or any other kinetic data [24]. In addition, the study revealed that the calcite leaching from the concrete will coat the co-precipitates and provide an additional increase in the reactive surface sites that can increase the life service of the barrier. The leaching of the heavy metals from the prepared concrete was reported after 72 h and was found to be less than 5%. All of the studied samples were found to have a considerable buffering effect on the acid mine drainage solution that increased the initial pH from 5.6 to values higher than 11.4 after 3 days of contact between the samples and the solutions [24].

4.2. Use of Supplementary Cementitious Materials

Red mud, fly ash, nano silica, and a mixture of silica fume, zeolite, and iron oxide were tested to investigate their effects on the pollutant removal performance of permeable concrete samples. Table 5 summarizes these effects. Permeable concrete samples were prepared using OPC, coarse aggregates, water, superplasticizer, and red mud and were investigated to evaluate their performance in the removal of the contaminants from acid mine drainage [30]. The strength, porosity, and permeability of the samples were measured, the strength was found to be correlated negatively with the increase in the red mud content in the sample, and the permeability and porosity had a positive correlation with the red mud content. These results are attributed to the highly alkaline nature of the red mud and higher Na_2O content that led to a fast reaction with the water at the onset of the hydration reaction. In addition, the use of red mud was reported to improve the removal reaction kinetics. At an optimum influent pH = 4 and hydraulic retention time = 24 h, the complete removal of Cu, Mn, Cd, and Zn was achieved, and the removal mechanism was explained by the hydrolysis of the portlandite upon its reaction with the sulfate ions to produce gypsum that led to the precipitation of these ions accompanied by the sorption onto the C-S-H and hematite in the red mud. The prepared material was reported to have had its efficiency in the remediation of heavy metal from acid mine drainage samples collected from a mining area in China proven.

Table 5. Effect of using SCMs on the performance of permeable reactive concrete barriers.

Material	Mix Design			Performance	Refs.
	W/C	A/C	Supplemented Material		
Red mud	0.21–0.23	3.25	Red mud 25, 50% superplasticizer	Reduced the compressive strength, effective at pH = 4 hydraulic retention time of 20 h	[30]
FA	0.27	4.02	Fly ash 30% superplasticizer	Enhanced the real acid mine drainage treatment	[31]
FA+ nano silica	0.26	5	Fly ash 20%, Nano silica 6%	Enhanced the nitrate removal and compressive strength	[157]
SF+ zeolite+ iron oxide	0.25	4	SF 5.05% zeolite 5.45% iron oxide 0.5%	Enhanced the heavy metal removal	[155]

A mix design containing type 1 OPC, granite (95 mm) coarse aggregates, and fly ash as SCM was tested to treat acid mine drainage collected from a gold mine and coalfield [31]. The study investigated the role of fly ash in enhancing the reactive performance of the barrier by improving the pollutant removal. The hardened concrete with the supplementary material was found to enhance the pollutant removal performance for Al, Fe, Mn, Co, and Ni, and this behavior was attributed to the high buffering capacity of the material that raised the acid mine drainage pH to 12 and, subsequently, led to the precipitation of the pollutants' hydroxides. The portlandite in the permeable concrete reacted with the sulfate ions in the contaminated solutions leading to the formation of the expansive gypsum. The presence of the fly ash was concluded to mitigate the damage that could be initiated by the formation of the gypsum. The combined effect of the use of fly ash and nano silica on the properties of the hardened permeable reactive concrete was investigated [157]. In this regard, the optimization of the mix design of permeable concrete consists of OPC type 2, coarse aggregates (No. 4), fine aggregates (No.6 and8), nano silica and fly ash was conducted according to Taguchi multivariant procedure (L9). The compressive strength, density, and void ratio, and permeability were measured according to the ASTM C39, ASTM C 1754, and ACI522, respectively. The enhanced nitrate removal was attributed to the provision of new surface functional groups, Si^{2+} and Si oxide, due to the addition of the SCM, and FTIR investigations indicated that the hydroxyl stretching peak at 3430 cm^{-1}, silicon tetrahedral peak at 1050 cm^{-1}, and Si-O bending vibration peak at 446 cm^{-1} were affected by the sorption. Another study investigated the effect of the flow configuration—gravity and down–up configuration—on the pollutant removal kinetics of permeable concrete supplemented with fly ash and silica fume and concluded that the gravity flow had two orders of magnitude less liquid–concrete contact time to have a similar acid mine drainage treatment quality [179]. This conclusion should be considered carefully, as the use of pumps in the down–up configuration will change the nature of the application from a passive to active mode.

4.3. Performance Comparison with Zero-Valent Iron

A batch experiments was conducted to compare the pollutant removal performance of permeable concrete materials against that of zero-valent iron [176]. The permeable concrete was prepared from type 1 OPC and 67 mm granite aggregates at a water-to-cement ratio equal to 0.27. That study investigated the performance of three samples, namely, permeable concrete (CEM1), permeable concrete with 30% FA SCM (30%FA), and zero-valent iron (ZVI), in the treatment of two types of acid mine drainage: from a gold mine and from a coal mine. The ZVI sample was found to buffer the acid mine drainage solutions from 4.15–5.79 to 6–8, whereas the cement-based samples buffered the solutions to higher pH values of 9–12. The results of the batch reactor tests conducted for 43 days, as shown in Figure 12a,b, indicate that the removal rates for Al, Fe, Ni, Co, Pb, and Zn from both acid mine solutions were higher than 80% for all of the studied permeable reactive barrier samples. The permeable-concrete-based samples had higher removal performances for the removal of Mg and Mn from both solutions. The high removal performances of these samples are attributed to the pH-driven metal precipitation of the pollutants on the surface of the permeable concrete and the formed gypsum. It should be noted that all of the studied materials led to the release of sulfate, which was higher in the ZVI sample [176]. In another batch study, the two permeable concrete samples and the zero-valent iron sample were tested to investigated their pollutant removal performance in the treatment of acid mine drainage collected from the coal and gold mines [178]. The first permeable concrete sample was non supplemented, i.e., type 1 OPC and granite aggregates, and the second was supplemented permeable concrete with fly ash (30%). The results of the batch experiment were compared against effluent discharge standards and revealed that Zn, Fe, Ni, Co, Pb, and Al were effectively removed by the three tested material. Both permeable concrete materials had a better performance at removing Mn and Mg. However, the treated solutions with the permeable concrete were buffered to a high pH and contained higher Cr^{6+} concentrations that affected their overall quality. In a third study, the hydraulic and pollutant removal performances of the permeable concrete and zero-valent iron were evaluated under dynamic flow conditions in a column experiment [167]. In this work, the performance concrete barrier buffered the initial acid mine pH from 2.99 to 11, whereas the ZVI barrier buffered the pH to 9. The permeable concrete containing 30% fly ash was reported to improve the retardation factors over that of the zero-valent iron. In terms of the hydraulic performance, the hydrodynamic dispersion coefficient for the 30-FA permeable concrete was higher than that of the zero-valent iron. Finally for a barrier of 1.5 m thickness, the estimated life service of the permeable concrete was double that of the zero-valent iron.

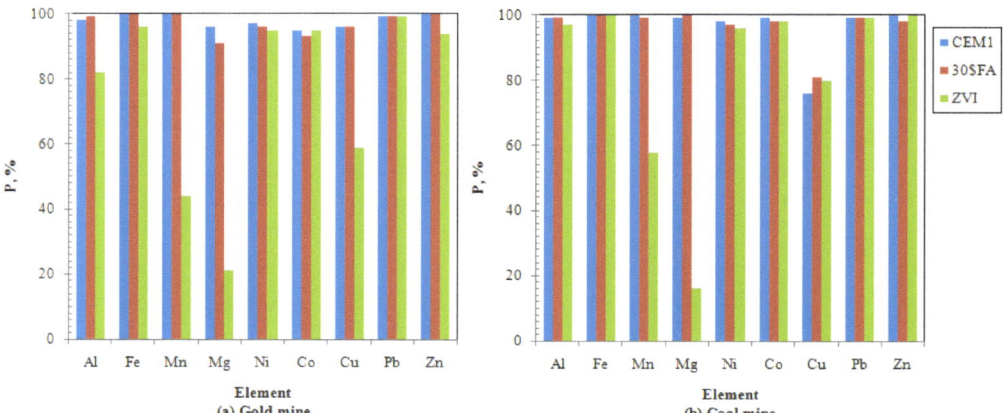

Figure 12. Removal performance of the different metals in the acid mine drainage from (**a**) gold and (**b**) coal mines (source data extracted from [176]).

4.4. Combined Treatment Method

A combination of active and passive processes were proposed for the treatment of acid mine drainage, and these processes included the use of permeable concrete as a pretreatment passive barrier and in active anaerobic digestion for the treatment of effluent collected from a coal mine in South Africa [175]. The mix design of the permeable concrete barrier was as follows: 4% silica fume, 8% fly ash, and granite coarse aggregate (13.2 and 9.5 mm). The use of the passive concrete barrier enabled the efficient removal of iron (99%), potassium (94%), and alumina (42%), whereas the bioreactor enabled the removal of COD (89.7%) and sulfate (99%). This study revealed that the iron precipitated in a respectively short contact time of 37 s and the satiability of the suspended ions presented an opportunity of its removal from the sludge.

4.5. Gaps in Investigating Permeable Reactive Concrete Barriers

As indicated above, a limited number of published research papers have been directed toward the investigation of the potential use of permeable concrete materials as permeable reactive barriers for ground water remediation. These efforts have mainly focused on studying their pollutant removal performance under the challenging operating conditions of the remediation of the mine acid drainage plume characterized by high sulfate content and acidity. In general, these conditions are known to affect the durability of cement-based materials, yet the obtained results from these research efforts reflect the potential feasibility of their application. This feasibility can be emphasized by addressing some gaps that exist in this field as follows:

- The beneficial use of fly ash to improve the mechanical strength of permeable reactive concrete barriers and to provide more active sites for the removal of pollutants is promising [31,174,178,179]. Yet, it was concluded the cement can contribute to an increase in the Cr concentration in the treated plume [178,180]. This point needs to be addressed in depth by investigating the stability of fly ash using a standardized test (e.g., TCLP) to assess the feasibility of this material's use. In addition, an in-depth analysis of the stability of the hardened material and the deduction of the chromium-leaching mechanism are needed.

- The reported mechanisms of pollutant removal were precipitation due to the alkalinity of the permeable reactive concrete media combined with physical and chemical sorption onto the hardened cement phases, the aggregates, and the SCM [30,173,176,178,179,181]. In particular, the reaction of the sulfate with the portlandite was identified as enhancing the precipitation of the pollutants. The effect of the portlandite reaction with the sulfate on

- the permeability of the barrier needs to be assessed, as the formation of expansive phases can affect the long-term hydraulic performance of the barrier.
- Concrete is known to have a high alkaline capacity, which has been reported to buffer the treated solution's pH to an unacceptable value, i.e., 11 [178]. This point can be addressed by using low-pH cement that can buffer the treated solutions to pH values comparable to that of the ZVI. Different types of additives can be used in this respect that should be studied in depth to ensure that the hardened material will meet the hydraulic and pollutant removal performances.
- Innovative permeable concretes have not been tested extensively for their applications in groundwater remediation and acid mine drainage treatment. In particular, magnesium phosphate cement is known for its fast hardening, near neutral pH, low water requirements, and high adhesive strength [5,6]. These materials have been tested for their application in low-development practices but not in permeable reactive barriers.
- The effect of the rheological characteristics of permeable concrete paste and their effects on the hydraulic and pollutant removal performances have been not investigated for permeable reactive concrete barriers. Most of the conducted studies relied on the use of the minimum amount of water (0.21–0.33) in conventional permeable concrete [30,31,156,179,180]. This limited range compared to that used in the low-development impact (0.25–0.45) practices reduces the porosity of the hardened material. This can affect the hydraulic performance of the barriers used in the remediation of the acid mine drainage over the long term, and the formation of expansive phases can lead to a further reduction in the porosity.
- The available experimental data on the performance of permeable concrete reactive materials were conducted within limited time frames of less than a year [21,24,176,178,180,183]. The long-term performances of these materials are required to be studied in depth to ensure the sustainable performance of these barriers throughout their service life and to identify threshold values for their reduced performances affecting their efficiency as a barrier.
- The durability of permeable reactive concrete barriers needs to be addressed to evaluate the effect of harsh operating conditions (i.e., high sulfate and acidic solutions) on the durability of such barriers. Future research needs to address the feasibility of performing scheduled maintenance or substitutions of the barrier material if the barrier does not reach its target remediation prior to reaching unacceptable reduced performances.
- Current research efforts in this field are focused at the lab-scale testing and mainly focus on conducting static experiments, with limited research studying dynamic conditions [30,31,36,181]. There is a need to address the performance of these materials under more realistic conditions that address upscale practical applications of these materials.
- The cost of the remediation technology is an important aspect that affects the decision-making process [21,31,177,180]. The cost estimate for these materials are limited, and these studies have addressed the costs of these materials and trench-type installation costs [24,30,179]. Detailed cost analyses have not been conducted; subsequently, there is a need to address the life cycle costs of the large-scale application of these materials.
- The environmental impacts of the use of permeable reactive concrete barriers need to be addressed. In this respect, the effect of the beneficial use of recycled wastes as SCMs and aggregates in conventional performance concrete should be quantified in terms of the reduction of the environmental footprint. In addition, the impacts of the buffer pH after treatment should be addressed.

5. Conclusions

The feasibility of using permeable concrete as a reactive barrier for the remediation of contaminated groundwater and the treatment of the acid mine drainage was addressed. This class of materials has been long implemented in low-impact development practices and their efficiency has been proven, but these materials are still considered innovative for their application as permeable reactive barriers, with a limited number of studies that have been published. These published studies have proved the efficiency of this class of materials for the removal of pollutants under the harsh acidic conditions of acid mine drainage, but some problems that affect the quality of the treated water have been identified. On the basis of the conducted literature review in this work on the mix design of conventional and innovative permeable concretes; their characterization and performance and durability assessment tests; and the feasibility assessment for their application as permeable reactive barriers, the gaps and recommended future research direction were identified to emphasize their feasibility toward pilot-scale applications. In this respect, the effect of the variation of the mix design of conventional and innovative permeable concretes on the rheology of the paste and the hydraulic, mechanical, and removal performances of the hardened material were identified. The practicality of the application of these materials in the remediation of contaminated groundwater and acid mine drainage were evaluated by reviewing and analyzing the published research in the field and identifying the gaps and suggestion of future research directions. The main conclusions that can be drawn from this work can be summarized as follows:

- The optimization of the mix design of the permeable reactive concrete barrier needs to be guided not only according to the required hydraulic and removal performances but also by the durability of the hardened materials over the designed service life;
- Identifying the environmental impacts of the optimized permeable reactive concrete barrier is crucial to ensure the sustainability of these materials. These impacts should consider the dynamics of the attenuation of the alkaline plume downstream of the barrier of the affected environment.

Author Contributions: Conceptualization, R.O.A.R.; methodology, R.O.A.R.; resources, R.O.A.R. and Y.-T.H.; writing—review and editing, R.O.A.R., A.M.E.-K. and Y.-T.H. All authors have read and agreed to the published version of the manuscript.

Funding: This research received no external funding.

Data Availability Statement: Not applicable.

Conflicts of Interest: The authors declare no conflict of interest.

References

1. LiberTexts. Functions of Water. Available online: https://med.libretexts.org/Courses/Metropolitan_State_University_of_Denver/Introduction_to_Nutrition_(Diker)/07%3A_Nutrients_Important_to_Fluid_and_Electrolyte_Balance/7.02%3A_Waters_Importance_to_Vitality (accessed on 26 July 2022).
2. Abdel Rahman, R.O.; El-Kamash, A.M.; Hung, Y.-T. Applications of Nano-Zeolite in Wastewater Treatment: An Overview. *Water* **2022**, *14*, 137. [CrossRef]
3. Beaudoin, J.; Odler, I. Hydration, setting and hardening of Portland cement. In *Lea's Chemistry of Cement and Concrete*; Elsevier Science: Amsterdam, The Netherlands, 2019; Volume 5, pp. 157–250.
4. Odler, I. Setting and hardening of Portland cement. In *Lea's Chemistry of Cement and Concrete*; Hewlett, P.C., Ed.; Elsevier Science: Amsterdam, The Netherlands, 2006; pp. 241–297.
5. Rahman, R.O.A.; Ojovan, M.I. Hydration process: Kinetics and thermodynamics. In *Sustainability of Life Cycle Management for Nuclear Cementation-Based Technologies*; Woodhead Publishing: Sawston, UK, 2021; pp. 125–160.
6. Rahman, R.O.A.; Rakhimov, R.Z.; Rakhimova, N.R.; Ojovan, M.I. *Cementitious Materials for Nuclear Waste Immobilization*; Wiley: New York, NY, USA, 2014; ISBN 9781118512005. [CrossRef]
7. Paine, K.A. Physicochemical and mechanical properties of Portland cement. In *Lea's Chemistry of Cement and Concrete*; Elsevier Science: Amsterdam, The Netherlands, 2019; pp. 285–340.
8. Saleh, H.M.; Rahman, R.O.A. (Eds.) Introductory chapter: Properties and application of cement based materials. In *Cement Based Materials*; Intech: London, UK, 2018; ISBN 978-1-78984-154-1. [CrossRef]
9. Wu, Y.; Qiao, W.G.; Li, Y.Z.; Zhang, S.; Sun, D.K.; Tang, C.; Liu, H.N.; Wang, T.J. Development and validation of environmentally friendly similar surrounding rock materials and cement slurry for surrounding rock repair and reinforcement. *J. Clean. Prod.* **2022**, *347*, 131288. [CrossRef]
10. Saleh, H.M.; Eskander, S.B. Innovative cement-based materials for environmental protection and restoration. In *New Materials in Civil Engineering*; Butterworth-Heinemann: Oxford, UK, 2020; pp. 613–641.
11. Tian, Y.; Themelis, N.J.; Zhao, D.; Bourtsalas, A.T.; Kawashima, S. Stabilization of Waste-to-Energy (WTE) fly ash for disposal in landfills or use as cement substitute. *Waste Manag.* **2022**, *150*, 227–243. [CrossRef] [PubMed]

12. Abdel Rahman, R.O.; Metwally, S.S.; El-Kamash, A.M. Improving the Performance of Engineering Barriers in Radioactive Waste Disposal Facilities: Role of Nano-Materials. In *Handbook of Nanomaterials and Nanocomposites for Energy and Environmental Applications*; Kharissova, O., Martínez, L., Kharisov, B., Eds.; Springer Nature: Cham, Switzerland, 2021, pp. 1183–1200. [CrossRef]
13. Chen, B.; Wu, F.; Qu, G.; Ning, P.; Ren, Y.; Liu, S.; Jin, C.; Li, H.; Zhao, C.; Liu, X.; et al. Waste control by waste: A comparative study on the application of carbide slag and quicklime in preparation of phosphogypsum-based ecological restoration materials. *Chem. Eng. Process. Process Intensif.* **2022**, *178*, 109051. [CrossRef]
14. Lai, H.; Du, J.; Zhou, C.; Liu, Z. Experimental study on ecological performance improvement of sprayed planting concrete based on the addition of polymer composite material. *Int. J. Environ. Res. Public Health* **2022**, *19*, 12121. [CrossRef]
15. Conner, J.R.; Hoeffner, S.L. The history of stabilization/solidification technology. *Crit. Rev. Environ. Sci. Technol.* **1998**, *28*, 325–396. [CrossRef]
16. Luhar, I.; Luhar, S.; Abdullah, M.M.A.B.; Sandu, A.V.; Vizureanu, P.; Razak, R.A.; Burduhos-Nergis, D.D.; Imjai, T. Solidification/Stabilization Technology for Radioactive Wastes Using Cement: An Appraisal. *Materials* **2023**, *15*, 954. [CrossRef]
17. Tyagi, S.; Annachhatre, A.P. A review on recent trends in solidification and stabilization techniques for heavy metal immobilization. *J. Mater. Cycles Waste Manag.* **2023**, *25*, 733–757. [CrossRef]
18. Tran, H.S.; Viet, N.T.T.; Duong, T.H.; Nguyen, L.H.; Kawamoto, K. Autoclaved aerated concrete grains as alternative absorbent and filter media for phosphorus recovery from municipal wastewater: A case study in Hanoi, Vietnam. *Environ. Technol. Innov.* **2023**, *31*, 103175. [CrossRef]
19. Abou-Elela, S.I.; Abo-El-Enein, S.A.; Hellal, M.S. Utilization of autoclaved aerated concrete solid waste as a bio-carrier in immobilized bioreactor for municipal wastewater treatment. *Desalination Water Treat.* **2019**, *168*, 108–116. [CrossRef]
20. Teymouri, E.; Wong, K.S.; Tan, Y.Y.; Pauzi, N.N.M. Mechanical behaviour of adsorbent pervious concrete using iron slag and zeolite as coarse aggregates. *Constr. Build. Mater.* **2023**, *388*, 131720. [CrossRef]
21. Shabalala, A.; Masindi, V. Insights into mechanisms governing the passive removal of inorganic contaminants from acid mine drainage using permeable reactive barrier. *J. Environ. Manag.* **2022**, *321*, 115866. [CrossRef] [PubMed]
22. Singh, A.; Sampath, P.V.; Biligiri, K.P. A review of sustainable pervious concrete systems: Emphasis on clogging, material characterization, and environmental aspects. *Constr. Build. Mater.* **2020**, *261*, 120491. [CrossRef]
23. Yun, C.M.; Rahman, M.R.; Kuok, K.K.; Sze, A.C.; Seng, A.L.; Bakri, M.K. Pervious Concrete Properties and Its Applications. In *Waste Materials in Advanced Sustainable Concrete: Reuse, Recovery and Recycle*; Springer International Publishing: Cham, Switzerland, 2022; pp. 1–23.
24. Holmes, R.R.; Hart, M.L.; Kevern, J.T. Heavy metal removal capacity of individual components of permeable reactive concrete. *J. Contam. Hydrol.* **2017**, *196*, 52–61. [CrossRef]
25. Park, S.-B.; Tia, M. An experimental study on the water-purification properties of porous concrete. *Cem. Concr. Res.* **2004**, *34*, 177–184. [CrossRef]
26. Huang, W.; Wang, H. Multi-aspect engineering properties and sustainability impacts of geopolymer pervious concrete. *Compos. Part B Eng.* **2022**, *242*, 110035. [CrossRef]
27. Shi, C. Hydraulic cement systems for stabilization/solidification. In *Stabilization and Solidification of Hazardous, Radioactive, and Mixed Wastes*; Spence, R.D., Shi, C., Eds.; CRC Press: Boca Raton, FL, USA, 2004; pp. 49–77.
28. Rahman, R.A.; El Abidin, D.Z.; Abou-Shady, H. Assessment of strontium immobilization in cement–bentonite matrices. *Chem. Eng. J.* **2013**, *228*, 772–780. [CrossRef]
29. Rahman, R.O.A.; Ojovan, M.I. Recent trends in the evaluation of cementitious material in radioactive waste disposal. In *Natural Resources and Control Processes*; Springer: Cham, Switzerland, 2016; pp. 401–448.
30. Xu, W.; Yang, H.; Mao, Q.; Luo, L.; Deng, Y. Removal of Heavy Metals from Acid Mine Drainage by Red Mud–Based Geopolymer Pervious Concrete: Batch and Long–Term Column Studies. *Polymers* **2022**, *14*, 5355. [CrossRef]
31. Shabalala, A.N.; Ekolu, S.O.; Diop, S.; Solomon, F. Pervious concrete reactive barrier for removal of heavy metals from acid mine drainage—Column study. *J. Hazard. Mater.* **2017**, *323*, 641–653. [CrossRef]
32. Chandrappa, A.K.; Biligiri, K.P. Pervious concrete as a sustainable pavement material–Research findings and future prospects: A state-of-the-art review. *Constr. Build. Mater.* **2016**, *111*, 262–274. [CrossRef]
33. Anwar, F.H.; El-Hassan, H.; Hamouda, M.; Hinge, G.; Mo, K.H. Meta-Analysis of the Performance of Pervious Concrete with Cement and Aggregate Replacements. *Buildings* **2022**, *12*, 461. [CrossRef]
34. Zhong, R.; Leng, Z.; Poon, C.-S. Research and application of pervious concrete as a sustainable pavement material: A state-of-the-art and state-of-the-practice review. *Constr. Build. Mater.* **2018**, *183*, 544–553. [CrossRef]
35. Ab Latif, A.; Putrajaya, R.; Ing, D.S. A Review of Porous Concrete Pavement: Compressive Strength and Clogging Investigation. *J. Adv. Res. Appl. Sci. Eng. Technol.* **2023**, *29*, 128–138.
36. Muthu, M.; Ramakrishnan, K.C.; Santhanam, M.; Rangarajan, M.; Kumar, M. Heavy metal removal and leaching from pervious concrete filter: Influence of operating water head and reduced graphene oxide addition. *J. Environ. Eng.* **2019**, *145*, 04019049. [CrossRef]
37. Clementino, F.d.S.; Santiago, J.M.; de Sousa, H.F.; da Conceição, I.G.C.; dos Santos, H.C. Use of permeable concrete with additives in wastewater treatment, focusing on biochar: A review. *Res. Soc. Dev.* **2021**, *10*, e562101019111. [CrossRef]
38. Rahman, R.O.A.; Ojovan, M.I. Toward Sustainable Cementitious Radioactive Waste Forms: Immobilization of Problematic Operational Wastes. *Sustainability* **2021**, *13*, 11992. [CrossRef]

39. Zhang, X.; Bai, C.; Qiao, Y.; Wang, X.; Jia, D.; Li, H.; Colombo, P. Porous geopolymer composites: A review. *Compos. Part A Appl. Sci. Manuf.* **2021**, *150*, 106629. [CrossRef]
40. Chen, X.; Guo, Y.; Ding, S.; Zhang, H.Y.; Xia, F.Y.; Wang, J.; Zhou, M. Utilization of red mud in geopolymer-based pervious concrete with function of adsorption of heavy metal ions. *J. Clean. Prod.* **2019**, *207*, 789–800. [CrossRef]
41. Yu, H.; Xu, M.; Chen, C.; He, Y.; Cui, X. A review on the porous geopolymer preparation for structural and functional materials applications. *Int. J. Appl. Ceram. Technol.* **2022**, *19*, 1793–1813. [CrossRef]
42. Kočí, V.; Černý, R. Directly foamed geopolymers: A review of recent studies. *Cem. Concr. Compos.* **2022**, *130*, 104530. [CrossRef]
43. Radina, L.; Sprince, A.; Pakrastins, L.; Gailitis, R.; Sakale, G. Foamed geopolymers: A review of recent studies. *J. Phys. Conf. Ser.* **2023**, *2423*, 012032. [CrossRef]
44. ACI. *522R-10. Report on Pervious Concrete (Reapproved 2011)*; American Concrete Institute: Farmington Hills, MI, USA, 2010. Available online: https://www.concrete.org/store/productdetail.aspx?ItemID=52210&Format=PROTECTED_PDF&Language=English&Units=US_AND_METRIC (accessed on 16 October 2023).
45. Singh, A.; Jagadeesh, G.S.; Sampath, P.V.; Biligir, K.P. Rational Approach for Characterizing In Situ Infiltration Parameters of Two-Layered Pervious Concrete Pavement Systems. *J. Mater. Civ. Eng.* **2019**, *31*, 04019258. [CrossRef]
46. NRMCA. *Previous in Practice Methods Materials Admixtures Guide to Specifying Pervious Concrete*; National Ready Mixed Concrete Association: Silver Spring, MD, USA. Available online: https://www.perviouspavement.org/downloads/PiP1.pdf (accessed on 16 October 2023).
47. Xie, H.-Z.; Li, L.G.; Ng, P.-L.; Liu, F. Effects of Solid Waste Reutilization on Performance of Pervious Concrete: A Review. *Sustainability* **2023**, *15*, 6105. [CrossRef]
48. Wijeyawardana, P.; Nanayakkara, N.; Gunasekara, C.; Karunarathna, A.; Law, D.; Pramanik, B.K. Improvement of heavy metal removal from urban runoff using modified pervious concrete. *Sci. Total Environ.* **2022**, *815*, 152936. [CrossRef] [PubMed]
49. Elizondo-Martinez, E.J.; Andres-Valeri, V.C.; Jato-Espino, D.; Rodriguez-Hernandez, J. Review of porous concrete as multifunctional and sustainable pavement. *J. Build. Eng.* **2020**, *27*, 100967. [CrossRef]
50. Teymouri, E.; Pauzi, N.N.M.; Wong, K.S. Developing Lignite Pervious Concrete for Application in Pedestrian Walkways and Urban Runoff Treatment. *Iran. J. Sci. Technol. Trans. Civ. Eng.* **2023**, *47*, 2949–2967. [CrossRef]
51. Ozel, B.F.; Sakallı, Ş.; Şahin, Y. The effects of aggregate and fiber characteristics on the properties of pervious concrete. *Constr. Build. Mater.* **2022**, *356*, 129294. [CrossRef]
52. Soto-Pérez, L.; Hwang, S. Mix design and pollution control potential of pervious concrete with non-compliant waste fly ash. *J. Environ. Manag.* **2016**, *176*, 112–118. [CrossRef]
53. Saboo, N.; Shivhare, S.; Kori, K.K.; Chandrappa, A.K. Effect of fly ash and metakaolin on pervious concrete properties. *Constr. Build. Mater.* **2019**, *223*, 322–328. [CrossRef]
54. Kim, H.-H.; Kim, C.-S.; Jeon, J.-H.; Park, C.-G. Effects on the physical and mechanical properties of porous concrete for plant growth of blast furnace slag, natural jute fiber, and styrene butadiene latex using a dry mixing manufacturing process. *Materials* **2016**, *9*, 84. [CrossRef]
55. Aoki, Y.; Sri Ravindrarajah, R.; Khabbaz, H. Properties of pervious concrete containing fly ash. *Road Mater. Pavement Des.* **2012**, *13*, 1–11. [CrossRef]
56. Wang, H.; Li, H.; Liang, X.; Zhou, H.; Xie, N.; Dai, Z. Investigation on the mechanical properties and environmental impacts of pervious concrete containing fly ash based on the cement-aggregate ratio. *Constr. Build. Mater.* **2019**, *202*, 387–395. [CrossRef]
57. Opiso, E.M.; Supremo, R.P.; Perodes, J.R. Effects of coal fly ash and fine sawdust on the performance of pervious concrete. *Heliyon* **2019**, *5*, e02783. [CrossRef] [PubMed]
58. Carmichael, M.J.; Arulraj, G.P.; Meyyappan, P.L. Effect of partial replacement of cement with nano fly ash on permeable concrete: A strength study. *Mater. Today Proc.* **2021**, *43*, 2109–2116. [CrossRef]
59. López-Carrasquillo, V.; Hwang, S. Comparative assessment of pervious concrete mixtures containing fly ash and nanomaterials for compressive strength, physical durability, permeability, water quality performance and production cost. *Constr. Build. Mater.* **2017**, *139*, 148–158. [CrossRef]
60. Arifi, E.; Cahya, E.N. Effect of fly ash on the strength of porous concrete using recycled coarse aggregate to replace low-quality natural coarse aggregate. *AIP Conf. Proc.* **2017**, *1887*, 020055. [CrossRef]
61. Bilal, H.; Chen, T.; Ren, M.; Gao, X.; Su, A. Influence of silica fume, metakaolin & SBR latex on strength and durability performance of pervious concrete. *Constr. Build. Mater.* **2021**, *275*, 122124.
62. Ortega-Villar, R.; Lizarraga-Mendiola, L.; Coronel-Olivares, C.; Lopez-Leon, L.D.; Bigurra-Alzati, C.A.; Vazquez-Rodriguez, G.A. Effect of photocatalytic Fe_2O_3 nanoparticles on urban runoff pollutant removal by permeable concrete. *J. Environ. Manag.* **2019**, *242*, 487–495. [CrossRef]
63. Liang, X.; Cui, S.; Li, H.; Abdelhady, A.; Wang, H.; Zhou, H. Removal effect on stormwater runoff pollution of porous concrete treated with nanometer titanium dioxide. *Transp. Res. D Transp. Environ.* **2019**, *73*, 34–45. [CrossRef]
64. Jian, S.; Wei, B.; Zhi, X.; Tan, H.; Li, B.; Li, X.; Lv, Y. Abrasion resistance improvement of recycled aggregate pervious concrete with granulated blast furnace slag and copper slag. *J. Adv. Concr. Technol.* **2021**, *19*, 1088–1099. [CrossRef]
65. Dahiru, D.; Ibrahim, M.; Gado, A.A. Evaluation of the effect of volcanic ash on the properties of concrete. *ATBU J. Environ. Technol.* **2019**, *12*, 79–100.

66. El-Hassan, H.; Kianmehr, P.; Zouaoui, S. Properties of pervious concrete incorporating recycled concrete aggregates and slag. *Constr. Build. Mater.* **2019**, *212*, 164–175. [CrossRef]
67. El-Hassan, H.; Kianmehr, P. Pervious concrete pavement incorporating GGBS to alleviatepavement runoff and improve urban sustainability. *Road Mater. Pavement Des.* **2016**, *19*, 167–181. [CrossRef]
68. Kim, I.T.; Park, C.; Kim, S.; Cho, Y.-H. Evaluation of field applicability of pervious concrete materials for airport pavement cement treated drainage base course. *Mater. Res. Innov.* **2015**, *19*, 378–388. [CrossRef]
69. Divsholi, B.S.; Lim, T.Y.D.; Teng, S. Durability properties and microstructure of ground granulated blast furnace slag cement concrete. *Int. J. Concr. Struct. Mater.* **2014**, *8*, 157–164. [CrossRef]
70. Güneyisi, E.; Gesŏglu, M.; Kareem, Q.; İpek, S. Effect of different substitution of natural aggregate by recycled aggregate on performance characteristics of pervious concrete. *Mater. Struct.* **2014**, *49*, 521–536. [CrossRef]
71. Zaetang, Y.; Sata, V.; Wongsa, A.; Chindaprasirt, P. Properties of pervious concrete containing recycled concrete block aggregate and recycled concrete aggregate. *Constr. Build. Mater.* **2016**, *111*, 15–21. [CrossRef]
72. Zhang, Z.; Zhang, Y.; Yan, C.; Liu, Y. Influence of crushing index on properties of recycled aggregates pervious concrete. *Constr. Build. Mater.* **2017**, *135*, 112–118. [CrossRef]
73. Zou, D.; Wang, Z.; Shen, M.; Liu, T.; Zhou, A. Improvement in freeze-thaw durability of recycled aggregate permeable concrete with silane modification. *Constr. Build. Mater.* **2020**, *268*, 121097. [CrossRef]
74. Cai, X.; Wu, K.; Huang, W.; Yu, J.; Yu, H. Application of recycled concrete aggregates and crushed bricks on permeable concrete road base. *Road Mater. Pavement Des.* **2020**, *22*, 2181–2196. [CrossRef]
75. Liu, W.J. Performance of new permeable concrete materials based on mechanical strength. *Nat. Environ. Pollut. Technol.* **2019**, *18*, 1683–1689.
76. Debnath, B.; Sarkar, P.P. Quantification of random pore features of porous concrete mixes prepared with brick aggregate: An application of stereology and mathematical morphology. *Constr. Build. Mater.* **2021**, *294*, 123594. [CrossRef]
77. Debnath, B.; Sarkar, P.P. Characterization of pervious concrete using over burnt brick as coarse aggregate. *Constr. Build. Mater.* **2020**, *242*, 11815. [CrossRef]
78. Li, L.G.; Ng, P.L.; Zeng, K.L.; Xie, H.Z.; Cheng, C.M.; Kwan, A.K.H. Experimental study and modelling of fresh behaviours of basalt fibre-reinforced mortar based on average water film thickness and fibre factor. *Materials* **2023**, *16*, 2137. [CrossRef]
79. Wang, S.; Zhang, G.; Wang, B.; Wu, M. Mechanical strengths and durability properties of pervious concretes with blended steel slag and natural aggregate. *J. Clean. Prod.* **2020**, *271*, 122590. [CrossRef]
80. Adil, G.; Kevern, J.T.; Mann, D. Influence of silica fume on mechanical and durability properties of pervious concrete. *Constr. Build. Mater.* **2020**, *247*, 118453. [CrossRef]
81. Toghroli, A.; Mehrabi, P.; Shariati, M.; Trung, N.T.; Jahandari, S.; Rasekh, H. Evaluating the use of recycled concrete aggregate and pozzolanic additives in fiber-reinforced pervious concrete with industrial and recycled fibers. *Constr. Build. Mater.* **2020**, *252*, 118997. [CrossRef]
82. Mondal, S.; Biligiri, K.P. Crumb Rubber and Silica Fume Inclusions in Pervious Concrete Pavement Systems: Evaluation of Hydrological, Functional, and Structural Properties. *J. Test. Eval.* **2018**, *46*, 20170032. [CrossRef]
83. Qin, Y.; Pang, X.; Tan, K.; Bao, T. Evaluation of pervious concrete performance with pulverized biochar as cement replacement. *Cem. Concr. Compos.* **2021**, *119*, 104022. [CrossRef]
84. Adamu, M.; Ayeni, K.O.; Haruna, S.I.; Ibrahim Mansour, Y.E.-H.; Haruna, S. Durability performance of pervious concrete containing rice husk ash and calcium carbide: A response surface methodology approach. *Case Stud. Constr. Mater.* **2021**, *14*, e00547. [CrossRef]
85. Hesami, S.; Ahmadi, S.; Nematzadeh, M. Effects of rice husk ash and fiber on mechanical properties of pervious concrete pavement. *Constr. Build. Mater.* **2014**, *53*, 680–691. [CrossRef]
86. Hari, R.; Mini, K. Mechanical and durability properties of basalt-steel wool hybrid fibre reinforced pervious concrete—A Box Behnken approach. *J. Build. Eng.* **2023**, *70*, 106307. [CrossRef]
87. Mohammed, B.S.; Liew, M.S.; Alaloul, W.S.; Khed, V.C.; Hoong, C.Y.; Adamu, M. Properties of nano-silica modified pervious concrete. *Case Stud. Constr. Mater.* **2018**, *8*, 409–422. [CrossRef]
88. Tarangini, D.; Sravana, P.; Rao, P.S. Effect of nano silica on frost resistance of pervious concrete. *Mater. Today Proc.* **2022**, *51*, 2185–2189. [CrossRef]
89. Nazeer, M.; Kapoor, K.; Singh, S. Strength, durability and microstructural investigations on pervious concrete made with fly ash and silica fume as supplementary cementitious materials. *J. Build. Eng.* **2023**, *69*, 106275. [CrossRef]
90. Mehrabi, P.; Shariati, M.; Kabirifar, K.; Jarrah, M.; Rasekh, H.; Trung, N.T.; Shariati, A.; Jahandari, S. Effect of pumice powder and nano-clay on the strength and permeability of fiber-reinforced pervious concrete incorporating recycled concrete aggregate. *Constr. Build. Mater.* **2021**, *287*, 122652. [CrossRef]
91. Liu, R.; Xiao, H.; Pang, S.D.; Geng, J.; Yang, H. Application of Sterculia foetida petiole wastes in lightweight pervious concrete. *J. Clean. Prod.* **2020**, *246*, 118972. [CrossRef]
92. Khankhaje, E.; Razman, M.; Mirza, J.; Warid, M.; Rafieizonooz, M. Properties of sustainable lightweight pervious concrete containing oil palm kernel shell as coarse aggregate. *Constr. Build. Mater.* **2016**, *126*, 1054e1065. [CrossRef]
93. Krishnan, C.; Santhanam, M.; Kumar, M.; Rangarajan, M. Iron oxide-modified pervious concrete filter for lead removal from wastewater. *Environ. Technol. Innov.* **2022**, *28*, 102681. [CrossRef]

94. Chen, Y.; Wang, K.; Wang, X.; Zhou, W. Strength, fracture and fatigue of pervious concrete. *Constr. Build. Mater.* **2013**, *42*, 97–104. [CrossRef]
95. Muthukumar, S.; Saravanan, A.J.; Raman, A.; Sundaram, M.S.; Angamuthu, S.S. Investigation on the mechanical properties of eco-friendly pervious concrete. *Mater. Today Proc.* **2021**, *46*, 4909–4914. [CrossRef]
96. Giustozzi, F. Polymer-modified pervious concrete for durable and sustainable transportation infrastructures. *Constr. Build. Mater.* **2016**, *111*, 502–512. [CrossRef]
97. AlShareedah, O.; Nassiri, S. Pervious concrete mixture optimization, physical, and mechanical properties and pavement design: A review. *J. Clean. Prod.* **2020**, *288*, 125095. [CrossRef]
98. Lang, L.; Duan, H.; Chen, B. Properties of pervious concrete made from steel slag and magnesium phosphate cement. *Constr. Build. Mater.* **2019**, *209*, 95–104. [CrossRef]
99. Gowda, S.B.; Goudar, S.K.; Thanu, H.; Monisha, B. Performance evaluation of alkali activated slag based recycled aggregate pervious concrete. *Mater. Today Proc.* **2023**. [CrossRef]
100. Lai, Z.; Hu, Y.; Fu, X.; Lu, Z.; Lv, S. Preparation of porous materials by magnesium phosphate cement with high permeability. *Adv. Mater. Sci. Eng.* **2018**, *2018*, 5910560.
101. Zhao, S.; Zhang, D.; Li, Y.; Gao, H.; Meng, X. Physical and Mechanical Properties of Novel Porous Ecological Concrete Based on Magnesium Phosphate Cement. *Materials* **2022**, *15*, 7521. [CrossRef] [PubMed]
102. Kim, G.; Jang, J.; Khalid, H.R.; Lee, H. Water purification characteristics of pervious concrete fabricated with CSA cement and bottom ash aggregates. *Constr. Build. Mater.* **2017**, *136*, 1–8. [CrossRef]
103. Phillip, E.; Choo, T.F.; Khairuddin, N.W.A.; Abdel Rahman, R.O. On the Sustainable Utilization of Geopolymers for Safe Management of Radioactive Waste: A Review. *Sustainability* **2023**, *15*, 1117. [CrossRef]
104. Luhar, I.; Luhar, S. A comprehensive review on fly ash-based geopolymer. *J. Compos. Sci.* **2022**, *6*, 219. [CrossRef]
105. Garg, M.; Valeo, C.; Gupta, R.; Prasher, S.; Sharma, N.R.; Constabel, P. Integrating natural and engineered remediation strategies for water quality management within a low-impact development (LID) approach. *Environ. Sci. Pollut. Res.* **2018**, *25*, 29304–29313. [CrossRef] [PubMed]
106. Liang, X.; Ji, Y. Mechanical properties and permeability of red mud-blast furnace slag-based geopolymer concrete. *SN Appl. Sci.* **2021**, *3*, 23. [CrossRef]
107. Liang, X.; Ji, Y. Experimental study on durability of red mud-blast furnace slag geopolymer mortar. *Constr. Build. Mater.* **2020**, *267*, 120942. [CrossRef]
108. Sun, Z.; Lin, X.; Vollpracht, A. Pervious concrete made of alkali activated slag and geopolymers. *Constr. Build. Mater.* **2018**, *189*, 797–803. [CrossRef]
109. Walling, S.A.; Provis, J.L. Magnesia-based cements: A journey of 150 years, and cements for the future? *Chem. Rev.* **2016**, *116*, 4170–4204. [CrossRef]
110. Rahman, R.O.A.; Ojovan, M.I. Life cycle of nuclear cementitious structures, systems, and components. In *Sustainability of Life Cycle Management for Nuclear Cementation-Based Technologies*; Rahman, R.O.A., Ojovan, M.I., Eds.; Elsevier-Woodhead Publishing: Sawston, UK, 2021; pp. 89–121. [CrossRef]
111. Zheng, X.; Pan, J.; Easa, S.; Fu, T.; Liu, H.; Liu, W.; Qiu, R. Utilization of copper slag waste in alkali-activated metakaolin pervious concrete. *J. Build. Eng.* **2023**, *76*, 107246. [CrossRef]
112. Lo, F.-C.; Lee, M.-G.; Lo, S.-L. Effect of coal ash and rice husk ash partial replacement in ordinary Portland cement on pervious concrete. *Constr. Build. Mater.* **2021**, *286*, 122947. [CrossRef]
113. Cai, J.; Liu, Z.; Xu, G.; Tian, Q.; Shen, W.; Li, B.; Chen, T. Mix design methods for pervious concrete based on the mesostructure: Progress, existing problems and recommendation for future improvement. *Case Stud. Constr. Mater.* **2022**, *17*, e01253. [CrossRef]
114. Rahman, R.O.A.; Ojovan, M.I. *Techniques to test cementitious systems through their life cycles, In Sustainability of Life Cycle Management for Nuclear Cementation-Based Technologies*; Rahman, R.O.A., Ojovan, M.I., Eds.; Woodhead Publishing: Sawston, UK, 2021; pp. 407–430. [CrossRef]
115. Li, K. *Durability Design of Concrete Structures: Phenomena, Modeling, and Practice*, 1st ed.; John Wiley & Sons Singapore Pte. Ltd.: Singapore, 2016.
116. Malhotra, V.M.; Carino, N.J. *Handbook on Non-Destructive Testing of Concrete*; CRC Press, LLC: Boca Raton, FL, USA, 2004.
117. Martins Filho, S.T.; Bosquesi, E.M.; Fabro, J.R.; Pieralisi, R. Characterization of pervious concrete focusing on non-destructive testing. *Rev. IBRACON Estrut. Mater.* **2020**, *13*, 483–500. [CrossRef]
118. Park, S.; Ju, S.; Kim, H.-K.; Seo, Y.-S.; Pyo, S. Effect of the rheological properties of fresh binder on the compressive strength of pervious concrete. *journal of materials research and technology. J. Mater. Res. Technol.* **2022**, *17*, 636–648. [CrossRef]
119. Kim, H.; Lee, H. Influence of cement flow and aggregate type on the mechanical and acoustic characteristics of porous concrete. *Appl. Acoust.* **2010**, *71*, 607–615. [CrossRef]
120. Nambiar, E.K.K.; Ramamurthy, K. Fresh state characteristics of foam concrete. *J. Mater. Civ. Eng.* **2008**, *20*, 111–117. [CrossRef]
121. Harshith, S.D.; Ahmad, E. Experimental Investigation of Porous Concrete for Concrete Pavement. *Int. J. Eng. Res.* **2020**, *9*, 657–660. [CrossRef]
122. Juradin, S.; Mihanović, F.; Ostojić-Škomrlj, N.; Rogošić, E. Pervious Concrete Reinforced with Waste Cloth Strips. *Sustainability* **2022**, *14*, 2723. [CrossRef]

123. da Costa, F.B.; Haselbach, L.M.; da Silva Filho, L.C. Pervious concrete for desired porosity: Influence of w/c ratio and a rheology-modifying admixture. *Constr. Build. Mater.* **2021**, *268*, 121084. [CrossRef]
124. Xie, X.; Zhang, T.; Yang, Y.; Lin, Z.; Wei, J.; Yu, Q. Maximum paste coating thickness without voids clogging of pervious concrete and its relationship to the rheological properties of cement paste. *Constr. Build. Mater.* **2018**, *168*, 732–746. [CrossRef]
125. Wang, Z.; Zou, D.; Liu, T.; Zhou, A.; Shen, M. A novel method to predict the mesostructure and performance of pervious concrete. *Constr. Build. Mater.* **2020**, *263*, 120117. [CrossRef]
126. Jimma, B.E.; Rangaraju, P.R. Film-forming ability of flowable cement pastes and its application in mixture proportioning of pervious concrete. *Constr. Build. Mater.* **2014**, *71*, 273–282. [CrossRef]
127. Risson, K.D.B.D.S.; Sandoval, G.F.B.; Pinto, F.S.C.; Camargo, M.; De Moura, A.C.; Toralles, B.M. Molding procedure for pervious concrete specimens by density control. *Case Stud. Constr. Mater.* **2021**, *15*, e00619. [CrossRef]
128. Banevičienė, V.; Malaiškienė, J.; Boris, R.; Zach, J. The Effect of Active Additives and Coarse Aggregate Granulometric Composition on the Properties and Durability of Pervious Concrete. *Materials* **2022**, *15*, 1035. [CrossRef]
129. Tang, C.-W.; Cheng, C.-K.; Ean, L.-W. Mix Design and Engineering Properties of Fiber-Reinforced Pervious Concrete Using Lightweight Aggregates. *Appl. Sci.* **2022**, *12*, 524. [CrossRef]
130. Muda, M.M.; Legese, A.M.; Urgessa, G.; Boja, T. Strength, Porosity and Permeability Properties of Porous Concrete Made from Recycled Concrete Aggregates. *Constr. Mater.* **2023**, *3*, 81–92. [CrossRef]
131. Geng, H.; Xu, Q.; Duraman, S.B.; Li, Q. Effect of Rheology of Fresh Paste on the Pore Structure and Properties of Pervious Concrete Based on the High Fluidity Alkali-Activated Slag. *Crystals* **2021**, *11*, 593. [CrossRef]
132. Laskar, A.I.; Bhattacharjee, R. Effect of Plasticizer and Superplasticizer on Rheology of Fly-Ash-Based Geopolymer Concrete. *ACI Mater. J.* **2013**, *110*, 513–518.
133. Amini, K.; Asce, S.M.; Wang, X.; Delatte, N.; Asce, F. Statistical modeling of hydraulic and mechanical properties of pervious concrete using nondestructive tests. *J. Mater. Civ. Eng.* **2018**, *30*, 04018077. [CrossRef]
134. Bordelon, A.C.; Roesler, J.R. Spatial distribution of synthetic fibers in concrete with X-ray computed tomography. *Cem. Concr. Compos.* **2014**, *53*, 35–43. [CrossRef]
135. Zhang, J.; Ma, G.; Ming, R.; Cui, X.; Li, L.; Xu, H. Numerical study on seepage flow in pervious concrete based on 3D CT imaging. *Constr. Build. Mater.* **2018**, *161*, 468–478. [CrossRef]
136. Yu, F.; Sun, D.; Hu, M.; Wang, J. Study on the pores characteristics and permeability simulation of pervious concrete based on 2D/3D CT images. *Constr. Build. Mater.* **2019**, *200*, 687–702. [CrossRef]
137. Wang, G.; Chen, X.; Dong, Q.; Yuan, J.; Hong, Q. Mechanical performance study of pervious concrete using steel slag aggregate through laboratory tests and numerical simulation. *J. Clean. Prod.* **2020**, *262*, 121208. [CrossRef]
138. Zhong, R.; Wille, K. Linking pore system characteristics to the compressive behavior of pervious concrete. *Cem. Concr. Compos.* **2016**, *70*, 130–138. [CrossRef]
139. Qin, Y.; Yang, H.; Deng, Z.; He, J. Water permeability of pervious concrete is dependent on the applied pressure and testing methods. *Ann. Mater. Sci. Eng.* **2015**, *2015*, 404136. [CrossRef]
140. Lederle, R.; Shepard, T.; Meza, V.D.L.V. Comparison of methods for measuring infiltration rate of pervious concrete. *Constr. Build. Mater.* **2020**, *244*, 118339. [CrossRef]
141. Sandoval, G.F.; Galobardes, I.; Teixeira, R.S.; Toralles, B.M. Comparison between the falling head and the constant head permeability tests to assess the permeability coefficient of sustainable Pervious Concretes. *Case Stud. Constr. Mater.* **2017**, *7*, 317–328. [CrossRef]
142. *ASTM C109*; Standard Test Method for Compressive Strength of Hydraulic Cement Mortars (Using 2-in. or [50-mm] Cube Specimens. ASTM: West Conshohocken, PA, USA, 2020. [CrossRef]
143. *ASTM C39*; Standard Test Method for Compressive Strength of Cylindrical Concrete Specimens. ASTM: West Conshohocken, PA, USA, 2021. [CrossRef]
144. *ASTM C496*; Standard Test Method for Splitting Tensile Strength of Cylindrical Concrete Specimens. ASTM: West Conshohocken, PA, USA, 2017. [CrossRef]
145. *ASTM C293*; Standard Test Method for Flexural Strength of Concrete (Using Simple Beam With Center-Point Loading). ASTM: West Conshohocken, PA, USA, 2016. [CrossRef]
146. Rahman, R.O.A.; Ojovan, M.I. *Sustainability of cementitious structures, systems, and components (SSC's): Long-term environmental stressors, In Sustainability of Life Cycle Management for Nuclear Cementation-Based Technologies*; Rahman, R.O.A., Ojovan, M.I., Eds.; Elsevier-Woodhead Publishing: Sawston, UK, 2021; pp. 181–232. [CrossRef]
147. *ASTM C452-21*; Standard Test Method for Potential Expansion of Portland-Cement Mortars Exposed to Sulfate. ASTM: West Conshohocken, PA, USA, 2021. [CrossRef]
148. *ASTM C1778*; Standard Guide for Reducing the Risk of Deleterious Alkali-Aggregate Reaction in Concrete. ASTM: West Conshohocken, PA, USA, 2022. [CrossRef]
149. *ASTMC666*; Standard Test Method for Resistance of Concrete to Rapid Freezing and Thawing. ASTM: West Conshohocken, PA, USA, 2015. [CrossRef]
150. *ASTM C88*; Standard Test Method for Soundness of Aggregates by Use of Sodium Sulfate or Magnesium Sulfate. ASTM: West Conshohocken, PA, USA, 2018. [CrossRef]

151. Teymouri, E.; Wong, K.S.; Rouhbakhsh, M.; Pahlevani, M.; Forouzan, M. Evaluating the Clogging Phenomenon in Pervious Concrete from January 2015 to December 2022. *Civ. Sustain. Urban Eng.* **2023**, *3*, 70–80. [CrossRef]
152. Zhang, X.; Li, H.; Harvey, J.T.; Liang, X.; Xie, N.; Jia, M. Purification effect on runoff pollution of porous concrete with nano-TiO$_2$ photocatalytic coating. *Transp. Res. Part D Transp. Environ.* **2021**, *101*, 103101. [CrossRef]
153. Pilon, B.S.; Tyner, J.S.; Yoder, D.C.; Buchanan, J.R. The effect of pervious concrete on water quality parameters: A case study. *Water* **2019**, *11*, 263. [CrossRef]
154. Muthu, M.; Santhanam, M.; Kumar, M. Pb removal in pervious concrete filter: Effects of accelerated carbonation and hydraulic retention time. *Constr. Build. Mater.* **2018**, *174*, 224–232. [CrossRef]
155. Yousefi, A.; Matavos-Aramyan, S. Mix Design Optimization of Silica Fume-Based Pervious Concrete for Removal of Heavy Metals from Wastewaters. *Silicon* **2018**, *10*, 1737–1744. [CrossRef]
156. Azad, A.; Saeedian, A.; Mousavi, S.-F.; Karami, H.; Farzin, S.; Singh, V.P. Effect of zeolite and pumice powders on the environmental and physical characteristics of green concrete filters. *Constr. Build. Mater.* **2019**, *240*, 117931. [CrossRef]
157. Alighardashi, A.; Mehrani, M.J.; Ramezanianpour, A.M. Pervious concrete reactive barrier containing nano-silica for nitrate removal from contaminated water. *Environ. Sci. Pollut. Res. Int.* **2018**, *25*, 29481–29492. [CrossRef]
158. Medawela, S.; Indraratna, B.; Athuraliya, S.; Lugg, G.; Nghiem, L.D. Monitoring the performance of permeable reactive barriers constructed in acid sulfate soils. *Eng. Geol.* **2021**, *296*, 106465. [CrossRef]
159. Rad, P.R.; Fazlali, A. Optimization of permeable reactive barrier dimensions and location in groundwater remediation contaminated by landfill pollution. *J. Water Process Eng.* **2020**, *35*, 101196. [CrossRef]
160. Rahman, R.A.; Moamen, O.A.; Hanafy, M.; Monem, N.A. Preliminary investigation of zinc transport through zeolite-X barrier: Linear isotherm assumption. *Chem. Eng. J.* **2012**, *185–186*, 61–70. [CrossRef]
161. Day, S.R.; O'Hannesin, S.F.; Marsden, L. Geotechnical techniques for the construction of reactive barriers. *J. Hazard. Mater.* **1999**, *67*, 285–297. [CrossRef] [PubMed]
162. Thakur, A.K.; Vithanage, M.; Das, D.B.; Kumar, M. A review on design, material selection, mechanism, and modelling of permeable reactive barrier for community-scale groundwater treatment. *Environ. Technol. Innov.* **2020**, *19*, 100917. [CrossRef]
163. Faisal, A.A.H.; Sulaymon, A.H.; Khaliefa, Q.M. A review of permeable reactive barrier as passive sustainable technology for groundwater remediation. *Int. J. Environ. Sci. Technol.* **2017**, *15*, 1123–1138. [CrossRef]
164. Singh, R.; Chakma, S.; Birke, V. Performance of field-scale permeable reactive barriers: An overview on potentials and possible implications for in-situ groundwater remediation applications. *Sci. Total. Environ.* **2023**, *858*, 158838. [CrossRef]
165. Song, J.; Huang, G.; Han, D.; Hou, Q.; Gan, L.; Zhang, M. A review of reactive media within permeable reactive barriers for the removal of heavy metal(loid)s in groundwater: Current status and future prospects. *J. Clean. Prod.* **2021**, *319*, 128644. [CrossRef]
166. Sakr, M.; El Agamawi, H.; Klammler, H.; Mohamed, M.M. Permeable reactive barriers as an effective technique for groundwater remediation: A review. *Groundw. Susain. Dev.* **2023**, *21*, 100914. [CrossRef]
167. O'Hannesin, S.F.; Gillham, R.W. Long-term performance of an in situ "iron wall" for remediation of VOCs. *Groundwater* **1998**, *36*, 164–170. [CrossRef]
168. Smith, J.; Boshoff, G.; Bone, B. Good practice guidance on permeable reactive barriers for remediating polluted groundwater, and a review of their use in the UK. *Land Contam. Reclam.* **2003**, *11*, 411–418. [CrossRef]
169. Flury, B.; Eggenberger, U.; Mäder, U. First results of operating and monitoring an innovative design of a permeable reactive barrier for the remediation of chromate contaminated groundwater. *Appl. Geochem.* **2009**, *24*, 687–696. [CrossRef]
170. Wanner, C.; Zink, S.; Eggenberger, U.; Mäder, U. Assessing the Cr(VI) reduction efficiency of a permeable reactive barrier using Cr isotope measurements and 2D reactive transport modeling. *J. Contam. Hydrol.* **2012**, *131*, 54–63. [CrossRef] [PubMed]
171. Wang, Y.; Pleasant, S.; Jain, P.; Powell, J.; Townsend, T. Calcium carbonate-based permeable reactive barriers for iron and manganese groundwater remediation at landfills. *Waste Manag.* **2016**, *53*, 128–135. [CrossRef]
172. Budania, R.; Dangayach, S. A comprehensive review on permeable reactive barrier for the remediation of groundwater contamination. *J. Environ. Manag.* **2023**, *332*, 117343. [CrossRef] [PubMed]
173. Mehrani, M.J.; Mehrani, A.A.; Alighardashi, A.; Ramezanianpour, A.M. An experimental study on the nitrate removal ability of aggregates used in pervious concrete. *Desalination Water Treat* **2017**, *86*, 124–130. [CrossRef]
174. Ekolu, S.O.; Bitandi, L.K. Prediction of Longevities of ZVI and Pervious Concrete Reactive Barriers Using the Transport Simulation Model. *J. Environ. Eng.* **2018**, *144*, 04018074. [CrossRef]
175. Thisani, S.K.; Von Kallon, D.V.; Byrne, P. Co-Remediation of Acid Mine Drainage and Industrial Effluent Using Passive Permeable Reactive Barrier Pre-Treatment and Active Co-Bioremediation. *Minerals* **2022**, *12*, 565. [CrossRef]
176. Shabalala, A. *Efficacies of Pervious Concrete and Zero-Valent Iron as Reactive Media for Treating Acid Mine Drainage*; Mine Water Solutions; Pope, J., Wolkersdorfer, C., Weber, A., Sartz, L., Wolkersdorfer, K., Eds.; IMWA: Wendelstein, Germany, 2020; pp. 83–87.
177. Shabalala, A.N. Utilisation of Pervious Concrete for Removal of Heavy Metals in Contaminated Waters: Opportunities and Challenges. In Proceedings of the 6th World Congress on Civil, Structural, and Environmental Engineering (CSEE'21), Virtual Conference, 21–23 June 2021. Paper No. ICEPTP 302. [CrossRef]
178. Shabalala, A.; Ekolu, S. Quality of water recovered by treating acid mine drainage using pervious concrete adsorbent. *Water SA* **2019**, *45*, 638–647. [CrossRef]
179. Thisani, S.K.; Von Kallon, D.V.; Byrne, P. Effects of Contact Time and Flow Configuration on the Acid Mine Drainage Remediation Capabilities of Pervious Concrete. *Sustainability* **2021**, *13*, 10847. [CrossRef]

180. Ekolu, S.O.; Solomon, F.; de Beer, F.; Bitandi, L.; Kilula, R.N.; Maseko, K.T.; Mahlangu, F.G. Measurement of pore volume, connectivity and clogging of pervious concrete reactive barrier used to treat acid mine drainage. *Environ. Sci. Pollut. Res.* **2022**, *29*, 55743–55756. [CrossRef] [PubMed]
181. Solomon, F.H.; Ekolu, S.O.; Musonda, I. Gravity-Fed Column Configuration for Acid Mine Drainage Experiment. *Int. J. Eng. Technol.* **2019**, *11*, 348–354. [CrossRef]
182. Ekolu, S.O.; Azene, F.Z.; Diop, S. A concrete reactive barrier for acid mine drainage treatment. In Proceedings of the Institution of Civil Engineers-Water Management; Thomas Telford Ltd.: London, UK, 2014; Volume 167, pp. 373–330.
183. Holmes, R.R.; Hart, M.L.; Kevern, J.T. Removal and breakthrough of lead, cadmium, and zinc in permeable reactive concrete. *Environ. Eng. Sci.* **2018**, *35*, 408–419. [CrossRef]

Disclaimer/Publisher's Note: The statements, opinions and data contained in all publications are solely those of the individual author(s) and contributor(s) and not of MDPI and/or the editor(s). MDPI and/or the editor(s) disclaim responsibility for any injury to people or property resulting from any ideas, methods, instructions or products referred to in the content.

Article

Growing an Enhanced Culture of Polyphosphate-Accumulating Organisms to Optimize the Recovery of Phosphate from Wastewater

Njabulo Thela [1,*], David Ikumi [1], Theo Harding [1] and Moses Basitere [2]

1. Water Research Group, New Engineering Building, University of Cape Town, Rondebosch, Cape Town 7700, South Africa
2. Academic Support Program for Engineering in Cape Town, Centre for Higher Education Development, University of Cape Town, Rondebosch, Cape Town 7700, South Africa
* Correspondence: njabulo.thela@uct.ac.za; Tel.: +27-216502603

Citation: Thela, N.; Ikumi, D.; Harding, T.; Basitere, M. Growing an Enhanced Culture of Polyphosphate-Accumulating Organisms to Optimize the Recovery of Phosphate from Wastewater. *Water* 2023, *15*, 2014. https://doi.org/10.3390/w15112014

Academic Editors: Yung-Tse Hung, Hamidi Abdul Aziz, Issam A. Al-Khatib, Rehab O. Abdel Rahman, Tsuyoshi Imai and Laura Bulgariu

Received: 20 March 2023
Revised: 18 May 2023
Accepted: 19 May 2023
Published: 26 May 2023

Copyright: © 2023 by the authors. Licensee MDPI, Basel, Switzerland. This article is an open access article distributed under the terms and conditions of the Creative Commons Attribution (CC BY) license (https://creativecommons.org/licenses/by/4.0/).

Abstract: Having certain bacteria called phosphorus-accumulating organisms (PAOs) is important for getting rid of phosphorus (P) in wastewater from homes. This happens in a process called enhanced biological phosphorus removal (EBPR), where PAOs are active in activated sludge. To design and make EBPR processes work better, we need to have an in-depth understanding of how PAOs work. The best way to learn about them is by studying them in a laboratory. This study undertook to culture these microorganisms in the laboratory. A University of Cape Town membrane bioreactor (UCTMBR) activated sludge (AS) system was used to grow the microorganisms and see how well it worked. This paper looked at what type of substrate PAOs like best, either acetate or propionate, and how providing them with more of their preferred substrate affects how they grow. During the process, it was observed that P was not released or taken up significantly when acetate was added to the influent. The levels were consistently low at around 5.74 ± 4.47 mgP/L infl (release) and 19.9 ± 7.17 mgP/L infl (uptake). The signs become much better when propionate was used instead of acetate. When the amount of propionate in the influent was increased from 50% to 76% (as a percentage of influent total chemical oxygen demand), the amount of P released went up to 155 ± 17.7 mgP/L infl, and the amount of P taken up went up to 213.7 ± 11.4 mgP/L infl. The proof given indicated that propionate is preferred by PAOs. This study found that when more propionate was added to the wastewater, the concentration of PAO biomass went up. This was shown by certain signs that PAOs display when they are present. Results presented in this journal article emanate from an MSc Thesis (Thela, 2022) published in open-source UCT.

Keywords: polyphosphate-accumulating organisms (PAOs); enhanced biological phosphorus removal (EBPR); acetate; propionate

1. Introduction

Phosphorus (P), or phosphate, as found in its oxidized form in nature, is an essential element required within living cells as part of the energy carrier adenosine triphosphate (ATP) and within deoxyribonucleic acid (DNA). Phosphate rock, a nonrenewable resource that is increasingly becoming scarce and expensive, is the main source of P [1]. Most of the globally mined phosphate is used in agricultural applications, mainly, in the fertilizer industry, but also to produce P-based pesticides and animal supplements [2]. However, most plants only absorb a small quantity of phosphate into cells, and animals and human beings excrete large quantities of the phosphate consumed via food sources. In addition, most of the phosphate introduced to agricultural soil via fertilizers is not consumed by crops but gets washed into the ground and surface water sources [3]. Using the current rate of mining, Desmidt et al. [2] estimated that the current phosphate rock reserves will be depleted in around 372 years.

Given this background, the development of a sustainable P recovery strategy is of critical importance. Conventional wastewater treatment plants (WWTPs) remove P biologically using enhanced biological P removal (EBPR) system or chemical precipitation with metal salts or a combination of both [4]. The most common feature of an EBPR system is the presence of an anaerobic cell, prior to the aerobic cell, with the influent fed into the anaerobic cell. Examples of EBPR processes include, but are not limited to, the Phoredox system, 5-Stage Bardenpho and the University of Cape Town (UCT) system. The chemical precipitants of metal-phosphate salts referred to above are commonly iron (Fe) or aluminum (Al) phosphate salts. In some instances, these inorganic metal phosphate salts are very stable chemical compounds and thus largely insoluble in water, making the uptake of phosphate from these salts difficult for plant life that absorbs phosphate in the dissolved forms: HPO_4^{2-} and $H_2PO_4^-$ [5]. For example, where Fe^{3+} is used to form inorganic phosphate salts (i.e., $FePO_4$) the P is no longer available for direct use as a slow-release fertilizer under normal pH conditions [2], thereby limiting the application of sludge produced from these types of chemical precipitants in agriculture. However, in other instances, where Fe^{2+} is available to react with P, vivianite [$Fe_3(PO_4)_2 \cdot 8H_2O$] can be formed, which is a good product for P recovery. With that said, in cases where Fe salts are not relied upon for P removal, the biological removal of P from wastewater using an EBPR process opens opportunities for recovering most of the P from sewage streams.

Enhanced biological P removal (EBPR) activated sludge (AS) system is one of the methods used to biologically remove P from wastewater. It uses bacteria called polyphosphate-accumulating organisms (PAOs) which can store P in the form of polyphosphate. These organisms are found in wastewater treatment plants where AS is recirculated between anaerobic and aerobic cells, with wastewater containing lots of volatile fatty acids (VFAs) sent to the anaerobic cell. A better understanding of PAO biomass metabolism can assist in the optimization of mathematical models used in the design of EEPR-configured WWTPs. One of the approaches to achieve this understanding is to culture these microorganisms in a controlled environment and study them. Wentzel, Ekama and Marais [6] described the inner cellular processes of PAOs: They take up VFAs from the bulk liquid in the anaerobic cell and store them internally as poly-β-hydroxy alkanoates (PHAs), using polyphosphate polymers and glycogen as energy sources. Because of polyphosphate hydrolysis, P, together with counterions (magnesium [Mg^{2+}], potassium [K^+] and calcium [Ca^{2+}]) are released to the bulk liquid in the anaerobic cell. In the aerobic cell, the PHA in PAO cells is used for energy, growth and to restore glycogen that was used up in the anaerobic cell. The energy is also used to replenish polyphosphate polymers depleted in the anaerobic cell [7], resulting in the uptake of P in the aerobic cell, together with the uptake of Mg^{2+}, K^+ and Ca^{2+} as counterions for the negatively charged polyphosphate polymers.

Treating AS that contains PAOs via an anaerobic digester (AD) breaks down polyphosphate into its components, resulting in the release of P, Mg^{2+}, K^+ and Ca^{2+} into the bulk liquid [8]. Nutrient-rich dewatering liquor resulting from the thickening of the AD sludge can be used to produce struvite ($MgNH_4PO_4 \cdot 6H_2O$). The production of struvite makes it possible to recover P since it is often used as a slow-release fertilizer, something not possible for Fe and Al phosphates that are known to, like in the example given above, lock the P with the chemical precipitate [2]. Conditions that promote the growth of PAO biomass within EBPR systems have been reported in numerous studies as outlined below:

- Wentzel et al. [9] reported on the process that can be used to grow PAOs and recommended that a mass fraction of 20% or more of the active biomass should be maintained within the anaerobic cell of the EBPR AS system.
- To make the anaerobic cell work well, it is important to try and ensure that as little as possible oxygen and/or nitrate is introduced to the anaerobic cell, either through recycle flows or other means. Many researchers agree that recycling nitrate or oxygen to the anaerobic cell makes EBPR not work as well as it should [9–13]. It results in a decreased EBPR performance.

- The presence of glycogen-accumulating organisms (GAOs) leads to reduced EBPR performance because, like PAOs, they take up VFAs under anaerobic conditions of the anaerobic cell of the EBPR AS system and store them as PHAs. However, GAOs only use glycogen to get energy, unlike PAOs which also use polyphosphate [14,15]. Studies have shown that it is better to use propionate instead of acetate to reduce competition between the two types of microorganisms [15,16].
- In agreement with [15,16], findings from Wang et al. [17] suggest that PAOs prefer more complex VFAs such as the butyrate used in the study. This is contrary to those findings made by Qui et al. [18], whose findings suggested that a simpler VFA (i.e., acetate) is preferred by PAOs.
- Numerous studies, including that performed by Jonsson et al. [19] and Choi et al. [20], showed that Ca^{2+} plays a small role in the makeup of polyphosphate polymers. However, it has been shown that high influent Ca^{2+} concentration indirectly impacts the metabolism of PAOs [21,22]. Barat et al. [21] found that high influent Ca^{2+} concentrations (>30 mgCa/L) led to a decrease in PAO activity and attributed this to the precipitation of amorphous calcium phosphate (ACP). The authors [21] concluded that the precipitation of ACP changed the metabolic pathway of PAOs from the polyphosphate-accumulating metabolic (PAM) pathway to the glycogen-accumulating metabolic (GAM) pathway. Zhang et al. [22], on the other hand, found 28–42 mgCa/L to be the ideal influent Ca^{2+} concentration where Ca-bound polyphosphate is synthesized, without the formation of Ca-phosphate precipitates.
- Studies performed on the effect of pH on the growth of PAOs show that keeping the pH level in the aerobic cell above 6.5 improves the rate at which they grow [23,24]. This concurs with findings by numerous other investigators who also observed that PAO activity improved as the pH went up [25–28].

The current study aimed to grow an enhanced culture of PAOs to significantly high concentrations, using an AS system with a University of Cape Town (UCT) configuration and having a membrane bioreactor (MBR) for solid-effluent separation (i.e., a UCTMBR AS system). Using membranes to separate solids and liquids meant that a higher concentration of biomass can be achieved, which is not achievable with conventional wastewater treatment plants fitted with secondary settling tanks for solid-effluent separation [29].

Growing an enhanced culture of PAOs implied that influent to the UCTMBR AS system will have characteristics different to those of typical municipal wastewater but will require a synthetic component to it [9]. The objective of the synthetic component is to supplement the influent with those nutrients that PAOs require but are insufficient in typical municipal wastewater. These include VFAs, P, Mg^{2+}, K^+ and sometimes Ca^{2+}, where the water is soft. The objective of this study is to investigate the influent conditions that would result in the development of an enhanced culture of PAOs in MBR EBPR systems. Specifically, the study evaluates (i) the preferred substrate between acetate and propionate, since there is not consensus in the literature reviewed on the preferred substrate [15–18] and (ii) the impact of increased influent Ca^{2+} concentration on P removal.

2. Materials and Methods

This study considered the operation of a University of Cape Town (UCT) configured activated sludge (AS) system, fitted with a submerged plate-and-frame membrane system, while changes in the influent characteristics were made to promote the growth of an enhanced culture of polyphosphate-accumulating organisms (PAOs). These changes were made periodically over a 6-month period and their impact on phosphorus (P) release and uptake within the anaerobic and aerobic cells, respectively, of the UCT Membrane Bioreactor (UCTMBR) (UCT Civil engineering workshop, Cape Town, SA) AS system were monitored. These were considered in relation to the growth of PAOs within the system.

2.1. Description and Operation of the UCTMBR AS System

A laboratory-scale nitrification-denitrification enhanced biological P removal (ND EBPR) AS system based on a UCTMBR process configuration was set up and operated within the UCT Water Research Group (WRG) laboratory. This system was seeded with waste activated sludge (WAS) biomass from a nearby wastewater treatment plant (WWTP), in Cape Town, South Africa, where this operation plant holds a UCT-configured process layout. The experimental system used in this study was configured as shown in Figure 1.

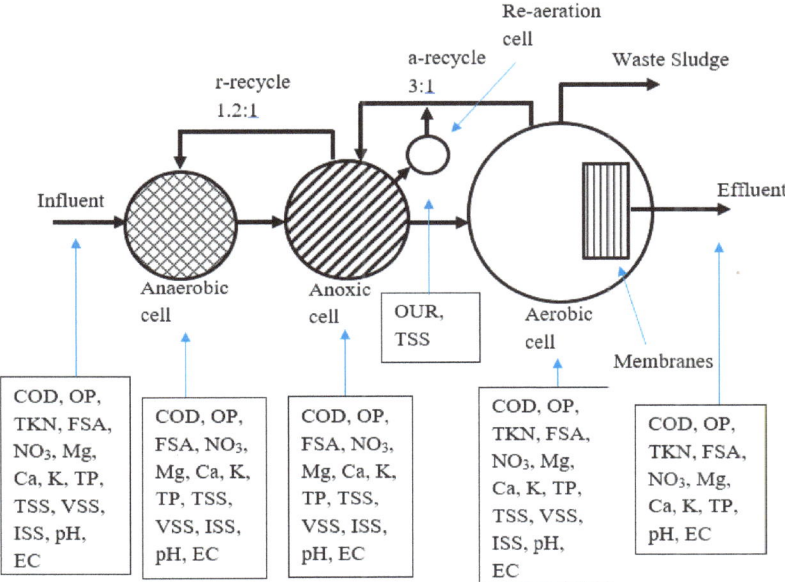

Figure 1. Schematic layout of the laboratory setup, including sampling points and parameters analyzed on the sample.

The way the system was set up and the membranes used to separate solids from liquids meant that (i) a lot of biomass can be retained (the system operates at high biomass concentration) and (ii) recycle flows can be changed to increase P removal, resulting in improved PAO biomass growth [29]. The UCTMBR AS system (Figure 1) had anerobic (29 L), anoxic (4.4 L) and aerobic (32 L) cells. Two cells made up the aerobic cell; it was a 29 L cell, fitted with membranes, and a 3 L reaeration cell fitted with a UCT Chemical Engineering DO/OUR meter (Hitech Micro Systems, Cape Town, SA) for oxygen utilization rate (OUR) measurements. Oxygen utilization rate data collected from the reaeration reactor were used in chemical oxygen demand (COD) balance calculations All the cells were made from a clear Perspex material. Except for the rectangular-shaped aerobic cell, all cells were cylindrical and were mixed continuously using mechanical stirrers. The rectangular-shaped aerobic cell was mixed with course bubbles. The influent flowrate was set at 60 L/d with its composition as detailed in Table 1 below. Mixed liquor recycle ratios were set at 3:1 for the a-recycle, resulting in a hydraulic retention time of 5.27 h in the anaerobic cell and 1.2:1 for the r-recycle, resulting in the hydraulic retention time of 0.338 h (20 min) in the anoxic cell. Ikumi [30] set the recycle ratio, AS flow from the anoxic cell to the reaeration cell, at 2:1 relative to the influent. The same recycle ratio was used in the current study. To make sure that the solids retention time was kept at 10 days sludge age, a volume of 4.5 L of AS was removed from the rectangular-shaped aerobic cell daily.

Table 1. Macronutrients added per 1200 mgCOD/L acetate to the influent.

	Influent Sewage Characteristics				
Parameter	Phase 1	Phase 2	Phase 3	Phase 4	Units
Influent flowrate	60	60	60	60	L/d
Total influent COD	1200	1200	1200	1200	mgCOD/L
Percentage COD supplied by settled WW	50	50	50	24 *	%
Percentage COD supplied by acetate	50	50	0 *	0	%
Percentage COD supplied by propionate	0	0	50 *	76 *	%
N requirements	50	50	50	50	mgN/L
TP	30	60 *	80 *	80	mgP/L
Magnesium	12.5	25 *	25	25	mgMg/L
Potassium	50	50	50	50	mgK/L
Calcium	20	40 *	26 *	14.5 *	mgCa/L
Yeast extract	10	10	10	10	mg/L

* Represents the parameter that was changed from the previous phase.

A peristaltic pump was used for the pumping requirements of the influent and all recycle flows. An Easy-load Masterflex (Model: 7518-00) peristaltic pump was used for pumping AS from the anoxic cell to the reaeration cell. Five A4 size, 0.45 μm membrane modules (Kubota Membrane Europe, London, UK), submerged in the rectangular-shaped aerobic cell, were used for solid–liquid separation. The experiment was operated at 21 ± 1 °C.

2.2. Analytical Methods

The UCTMBR AS system was checked regularly by doing tests at certain sampling points, as shown in Figure 1. To confirm the steady state, the frequency of sampling and testing was increased to every second day over a period of 10 days. The analyses conducted, more specifically during steady state, provided sufficient insight into the performance of the AS system. The data we gathered helped us understand the characteristics of the influent and those of the AS. All the COD, total kjeldahl nitrogen (TKN), total P (TP), total suspended solids (TSS), volatile suspended solids (VSS) and inorganic suspended solids (ISS) were tested following instructions detailed in the Standard Methods for the examination of water and wastewater [31]. Colorimetric methods were used to determine free and saline ammonia (FSA), nitrate (NO_3^-), nitrite (NO_2^-), orthophosphate (OP), Mg^{2+}, Ca^{2+} and K^+ concentrations. This process was automated with a Gallery™ Discrete Analyzer (Thermo Fischer Scientific, Waltham, Massachusetts, USA) using standard methods of the instrument. The instrument was calibrated before any analysis was performed. To ensure that the methods were still accurate, a standard solution, with known concentrations of the chemical compounds listed above, was analyzed. Where pH and electrical conductivity (EC) measurements were required, an Accsen pH Meter (Model: pH 8) and Jenway (Model 4510) were used, respectively.

2.3. Influent Feed Composition and MBR Operational Changes

The changes to the biological phosphate storage capacity of the UCTMBR AS system biomass, evaluated in this study, relate to changes made to the influent wastewater characteristics fed to this AS system. Within this study, these influent wastewater characteristic changes were evaluated during 4 phases and described as such in this paper (see Table 1).

Changes to the influent characteristics fed to the AS system resulted from the changes in the make-up of organic components that contributes to the influent COD concentration, where in the case of this study the influent was made up of settled sewage, acetate or propionate with added tap water, for dilution to the desired COD concentration. Further nutrients (calcium chloride dihydrate, for Ca^{2+}, magnesium chloride hexahydrate, for Mg^{2+} and dipotassium hydrogen phosphate, for P) were added to supplement those nutrients that were in short supply as recommended by Wentzel et al. [9].

Wentzel et al. [9] found that if one adds more VFAs to the influent directed to the anaerobic cell of an EBPR system, PAOs will perform better than other microorganisms present in the EBPR system, including ordinary heterotrophic organisms (OHOs). One of the objectives of this study was to operate the system at elevated biomass concentrations, hence the high COD flux to the system (72 gCOD/d). Changes in the make-up of the influent wastewater for Phases 1, 2, 3 and 4 were described in Table 1.

The influent feed characteristic changes and corresponding phases can be summarized as follows:

- Phase 1: refers to the period from Day 0 to Day 22. During this period, the concentration of total phosphorus (TP), Mg^{2+} and Ca^{2+} was fixed at half the concentrations recommended in the literature [9,15,16,21,22].
- These concentrations were, on Day 22, increased to those used in numerous investigations that involved PAOs [9,15,16,21,22] and this was the beginning of Phase 2, which ran from Day 22 to Day 74.
- Phase 3: (Period from Day 74 to Day 145) As per the findings from recent studies [15,16], the acetate that made up 50% of the total influent COD was replaced with propionate, while the percentage remained unchanged at 50% of the total influent COD. To reduce chemical precipitation potential as per the findings from recent investigations [20–22], the influent Ca^{2+} concentration was reduced. These changes to the influent characteristics resulted in increased PAO activity as discussed in Section 3.1.
- Phase 4: (Period from Day 145 to Day 193) The influent COD contributed by propionate was increased to 76% after Phase 3 steady-state analyses were performed. Due to the reduction in settled sewage, which contributed a significant amount of influent Ca^{2+}, the overall influent Ca^{2+} concentration decreased further as shown in Table 1.

Throughout the experiment, the influent TKN and K^+ concentrations were higher than those shown in Table 1, because of (i) a high settled sewage TKN concentration and (ii) the introduction of K^+ by dipotassium hydrogen phosphate (K_2HPO_4) used to supplement influent TP.

Once the influent acetate COD concentration reached 400 mgCOD/L, representing 80% of the total influent COD, Wentzel et al. [9] started to add macronutrients. Now it is usual to do this when studying PAO cultures in AS systems [9,15,16,22]. To be safe during the investigation, micronutrients were added when the percentage of COD from synthetic sewage (i.e., propionate) was increased to 76% (see Table 2).

Table 2. Micronutrient's stock solution.

Chemical	Compound	Quantity	Units
Solution A			
Boric acid	H_3BO_3	0.2498	g
Copper sulphate pentahydrate	$CuSO_4 \cdot 5H_2O$	0.2508	g
Potassium iodide	KI	0.0626	g
Manganese chloride tetrahydrate	$MnCl_2 \cdot 4H_2O$	1.6643	g
Sodium molybdate dihydrate	$Na_2MoO_4 \cdot 2H_2O$	0.1250	g
Zinc sulphate heptahydrate	$ZnSO_4 \cdot 7H_2O$	1.2561	g
Cobalt chloride hexahydrate	$CoCl_2 \cdot 6H_2O$	0.4775	g
	Feed for	14	Days
	Volume	2	L
	Feed/day	143	mL
Solution B			
Iron (II) sulphate heptahydrate	$FeSO_4 \cdot 7H_2O$	2,195	g
	Feed for	7	Days
	Volume	L	L
	Fed/day	143	mL

2.4. Data Collection and Evaluation

Measurements were conducted and data collected from (1) the influent to the AS system, made up of settled sewage, VFAs, i.e., acetate or propionate and tap water from dilution and (2) at the sampling points of the experimental AS system as shown in Figure 1. The settled sewage collected from a nearby WWTP in Cape Town, South Africa, was analyzed for COD, TP, K^+, Ca^{2+}, Mg^{2+}, TKN and FSA concentrations. Tap water, also used to make up a feed, was analyzed for three minerals: K^+, Ca^{2+} and Mg^{2+} concentrations only. These tests were performed to figure out how much extra nutrients needed to be added to the feed to make sure that PAOs were getting all the nutrients they require. Calcium chloride dihydrate was used for Ca^{2+}, dipotassium hydrogen phosphate was used to supplement P and magnesium chloride hexahydrate was used for Mg^{2+}. A schematic representation of the feed preparation process is shown in Figure 2. How well the UCTMBR system was working was checked by doing tests in sampling points marked in Figure 1.

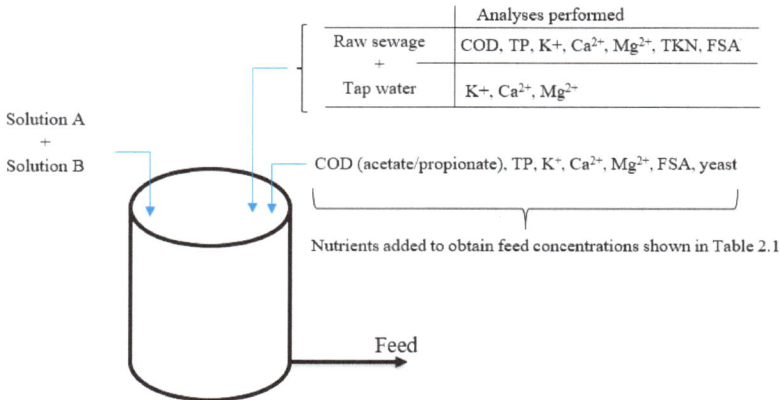

Figure 2. Schematic representation of the feed preparation process.

The continuity of the measured datasets over the experimental system was evaluated using a material mass balance method for COD, nitrogen (N) and P components [32]. The principle of conservation of mass was used to check steady-state analysis results for reliability and accuracy. Material balances with a 20% range (i.e., 90% to 110%) are indicative of accurate and reliable experimental data [30]. Given that experimental work can be influenced by systematic and random errors, including instrument calibration, poor equipment calibration and even inaccurate readings from the user, obtaining a 100% balance can be challenging. In some cases, mass balances in the 80% to 120% range are also acceptable if they can be justified [30].

The COD, N and P mass balances obtained (see Table 3) for Phases 3 and 4 of the system, at steady state, showed that a significant level of confidence can be placed on the data collected during the two phases. Due to their limited evidence of PAO activity, Phases 1 and 2 were not allowed to reach a steady state, hence the exclusion of these periods, as shown in Table 3.

Table 3. Mass balances during Phase 3 and Phase 4 steady-state periods.

Parameter	Phase 3	Phase 4
COD mass balance	119%	99%
N mass balance	89%	105%
P mass balance	111%	96%

3. Results and Discussion

3.1. Changes in OP Concentrations Due to Influent Characteristics Changes

Figure 3, which shows OP concentration variations in the influent, anerobic and aerobic cells (i.e., effluent and aerobic cell OP concentrations are the same), shows that the increase in influent TP, Mg^{2+} and Ca^{2+} (see Table 1) leads to an increase in the OP concentration in both anerobic and aerobic cells. This means that adding more nutrients to the feed did not affect PAO activity. If it did, OP concentration would have been much higher in the anerobic than in the aerobic cell, due to anerobic P release. Concurrent with the OP concentration increase on both the anerobic and aerobic cells, there was an increase in the inorganic suspended solids to volatile suspended solids ratio (ISS/VSS ratio) on both cells, accompanied by an increase in both P and Ca^{2-} removals (see Table 4). The P removal increased from 7.1 mgP/L (Phase 1) to 24.1 mgP/L (Phase 2), while that of Ca^{2+} increased from 7.8 mgCa/L to 29.4 mgCa/L. These observations could have meant that the activity of PAOs increased. However, because of the buildup of white precipitates in the a-recycle lines and the increase in removing P and Ca^{2+}, it seemed likely that P and Ca^{2+} were not removed biologically, but rather via mineral precipitation. Barat et al. [21] showed that high influent Ca^{2+} concentrations (>30 mgCa/L) led to a decrease in PAO activity and an increase in the formation of Ca precipitates. As mentioned earlier, this was attributed to the precipitation of amorphous calcium phosphate (ACP or $Ca_3[PO_4]_2$), and the conclusion reached was that the precipitation of ACP changed the metabolic pathway of PAOs from polyphosphate-accumulating metabolic (PAM) pathway to glycogen-accumulating metabolic (GAM) pathway [19]. Hence, given the experimental evidence, it was likely that the precipitate observed during Phase 2 may have been ACP. Moreover, numerous studies support this conclusion [20,33,34], where each investigation reported that although Ca^{2+} does help in the biological P removal through the formation of polyphosphate, the amount required to form polyphosphate is little. Therefore, the elevated Ca^{2+} removal observed could only be attributed to the precipitation of calcium phosphate.

Figure 3. Orthophosphate (OP) concentration in the feed tank, anerobic cell and effluent.

Table 4. P, Ca^{2+} and Mg^{2+} removal during Phase 1 and Phase 2.

| | Phase 1 | | | Phase 2 | | | |
Component	In	Out	Removal	In	Out	Removal	Units
P	30.1 ± 0.32	23 ± 2.2	7.1	57.9 ± 2.8	33.8 ± 5.26	24.1	mgP/L
Ca^{2+}	31.7 ± 2.14	23.9 ± 2.31	7.8	54.8 ± 4.99	25.4 ± 0.49	29.4	mgCa/L
Mg^{2+}	11.9 ± 1.23	11.3 ± 1.23	0.6	25.3 ± 2.27	22.3 ± 3.27	3.0	mgMg/L

An increase in the anerobic cell OP concentration was observed a few days after the commencement of Phase 3 (see Figure 3), which indicated an increase in anerobic P release. This meant that either there were more PAOs or their metabolic pathway had changed from the GAM pathway to the PAM pathway. Either way, this was taken as evidence that PAO activity was on the rise in the enhanced biological P removal (EBPR) activated sludge (AS) system. An increase in the influent propionate COD from 50% (Phase 3) to 76% (Phase 4) of the total influent COD resulted in a further increase in OP release as illustrated by the increased anerobic cell OP concentration (see Figure 3), and this meant that there was a further increase in PAO activity.

For the current discussion, Phase 1 and Phase 2 were grouped together and called the Start-up period. This was possible since they displayed approximately the same magnitude of P release and P uptake in the anerobic and aerobic cells, respectively. Together with OP concentrations measured from all the sampling points shown in Figure 1, measured flowrates (influent, a-recycle and r-recycle flows) were used to calculate the P released and the P taken up for the Start-up, Phase 3, and Phase 4 periods. Results obtained for P release and P uptake are shown in Figure 4 a, b, respectively. In simpler terms, the amount of P released in the anerobic cell increased from a low level during the Start-up phase (5.74 ± 4.47 mgP/L infl) to a higher level (54.5 ± 8.15 mgP/L infl) during Phase 3, as shown in Figure 4 a. During Phase 4, the release of P in the anerobic cell increased to 155 ± 17.7 mgP/L infl. During the Start-up phase, only a small amount of P was taken up in the aerobic cell (19.9 ± 7.17 mgP/L infl). However, in Phase 3, a lot more P was taken up as shown in Figure 4 b. During Phase 4, when the amount of propionate in the influent was increased by almost one-third, the amount of P taken up went up to 213.7 ± 11.4 mgP/L infl. The discussed observations are proof of PAO activity, which began when acetate in the influent was replaced with propionate (i.e., the beginning of Phase 3).

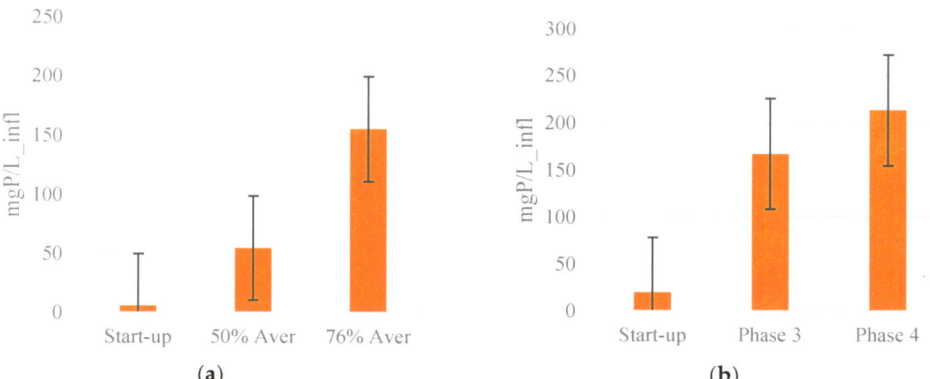

Figure 4. (**a**) OP release in the anerobic cell and (**b**) OP uptake in the aerobic cell.

3.2. Changes in Phosphate Concentrations Due to Influent Characteristics Changes

Acevedo et al. [35] studied how polyphosphate storage levels affected PAO metabolism in a sequencing batch reactor (SBR). During the study, Equation (1) was used to calculate the concentration of polyphosphate (PP) within the PAO biomass. This equation was modified to account for the OP observed in both anerobic and aerobic cells in the current investigation (see Equation (2)).

$$PP = TP - (f_{P,VSS} \times X_{VSS}) \qquad (1)$$

$$PP = TP - OP - (f_{P,VSS} \times X_{VSS}) \qquad (2)$$

Figure 5 shows how much PP was in the AS biomass in the aerobic and anerobic cells during Phase 3 and Phase 4. As discussed above, there was no PAO activity at the beginning, so the Start-up phase is not shown in Figure 5. In Phase 3, the concentration of PP in the anerobic cell was less than half of the concentration in the aerobic cell, as seen in Figure 5. Again, in Phase 4, the concentration of PP in the anerobic cell was less than that of the aerobic cell and the difference was bigger during this phase. It was 0.023 ± 0.008 mgPP/mgVSS in the anerobic cell and 0.07 ± 0.003 mgPP/mgVSS in the aerobic cell. These observations have shown that PP is utilized as propionate is sequestered by PAOs and replenished as P is taken up. This concurs with findings from numerous studies [7,36,37]. Moreover, in Phase 4, the concentration of organic P of 0.0901 ± 0.003 mgP/mgVSS in the biomass is greater than 0.02 mgP/mgVSS in fully aerated AS system biomass. The increase in stored PP concentration in the aerobic cell provided additional proof that the population of PAOs is directly proportional to the VFA influent concentration. As shown in Figure 5, PP went up during Phase 4 compared to Phase 3.

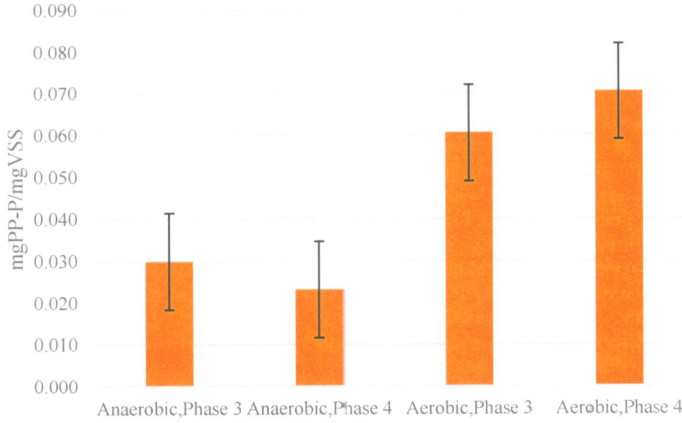

Figure 5. Polyphosphate to volatile suspended solids ratio in the anerobic and aerobic cells during Phase 3 and Phase 4 steady-state periods.

4. Conclusions

This study has demonstrated a strategy to grow an enhanced culture of PAOs in a UCTMBR AS system. This strategy can be used by researchers working on advancing the understanding of PAO metabolism. This study has also shown that PAOs prefer propionate, a more complex VFA, over acetate, which is simpler, consistent with previous findings [15–17]. Though not explicit in the current study and therefore requiring further investigation, this study has also shown that an elevated influent Ca^{2+} concentration can form precipitates, hindering the EBPR performance, consistent with findings made in the past [21,22]. In addition, when membranes are used for solid–liquid separation, as was performed in this investigation, precipitates can also form on the surface of the membrane, resulting in reduced membrane performance. This phenomenon also requires further investigation since it can affect the performance of membranes in real WWTP applications treating hard water (i.e., high influent Ca^{2+} concentration).

Author Contributions: Conceptualization, N.T. and D.I.; methodology, N.T. and D.I.; validation, D.I. and T.H.; formal analysis, N.T., T.H. and D.I.; investigation, N.T.; resources, D.I.; data curation, N.T.; writing—original draft preparation, N.T. and D.I.; writing—review and editing, N.T., T.H. and M.B.; visualization, N.T. and T.H.; supervision, D.I.; project administration, N.T. and M.B.; funding acquisition, D.I. All authors have read and agreed to the published version of the manuscript.

Funding: This research was funded by the Water Research Commission, grant number K5/2839//3.

Data Availability Statement: Data will be available on request.

Acknowledgments: Not applicable.

Conflicts of Interest: The authors declare no conflict of interest. The funder had no role in the design of the study; in the collection, analyses, or interpretation of data; in the writing of the manuscript; or in the decision to publish the results.

References

1. Cordell, D.; Rosemarin, A.; Schröder, J.J.; Smit, A.L. Towards global phosphorus security: A systems framework for phosphorus recovery and reuse options. *Chemosphere* **2011**, *84*, 747–758. [CrossRef] [PubMed]
2. Desmidt, E.; Ghyselbrecht, K.; Zhang, Y.; Pinoy, L.; van der Bruggen, B.; Verstraete, W.; Rabaey, K.; Meesschaert, B. Global phosphorus scarcity and full-scale P-recovery techniques: A review. *Crit. Rev. Environ. Sci. Technol.* **2015**, *45*, 336–384. [CrossRef]
3. Wang, R.; Cai, C.; Zhang, J.; Sun, S.; Zhang, H. Study on phosphorus loss and influencing factors in the water source area. *Int. Soil Water Conserv. Res.* **2022**, *10*, 324–334. [CrossRef]
4. Tchobanoglous, G.; Stensel, H.D.; Tsuchihashi, R.; Burton, F.; Abu-Orf, M.; Bowden, G.; Pfrang, W. *Wastewater Engineering: Treatment and Resource Recovery—Vol. 1*, 5th ed.; McGraw-Hill: New York, NY, USA, 2014; p. 648.
5. Turner, B.L.; McKelvie, I.D.; Haygarth, P.M. Characterisation of water-extractable soil organic phosphorus by phosphatase hydrolysis. *Soil Biol. Biochem.* **2002**, *34*, 27–35. [CrossRef]
6. Wentzel, M.C.; Ekama, G.A.; Marais, G.V.R. Processes and modelling of nitrification denitrification biological excess phosphorus removal systems—A review. *Water Sci. Technol.* **1992**, *25*, 59–82. [CrossRef]
7. Smolders, G.; van der Meij, J.; van Loosdrecht, M.; Heijnen, J. A structured metabolic model for anaerobic and aerobic stoichiometry and kinetics of the biological phosphorus removal process. *Biotechnol. Bioeng.* **1995**, *47*, 277–287. [CrossRef]
8. Harding, T. A steady state stoichiometric model describing the anaerobic digestion of biological excess phosphorus removal waste activate sludge. Master's Thesis, University of Cape Town, Cape Town, South Africa, 2009.
9. Wentzel, M.C.; Loewenthal, R.E.; Ekama, G.A.; Marais, G. Enhanced polyphosphate organism cultures in activated sludge systems-Part 1: Enhanced culture development. *Water SA* **1988**, *14*, 81–92.
10. Barnard, J.L. A review of biological phosphorus removal in the activated sludge process. *Water SA* **1976**, *2*, 136–144.
11. Hascoet, M.; Florentz, M.; Granger, P. Biochemical aspects of enhanced biological phosphorus removal from wastewater. *Water Sci. Technol.* **1985**, *17*, 23–41. [CrossRef]
12. Rabinowitz, B. Chemical and biological phosphorus removal in the activated sludge process. Master's Thesis, University of Cape Town, Cape Town, South Africa, 1980.
13. du Toit, G.J.; Ramphao, M.; Parco, V.; Wentzel, M.; Ekama, G.A. Design and performance of BNR activated sludge systems with flat sheet membranes for solid-liquid separation. *Water Sci. Technol.* **2007**, *56*, 105–113. [CrossRef]
14. Mino, T.; Satoh, H.; Matsuo, T. Metabolisms of different bacterial populations in enhanced biological phosphate removal processes. *Water Sci. Technol.* **1994**, *29*, 67–70. [CrossRef]
15. Oehmen, A.; Yuan, Z.; Blackall, L.L.; Keller, J. Comparison of acetate and propionate uptake by polyphosphate accumulating organisms and glycogen accumulating organisms. *Biotechnol. Bioeng.* **2005**, *91*, 162–168. [CrossRef]
16. Carvalheira, M.; Oehmen, A.; Carvalho, G.; Reis, M.A. The effect of substrate competition on the metabolism of polyphosphate accumulating organisms (PAOs). *Water Res.* **2014**, *64*, 149–159. [CrossRef]
17. Wang, L.; Liu, J.; Oehmen, A.; Le, C.; Geng, Y.; Zhou, Y. Butyrate can support PAOs but not GAOs in tropical climates. *Water Res.* **2021**, *193*, 116884. [CrossRef]
18. Qiu, G.; Zuniga-Montanez, R.; Law, Y.; Thi, S.S.; Nguyen, T.Q.N.; Eganathan, K.; Liu, X.; Nielsen, P.H.; Williams, R.B.; Wuertz, S. Polyphosphate-accumulating organisms in full-scale tropical wastewater treatment plants use diverse carbon sources. *Water Res.* **2019**, *149*, 496–510. [CrossRef]

19. Jonsson, K.; Johansson, P.; Christensson, M.; Lee, N.; Lie, E.; Welander, T. Operational factors affecting enhanced biological phosphorus removal at the wastewater treatment plant in Helsingborg, Sweden. *Water Sci. Technol.* **1996**, *34*, 67–74. [CrossRef]
20. Choi, H.J.; Yu, S.W.; Lee, S.M.; Yu, S.Y. Effects of potassium and magnesium in the enhanced biological phosphorus removal process using a membrane bioreactor. *Water Environ. Res.* **2011**, *83*, 613–621. [CrossRef]
21. Barat, R.; Montoya, T.; Borrás, L.; Ferrer, J.; Seco, A. Interactions between calcium precipitation and the polyphosphate-accumulating bacteria metabolism. *Water Res.* **2008**, *42*, 3415–3424. [CrossRef]
22. Zhang, H.L.; Sheng, G.P.; Fang, W.; Wang, Y.P.; Fang, C.Y.; Shao, L.M.; Yu, H.Q. Calcium effect on the metabolic pathway of phosphorus accumulating organisms in enhanced biological phosphorus removal systems. *Water Res.* **2015**, *84*, 171–180. [CrossRef]
23. Romanski, J.; Heider, M.; Wiesmann, U. Kinetics of anaerobic orthophosphate release and substrate uptake in enhanced biological phosphorus removal from synthetic wastewater. *Water Res.* **1997**, *31*, 3137–3145. [CrossRef]
24. Filipe, C.D.; Daigger, G.T.; Grady Jr, C.L. Effects of pH on the rates of aerobic metabolism of phosphate-accumulating and glycogen-accumulating organisms. *Water Environ. Res.* **2001**, *73*, 213–222. [CrossRef] [PubMed]
25. Bond, P.L.; Keller, J.; Blackall, L.L. Anaerobic phosphate release from activated sludge with enhanced biological phosphorus removal. A possible mechanism of intracellular pH control. *Biotechnol. Bioeng.* **1999**, *63*, 507–515. [CrossRef]
26. Liu, W.T.; Mino, T.; Matsuo, T.; Nakamura, K. Biological phosphorus removal processes-effect of pH on anaerobic substrate metabolism. *Water Sci. Technol.* **1996**, *34*, 25–32.
27. Oehmen, A.; Vives, M.T.; Lu, H.; Yuan, Z.; Keller, J. The effect of pH on the competition between polyphosphate-accumulating organisms and glycogen-accumulating organisms. *Water Res.* **2005**, *39*, 3727–3737. [CrossRef] [PubMed]
28. Smolders, G.J.F.; van der Meij, J.; van Loosdrecht, M.C.M.; Heijnen, J.J. Model of the anaerobic metabolism of the biological phosphorus removal process: Stoichiometry and pH influence. *Biotechnol. Bioeng.* **1994**, *43*, 461–470. [CrossRef]
29. Ramphao, M.; Wentzel, M.C.; Merritt, R.; Ekama, G.A.; Young, T.; Buckley, C.A. Impact of membrane solid–liquid separation on design of biological nutrient removal activated sludge systems. *Biotechnol. Bioeng.* **2005**, *89*, 630–646. [CrossRef]
30. Ikumi, D.S. The development of a three-phase plant-wide mathematical model for sewage treatment. Ph.D. Thesis, University of Cape Town, Cape Town, South Africa, 2009.
31. American Public Health Association; American Public Health Association; American Water Works Association. Water Pollution Contorl Federation. In *Standard Methods for the Examination of Water and Wastewater*; American Public Health Association: Washington, DC, USA, 1975.
32. Henze, M.; van Losdrecht, M.C.M.; Ekama, G.A.; Brdjanovic, D.; Amy, G.; Comeau, Y.; Mahmoud, N.; Stenstrom, M.K.; Wentzel, M.C.; Zeeman, G.; et al. *Biological Wastewater Treatment: Principles, Modelling and Design*; IWA Publishing: London, UK, 2008; pp. 156–220.
33. Arvin, E.; Holm-Kristensen, G. Exchange of organics, phosphate and cations between sludge and water in biological phosphorus and nitrogen removal processes. *Water Sci. Technol.* **1985**, *17*, 147–162. [CrossRef]
34. Comeau, Y.; Hall, K.J.; Hancock, R.E.; Oldham, W.K. Biochemical model for enhanced biological phosphorus removal. *Water Res.* **1986**, *20*, 1511–1521. [CrossRef]
35. Acevedo, B.; Oehmen, A.; Carvalho, G.; Seco, A.; Borrás, L.; Barat, R. Metabolic shift of polyphosphate-accumulating organisms with different levels of polyphosphate storage. *Water Res.* **2012**, *46*, 1889–1900. [CrossRef]
36. Wentzel, M.C.; Ekama, G.A.; Dold, P.L.; Marais, G.V. Biological excess phosphorus removal—Steady state process design. *Water SA* **1990**, *16*, 29–48.
37. Brdjanovic, D.; van Loosdrecht, M.C.M.; Hooijmans, C.M.; Mino, T.; Alaerts, G.J.; Heijnen, J.J. Effect of polyphosphate limitation on the anaerobic metabolism of phosphorus-accumulating microorganisms. *Appl. Microbiol. Biotechnol.* **1998**, *50*, 273–276. [CrossRef]

Disclaimer/Publisher's Note: The statements, opinions and data contained in all publications are solely those of the individual author(s) and contributor(s) and not of MDPI and/or the editor(s). MDPI and/or the editor(s) disclaim responsibility for any injury to people or property resulting from any ideas, methods, instructions or products referred to in the content.

Utilizing Electricity-Producing Bacteria Flora to Mitigate Hydrogen Sulfide Generation in Sewers through an Electron-Pathway Enabled Conductive Concrete

Huy Thanh Vo [1], Tsuyoshi Imai [2,*], Masato Fukushima [3], Tasuma Suzuki [4], Hiraku Sakuma [5], Takashi Hitomi [5] and Yung-Tse Hung [6]

[1] Faculty of Urban Engineering, Mientrung University of Civil Engineering, Tuy Hoa 620000, Vietnam; huypocrisy@gmail.com
[2] Graduate School of Sciences and Technology for Innovation, Yamaguchi University, Yamaguchi 755-8611, Japan
[3] Fuso Cooperation, Takamatsu 761-8551, Japan; m.fukushima@fuso-inc.co.jp
[4] Kajima Technical Research Institute, Kajima Cooperation, Tokyo 182-0036, Japan; suzuktas@kajima.com
[5] Nakagawa Humepipe Industry Co., Ltd., Ibaraki 300-0051, Japan
[6] Department of Civil and Environmental Engineering, Cleveland State University, Cleveland, OH 44115, USA; y.hung@csuohio.edu
* Correspondence: imai@yamaguchi-u.ac.jp

Abstract: This study aims to demonstrate the effectiveness of using biological oxidation for hydrogen sulfide (H_2S) control. A long-term experiment was conducted using a rod-shaped electrode made of highly conductive concrete, which provided an electron pathway for H_2S mitigation. Bacterial flora analysis was conducted using PCR-DGGE and metagenomic analysis by next-generation sequencing to identify electricity-producing bacteria. Results showed that H_2S was effectively mitigated, and electricity-producing bacteria, including *Geobacter* sp. and *Pelobacter* sp., were found around the inner surface of the anode. The study found that highly conductive concrete can create an electron pathway for biological oxidation of H_2S. Oxygen from the air layer near the surface of the water can act as an electron acceptor, even under anaerobic conditions, enabling effective H_2S control in sewer systems.

Keywords: hydrogen sulfide; sewer pipe; conductive concrete; electron pathway; electricity-producing bacteria; PCR-DGGE; next-generation sequencer

1. Introduction

Microbiologically influenced corrosion (MIC) has emerged as a significant concern for civil engineers to protect construction materials, such as pipeline systems, sewers, and underground water systems. Corrosion by hydrogen sulfide can occur in wastewater treatment systems where anaerobic microorganism convert sulfates in the wastewater to aqueous H_2S. Sulfate-reducing bacteria (SRB) are known to generate a variety of destructive substances as organic acids, hydrogen sulfide, and other sulfur-containing compounds that can initiate degradation of material surfaces. Sewage pipeline corrosion caused by hydrogen sulfide is a common problem that can lead to the deterioration of sewer infrastructure, causing high maintenance costs and potential structure failures [1,2]. In anaerobic conditions and moisture, hydrogen sulfide formed by SRB communities of wastewater can react with sulfur oxidizing bacteria (SOB) to form sulfuric acid. This biogenic acid can then dissolute with calcium-silicate-hydrate as hydrate products in the concrete, developing calcium sulfate and water [3–5]. This degradation process can be accelerated by time to lead to structural failures and collapsing concrete and posing a risk to safety.

Most sewer systems are located underground and directly exposed to wastewater, making them highly susceptible and prone to microbiologically induced corrosion. Consequently, significant financial resources are often required to repair and maintain these systems. Governments worldwide are increasingly being compelled to seek solutions to address this ongoing challenge, which may involve implementing preventive measures or replacing aging infrastructure [6]. The costs associated with corrosion prevention methods can vary significantly depending on the size and complexity of the sewer system.

To prevent corrosion by H_2S, using oxidant chemicals can be used to minimize the amount of sulfate in the wastewater and to maintain adequate levels of dissolved oxygen to prevent the growth of anaerobic bacteria. Protective coatings and liners can also be consumed to reduce the exposure of concrete surfaces to biogenic acid and corrosive compounds. The use of oxidant as an electron acceptor to oxidize sulfide through biological means has been investigated in sewer systems. Sewer aeration or bio-oxidation has been identified as an effective measure for mitigating sulfide generation and corrosion in sewer systems [7,8]. Adjoining nitrate/nitrite into sewers can be an effective solution for preventing the occurrence of sulfide. This is because SRBs, which typically produce sulfide as a byproduct of their metabolism, can use nitrate as an alternative electron acceptor [9]. In addition to reducing sulfide production, nitrate injection can also benefit sewer infrastructure by preventing the formation of sulfuric acid. Sulfuric acid can corrode concrete and metal pipes, so by reducing its formation, nitrate injection can help to extend the lifespan of sewer infrastructure. Microbial fuel cells (MFCs) have been investigated as a green potential approach to mitigate hydrogen sulfide emissions in sewer systems by utilizing the microorganisms in the MFCs to degrade organic matter in wastewater and produce electrons, which can be harnessed by an electrode to generate electricity [10,11]. By contrast, to protect sewer concrete materials, several recent studies have explored the use of protective coatings and liners, such as polysiloxane, epoxy coatings, or polyurea inners to prevent exposure of biogenic acid and H_2S [12,13]. Recent studies have emphasized the importance of microorganisms surrounding the electrodes of MFCs in the removal of hydrogen sulfide by utilizing it as a substrate for biological oxidation. MFCs utilize microorganisms to generate electrical energy while simultaneously removing organic matter from wastewater.

Research into MFCs for biocorrosion mitigation is ongoing, with promising results thus far. Researchers are working toward developing practical and effective and sustainable solutions for the long-term protection of biocorrosion of concrete structures. In previous studies, the authors applied the principle of microbial fuel cells and conducted experiments on suppressing hydrogen sulfide generation using conductive concrete [14]. Based on their demonstration, it has been shown that the conductive substance within the concrete is capable of absorbing hydrogen sulfide. Furthermore, that study also confirmed that hydrogen sulfide can be biologically oxidized through the inoculation of electricity-producing bacteria (EPB). While conductive concrete has been recognized as a useful technology for deicing snow and heated pavements [15], its potential as a MFC in sewer systems for treating wastewater has been largely unexplored. Further research is needed to investigate this application and its implications for sustainable infrastructure. Figure 1 illustrates the proposed mechanism of this biological oxidation process as previously researched. However, despite this, there is currently no molecular biological evidence supporting the claim that EPB present in sewage sludge can biologically oxidize hydrogen sulfide even when the inoculation of EPB is not performed.

The present study aims to demonstrate the mechanisms behind biocorrosion by identifying the microbial communities responsible for the process. In addition, we will explore the efficacy of using conductive concrete as a microbial fuel cell system to target these organisms. Specifically, we will test the ability of electrically conductive concrete to suppress the generation of hydrogen sulfide, a key contributor to biocorrosion, through experimental trials. Furthermore, we will use molecular biological methods to analyze the microbial community in the sludge and determine the changes resulting from the use of conductive concrete.

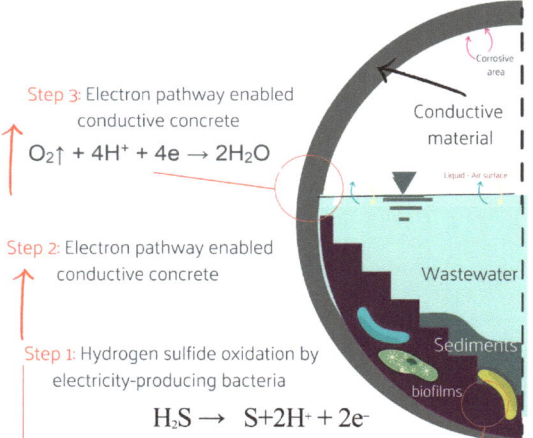

Figure 1. Mechanism of hydrogen sulfide generation inhibition by EPB.

2. Materials and Methods

The experiment was conducted under controlled laboratory conditions with a temperature of 25 ± 1 °C. The aqueous solution was prepared using distilled water (SA-2100A·A type, Tokyo Rikakikai Co., Ltd., Tokyo, Japan), which was of high purity and ensured that the solution was free from any impurities. Additionally, reagents of high grade were used in the experiment from reliable suppliers such as Fujifilm Wako Pure Chemical Co., Ltd. (Osaka, Japan), Kishida Chemical Co., Ltd. (Osaka, Japan), and Nakarai Tesc Co., Ltd. (Kyoto, Japan). The selection of high-grade reagents ensured the accuracy and reliability of the experiment results.

2.1. Preparation of Conductive Material and Electrodes

In this study, San-Earth M5C (referred to as San-Earth) was used as a type of conductive substance that has been shown in previous study to inhibit the generation of sulfides (including sulfide ion S^{2-}, hydrogen sulfide ion HS^-, and hydrogen sulfide H_2S) in water [14]. San-Earth contains amorphous carbon, a byproduct of oil refining processes, with a maximum particle size of 0.3 mm and a specific surface area of 1.9 m^2/g (as shown in Figure 2A). The electrodes were made in a cylindrical shape, with a diameter of 16 mm and a length of 75 mm, and the water-to-powder ratio was set at 42% in accordance with a previous study [14]. After undergoing treatment with an alum-based scour remover, the electrodes retained their specific characteristics of not expanding, shrinking, or cracking during curing [14]. The electrode shown in Figure 2B is positioned within a simulated sewer pipe as Figure 1, allowing for electrochemical reaction simulation-like real field conditions (described in Section 2.2).

2.2. Hydrogen Sulfide Suppression Experiment with Conductive Concrete

The objective of this experiment was to gather molecular biological data that would serve as the basis for the biological oxidation of sulfide by providing electron transfer pathways. To accomplish this goal, two sets of systems were prepared (referred to Figure 2B). The first set included an anode electrode (a bar-shaped concrete electrode manufactured by San Earth) positioned 8.5 cm deep at the bottom of an aquarium (18.5 cm tall × 29.0 cm wide × 13.0 cm high, with a total volume of 6.98 L) and a cathode electrode (a bar-shaped concrete electrode, identical to the anode electrode) that was partially submerged in water with a 100 Ω external resistor. The connection between the two electrodes created electron transfer pathways. The second set of systems did not include this connection.

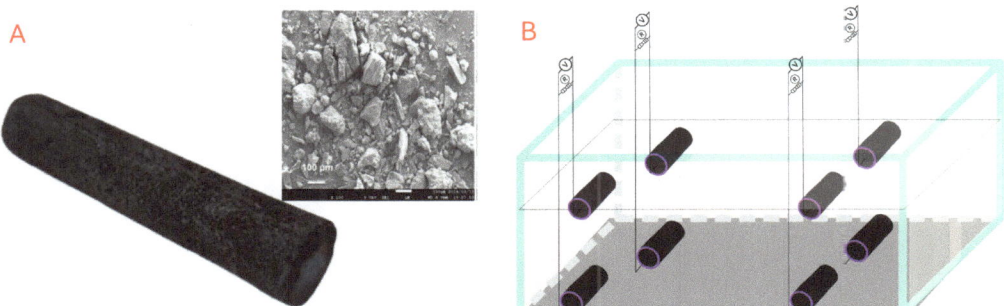

Figure 2. Illustration of the conductive concrete and electrodes used in the experiment. (**A**) a scanning electron microscopy image of the conductive substance (amorphous carbon) of the electrode made from conductive concrete; (**B**) the experimental setup with a rod-shaped electrode (V: Voltmeter, R: Resistor, Anode electrode is placed at the bottom of the water tank, cathode electrode is placed near the water surface with half of it submerged). The diagram depicts a closed circuit with an electron pathway.

The copper wire was used to connect the bar-shaped concrete electrodes, with the connection embedded in concrete and equipped with anticorrosion measures. The system had an internal resistance of 680 Ω. To evaluate the targeted effect of biological oxidation and prevent the prolonged suppression of hydrogen sulfide adsorption, the surface area of the anode electrode was significantly reduced to less than one-tenth of the previous study [14]. This modification allowed for a more focused evaluation of biological oxidation while reducing the surface area available for adsorption of hydrogen sulfide.

Using the experimental systems described above, biological solids (referred to as sludge) near the surface of the anode electrode were collected from one group of four bar-shaped conductive concrete electrodes at regular intervals. Bacterial analysis was then performed on the collected samples. To facilitate the interpretation of the bacterial analysis results, a tank with a flat concrete specimen was also prepared for comparison.

2.3. Preparation and Analysis Instruction of Wastewater Samples

The experiment was initiated by mixing 0.48 L of excess sludge (with SS of 8540 mg/L and VSS of 7410 mg/L) and digested sludge (with SS of 8040 mg/L and VSS of 7240 mg/L) obtained from the Ube City Wastewater Treatment Plant (Ube WWTP) with 3.84 L of artificial wastewater. The composition of the synthetic wastewater was as follows: in 1 L of distilled water, $NaHCO_3$ (2.0 g), K_2HPO_4 (2.0 g), yeast extract (0.02 g), glucose (2.0 g), $(NH_4)_2HPO_4$ (0.70 g), KCl (0.75 g), NH_4Cl (0.85 g), $FeCl_3·6H_2O$ (0.42 g), $MgCl_2·6H_2O$ (0.81 g), $MgSO_4·7H_2O$ (0.25 g), $CoCl_2·6H_2O$ (0.018 g), and $CaCl_2·6H_2O$ (0.15 g) were added. This mixture was stirred until all the components were completely dissolved. The resulting synthetic wastewater was then used for the experimental trials. Following the mixing, the system was allowed to settle in a static state before starting the experiment.

The pH, sulfate ion, and sulfide concentrations in the aqueous phase were continuously monitored. Sulfate ions in the water were measured using the barium sulfate turbidimetric method in accordance with USEPA method 375.4, following filtration through a 0.45 μm membrane filter. Sulfides were quantified using the methylene blue method according to USEPA method 376.2. On days 20, 40, and 68 after the start of the experiment, glucose (at a concentration of 2000 mg/L) and magnesium sulfate heptahydrate (at a concentration of 34 mg S/L) were added, once SRB had significantly reduced the sulfate ion concentration to almost 0 mg S/L. The concentration of sulfate ion added was set according to the concentration of sulfate ion typically observed in actual sewage (approximately 100 $mgSO_4^{2-}$/L).

2.4. Analysis of the Microbial Community Involved in Inhibition of Sulfide Generation

The microbial community involved in sulfide generation inhibition was investigated by analyzing sludge samples collected from the wastewater tank and anode electrode. The analysis utilized Polymerase Chain Reaction—Denaturing Gradient Gel Electrophoresis (PCR-DGGE) and next-generation sequencing techniques to identify microbial species and assess their abundance. The aim of the analysis was to determine differences in the microbial species that were present or absent in relation to the presence of electron transfer pathways and growth conditions. PCR-DGGE was utilized to identify dominant microbial species and monitor changes in their abundance over time, while next-generation sequencing provided a comprehensive profile of the microbial community and identified the full range of microbial species present in the samples.

2.4.1. Sample Collection Method of Sludge and Types of Experimental Systems

Sludge samples were collected using a spatula from the conductive concrete electrode surface and a sterile pipette from the bottom of the tank to assess temporal changes. The surface of the concrete was cut off to a thickness of 0.5–1 mm to investigate the possibility of electron-emitting bacteria growing inside. The collected samples were stored in sterile plastic tubes at $-20\ ^\circ$C until further analysis. Table 1 summarizes the experimental systems and sludge sampling methods used in the analysis of the bacterial community involved in the suppression of sulfide generation by PCR-DGGE.

Table 1. Types of experimental systems and methods for sludge collection.

Sample	Experimental System	Methods for Sludge Collection
①	Open circuit (Without electron pathway) conductive concrete	Collect sludge on electrode surfaces with a medicine spoon
②		Sludge is scraped from the electrode surface using a cutter knife, enabling the accumulation and growth of microbial communities for subsequent analysis
③		Sludge obtained from the bottom of the tank
④	Closed circuit (With electron pathway) conductive concrete	Collect sludge on electrode surfaces with a medicine spoon
⑤		Sludge is scraped from the electrode surface using a cutter knife, enabling the accumulation and growth of microbial communities for subsequent analysis
⑥		Sludge obtained from the bottom of the tank
⑦	Normal concrete	The biofilm on the surface of the concrete was removed using a cutter knife
⑧		Sludge sample obtained from the bottom of the tank

2.4.2. Analyzing the Bacterial Community by PCR-DGGE Method

PCR-DGGE was used to analyze the bacterial community present in the sludge based on the 16S rRNA gene V3 region sequence [16–19]. DNA extraction was carried out using the DNA Extraction Kit (Nippon Gene Co., Ltd., Tokyo, Japan). The PCR protocol utilized a two-step nested PCR to increase the specificity and yield of the desired amplicon. The first-step primer used for the 16S rRNA gene was 27f/1492r for bacteria and 21f/958r for archaea [16,18]. The second step amplified the V3 region using 341f-GC/518r for bacteria and 340f-GC/519r for archaea [18,19]. The reaction conditions for bacteria and archaea PCR followed by PCR amplification of the 16S V3-V4 region [18]. Each PCR reaction (25 µL) contained Emerald Amp Max PCR Master, 10 µM of each primer, and 1.5 µL or 3.0 µL of template DNA. The PCR products were confirmed by agarose gel electrophoresis.

DGGE analysis was conducted using the DCode Microbial Community Analysis System (Bio-Rad, Berkeley, CA, USA) on 8% polyacrylamide gels containing a denaturing gradient of 40–70%. Electrophoresis was performed at a holding temperature of 60 $^\circ$C and 20 V for 10 min, followed by 16 h of running at 70 V and the same temperature.

After electrophoresis, the gels were stained with SYBR Gold for 1 h and DNA bands were visualized using a Chemi Doc XRS UV imaging system (Bio-Rad). DNA bands were excised and a PCR targeting the V3 region was carried out using the primers 341f/518r for bacteria and 340f/519r for archaea. DNA base sequences were determined using an Ion S5 DNA sequencer (Thermo Fisher Scientific Inc., Waltham, MA, USA). MEGA X software was used for DNA base sequence analysis, and the BLAST program (NCBI) was used to search the 16S rRNA gene (194 bp) database.

2.4.3. Next-Generation Sequencing-Based 16S Metagenomic Analysis

The 16S metagenomic analysis was performed using next-generation sequencing to analyze DNA extracted from sludge collected in Section 2.4.1. The DNA extraction was carried out using the NucleoSpin® Soil Kit (Takara Bio Inc., Kusatsu, Japan). The extracted DNA was then amplified using PCR with two primer sets targeting the 16S V3-V4 region [20]. The forward and reverse primers used for the amplification were CGTCGGCAGCGTCAGATGT-GTATAAGAGACAGCCTACGGGNGGCWGCAG and GTCTCGTGGGCTCGGAGATGTG-TATAAGAGACAGGACTACHVGGGTATCTAATCC, respectively. The PCR amplification conditions were as follows: initial denaturation at 95 °C for 3 min, followed by 25 cycles of denaturation at 95 °C for 30 s, annealing at 55 °C for 30 s, extension at 72 °C for 30 s, and a final extension at 72 °C for 5 min.

The amplified sample was further amplified using the Nextera XT Index Kit (Illumina, San Diego, CA, USA), which added barcodes and adapter sequences to both ends of the amplicons. The amplification conditions were as follows: initial denaturation at 95 °C for 3 min, followed by 8 cycles of denaturation at 95 °C for 30 s, annealing at 55 °C for 30 s, extension at 72 °C for 30 s, and a final extension at 72 °C for 5 min.

The V3-V4 region was then sequenced using the MiSeq Reagent Kit v3. Finally, 16S metagenomic analysis was performed on the sequence data of each sample obtained by sequencing using the Base Space analysis software. The Green Genes database was used for the analysis.

3. Results

3.1. Inhibition of Hydrogen Sulfide Generation Using Conductive Concrete

Figure 3 displays the change in sulfate and sulfide ion concentrations throughout time. The initial addition of substrate, marked by the arrow in the diagram, occurred on the 20th day after the experiment began and was designated as the first cycle. The intervening days between each consecutive addition of substrate were classified as the second, third, and fourth cycles, respectively. Voltage recordings confirmed power generation during the experiment. However, the data logger malfunctioned during the second cycle, and no further data could be obtained. Sulfate ions were consumed within about five days after the start of the experiment in both cycles. For sulfide ions, no significant difference was observed in the first and second cycles, but in the third and fourth cycles, the maximum sulfide ion concentration was higher in the open circuit (without an electron pathway). This may be attributed to the insufficient growth of microorganisms responsible for the biological oxidation of sulfide during the early stages of the experiment. The main mechanism for sulfide removal was thought to be adsorption by the amorphous carbon contained in San-Earth, and sulfide was removed from the water without being affected by the presence of electron transfer pathways. However, as the cycles progressed, the adsorption sites became saturated, and biological oxidation was considered to be the primary mechanism for sulfide suppression. This led to a significant difference in the suppression effect due to the presence or absence of an electron transfer pathway. The average reduction rate of sulfide ion concentration in the closed circuit (with an electron pathway) compared to the open circuit (without an electron pathway) was calculated and is presented in Figure 3B. The reduction rates were 41.5% and 27.0% in the third and fourth cycles, respectively, and the difference was evident between the two circuits.

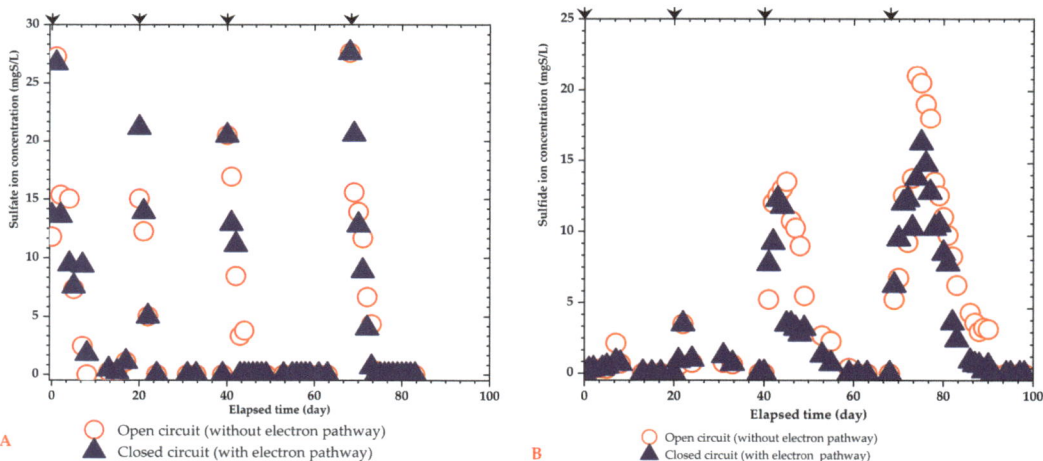

Figure 3. Illustration of the results of a 91-day experiment on inhibiting hydrogen sulfide generation using conductive concrete. (**A**) changes in sulfate ion concentration, (**B**) changes in sulfide ion concentration. Glucose at 2000 mg/L and magnesium sulfate at 34 mg S/L (equivalent to 100 mg SO_4^{2-}/L) were added on day 20, 40, and 68.

3.2. Analyzing the Bacterial Community Involved in the Suppression of Sulfide Generation by PCR-DGGE Method

Figure 4 displays the DGGE images of the bacteria at the end of each cycle, while Figure 5 shows an enlarged view of the blue-framed section after the first cycle in Figure 4. The samples in Table 1 correspond to the numbers in the DGGE results shown in Figures 4 and 5. Table 2 presents information on the bacteria obtained from sequencing data after the first cycle. The symbols in Figure 5 correspond to those in Table 2, and the arrows in the image indicate the bacteria with the highest homology, identified by matching the DNA sequences with the gene database.

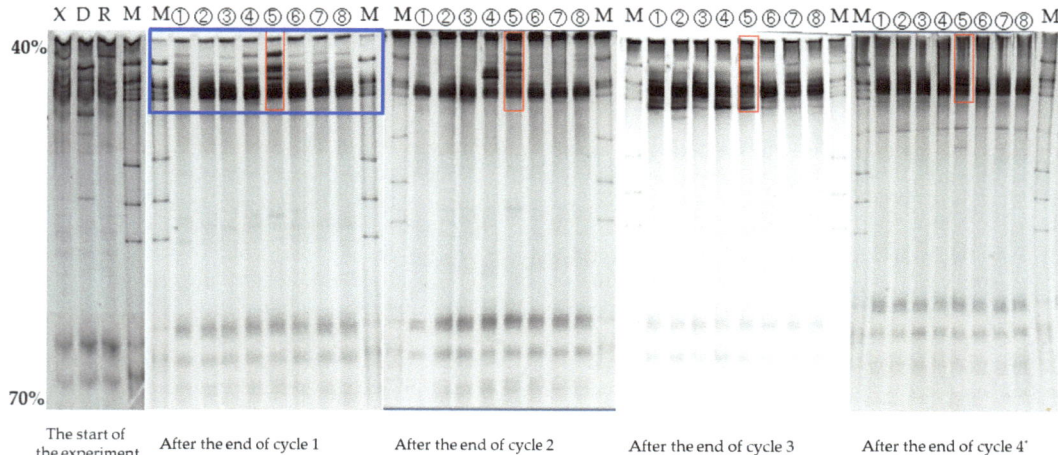

Figure 4. Results of DGGE analysis of bacteria at the end of each cycle. (*: left for about one month after the substrate was completely consumed, M: marker, X: mixed sludge, D: digested sludge, R: excess sludge).

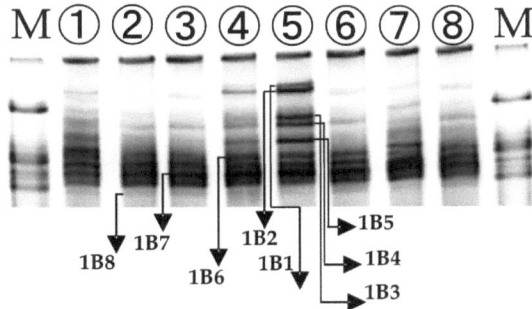

Figure 5. DGGE analysis results of bacteria at the end of cycle 1.

Table 2. Information on bacteria obtained from community analysis.

Symbol	Cycle	Bacteria	Similarity (%)
1B1		*Geobacter uraniireducens* Rf4	98.28
1B2		*Geobacter psychrophilus* strain P35	100.00
1B3		*Pelobacter carbinolicus* DSM 2380	96.88
1B4	After the end of the cycle 1	*Pelobacter carbinolicus* DSM 2380	95.77
1B5		*Desulfuromonas acetexigens* strain 2873	97.56
1B6		*Macrococcus epidermidis* strain CCN 7099	100.00
1B7		*Fusibacter fontis* strain KhaIAKB1	98.59
1B8		*Fusibacter fontis* strain KhaIAKB1	98.63

Bacteria of 1B1–1B4 in Table 2, which were thinly banded in the excess and digested sludge at the start of the experiment in Figure 4, were confirmed to be electricity-producing bacteria by analysis of sequencing data [21–23]. As clearly shown in Figure 5, these 1B1 to 1B4 bands were densest in the sludge inside the anode electrode (Figure 5, ⑤). Next, the sludge on the electrode surface (Figure 5, ④) was the second densest with 1B1 to 1B4 bands. These results suggest that electricity-producing bacteria grow predominantly inside the anode electrode rather than on its surface, and the presence or absence of an electron transfer pathway is significantly related to their growth.

The temporal changes in bands 1B1 to 1B4 were analyzed (Figure 4) to investigate the impact of the electron transfer pathways on the growth of electricity-producing bacteria in experimental systems. Dense bands were detected in the first, second, and third cycles of samples ④ and ⑤. Although the band in the third cycle appeared lighter, subsequent analysis of the 16S metagenome data using a next-generation sequencer revealed no changes in the electron-releasing bacteria. Furthermore, the overexposure of the image was evidenced by the photography conditions of the marker. Conversely, bands that were only slightly detected (faint) in cycles 1 (samples of ①–③, open circuit without electron transfer pathway) and ⑦, ⑧ (normal concrete) became even fainter and decreased in cycles 2 and 3, indicating that electricity-producing bacteria did not proliferate in the absence of an electron pathway.

Upon integration of the results presented in Section 3.1 with the current findings, it can be inferred that the biological oxidation of sulfide by electricity-producing bacteria is taking place. Following the completion of cycle 4, as denoted by an asterisk in Figure 4, the substrate was left for approximately one month to assess the decline in electron-producing bacteria. The boxed region demonstrates a noticeable reduction in the intensity of bands 1B1–1B4, signifying a substantial decrease in electron-producing bacteria during substrate consumption within a month. This observation is also supported by the 16S metagenomic sequencing results presented in the subsequent section.

3.3. Next-Generation Sequencing-Based 16S Metagenomic Analysis: Quantitative Evaluation of Bacteria Involved in Sulfide Generation Inhibition

The abundance of electricity-producing and sulfate-reducing bacteria during each cycle is illustrated in Figure 6, which is based on 16S metagenomic analysis using next-generation sequencing. Table 3 provides the total number of reads obtained for each sample, and Table 4 illustrates the top 20 genera identified in the analysis. The "% hits" in Figure 6 represents the percentage of the total number of reads accounted for by a particular species, such as *Geobacter* sp. *Geobacter* sp. Was the most frequently detected representative electron-emitting bacteria identified by both the PCR-DGGE method and 16S metagenomic analysis using next-generation sequencing (see Table 4). In Figure 6A, all species belonging to the *Geobacter* genus, including those that were not identified, are collectively represented as the genus. This is also the case for Figure 6B–D. The figure displays the percentage of reads relative to the total number of reads, as can be seen from Table 3, and the number of reads per sample varies slightly, requiring accurate determination of microorganism evolution between cycles. The electron-emitting bacteria were detected in the following order of decreasing detection rate: inside electrode with electron transfer pathway > electrode surface with electron transfer pathway > inside electrode without electron transfer pathway > electrode surface without electron transfer pathway, as shown in the Figure 6A. This order was consistent for all cycles. These findings support the results obtained by the PCR-DGGE method (Section 3.2) and suggest that the presence or absence of the electron transfer pathways has a significant effect on the growth of *Geobacter* sp., a representative electricity-producing bacterium. Notably, *Geobacter* sp. Growth was most prominent in the electrode interior with the electron transfer pathways. Thus, environmental conditions that promote the electron transfer pathways can enhance the growth and accumulation of *Geobacter* sp., a representative EPB.

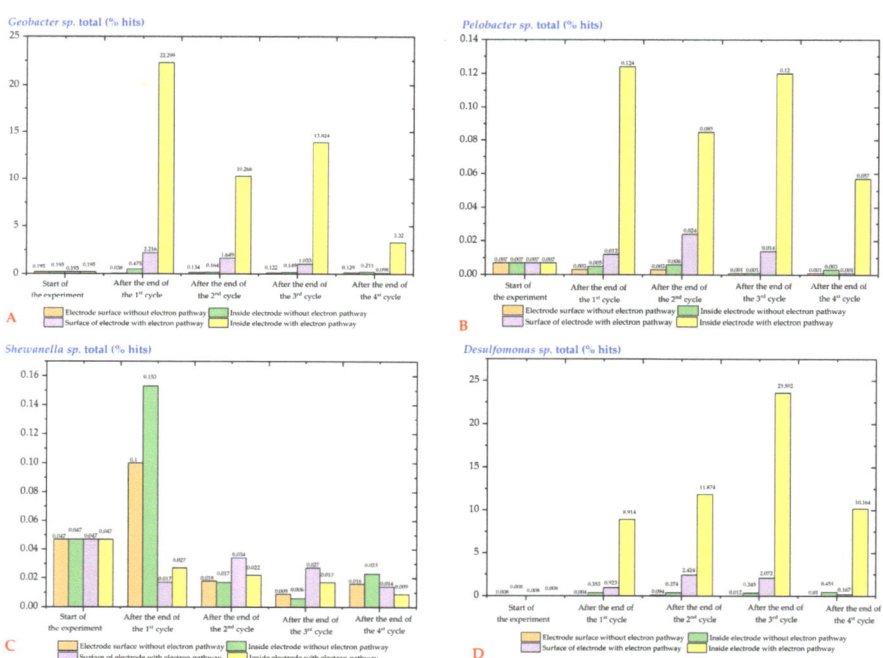

Figure 6. Changes in representative electricity-producing bacteria (*Geobacter* sp., *Pelobacter* sp., *Shewanella* sp., *and Desulfomonas* sp.) in each cycle extracted and organized from the analysis results of 16S metagenomes using next-generation sequencer. (**A**) *Geobacter* sp., (**B**) *Pelobacter* sp., (**C**) *Shewanella* sp., (**D**) *Desulfomonas* sp. Units in each graph represent the proportion of detected bacteria.

Table 3. Total number of reads for each sample.

	Number of Reads			
	Electrode Surfaces without Electron Pathways	Inside the Electrode without Electron Pathways	Electrode Surfaces with Electron Pathways	Inside the Electrode with Electron Pathways
At the start of the experiment	217.894	217.894	217.894	217.894
1st cycle	173.389	209.664	193.306	172.297
2nd cycle	163.275	171.655	192.780	216.892
3rd cycle	200.233	223.374	202.819	193.217
4th cycle	203.301	197.302	207.014	217.431

Table 4. Illustration of top 20 Genera from 16S Metagenome Analysis. (Other: percentage of total reads attributed to all genera ranked 21st and below, including unidentified species).

Before the Experiment Started		After the 1st Cycle								After the 2nd Cycle							
		Electrode Surface without Electron Pathway		Inside Electrode without Electron Pathway		Electrode Surface without Electron Pathway		Inside Electrode without Electron Pathway		Electrode Surface without Electron Pathway		Inside Electrode without Electron Pathway		Electrode Surface without Electron Pathway		Inside the Electrode with Electron Pathway	
Genus	%_hits	Genus	%_hits	Genus	%_hits	Genus	%_hits	Genus	%_hits	Genus	%_hits	Genus	%_hits	Genus	%_hits	Genus	%_hits
Dechloromonas	5.71	Pseudomonas	14.49	Clostridium	6.28	Trichococcus	10.41	Geobacter	22.3	Trichococcus	9.17	Trichococcus	11.82	Thauera	12.21	Desulfuromonas	11.88
Ferribacterium	2.82	Comamonas	9.02	Parabacteroides	5.77	Clostridium	7.57	Desulfuromonas	8.91	Clostridium	7.6	Clostridium	9.49	Clostridium	6.4	Clostridium	10.5
Nitrospira	2.39	Stenotrophomonas	3.92	Trichococcus	5.05	Alkaliphilus	5.47	Parabacteroides	4.82	Parabacteroides	4.55	Parabacteroides	3.87	Trichococcus	5.46	Geobacter	10.27
Anaerobaculum	1.99	Ochrobactrum	3.34	Comamonas	4.79	Parabacteroides	3.9	Clostridium	3.6	Cystobacter	4.15	Alkaliphilus	3.54	Chthoniobacter	3.32	Trichococcus	6.4
Clostridium	1.75	Thauera	3.12	Fusibacter	3.91	Fusibacter	2.68	Trichococcus	3.55	Alkaliphilus	2.71	Cystobacter	2.91	Parabacteroides	2.84	Thauera	5.07
Saccharopolyspora	1.69	Azospirillum	2.99	Pseudomonas	3.55	Geobacter	2.22	Alkaliphilus	2.6	Fusibacter	2.25	Pedobacter	2.73	Pedobacter	2.83	Alkaliphilus	2.46
Levinella	1.45	Campylobacter	2.37	Alkaliphilus	3.13	Sedimentibacter	2.02	Pedobacter	2.13	Pedobacter	2.07	Sedimentibacter	2.12	Desulfuromonas	2.42	Pedobacter	2.37
Caldilinea	1.31	Diaphorobacter	2.1	Brevundimonas	2.47	Enterococcus	1.84	Fusibacter	2.09	Sedimentibacter	2.02	Fusibacter	1.92	Cystobacter	1.81	Parabacteroides	2.32
Tepidanaerobacter	1.27	Shinella	1.82	Arcobacter	1.71	Nitrospira	1.42	Desulfobulbus	1.6	Chryseobacterium	1.86	Sphingobacterium	1.9	Arcobacter	1.77	Sedimentibacter	1.58
Candidatus Scalindua	1.25	Delftia	1.55	Lactococcus	1.55	Lactococcus	1.4	Sedimentibacter	1.27	Sphingobacterium	1.61	Candidatus Tammella	1.32	Geobacter	1.65	Fusibacter	1.46
Thauera	1.16	Clostridium	1.53	Sedimentibacter	1.53	Heliorestis	1.37	Dechloromonas	1.14	Chthoniobacter	1.48	Chryseobacteruim	1.25	Sphaerochaeta	1.53	Desulfovibrio	0.95
Bifidobacterium	1.13	Trichococcus	1.63	Desulfobulbus	1.31	Saccharopolyspora	1.3	Lactococcus	1.11	Candidatus Tammella	1.11	Heliorestis	1.2	Fusibacter	1.5	Candidatus Tammella	0.93
Vogesella	1.05	Flavobacterium	1.57	Pedobacter	1.22	Desulfovibrio	1.28	Thauera	0.99	Desulfovibrio	0.87	Desulfovibrio	0.78	Sphingobacterium	1.49	Cystobacter	0.86
Rhodobacter	1	Acidovorax	1.54	Saccharopolyspora	1.09	Candidatus Tammella	0.95	Enterococcus	0.77	Heliorestis	0.76	Myroides	0.76	Azoarcus	1.36	Lactococcus	0.81
Thermodesulfovibrio	0.97	Uliginosibacterium	1.53	Desulfomicrobium	1.07	Desulfuromonas	0.92	Tolumonas	0.65	Aminiphilus	0.64	Desulfobulbus	0.71	Alkaliphilus	1.31	Sphingobacterium	0.77
Dokdonella	0.89	Deceoia	1.52	Stenotrophomonas	1.02	Dechloromonas	0.85	Bacteroides	0.64	Treponema	0.56	Thauera	0.68	Desulfovibrio	1.3	Arcobacter	0.7

Table 4. Cont.

Before the Experiment Started

Genus	%_hits
Hyphomicrobium	0.86
Aminiphilus	0.83
Megasphaera	0.82
Azospirillum	0.82
other	68.86

After the 1st Cycle

Electrode Surface without Electron Pathway		Inside Electrode without Electron Pathway		Electrode Surface without Electron Pathway		Inside Electrode without Electron Pathway	
Genus	%_hits	Genus	%_hits	Genus	%_hits	Genus	%_hits
Snowella	1.36	Heliorestis	0.94	Azospirillum	0.79	Desulfovibrio	0.63
Brevundimonas	1.26	Enterococcus	0.91	Pedobacter	0.74	Cystobacter	0.6
Xenophilus	1.21	Lactococcus	0.9	Candidatus Scalindua	0.71	Desulfosarcina	0.56
Bdellovibrio	1.16	Bacteroides	0.87	Holdemania	0.7	Saccharopolyspora	0.54
other	40.51	other	50.94	other	51.48	other	39.49

After the 2nd Cycle

Electrode Surface without Electron Pathway		Inside Electrode without Electron Pathway		Electrode Surface without Electron Pathway		Inside Electrode without Electron Pathway	
Genus	%_hits	Genus	%_hits	Genus	%_hits	Genus	%_hits
Synergistes	0.54	Flavobacterium	0.64	Pseudomonas	1.23	Pseudomonas	0.7
Ferridobacterium	0.54	Treponema	0.61	Sedimentibacter	1.14	Myroides	0.69
Acholeplasma	0.53	Chthoniobacter	0.59	Synergistes	1.03	Comamonas	0.67
Tepidanaerobacter	0.52	Synergistes	0.56	Aequorivita	0.95	Acholeplasma	0.58
other	54.46	other	50.6	other	46.47	other	38.05

After the 3rd Cycle

Electrode Surface without Electron Pathway		Inside Electrode without Electron Pathway		Electrode Surface without Electron Pathway		Inside Electrode without Electron Pathway	
Genus	%_hits	Genus	%_hits	Genus	%_hits	Genus	%_hits
Clostridium	35.95	Clostridium	49.59	Clostridium	29.51	Clostridium	24.78
Trichococcus	17.1	Trichococcus	12.02	Trichococcus	6.24	Desulfuromonas	23.59
Parabacteroides	1.91	Parabacteroides	2.38	Pedobacter	4.57	Geobacter	13.82
Sedimentibacter	1.58	Pedobacter	1.8	Parabacteroides	3.08	Trichococcus	4.74
Alkaliphilus	1.56	Heliorestis	1.56	Desulfuromonas	2.07	Pedobacter	2.72
Pedobacter	1.38	Sedimentibacter	1.29	Tolumonas	1.75	Parabacteroides	1.92
Heliorestis	1.37	Alkaliphilus	1.23	Bacteroides	1.53	Treponema	0.76
Desulfomicrobium	0.89	Desulfomicrobium	0.73	Sedimentibacter	1.4	Sedimentibacter	0.73
Zoogloea	0.66	Candidatus Tammella	0.66	Hydrogenophaga	1.15	Sphingobacterium	0.65
Blautia	0.65	Desulfovibrio	0.65	Chthoniobacter	1.13	Alkaliphilus	0.6
Desulfovibrio	0.64	Zoogloea	0.6	Alkaliphilus	1.05	Tolumonas	0.56
Hydrogenophaga	0.59	Pseudomonas	0.57	Geobacter	1.03	Desulfobulbus	0.52
Candidatus Tammella	0.59	Enterococcus	0.55	Zoogloea	0.91	Desulfuromonas	0.48
Anaerostipes	0.57	Acetobacterium	0.45	Treponema	0.9	Bacteroides	0.4
Fusibacter	0.52	Myroides	0.44	Sphingobacterium	0.87	Lactococcus	0.39
Dechloromonas	0.48	Desulfonauticus	0.41	Desulfomicrobium	0.81	Desulfomicrobium	0.37
Pseudomonas	0.47	Flavobacterium	0.4	Candidatus Tammella	0.72	Myroides	0.35
Bacteroides	0.46	Fusibacter	0.39	Thiobacillus	0.68	Desulfovibrio	0.34
Enterococcus	0.45	Desulfobulbus	0.35	Flavobacterium	0.67	Cystobacter	0.33
Desulfonauticus	0.44	Desulfuromonas	0.35	Desulfovibrio	0.65	Desulfosarcina	0.31
other	31.74	other	23.61	other	39.28	other	21.66

After the 4th Cycle

Electrode Surface without Electron Pathway		Inside Electrode without Electron Pathway		Electrode Surface without Electron Pathway		Inside Electrode without Electron Pathway	
Genus	%_hits	Genus	%_hits	Genus	%_hits	Genus	%_hits
Clostridium	25.68	Clostridium	36.07	Clostridium	18.96	Trichococcus	28.13
Trichococcus	11.16	Trichococcus	19.03	Trichococcus	12.17	Clostridium	20.06
Anaerostipes	4.14	Pedobacter	4	Anaerostipes	9.99	Desulfuromonas	10.16
Blautia	3.81	Bacteroides	1.73	Blautia	8.69	Anaerostipes	4.89
Pedobacter	3.11	Desulfomicrobium	1.69	Pseudomonas	2.8	Blautia	4.04
Pseudomonas	3.08	Anaerostipes	1.6	Alkaliphilus	1.87	Geobacter	3.32
Desulfomicrobium	2.43	Blautia	1.43	Thauera	1.64	Pedobacter	1.84
Bacteroides	2.19	Acidaminococcus	1.22	Hydrogenophaga	1.59	Parabacteroides	1.8
Acidaminococcus	1.62	Treponema	1.19	Pedobacter	1.54	Desulfovibrio	1
Parabacteroides	1.31	Parabacteroides	1.16	Shinella	1.42	Acidaminococcus	1
Alkaliphilus	1.06	Desulfobulbus	0.85	Parabacteroides	1.16	Desulfomicrobium	0.84
Treponema	1.03	Sphingobacterium	0.82	Agrobacterium	0.97	Sedimentibacter	0.65
Sedimentibacter	0.99	Heliorestis	0.8	Acidaminococcus	0.89	Treponema	0.63
Heliorestis	0.75	Acholeplasma	0.78	Bacteroides	0.78	Bacteroides	0.63
Heliorestis	0.68	Sedimentibacter	0.65	Candidatus Tammella	0.75	Sphingobacterium	0.62
Bellilinea	0.67	Desulfovibrio	0.59	Arcobacter	0.74	Alkaliphilus	0.62
Sphingobacterium	0.66	Methyloversatilis	0.54	Comamonas	0.74	Candidatus Tammella	0.54
Candidatus Tammella	0.57	Alkaliphilus	0.51	Sphingobacterium	0.66	Desulfobulbus	0.52
Desulfotalea	0.53	Rhodobacter	0.5	Rhizobium	0.66	Anaeromusa	0.48
Methyloversatilis	0.52	Desulfotignum	0.49	Desulfomicrobium	0.62	Heliorestis	0.35
other	34.03	other	24.36	other	31.38	other	17.89

Figure 6B summarizes the results for *Pelobacter* sp., which was detected and identified by both the PCR-DGGE method and 16S metagenomic analysis using next-generation sequencing. Its detection frequency was lower than that of *Geobacter* sp., but the order of detection was consistent for all cycles: inside electrode with electron pathway >> electrode surface with electron pathway >> inside electrode without electron pathway >> electrode surface without electron pathway. These results suggest that the presence of the electron transfer pathways also influences the growth of *Pelobacter* sp. Figure 6C summarizes the results of *Shewanella* sp., a well-known electricity-producing bacteria in the field of microbial fuel cells. This bacterium was not detected or identified in the bacterial flora analysis by PCR-DGGE but was detected in the 16S metagenomic analysis using next-generation sequencing. Contrary to the results of *Geobacter* sp. And *Pelobacter* sp., *Shewanella* sp. Showed almost no growth with or without an electron transfer pathway. This indicates that although *Shewanella* sp. Is an electricity-producing bacteria, it was not involved in the biological oxidation of sulfide in this experimental system. Therefore, not all electron-emitting bacteria are necessarily involved in the biological oxidation of sulfide. *Desulfomonas* sp., a sulphate-reducing bacterium, was initially detected at low levels before the start of the experiment, but its detection rate increased as the cycle progressed, particularly inside the surface layer of the electrode, as shown in Figure 6D.

4. Discussion

The study found that conductive concrete can suppress H_2S through biological oxidation, with electricity-producing bacteria playing a significant role. The presence of the electron transfer pathways was found to be essential for the growth of these bacteria, with *Geobacter* sp. being the most frequently detected. *Pelobacter* sp. was also detected but had a lower detection frequency. *Shewanella* sp., another electricity-producing bacterium, was not involved in the biological oxidation of sulfide. *Desulfomonas* sp., a sulfate-reducing bacterium, showed significant growth near the electrode poles with EP3, suggesting symbiosis between them. The findings of this study indicate that the presence of an electron transfer pathway is critical for the growth and accumulation of electricity-producing bacteria such as *Geobacter* sp. and *Pelobacter* sp. The reduction rates observed in this study were lower than those reported in previous studies [14], possibly due to the smaller surface area of the anodes in this experimental system. This finding suggests that the biological oxidation effect of hydrogen sulfide was observed, and further bacterial flora analysis was conducted to gain a deeper understanding of the process. The importance of an electron transfer pathway was confirmed by the growth of *Geobater* sp. and *Pelobacter* sp. The results indicate that presence of an electron transfer pathway has a significant impact on the growth of these bacteria and agree with the previous studies [24,25]. The sulfate-reducing bacterium was found to be most concentrated in the sediment within the anode electrode of the closed circuit, similar to the electricity-producing bacteria. This suggests that sulfate-reducing bacteria and electron-excreting bacteria can coexist and thrive in anaerobic conditions. However, without an electron transfer pathway, the growth of sulfate-reducing bacteria was limited, as indicated by a considerably fainter band (Figure 6). Next-generation sequencing analysis revealed that *Geobacter* sp. proliferated rapidly inside the electrode when an electron transfer pathway was provided. In contrast, *Shewanella* sp. showed little to no growth regardless of the presence or absence of an electron transfer pathway. These findings suggest that not all electricity-producing bacteria are involved in the biological oxidation of sulfide. This could be explained that *Geobacter* sp. use a unique electron transfer pathway called the direct interspecies electron transfer (DIET) pathway, which allow their electrically conductive pili (e-pili) to plug into conductive carbon substances, such as San-Earth [26]. The molecular analysis of the microbial community in anaerobic environment is consistent with previous findings [27–29] that *Geobacter* species are the most numerous bacteria (Table 4) and exhibit a remarkable level of metabolic activity, especially in the presence of conductive materials. This property makes conductive concrete a supportive structure for EPB bacteria's e-pili to anchor, leading to their strong mobilization.

The presence of electricity-producing bacteria, particularly *Geobacter* sp. and *Pelobacter* sp., has been shown to significantly increase in response to biological oxidation of sulfide in experimental systems. These findings highlight the potential of utilizing electricity-producing bacteria for the removal of sulfide in industrial wastewater and other environmental systems [30–32]. Further research is needed to identify the specific bacteria responsible for sulfide oxidation and to optimize conditions that promote their growth and activity to enhance their performance in sulfide removal applications. Interestingly, *Desulfomonas* sp., a sulfate-reducing bacterium that does not produce electrons, was observed to exhibit significant growth near the electrode poles alongside the electricity-producing bacteria, suggesting symbiosis. Despite the accumulation of *Desulfomonas* sp. in the electrode of the conductive concrete, its presence had little effect on the rate of sulfate concentration decrease (Figure 3), indicating that it does not accelerate sulfide formation. However, the abundance of *Desulfomonas* sp. was significantly higher with an electron transfer pathway than without, as supported by the results of PCR-DGGE in Section 3.2. These results suggest that creating favorable environmental conditions for the growth of electricity-producing bacteria using conductive concrete, which provides an electron transfer pathway, can promote the growth and accumulation of *Geobacter* sp. and *Pelobacter* sp. near the surface of the concrete. The growth and accumulation of these electron-emitting bacteria, along with the inhibition of sulfide formation, provide evidence for the contribution of biological oxidation to the inhibition of hydrogen sulfide generation.

5. Conclusions

In this study, we aimed to investigate the potential of biological oxidation for controlling the formation of hydrogen through experiments involving the use of conductive concrete to provide an electron transfer pathway, and the analysis of bacterial flora through molecular biological methods such as PCR-DGGE and next-generation sequencing. The results revealed that *Geobacter* sp. and *Pelobacter* sp., which are known as typical electricity-producing bacteria, were found to grow and accumulate in the immediate vicinity of the conductive concrete surface. Further, the growth and accumulation of these electricity-producing bacteria were found to be associated with the suppression of sulfide formation. These findings provided compelling evidence that biological oxidation plays a critical role in inhibiting the generation of hydrogen sulfide. In the future, long-term demonstration tests are planned to be conducted using a new conductive concrete that has been separately developed, along with actual sewage water. Such a study will investigate whether the growth and accumulation of electricity-producing bacteria occurs in a similar manner to the present study. The findings from such experiments will provide insights into the effectiveness of the use of conductive concrete as a potential solution for controlling the generation of hydrogen sulfide in wastewater treatment plants.

Author Contributions: Conceptualization and writing—original draft preparation, H.T.V.; methodology, validation, supervision T.I.; methodology, visualization, software, H.T.V., M.F.; formal analysis, M.F., T.S.; investigation, project administration, T.I., H.S. and T.H.; writing—review and editing, T.I., Y.-T.H. All authors have read and agreed to the published version of the manuscript.

Funding: This research was funded in part by JSPS KAKENHI (20K04749); GAIA Project, Ministry of Land, Infrastructure, Transport and Tourism (No. 4).

Data Availability Statement: Data may be made available on request from the corresponding author.

Acknowledgments: We are gratefully acknowledged for the support from Yamaguchi University for providing the infrastructural facilities, grateful to Shuji Tanaka, a consultant for TNK Waterworks Co., Ltd., for his useful comments and advice.

Conflicts of Interest: The authors declare no conflict of interest.

References

1. Foorginezhad, S.; Mohseni-Dargah, M.; Firoozirad, K.; Aryai, V.; Razmjou, A.; Abbassi, R.; Garaniya, V.; Beheshti, A.; Asadnia, M. Recent Advances in Sensing and Assessment of Corrosion in Sewage Pipelines. *Process Saf. Environ. Prot.* **2021**, *147*, 192–213. [CrossRef]
2. Wang, Y.; Li, P.; Liu, H.; Wang, W.; Guo, Y.; Wang, L. The Effect of Microbiologically Induced Concrete Corrosion in Sewer on the Bearing Capacity of Reinforced Concrete Pipes: Full-Scale Experimental Investigation. *Buildings* **2022**, *12*, 1996. [CrossRef]
3. Chaudhari, B.; Panda, B.; Šavija, B.; Chandra Paul, S. Microbiologically Induced Concrete Corrosion: A Concise Review of Assessment Methods, Effects, and Corrosion-Resistant Coating Materials. *Materials* **2022**, *15*, 4279. [CrossRef] [PubMed]
4. Little, B.J.; Blackwood, D.J.; Hinks, J.; Lauro, F.M.; Marsili, E.; Okamoto, A.; Rice, S.A.; Wade, S.A.; Flemming, H.C. Microbially influenced corrosion—Any progress? *Corros. Sci.* **2020**, *170*, 108641. [CrossRef]
5. Zhang, L.; De Schryver, P.; De Gusseme, B.; De Muynck, W.; Boon, N.; Verstraete, W. Chemical and biological technologies for hydrogen sulfide emission control in sewer systems: A review. *Water Res.* **2008**, *42*, 1–12. [CrossRef] [PubMed]
6. Pikaar, I.; Sharma, K.R.; Hu, S.; Gernjak, W.; Keller, J.; Yuan, Z. Reducing sewer corrosion through integrated urban water management. *Science* **2014**, *345*, 812–814. [CrossRef]
7. Nielsen, A.H.; Vollertsen, J. Model Parameters for Aerobic Biological Sulfide Oxidation in Sewer Wastewater. *Water* **2021**, *13*, 981. [CrossRef]
8. Anwar, A.; Liu, X.; Zhang, L. Biogenic corrosion of cementitious composite in wastewater sewerage system—A review. *Process Saf. Environ. Prot.* **2022**, *165*, 545–585. [CrossRef]
9. Mohanakrishnan, J.; Gutierrez, O.; Meyer, R.L.; Yuan, Z. Nitrite effectively inhibits sulfide and methane production in a laboratory scale sewer reactor. *Water Res.* **2008**, *42*, 3961–3971. [CrossRef]
10. Chaturvedi, V.; Verma, P. Microbial fuel cell: A green approach for the utilization of waste for the generation of bioelectricity. *Bioresour. Bioprocess.* **2016**, *3*, 38. [CrossRef]
11. Venkatramanan, V.; Shah, S.; Prasad, R. A Critical Review on Microbial Fuel Cells Technology: Perspectives on Wastewater Treatment. *Open Biotechnol. J.* **2021**, *15*, 131–141. [CrossRef]
12. Van Dinh, C. Anticorrosion Behavior of the SiO_2/Epoxy Nanocomposite-Concrete Lining System under H_2SO_4 Acid Aqueous Environment. *ACS Omega* **2020**, *5*, 10533–10542. [CrossRef]
13. Stanaszek-Tomal, E.; Fiertak, M. Biological and chemical corrosion of cement materials modified with polymer. *Bull. Polish Acad. Sci. Tech. Sci.* **2015**, *63*, 591–596. [CrossRef]
14. Imai, T.; Vo, H.T.; Fukushima, M.; Suzuki, T.; Sakuma, H.; Hitomi, T. Application of Conductive Concrete as a Microbial Fuel Cell to Control H2S Emission for Mitigating Sewer Corrosion. *Water* **2022**, *14*, 3454. [CrossRef]
15. Sassani, A.; Ceylan, H.; Kim, S.; Arabzadeh, A.; Taylor, P.C.; Gopalakrishnan, K. Development of Carbon Fiber-modified Electrically Conductive Concrete for Implementation in Des Moines International Airport. *Case Stud. Constr. Mater.* **2018**, *8*, 277–291. [CrossRef]
16. Takahashi, S.; Tomita, J.; Nishioka, K.; Hisada, T.; Nishijima, M. Development of a prokaryotic universal primer for simultaneous analysis of Bacteria and Archaea using next-generation sequencing. *PLoS ONE* **2014**, *9*, e105592. [CrossRef]
17. Muyzer, G.; De Waal, E.C.; Uitterlinden, A.G. Profiling of complex microbial populations by denaturing gradient gel electrophoresis analysis of polymerase chain reaction-amplified genes coding for 16S rRNA. *Appl. Environ. Microbiol.* **1993**, *59*, 695–700. [CrossRef]
18. Kongjan, P.; O-Thong, S.; Kotay, M.; Min, B.; Angelidaki, I. Biohydrogen production from wheat straw hydrolysate by dark fermentation using extreme thermophilic mixed culture. *Biotechnol. Bioeng.* **2010**, *105*, 899–908. [CrossRef]
19. Schäfer, H.; Muyzer, M.G. Denaturing gradient gel electrophoresis in marine microbial ecology. *Methods Microbiol.* **2001**, *30*, 452–468.
20. Illumina, 16S Metagenomic Sequencing Library Preparation, Part # 150. Illumina.com. 2013. Available online: https://jp.support.illumina.com/content/dam/illumina-support/documents/documentation/chemistry_documentation/16s/16s-metagenomic-library-prep-guide-15044223-b.pdf (accessed on 24 May 2022).
21. Ewing, T.; Ha, P.T.; Beyenal, H. Evaluation of long-term performance of sediment microbial fuel cells and the role of natural resources. *Appl. Energy* **2017**, *192*, 490–497. [CrossRef]
22. Liu, L.; Tsyganova, O.; Lee, D.-J.; Su, A.; Chang, J.-S.; Wang, A.; Ren, N. Anodic biofilm in single-chamber microbial fuel cells cultivated under different temperatures. *Int. J. Hydrogen Energy* **2012**, *37*, 15792–15800. [CrossRef]
23. Rubaba, O.; Araki, Y.; Yamamoto, S.; Suzuki, K.; Sakamoto, H.; Matsuda, A.; Futamata, H. No TitleElectricity Producing Property and Bacterial Community Structure in Microbial Fuel Cells Equipped with Membrane Electrode Assembly. *J. Biosci. Bioeng.* **2013**, *116*, 106–113. [CrossRef] [PubMed]
24. Kondaveeti, S.; Lee, S.H.; Park, H.D.; Min, B. Specific enrichment of different *Geobacter* sp. in anode biofilm by varying interspatial distance of electrodes in air-cathode microbial fuel cell (MFC). *Electrochim. Acta* **2020**, *331*, 135388. [CrossRef]
25. Blanchet, E.; Desmond, E.; Erable, B.; Bridier, A.; Bouchez, T.; Bergel, A. Comparison of synthetic medium and wastewater used as dilution medium to design scalable microbial anodes: Application to food waste treatment. *Bioresour. Technol.* **2015**, *185*, 106–115. [CrossRef] [PubMed]
26. Lovley, D.R. Syntrophy Goes Electric: Direct Interspecies Electron Transfer. *Annu. Rev. Microbiol.* **2017**, *71*, 643–664. [CrossRef]

27. Baek, G.; Kim, J.; Cho, K.; Bae, H.; Lee, C. The biostimulation of anaerobic digestion with (semi)conductive ferric oxides: Their potential for enhanced biomethanation. *Appl. Microbiol. Biotechnol.* **2015**, *99*, 10355–10366. [CrossRef]
28. González, J.; Sánchez, M.; Gómez, X. Enhancing Anaerobic Digestion: The Effect of Carbon Conductive Materials. *C J. Carbon Res.* **2018**, *4*, 59. [CrossRef]
29. Zhao, Z.; Zhang, Y.; Woodard, T.L.; Nevin, K.P.; Lovley, D.R. Enhancing syntrophic metabolism in up-flow anaerobic sludge blanket reactors with conductive carbon materials. *Bioresour. Technol.* **2015**, *191*, 140–145. [CrossRef]
30. Nielsen, L.P.; Risgaard-Petersen, N.; Fossing, H.; Christensen, P.B.; Sayama, M. Electric currents couple spatially separated biogeochemical processes in marine sediment. *Nature* **2010**, *463*, 1071–1074. [CrossRef]
31. Clauwaert, P.; Rabaey, K.; Aelterman, P.; de Schamphelaire, L.; Pham, T.H.; Boeckx, P.; Boon, N.; Verstraete, W. Biological denitrification in microbial fuel cells. *Environ. Sci. Technol.* **2007**, *41*, 3354–3360. [CrossRef]
32. Yanuka-Golub, K.; Reshef, L.; Rishpon, J.; Gophna, U. Specific *Desulfuromonas* Strains Can Determine Startup Times of Microbial Fuel Cells. *Appl. Sci.* **2020**, *10*, 8570. [CrossRef]

Disclaimer/Publisher's Note: The statements, opinions and data contained in all publications are solely those of the individual author(s) and contributor(s) and not of MDPI and/or the editor(s). MDPI and/or the editor(s) disclaim responsibility for any injury to people or property resulting from any ideas, methods, instructions or products referred to in the content.

Article

Chromium Removal from Aqueous Solution Using Natural Clinoptilolite

Tonni Agustiono Kurniawan [1,*], Mohd Hafiz Dzarfan Othman [2], Mohd Ridhwan Adam [2], Xue Liang [3], Huihwang Goh [3], Abdelkader Anouzla [4], Mika Sillanpää [5], Ayesha Mohyuddin [6] and Kit Wayne Chew [7]

1. College of the Environment and Ecology, Xiamen University, Xiamen 361102, China
2. Advanced Membrane Technology Research Centre (AMTEC), Faculty of Chemical and Energy Engineering, Universiti Teknologi Malaysia (UTM), Skudai 81310, Malaysia
3. School of Electrical Engineering, Guangxi University, Nanning 530004, China
4. Department of Process Engineering and Environment, Faculty of Science and technology, University Hassan II of Casablanca, Mohammedia 28806, Morocco
5. Department of Chemical Engineering, School of Mining, Metallurgy and Chemical Engineering, University of Johannesburg, Doornfontein 2028, South Africa
6. Department of Chemistry, School of Science, University of Management and Technology, Lahore 54770, Pakistan
7. School of Chemistry, Chemical Engineering and Biotechnology, Nanyang Technological University (NTU), Singapore 637459, Singapore
* Correspondence: tonni@xmu.edu.cn

Citation: Kurniawan, T.A.; Othman, M.H.D.; Adam, M.R.; Liang, X.; Goh, H.; Anouzla, A.; Sillanpää, M.; Mohyuddin, A.; Chew, K.W. Chromium Removal from Aqueous Solution Using Natural Clinoptilolite. *Water* 2023, 15, 1667. https://doi.org/10.3390/w15091667

Academic Editors: Yung-Tse Hung, Hamidi Abdul Aziz, Issam A. Al-Khatib, Rehab O. Abdel Rahman and Tsuyoshi Imai

Received: 27 March 2023
Revised: 9 April 2023
Accepted: 23 April 2023
Published: 25 April 2023

Copyright: © 2023 by the authors. Licensee MDPI, Basel, Switzerland. This article is an open access article distributed under the terms and conditions of the Creative Commons Attribution (CC BY) license (https://creativecommons.org/licenses/by/4.0/).

Abstract: This work investigates the applicability of clinoptilolite, a natural zeolite, as a low-cost adsorbent for removing chromium from aqueous solutions using fixed bed studies. To improve its removal performance for the inorganic pollutant, the adsorbent is pretreated with NaCl to prepare it in the homoionic form of Na^+ before undertaking ion exchange with Cr^{3+} in aqueous solution. This work also evaluates if treated effluents could meet the required effluent discharge standard set by legislation for the target pollutant. To sustain its cost-effectiveness for wastewater treatment, the spent adsorbent is regenerated with NaOH. It was found that the clinoptilolite treated with NaCl has a two-times higher Cr adsorption capacity (4.5 mg/g) than the as-received clinoptilolite (2.2 mg/g). Pretreatment of the clinoptilolite with NaCl enabled it to treat more bed volume (BV) (64 BV) at a breakthrough point of 0.5 mg/L of Cr concentration and achieve a longer breakthrough time (1500 min) for the first run, as compared to as-received clinoptilolite (32 BV; 250 min). This suggests that pretreatment of clinoptilolite with NaCl rendered it in the homoionic form of Na^+. Although pretreated clinoptilolite could treat the Cr wastewater at an initial concentration of 10 mg/L, its treated effluents were still unable to meet the required Cr limit of less than 0.05 mg/L set by the US Environmental Protection Agency (EPA).

Keywords: adsorption; clinoptilolite; ion exchange; low-cost adsorbent; water pollution; zeolite

1. Introduction

Water is a fundamental part of life. Without water, there is no life. Within the scope of UN Sustainable Development Goal (SDG) #6 "Clean water and sanitation for all", the development of sustainable water treatment technologies serves as an enabler for water and sanitation equity. As the lack of clean water in different parts of the world results from water pollution caused by refractory pollutants such as inorganic contaminants, water is a deal breaker for accomplishing the SDGs [1]. However, the presence of heavy metals in the aquatic environment due to untreated industrial wastewater effluents in water bodies and their potential effects on living organisms has emerged as one of the major environmental concerns worldwide [2]. Water shortage and safety concerns, exacerbated by increasing water demand and water pollution, also represent major challenges in global efforts to contribute to the UN SDGs, while ensuring the provision of clean water as a basic human right for vulnerable communities [3].

To address the demand of our society for "clean water", various technologies have been developed to deal with the shortage of conventional water resources by harvesting it from non-conventional resources, including treated effluents [4]. As water treatment is crucial to a healthy community and a safe environment, wastewater needs to be treated thoroughly so that it does not harm the environment into which it is discharged. Therefore, any water technology has to meet stringent discharge standards for effluents required by environmental legislation [5]. The technology must also be robust to maintain its performance requirements [6]. Other factors such as the characteristics of wastewater, the legal requirements of residual effluent prior to their discharge, treatment performance, plant flexibility and reliability, and long-term environmental impacts need to be taken into account when selecting the most appropriate technology for wastewater treatment [7].

As environmental legislation imposing effluent limits for wastewater discharged from wastewater treatment plants has become increasingly strict, cost-effective water technologies have been in demand in the global market. New approaches need to be examined to supplement existing conventional treatments such as chemical precipitation [8]. The approaches cover avoiding consumption of excessive chemicals and reducing the generation of toxic sludge or secondary waste post-treatment, while simultaneously improving the ability of treated effluent to comply with the requirements of legislation and reducing energy consumption and treatment costs [9].

As traditional treatment alternatives cannot optimize the removal of target contaminants from industrial wastewater, there is a growing need to develop other environmentally sound technologies that could improve their performance for water treatment applications. For this reason, membrane filtration has been developed for the removal of refractory pollutants in wastewater. Unlike other separation technologies, membrane separation has key benefits such as low environmental pollutant emissions, as it represents a physical separation at moderate operating conditions [10]. Despite the ability of membrane filtration to remove target pollutants from wastewater, its limitations are attributed to its costly treatment costs due to massive energy consumption [11]. They are also not cost-effective to treat polluted wastewater due to heavy metals with concentrations over 100 mg/L [12]. Therefore, the search for alternative treatments has intensified in recent years.

Like membrane filtration, a polluted water environment can be restored using low-cost materials based on a physico-chemical process [13]. Through mass transfer, by which a target pollutant is relocated from the liquid phase to the surface of a solid through physico-chemical interactions [14], adsorption has been widely recognized as a novel strategy for treating wastewater laden with inorganic pollutants [15]. Due to its large surface area, adsorption using activated carbon (AC) can eliminate inorganic pollutants such as metals and other refractory pollutants [16]. Although treated effluents can meet the limit of metal effluent, the utilization of AC remains costly for a large-scale application.

The diverse applications of functional materials for adsorbents have recently responded to the need for cost-effective water treatment. Consequently, there is a growing motivation to use non-conventional materials for the removal of inorganic pollutants from polluted water [17]. Natural resources that are locally available in large quantities such as clinoptilolite can be chemically modified and used as low-cost adsorbents [18]. Conversion of the clinoptilolite into functional materials, which can be utilized for water purification, would add to their commercial value and help users minimize waste disposal costs while providing another option to costly AC [19].

The need for sustainable techniques that do not lead to the generation of hazardous by-products has resulted in the practical utilization of clinoptilolite as an adsorbent for environmental remediation. Natural clinoptilolite has gained popularity due to its ion exchange capability [20]. Large deposits of clinoptilolite in Greece and the UK provide industrial users with cost efficiency. This enables them to treat wastewater laden with heavy metals cost-effectively. The market price of clinoptilolite is about USD 0.4 per kg, depending on its quality [21].

Clinoptilolite, a high-silica member of the heulandite group of natural zeolite, is abundantly available in nature. As a crystalline aluminosilicate from natural resources, zeolite has high cation exchange capacities (CEC) with certain metal ions in the solution [22]. The exchange characteristics of clinoptilolite are attributed to the existence of its ne-gatively charged lattice, which is exchangeable with heavy metals [23]. Since alumunium has one less positive charge than silicon, the framework has a net negative charge of one at the site of each alumunium atom and is counterbalanced by the exchangeable cations such as Na^+, K^+, and Mg^{2+} [24]. The microporosity and high surface area of clinoptilolite make it widely utilized in applications as an ion exchanger, adsorbent, and separation media.

A preliminary study has been undertaken using clinoptilolite as an adsorbent for Cr removal from aqueous solutions using batch modes [25]. Although batch studies are convenient to assess the removal capability of low-cost adsorbents on target adsorbate, they only yield information on the capacity of the media for target metal ions and the rate of metal uptake [26]. Consequently, there is a growing need to perform fixed bed tests using a column prior to scaling up.

To demonstrate its novelty, this work investigates the applicability of clinoptilolite for the treatment of wastewater laden with Cr(VI) based on fixed bed studies. To enhance its treatment performance for the target pollutant, the clinoptilolite was pretreated with NaCl. Chemical pretreatment of clinoptilolite with NaCl was carried out to prepare the adsorbent in the homoionic form of Na^+ prior to ion exchange with Cr^{3+} at acidic conditions [27]. This work also evaluates if treated effluents could meet the required discharge standard imposed by legislation [28]. To sustain its cost-effectiveness for wastewater treatment, spent clinoptilolite was regenerated with NaOH [29]. The performance of clinoptilolite in this work for Cr removal is also compared to that of other studies using similar natural materials.

It is expected that contaminated water laden with Cr could be treated cost-effectively with clinoptilolite. This would assist users in minimizing the treatment cost of their wastewater, while meeting the requirement of discharge effluent standards set by local legislation [30].

2. Material and Methods

All the chemicals were of analytical grade, supplied by Merck (US), and used without purification. Deionized water was applied to prepare working solutions and reagents.

2.1. Cr(VI) Aqueous Solutions

$K_2Cr_2O_7$ was utilized as a source of Cr(VI) in aqueous solutions [31]. To ensure its purity, before being dissolved in deionized water, the chemical was dried in an oven at 100 °C overnight and cooled in a desiccator at ambient temperature. A stock solution of 50 mg/L was obtained by dissolving 0.1414 g of $K_2Cr_2O_7$ in 100 mL of deionized water, while the Cr concentration in working solution was varied by diluting the stock solution [31].

Prior to its use, the pH of the Cr solution was measured using a pH meter. pH adjustment was undertaken using 0.1 M NaOH and/or 0.1 M H_2SO_4, which represent strong alkaline and strong acid, respectively. To analyze the remaining Cr concentration in the samples after treatment, about 0.25 g of 1.5-diphenylcarbazide was dissolved in 50 mL of acetone and stored in a brown bottle [32].

2.2. Treatment of Clinoptilolite with NaCl

The adsorbent in this study is natural clinoptilolite. Its physical characteristics are listed in Table 1. Prior to experiments, the adsorbent was treated with 2 M NaCl. The suspension was continuously agitated for 24 h using a rotary shaker. The clinoptilolite was separated from the supernatant using GF/C filters, and the liquid was drained [33]. The washing process was repeated to remove excess NaCl from the surface of the clinoptilolite. Finally, the adsorbent was dried for 3 h and stored until required for further use [34]. To understand the change in its morphology before and after its treatment with NaCl, the clinoptilolite was characterized with TEM (transmission electron microscope), FTIR (Fourier transformation infrared), and XRD (X-ray diffraction).

Table 1. Physical properties of clinoptilolite.

Property	
Solid density (g/cm^3)	2.10
Particle size (mm)	0.68
Packing density (g/cm^3)	2.25
Total surface area (m^2/g)	800
Cation exchange capacity (meq/g)	2.50

2.3. Fixed Bed Study

In a fixed-bed study, a glass column, 50 cm in length with 1.0 cm of internal diameter, was packed with a known mass of an adsorbent. At the bottom of the column, a 1 cm layer of glass beads was fitted. The designated column was filled by the adsorbent. Feeding solutions containing Cr(VI) with a concentration of 10 mg/L were then prepared from the stock solution. After adjusting its pH to optimum based on the results of batch studies, the feeding solutions were introduced at the top of the column, and the column was operated with the feeding solutions flowing from top to bottom [35].

The pH of the effluents was monitored hourly to check if there was any change that might take place [36]. A flow rate of 5.0 mL/min was retained with a peristaltic pump. This flow rate might slightly vary between the runs [37]. The effluent samples were periodically collected by a fraction collector and then analyzed for residual Cr concentrations.

The column operations ended after the saturation point was attained, and there was no difference in concentration between influent and effluent ($C_e/C_o = 1$). In this condition, all surface sites of the zeolite were occupied by adsorbed Cr. The column was washed with deionized water to eliminate unadsorbed metal [38].

2.4. Column Regeneration

After complete exhaustion ($C_e/C_o = 1$), the column was desorbed by passing regenerants to recover the accumulated Cr on the saturated adsorbent. 0.1 M NaOH solution was used for the regeneration of spent clinoptilolite. Desorption was terminated as soon as the effluent metal concentration was negligible. Afterwards, the column was rinsed with deionized water at the same flow rate of 5 mL/min until the pH of the effluent was equal to its influent of 7.0–7.2 [39]. To quantify the regeneration efficiency (RE) of the spent adsorbent and evaluate its reusability, the following method was applied [40]:

$$(\%RE) = \frac{(A_r)}{(A_0)} \times 100 \quad (1)$$

where A_0 and A_r are the adsorption capacities of the adsorbent before and after regeneration, while the % of loss in the Cr adsorption capacity is the fraction of adsorbate that could no longer be adsorbed to that adsorbed during the first cycle was calculated as:

$$(\%LAC) = \frac{(A - I)}{A} \times 100 \quad (2)$$

where I and A are the amount adsorbed in each subsequent cycle and the amount adsorbed during the first cycle, respectively [41].

2.5. Chemical Analysis of Cr Concentration

Changes in the Cr(VI) concentrations in the solution after adsorption treatments were calculated colorimetrically based on the Standard Methods [42]. A purple-violet complex resulted from the reactions between 1.5-diphenylcarbazide and Cr^{6+} in acidic conditions. Absorbance was determined at wavelength (λ) 540 nm after 10 min of color development. The least detectable concentration based on this method is 0.005 mg/L as Cr(VI) [42]. In acidic solutions, both $HCrO_4^-$ and $Cr_2O_7^{2-}$ anions can be detected.

2.6. Statistical Analysis

To ensure the accuracy of the obtained data, all the studies were undertaken in duplicate. The relative standard deviation of Cr removal for the studies was less than 1.0%. When the relative error exceeded this criterion, a third experiment was carried out [43].

3. Results and Discussion

3.1. Effect of Chemical Pretreatment

Clinoptilolite contains a complement of exchangeable sodium, potassium, magnesium, and calcium ions, with the selectivity of metals as follows: $K^+ > Mg^{2+} > Ca^{2+} > Na^+$ [44]. To prepare the clinoptilolite in the homoionic form of Na^+, it was treated with NaCl before adsorption. This treatment was conducted based on the findings of previous studies that Na^+ was the most effective exchangeable ion for the ion exchange of heavy metals [45].

After treatment, certain cations such as K^+ and Ca^{2+} were strongly held by the clinoptilolite in preference to Na^+. Therefore, Na^+ is mostly involved in the ion exchange process [46]. The exposure of clinoptilolite to concentrated NaCl led to the production of the Na-rich sample, and this was attributed to the exchange of Ca^{2+} Mg^{2+} and K^+ were not exchangeable with other cations, as they were associated with impurities in the sample. For these reasons, the K^+, Ca^{2+}, and Mg^{2+} contents in the clinoptilolite could be ignored [47].

3.2. Characterization of Clinoptilolite before and after Pretreatment

To understand the difference in its morphology before and after its pretreatment with NaCl, the adsorbent was characterized using TEM, FTIR, and XRD. Figure 1 presents the morphology of clinoptilolite based on TEM analysis with scale bars of 1 μm (a), 500 nm (b), and 200 nm (c), respectively. Figure 1c indicates the lamellar-shaped particles of the clinoptilolite after metal adsorption, as compared to the natural clinoptlolite [48].

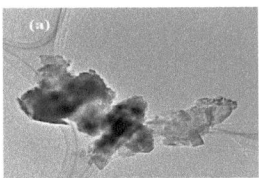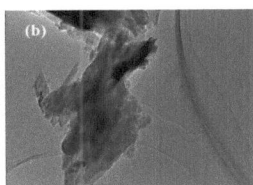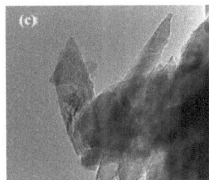

Figure 1. TEM characterization of clinoptilolite (**a**) before pretreatment; (**b**) after pretreatment; (**c**) after metal adsorption.

In addition, XRD was used to characterize the crystallinity of the adsorbent (Figure 2). The XRD characterization of the clinoptilolite was indicated by multi-diffraction peaks at 2θ of 9.768°, 11.105°, 13.220°, 16.796°, 18.894°, 20.762°, 22.247°, 22.615°, 25.930°, 26.547°, 28.040°, 29.885°, 31.861°, 32.580°, 36.451°, and 50.051°, as confirmed by the JCPDS card (01-079-1460). The pattern peaks confirmed that clinoptilolite was the main phase of this adsorbent.

To confirm the elemental composition of the clinoptilolite, FTIR was utilized to analyze the functional group on its surface (Figure 3).

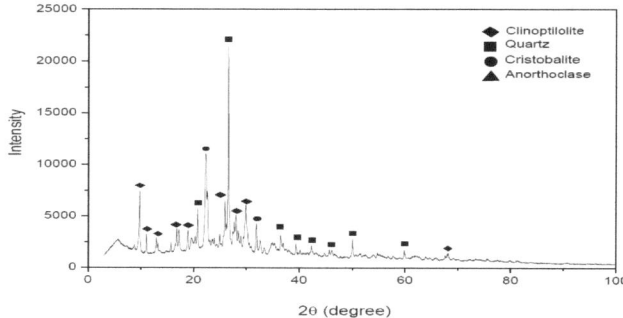

Figure 2. XRD analysis of clinoptilolite.

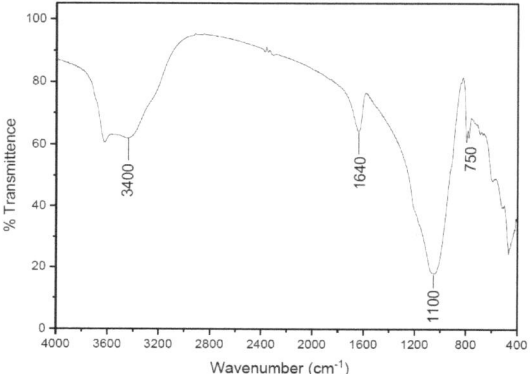

Figure 3. FTIR results of clinoptilolite.

Figure 3 shows the existence of spectra related to water molecules at 3400 and 1640 cm^{-1}. The asymmetric stretching of the spectrum at 1100 cm^{-1} was associated with the SiO$_4$ tetrahedral [49]. A weaker spectrum at 1000 cm^{-1} was related to the vibration that involves ≡Al−O due to the vacancies in Al^{3+}. The peak at 750−700 cm^{-1} was associated with the symmetric vibration of SiO$_4$, which indicated the presence of Si−O···HO−Si bonds [50].

To further identify the elemental composition of clinoptilolite, X-ray fluorescence (XRF) analysis was carried out. As presented in Table 2, the major elements of the clinoptilolite are SiO$_2$ (73%) (w/w), and Al$_2$O$_3$ (12%) (w/w). Several trace elements were also present. They included CaO (2.1% (w/w), K$_2$O (6.7% (w/w), Fe$_2$O$_3$ (4.3% (w/w)), and MgO (1.9% (w/w)). Their presence might contribute to Cr adsorption during water treatment [51].

Table 2. Composition of clinoptilolite.

Oxide	Al$_2$O$_3$	SiO$_2$	MgO	CaO	Fe$_2$O$_3$	K$_2$O
Composition (%)	12.0	73.0	1.9	2.1	4.3	6.7

In adsorption treatment, adsorbate and adsorbent interacted physically in the aqueous phase [52]. As no ·OH was involved in the degradation of the target pollutant, there was no change in the chemical composition of the starting compounds after treatment [53]. Hence, it is not necessary to prove the stability of the adsorbent before and after treatment [54].

3.3. Fixed Bed Studies

In practice, fixed-bed columns are widely used in chemical industries due to their simple and continuous operation [55]. Column operation is essential for the industrial-scale formulation of certain technical systems as it provides credible data on acceptable flow rate, breakthrough time, and loss of adsorption capacity from the first cycle to subsequent cycles [56]. In addition, column studies more accurately quantify the adsorption capacity of an adsorbent for an adsorbate [57]. By using a breakthrough, the practical applicability and feasibility of an adsorbent for Cr removal can be evaluated for industrial application [58].

In column studies, the regeneration of adsorbent and recovery of adsorbate material are the key factors in wastewater treatment applications [59]. To design such an adsorption/desorption process in column operations, the adsorption capacities and adsorption kinetics between adsorbent and adsorbate need to be clearly defined [60]. One way to obtain these characteristics is by examining the concentration of adsorbate in the effluent versus the number of bed volume (BV), which could be treated by adsorbent until reaching complete exhaustion [61]. By using a breakthrough technique, the behavior of metal adsorption on the adsorbent surface can be evaluated [62].

3.3.1. Breakthrough

Generally, breakthrough is defined by the point where a specified amount of the influent is detected in the effluent, while the number of bed volume (BV) represents the ratio between the volume of adorbate solution treated and the volume of adsorbent utilized [63]. Both parameters are widely employed to compare the removal performance of adsorbents for certain metal ions [64]. The Cr uptake at the 5% breakthrough point was chosen as the operational capacity of the fixed bed study [65].

To assess its practical utility for Cr removal, column studies were also performed for all types of clinoptilolite. A typical breakthrough curve, representing the ratio of effluent concentration over influent concentration versus the number of BV passed through the column until reaching complete exhaustion, is presented in Figure 4.

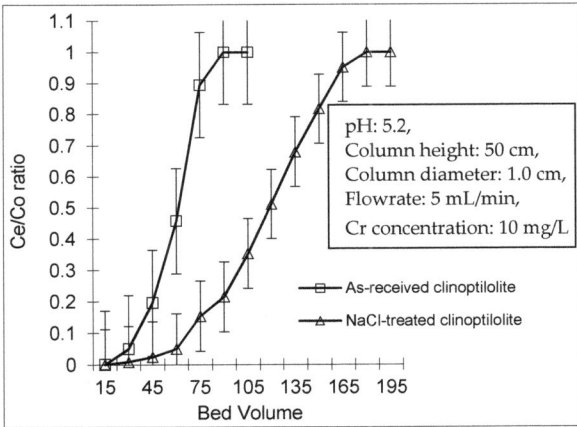

Figure 4. Breakthrough curve of as-received clinoptilolite and NaCl-treated clinoptilolite on Cr removal for the first run.

The breakthrough curves of all types of clinoptilolite for the first run, presented in Figure 4, are of the conventional "S" shape. It is interesting to note that the breakthrough point was accomplished when the Cr first appeared in the effluent ($C_e/C_o = 0.05$), while the saturation point was attained in equilibrium when no concentration difference was found between the influent and effluent ($C_e/C_o = 1$) [66].

The two breakthrough curves in the above figure demonstrate that the removal performance of clinoptilolite on Cr was strongly influenced by the way in which the clinoptilolite was treated prior to ion exchange [67]. Figure 4 shows that the complete breakthrough of the as-received clinoptilolite occurred at 32 BV (1.6 L of influent). This indicates that the breakthrough rapidly proceeded at the beginning of the adsorption process, but it started to decline steadily until becoming completely exhausted at 91 BV (5.2 L of feeding solution). At the initial stage of adsorption, when there was an excess of adsorption sites on the clinoptilolite's surface, Cr(III) ions were adsorbed rapidly [68]. However, when the surface sites of clinoptilolite were densely covered by adsorbate, the available sites for metal binding became saturated, resulting in a lower removal rate [69]. It is essential to note that the breakthrough curve of the as-received clinoptilolite in Figure 4 did not follow an ideal "S" shape profile, suggesting the inefficient use of adsorbent and that the large adsorption zone was within the clinoptilolite bed [70].

Compared to the as-received clinoptilolite, the NaCl-treated clinoptilolite achieved the breakthrough point of 0.5 mg/L of effluent Cr concentration remarkably later at 64 BV, corresponding to 3.6 L of influent, and became completely exhausted at 182 BV (about 10.4 L). The significant difference in terms of Cr removal performance between the two types of clinoptilolite suggests that chemical pretreatment of clinoptilolite with NaCl rendered it in the homoionic form of Na^+ ($p \leq 0.05$; paired t-test). Consequently, the Na^+ of the clinoptilolite could be easily replaced by Cr^{3+} in the solution [71]. This provides convincing evidence to explain the proposed adsorption mechanism: that Cr removal by the clinoptilolite resulted from ion exchange, although some might be due to passive physical adsorption [72].

The Cr adsorption capacities of clinoptilolite, determined based on the breakthrough curve area under complete exhaustion, were obtained from the dynamic study by divi-ding the total weight of solute adsorbed by the total weight of adsorbent used. For comparison purposes, the adsorption capacities of clinoptilolite, determined from the batch studies at the same Cr concentration of 10 mg/L, are also presented in Table 3.

Table 3. Comparison of the Cr adsorption capacity of clinoptilolite between column and batch studies for the first run.

Types of Adsorbent	Cr Adsorption Capacity (mg/g)		Difference of Cr Adsorption Capacity (%)
	Fixed Bed Studies	Batch Studies [25]	
As-received clinoptilolite	2.2	1.8	23
NaCl-treated clinoptilolite	4.5	3.2	41

Table 3 shows that the results of column operations were higher than those of batch studies at the same Cr concentration of 10 mg/L. This can be due to the inherent difference in the nature of both studies. In batch studies, the concentration gradient reduced with the longer contact time, while in column studies, the clinoptilolite continuously had phy-sico-chemical interactions with fresh adsorbate solution at the interface of the adsorption zone, as the feeding solution passed through the column [73]. Hence, the concentration gradient increased with a longer residence time. When running columns in series, the first run of column operations needs to be undertaken until attaining complete exhaustion [74].

Monitoring of the effluent pH of NaCl-treated clinoptilolite in column operation shows a substantial increase of pH from 5.0 to 9.0 during the first run; thus, indicating that Cr adsorption on the surface of clinoptilolite releases OH^- into the system [75]. Al-though the column performance of NaCl-treated clinoptilolite tended to deteriorate from the first cycle to the subsequent cycle, it is suggested that NaCl-treated clinoptilolite was a relatively good adsorbent for Cr removal, as this adsorbent exhibited a reasonable Cr adsorption capacity for the first two cycles [76].

3.3.2. Regeneration

Regeneration of a spent adsorbent is necessary when the adsorbent used is expensive or not always available in large quantities [77]. From an economical point of view, an adsorbent can be considered efficient and effective if it is easily regenerated and re-utilized as frequently as possible without altering its removal performance on certain metals [78]. Therefore, column regeneration using a selected chemical was undertaken to restore the removal performance of the same column to its original state [79].

When clinoptilolite in the same column becomes exhausted or when the effluent from an adsorbed bed reaches the allowable discharge level, the recovery of the adsorbed material as well as the regeneration of the adsorbent become necessary [80]. Chemical regeneration is a definite option for this purpose. Therefore, Cr desorption from the clinoptilolite surface was performed with NaOH (Figure 5).

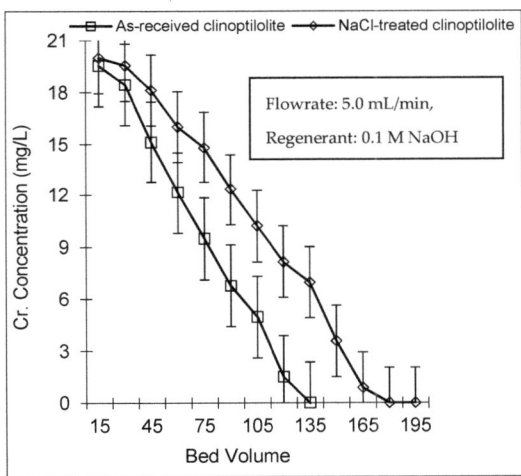

Figure 5. Regeneration curve of all types of clinoptilolite by 0.1 M NaOH.

It was found that complete Cr desorption from NaCl-treated clinoptilolite, which occurred at 10.86 L of NaOH (about 164 BV), accounted for the recovery of 93% of Cr from the adsorbent (Table 4); while Cr desorption from the as-received clinoptilolite, which occurred at 96 BV (corresponding to 5.4 L of the same regenerant), provided 91% of Cr recovery. Thus, this suggests that Cr desorption from the former needed an additional volume of regenerant, leading to a costly treatment cost [81].

Table 4. Comparison of total mass balance of Cr adsorbed on clinoptilolite before and after regeneration for the first cycle of all types of clinoptilolite.

Types of Clinoptilolite	Cr before Regeneration (mg/g)	Cr after Regeneration (mg/g)	Regeneration Efficiency (%) *
As-received clinoptilolite	2.2	2.0	91
NaCl-treated clinoptilolite	4.5	4.2	93

Note: * Remarks: % regeneration efficiency (RE) was calculated using Equation (1).

In the first regeneration cycle of the NaCl-treated clinoptilolite and the as-received clinoptilolite, a loss of Cr adsorption capacity of about 23 and 13% was found, respectively (Table 5). This indicates that NaOH is not an ideal regenerant for both types of clinoptilolite, as their ion exchange capacity with Cr^{3+} tended to decline. Other ion exchangers, such as NaCl, should be tested to desorb Cr from the clinoptilolite's surface, as Cr adsorption on clinoptilolite occurs due to ion exchange between the Na^+ of clinoptilolite and the Cr^{3+} in the solution [82]. Thus, this suggests that there is a need to examine the current regeneration procedure more rigorously in order for the regenerated clinoptilolite to adsorb more metals [83].

Table 5. Summary of column performance for Cr adsorption by all types of clinoptilolite.

Type of Clinoptilolite	BV Treated at 1st Run		Initial Cr Adsorption Capacity (mg/g)	Cr Adsorption Capacity (mg/g) At 2nd Run	Loss of Adsorption Capacity (%) *
	At Breakthrough	At Exhaustion			
As-received clinoptilolite	32	91	2.2	1.7	23
NaCl-treated clinoptilolite	64	181	4.5	3.9	13

Note: * Remarks: % loss of adsorption capacity was calculated using Equation (2).

Table 5 shows that the Cr adsorption capacity of clinoptilolite remarkably deteriorated over the two cycles. The rate of this deterioration decreased with the increasing number of successive cycles due to exposure to the alkaline regenerant [84]. Although the Cr removal performance of NaCl-treated clinoptilolite in column operation was not excellent, it is important to note that this adsorbent is technically capable of treating Cr-rich effluents at a low cost [85]. Therefore, it needs further consideration before being used for wastewater treatment on an industrial scale.

3.3.3. Cr Adsorption Capacity of Clinoptilolite

Clinoptilolite treated with NaCl has a two-times higher Cr adsorption capacity (4.5 mg/g) than as-received clinoptilolite (2.2 mg/g). Pretreatment of clinoptilolite with NaCl enabled it to treat more bed volume (64 BV) at the breakthrough point of 0.5 mg/L of Cr concentration and achieve a longer breakthrough time (1500 min) for the first run, as compared to the as-received clinoptilolite (32 BV; 250 min) because pretreatment of clinoptilolite with NaCl rendered it in the homoionic form of Na^+. Consequently, the Na^+ of the clinoptilolite could be replaced by Cr^{3+} in the solution via an ion exchange mechanism, and some might be due to passive adsorption [86]. Although the Cr removal performance of NaCl-treated clinoptilolite is not excellent, this adsorbent has a reasonable Cr adsorption capacity (4.5 mg/g). Statistically, the difference in terms of Cr adsorption capacities between treated and untreated clinoptilolite was negligible ($p > 0.05$; paired t-test).

To understand the performance of clinoptilolite, its adsorption capacity for Cr in this work was compared to that of previous works for a variety of heavy metals (Table 6). The table shows that the adsorption capacity of clinoptilolite for Cr(III) was comparable to those for Cd(II), Cr(VI), Co(II), Ni(II), Zn(II), and Cu(II). In spite of their low metal adsorption capacities, natural materials, including clinoptilolite, have the ability to remove inorganic pollutants through ion exchange with the target contaminant [87]. It is important to note that the adsorption capacity of an adsorbent varies depending on the initial concentration of adsorbate, the type of adsorbent, and chemical pretreatment [88].

The difference in Cr adsorption capacities between the two chemically treated adsorbents was attributed to the fact that clinoptilolite has fewer negatively charged adsorption sites for Coulombic forces with Cr^{3+}. Despite the fact that Cr removal by clinoptilolite occurred due to ion exchange, Cr removal might be due to adsorption on the clinoptilolite surface. Consequently, it resulted in a lower uptake of Cr by clinoptilolite [89].

Table 6. An overview of Cr adsorption capacity by different types of zeolite.

Material	Reference	Cd^{2+}	Cr^{3+}	Cr^{6+}	Co^{2+}	Ni^{2+}	Zn^{2+}	Cu^{2+}	Pb^{2+}
Clinoptilolite	[64]	2.4	0		1.4	0.5	0.5	1.6	1.6
	[69]	1.2							1.4
	[62]	3.7		2.4	1.5	0.9	2.7	3.8	6.0
	Present study		4.5						
Chabazite	[70]	137.0							175
	[62]	6.7		3.6	5.8	4.5	5.5	5.1	6.0
Chabazite–philipsite	[63]		7.1						
	[73]		0.3		0.6		0.04		0.4

Further column studies should be conducted using a flow rate of less than 5 mL/min. Since a low flow rate of feeding solution increases the physico-chemical interaction between clinoptilolite and the target pollutant in the column, the solution has more available residence time to diffuse into the adsorbent for adsorption before it is swept through the column. Hence, it might maximize the treated volume of feeding solutions until breakthrough, extend the lifespan of the bed, and result in a higher Cr removal.

3.3.4. Adsorption Mechanism of Cr Removal by Clinoptilolite

The first step is the reduction of Cr(VI) to Cr(III) (Equation (3)). Although $Cr_2O_7^{2-}$ was utilized as the source of Cr(VI) in aqueous solution, under pH < 6, Cr(VI) exists in the predominant form of $HCrO_4^-$ [90], with the hydrolysis reaction of $Cr_2O_7^{2-}$ as follows:

$$Cr_2O_7^{2-} + H_2O \longleftrightarrow 2\, HCrO_4^- \quad pK_3 = 14.56 \quad (3)$$

The second step controlling Cr removal by clinoptilolite is represented as follows:

$$[Cr(OH)]^{+2}_{(s)} + Na_n A_{(z)} + n\, H_2O_{(s)} \longleftrightarrow ([Cr(OH)]^{+2} - H_n - A)_{(z)} + n\, Na^+_{(s)} + n\, OH^-_{(s)} \quad (4)$$

where A and n represent the adsorption sites on the clinoptilolite's surface and the coefficient of the reaction component, respectively, while subscripts s and z denote the "solution" and "clinoptilolite" phases, respectively. Equation (4) shows that the negative charge of the clinoptilolite, which comes from the tetrahedrally coordinated aluminum, is balanced by the exchangeable Cr^{3+}, suggesting that the Cr uptake by the clinoptilolite occurred due to ion exchange and/or adsorption [91]. Cr adsorption is not a fundamentally different process from that of ion exchange. The mechanism of Cr removal by the clinoptilolite in the solution is facilitated by the ion exchange between Cr^{3+} and the Na^+ of the clinoptilolite network.

As pH increased to 5.0, the adsorption shifted from left to right, which led to the production of more surface complex ($[Cr(OH)]^{+2}-H_n-A$) on the clinoptilolite. The final pH of the solution slightly increased after adsorption because the hydrolysis reaction of the clinoptilolite caused more OH^- release into the solution, resulting in a higher Cr removal. The presence of OH^- in the solution caused Cr^{3+} to be accommodated in the surface lattice of the clinoptilolite, implying that Cr removal by the clinoptilolite is pH-dependent [92].

4. Conclusions

This fixed-bed study has revealed the engineering applicability of clinoptilolite as a low-cost adsorbent for treatment of Cr-laden wastewater [85]. The adsorbent pretreated with NaCl had a significantly higher Cr adsorption capacity (4.5 mg/g) as compared to the clinoptilolite in its as-received form (2.2 mg/g). Pretreatment of clinoptilolite with NaCl rendered it in the homoionic form of Na^+. Hence, this facilitated the pretreated clinoptilolite to treat more bed volume (64 BV) at the breakthrough point and accomplish a longer time to attain a breakthrough (1500 min) for the first run, as compared to clinoptilolite in its as-received form (32 BV; 250 min). Despite showing that the pretreated clinoptilolite could treat the Cr-laden wastewater at 10 mg/L of initial Cr concentration, the treated effluents still could not meet the required Cr limit of less than 0.05 mg/L set by the US Environmental Protection Agency (EPA).

Author Contributions: Conceptualization, A.A.; methodology, H.G.; validation, X.L. and M.R.A.; formal analysis, K.W.C.; investigation, M.S.; resources, M.H.D.O.; writing—original draft preparation, T.A.K.; writing—review and editing, T.A.K.; supervision, A.M. All authors have read and agreed to the published version of the manuscript.

Funding: This work received Research Grants No. Q.J130000.21A6.00P14 and No. Q.J130000.3809.22H07 from the Universiti Teknologi Malaysia (UTM).

Data Availability Statement: Not applicable.

Conflicts of Interest: The authors declare no conflict of interest. The funders had no role in the design of the study; in the collection, analyses, or interpretation of data; in the writing of the manuscript, or in the decision to publish the results.

References

1. Zhu, M.; Kurniawan, T.A.; Duan, L.; Song, Y.; Hermanowicz, S.W.; Othman, M.H.D. Advances in BiOX-based ternary photo-catalysts for water technology and energy storage applications: Research trends, challenges, solutions, and ways forward. *Rev. Environ. Sci. Bio/Technol.* **2022**, *21*, 331–370. [CrossRef]
2. Dun, F.; Huang, Y.; Zhang, X.; Kurniawan, T.A.; Ouyang, T. Uncovering potentials of integrated TiO_2(B) nanosheets and H_2O_2 for removal of tetracycline from aqueous solution. *J. Mol. Liq.* **2017**, *248*, 112–130. [CrossRef]
3. Liang, X.; Kurniawan, T.A.; Goh, H.H.; Zhang, D.; Dai, W.; Liu, H.; Goh, K.C.; Othman, M.H.D. Conversion of landfilled waste-to-electricity (WTE) for energy efficiency improvement in Shenzhen (China): A strategy to contribute to resource recovery of unused methane for generating renewable energy on-site. *J. Clean. Prod.* **2022**, *369*, 133078. [CrossRef]
4. Liang, X.; Goh, H.H.; Kurniawan, T.A.; Zhang, D.; Dai, W.; Liu, H.; Liu, J.; Goh, K.C. Utilizing landfill gas (LFG) to electrify digital data centers in China for accelerating energy transition in Industry 4.0 era. *J. Clean. Prod.* **2022**, *369*, 133297. [CrossRef]
5. Sniatala, B.; Kurniawan, T.A.; Sobotka, D.; Makinia, J.; Othman, M.H.D. Macro-nutrients recovery from liquid waste as a sustainable resource for production of recovered mineral fertilizer: Uncovering alternative options to sustain global food security cost-effectively. *Sci. Total Environ.* **2023**, *856*, 159283. [CrossRef] [PubMed]
6. Maiurova, A.; Kurniawan, T.A.; Kustikova, M.; Bykovskaia, E.; Othman, M.H.D.; Singh, D.; Goh, H.H. Promoting digital transformation in waste collection service and waste recycling in Moscow (Russia): Applying a circular economy paradigm to mitigate climate change impacts on the environment. *J. Clean. Prod.* **2022**, *354*, 131604. [CrossRef]
7. Kurniawan, T.A.; Liang, X.; O'Callaghan, E.; Goh, H.; Othman, M.H.D.; Avtar, R.; Kusworo, T.D. Transformation of solid waste management in China: Moving towards sustainability through digitalization-based circular economy. *Sustainability* **2022**, *14*, 2374. [CrossRef]
8. Huang, Y.; Zeng, X.; Guo, L.; Lan, J.; Zhang, L.; Cao, D. Heavy metal ion removal of wastewater by zeolite-imidazolate frameworks. *Sep. Purif. Technol.* **2018**, *194*, 462–469. [CrossRef]
9. Obaid, S.S.; Gaikwad, D.K.; Sayyed, M.I.; Khader, A.R.; Pawar, P.P. Heavy metal ions removal from waste water by the natural zeolites. *Mater. Today Proc.* **2018**, *5*, 17930–17934. [CrossRef]
10. Meng, Q.; Chen, H.; Lin, J.; Lin, Z.; Sun, J. Zeolite A synthesized from alkaline assisted pre-activated halloysite for efficient heavy metal removal in polluted river and industrial wastewater. *J. Environ. Sci.* **2017**, *56*, 254–262. [CrossRef]
11. Lahnafi, A.; Elgamouz, A.; Tijani, N.; Jaber, L.; Kawde, A.N. Hydrothermal synthesis and electrochemical characterization of novel zeolite membranes supported on flat porous clay-based microfiltration system and its application of heavy metals removal of synthetic wastewaters. *Microporous Mesoporous Mater.* **2022**, *334*, 111778. [CrossRef]
12. Isawi, H. Using zeolite/polyvinyl alcohol/sodium alginate nanocomposite beads for removal of some heavy metals from wastewater. *Arab. J. Chem.* **2020**, *13*, 5691–5716. [CrossRef]
13. Aloulou, W.; Aloulou, H.; Khemakhem, M.; Duplay, J.; Daramola, M.O.; Amar, R.B. Synthesis and characterization of clay-based ultrafiltration membranes supported on natural zeolite for removal of heavy metals from wastewater. *Environ. Technol. Innov.* **2020**, *18*, 100794. [CrossRef]
14. Yurekli, Y. Removal of heavy metals in wastewater by using zeolite nano-particles impregnated polysulfone membranes. *J. Hazard. Mater.* **2016**, *309*, 53–64. [CrossRef]
15. Grba, N.; Baldermann, A.; Dietzel, M. Novel green technology for wastewater treatment: Geo-material/geopolymer applications for heavy metal removal from aquatic media. *Int. J. Sediment. Res.* **2022**, *38*, 33–48. [CrossRef]
16. Janjhi, F.A.; Ihsanullah, I.; Bilal, M.; Castro-Muñoz, R.; Boczkaj, G.; Gallucci, F. MXene-based materials for removal of antibiotics and heavy metals from wastewater–A review. *Water Resour. Ind.* **2023**, *29*, 100202. [CrossRef]

17. Zanin, E.; Scapinello, J.; de Oliveira, M.; Rambo, C.L.; Franscescon, F.; Freitas, L.; de Mello, J.M.M.; Fiori, M.A.; Oliveira, J.V.; Dal Magro, J. Adsorption of heavy metals from wastewater graphic industry using clinoptilolite zeolite as adsorbent. *Process. Saf. Environ.* **2017**, *105*, 194–200. [CrossRef]
18. Li, Z.; Wu, L.; Sun, S.; Gao, J.; Zhang, H.; Zhang, Z.; Wang, Z. Disinfection and removal performance for Escherichia coli, toxic heavy metals and arsenic by wood vinegar-modified zeolite. *Ecotoxicol. Environ. Saf.* **2019**, *174*, 129–136. [CrossRef]
19. Luo, H.; Law, W.W.; Wu, Y.; Zhu, W.; Yang, E.H. Hydrothermal synthesis of needle-like nanocrystalline zeolites from metakaolin and their applications for efficient removal of organic pollutants and heavy metals. *Microporous Mesoporous Mater.* **2018**, *272*, 8–15. [CrossRef]
20. Egashira, R.; Tanabe, S.; Habaki, H. Adsorption of heavy metals in mine wastewater by Mongolian natural zeolite. *Procedia Eng.* **2012**, *42*, 49–57. [CrossRef]
21. Ajiboye, T.O.; Oyewo, O.A.; Onwudiwe, D.C. Simultaneous removal of organics and heavy metals from industrial wastewater: A review. *Chemosphere* **2021**, *262*, 128379. [CrossRef] [PubMed]
22. Wang, Z.; Tan, K.; Cai, J.; Hou, S.; Wang, Y.; Jiang, P.; Liang, M. Silica oxide encapsulated natural zeolite for high efficiency removal of low concentration heavy metals in water. *Colloid. Surf. A* **2019**, *561*, 388–394. [CrossRef]
23. Dun, F.; Kurniawan, T.A.; Li, H.; Wang, L.; Chen, Z.; Wang, H.; Li, W.; Wang, Y.; Li, Q. Applicability of HDPC-supported Cu nanoparticles composite synthesized from unused waste digestate for octocrylene degradation in aqueous solutions. *Chem. Eng. J.* **2019**, *355*, 650–660. [CrossRef]
24. Babel, S.; Kurniawan, T.A. Low-cost adsorbents for heavy metals uptake from contaminated water: A review. *J. Hazard. Mater.* **2003**, *97*, 219–243. [CrossRef]
25. Kurniawan, T.A.; Babel, S. Cr(VI) removal from contaminated wastewater using natural zeolite. *J. Ion Exch.* **2003**, *14*, 289–293. [CrossRef]
26. Kurniawan, T.A.; Avtar, R.; Singh, D.; Xue, W.; Dzarfan Othman, M.H.; Hwang, G.H.; Iswanto, I.; Albadarin, A.B.; Kern, A.O. Reforming MSWM in Sukunan (Yogjakarta, Indonesia): A case-study of applying a zero-waste approach based on circular economy paradigm. *J. Clean. Prod.* **2020**, *284*, 124775. [CrossRef]
27. Kurniawan, T.A.; Chan, G.; Lo, W.H.; Babel, S. Physico-chemical treatment techniques for wateswater laden with heavy metals. *Chem. Eng. J.* **2006**, *118*, 83–98. [CrossRef]
28. Kurniawan, T.A.; Chan, G.Y.S.; Lo, W.H. Comparisons of low-cost adsorbents for treating wastewaters laden with heavy metals. *Sci. Total Environ.* **2006**, *366*, 409–426. [CrossRef]
29. Kurniawan, T.A.; Lo, W.H.; Chan, G. Physico-chemical treatments for removal of recalcitrant contaminants from landfill leachate. *J. Hazard. Mater.* **2006**, *29*, 80–100. [CrossRef]
30. Dun, F.; Kurniawan, T.A.; Gui, H.; Li, H.; Wang, Y.; Li, Q. Role of Cu_xO-anchored pyrolyzed hydrochars on H_2O_2-activated degradation of tetracycline: Effects of pyrolysis temperature and pH. *Ind. Eng. Chem. Res.* **2022**, *61*, 8847–8857. [CrossRef]
31. Kurniawan, T.A.; Singh, D.; Avtar, R.; Dzarfan Othman, M.H.; Hwang, G.H.; Albadarin, A.B.; Rezakazemi, M.; Setiadi, T.; Shirazian, S. Resource recovery from landfill leachate: An experimental investigation and perspectives. *Chemosphere* **2021**, *274*, 129986. [CrossRef] [PubMed]
32. Kurniawan, T.A.; Liang, X.; Singh, D.; Othman, M.H.D.; Goh, H.H.; Gikas, P.; Kern, A.O.; Kusworo, T.D.; Shoqeir, J.A. Harnessing landfill gas (LFG) for electricity: A strategy to mitigate greenhouse gas emissions in Jakarta (Indonesia). *J. Environ. Manag.* **2022**, *301*, 113882. [CrossRef] [PubMed]
33. Dun, F.; Kurniawan, T.A.; Li, Q.; Gui, H.; Othman, M.H.D. Treatment of As(III)-contaminated water using iron-coated carbon fiber. *Materials* **2022**, *15*, 4365. [CrossRef]
34. Kurniawan, T.A.; Othman, M.H.D.; Singh, D.; Avtar, R.; Goh, H.H.; Setiadi, T.; Lo, W.H. Technological solutions for long-term management of partially used nuclear fuel: A critical review. *Ann. Nucl. Energy* **2022**, *166*, 108736. [CrossRef]
35. Kurniawan, T.A.; Othman, M.H.D.; Hwang, G.H.; Gikas, P. Unlocking digital technology in waste recycling industry in Industry 4.0 era: A transformation towards digitalization-based circular economy in Indonesia. *J. Clean. Prod.* **2022**, *357*, 131911. [CrossRef]
36. Huuha, T.; Kurniawan, T.A.; Sillanpää, M. Removal of silicon from pulping whitewater using integrated treatment of chemical precipitation & evaporation. *Chem. Eng. J.* **2010**, *158*, 584–592. [CrossRef]
37. Kurniawan, T.A.; Othman, M.H.D.; Adam, M.R.; Goh, H.H.; Mohyudin, A.; Avtar, R.; Kusworo, T.D. Treatment of pulping whitewater using membrane filtrations. *Chem. Papers* **2022**, *76*, 5001–5010. [CrossRef]
38. Kurniawan, T.A.; Maiurova, A.; Kustikova, M.; Bykovskaia, E.; Othman, M.H.D.; Goh, H.H. Accelerating sustainability transition in St. Petersburg (Russia) through digitalization-based circular economy in waste recycling industry: A strategy to promote carbon neutrality in era of Industry 4.0. *J. Clean. Prod.* **2022**, *363*, 132452. [CrossRef]
39. Kurniawan, T.A.; Singh, D.; Xue, W.; Avtar, R.; Othman, M.H.D.; Hwang, G.H.; Setiadi, T.; Albadarin, A.B.; Shirazian, S. Resource recovery toward sustainability through nutrient removal from landfill leachate. *J. Environ. Manag.* **2021**, *287*, 112265. [CrossRef]
40. Premakumara, D.G.J.; Canete, A.L.M.L.; Nagaishi, M.; Kurniawan, T.A. Policy implementation of the Republic Act (RA) No. 9003 in the Philippines on MSW management: A case study of Cebu City. *Waste Manag.* **2014**, *34*, 971–979. [CrossRef]
41. Ulfat, W.; Mohyuddin, A.; Amjad, M.; Kurniawan, T.A.; Mujahid, B.; Nadeem, S.; Javed, M.; Amjad, A.; Ashraf, A.Q.; Othman, M.H.D.; et al. Reuse of buffing dust-laden tanning waste hybridized with polystyrene for fabrication of thermal insulation materials. *Sustainability* **2023**, *15*, 1958. [CrossRef]

42. *Standard Methods for the Examination of Water and Wastewater*, 20th ed.; American Public Health Association (APHA): Washington, DC, USA, 2018.
43. Kurniawan, T.A.; Othman, M.H.D.; Liang, X.; Ayub, M.; Goh, H.H.; Kusworo, T.D.K.; Mohyuddin, A.; Chew, K.W. Microbial fuel cells (MFC): A potential game changer in renewable energy development. *Sustainability* **2023**, *14*, 16847. [CrossRef]
44. Kurniawan, T.A.; Othman, M.H.D.; Liang, X.; Goh, H.H.; Chew, K.W. From liquid waste to mineral fertilizer: Recovery, recycle and reuse of high-value macro-nutrients from landfill leachate to contribute to circular economy, food security, and carbon neutrality. *Process Saf. Environ. Prot.* **2023**, *170*, 791–807. [CrossRef]
45. Kurniawan, T.A.; Meidiana, C.; Othman, M.H.D.; Goh, H.H.; Chew, K.W. Strengthening waste recycling industry in Malang (Indonesia): Lessons from waste management in the era of Industry 4.0. *J. Clean. Prod.* **2023**, *382*, 135296. [CrossRef]
46. Kurniawan, T.A.; Lo, W.; Liang, X.; Goh, H.H.; Othman, M.H.D.; Chew, K.W. Remediation technologies for contaminated groundwater laden with arsenic (As), mercury (Hg), and/or fluoride (F): A critical review and ways forward to contribute to carbon neutrality. *Separ. Purif. Technol.* **2023**, *314*, 123474. [CrossRef]
47. Kurniawan, T.A.; Lo, W.H.; Liang, X.; Goh, H.H.; Othman, M.H.D.; Chong, K.K.; Mohyuddin, A.; Kern, A.O.; Chew, K.W. Heavy metal removal from aqueous solution using unused biomaterials and/or functional nanomaterials: Recent advances and ways forward in wastewater treatment. *J. Compos. Sci.* **2023**, *7*, 84. [CrossRef]
48. Lo, H.M.; Liu, M.; Pai, T.; Liu, W.; Wang, S.; Banks, C.; Hung, C.; Chiang, C.; Lin, K.; Chiu, H.; et al. Biostabilization assessment of MSW co-disposed with MSWI fly ash in anaerobic bioreactors. *J. Hazard. Mater.* **2009**, *162*, 1233–1242. [CrossRef]
49. Kurniawan, T.A.; Othman, M.H.D.; Liang, X.; Goh, H.H.; Gikas, P.; Chong, K.K.; Chew, K.W. Challenges and opportunities for biochar management to promote circular economy and carbon neutrality. *J. Environ. Manag.* **2023**, *332*, 117429. [CrossRef]
50. Lo, H.M.; Kurniawan, T.A.; Pai, T.Y.; Liu, M.H.; Chiang, C.F.; Wang, S.C.; Wu, K.C. Effects of spiked metals on MSW anaerobic digestion. *Waste Manag. Res.* **2012**, *30*, 32–48. [CrossRef]
51. Al-Daghistani, A.I.; Mohammad, B.T.; Kurniawan, T.A.; Singh, D.; Rabadi, A.D.; Xue, W.; Avtar, R.; Othman, M.H.D.; Shirazian, S. Characterization and applications of *Thermomonas hydrothermalis* isolated from Jordanian hot springs for biotechnological and medical purposes. *Process Biochem.* **2021**, *104*, 171–181. [CrossRef]
52. Mohyuddin, A.; Kurniawan, T.A.; Khan, Z.; Nadeem, S.; Javed, M.; Dera, A.A.; Iqbal, S.; Awwad, N.S.; Ibrahium, H.A.; Abourehab, M.A.S.; et al. Comparative insights into the antimicrobial, antioxidant, and nutritional potential of the *Solanum Nigrum* complex. *Processes* **2022**, *10*, 1455. [CrossRef]
53. Kurniawan, T.A.; Haider, A.; Ahmad, H.M.; Mohyuddin, A.; Aslam, H.M.U.; Nadeem, S.; Javed, M.; Othman, M.H.D.; Goh, H.H.; Chew, K.W. Source, occurrence, distribution and implications of microplastic pollutants in freshwater ecosystem on sustainability: A critical review and the way forward. *Chemosphere* **2023**, *325*, 138367. [CrossRef] [PubMed]
54. Ali, A.A.; Bishtawi, R.E. Removals of lead and nickel ions using zeolite tuff. *J. Chem. Biotechnol.* **1997**, *69*, 27–34. [CrossRef]
55. Blanchard, G.; Maunaye, M.; Martin, G. Removal of heavy metals from water by means of natural zeolites. *Water Res.* **1984**, *18*, 1501–1507. [CrossRef]
56. Curković, L.; Stefanovic, S.C.; Filipan, T. Metal ion exchange by natural and modified zeolites. *Water Res.* **1997**, *31*, 1379–1382. [CrossRef]
57. Grant, D.C.; Skriba, M.C. Removal of radioactive contaminants from West Valley waste streams using natural zeolites. *Environ. Prog.* **1987**, *6*, 104–109. [CrossRef]
58. Groffman, A.; Peterson, S.; Brookins, D. Removing lead from wastewater using zeolite. *Water Environ. Technol.* **1987**, *4*, 54–59.
59. Leinonen, H.; Lehto, J. Purification of metal finishing wastewaters with zeolites and activated carbons. *Water Manag. Res.* **2001**, *19*, 45–57. [CrossRef]
60. Meshko, V.; Markovska, L.; Mincheva, M.; Rodrigues, A.E. Adsorption of basic dyes on granular activated carbon and natural zeolite. *Water Res.* **2001**, *35*, 3357–3366. [CrossRef]
61. Mondales, K.D.; Carland, R.M.; Aplan, F.F. The comparative ion-exchange capacities of natural sedimentary and synthetic zeolites. *Min. Eng.* **1995**, *8*, 535–548. [CrossRef]
62. Ouki, S.K.; Kavanagh, M. Performance of natural zeolites for the treatment of mixed metal-contaminated effluents. *Waste Manag. Res.* **1997**, *15*, 383–394. [CrossRef]
63. Pansini, M.; Colella, C.; De'Gennaro, M. Chromium removal from water by ion exchange using zeolite. *Desalination* **1991**, *83*, 145–157. [CrossRef]
64. Zamzow, M.J.; Eichbaum, B.R.; Sandgren, K.R.; Shanks, D.E. Removal of heavy metal and other cations from wastewater using zeolites. *Separ. Sci. Technol.* **1990**, *25*, 1555–1569. [CrossRef]
65. Kurniawan, T.A. *Removal of Toxic Chromium from Wastewater*; Nova Science Publisher: New York, NY, USA, 2010; ISBN 978-1-60876-340-5.
66. Dun, F.; Kurniawan, T.A.; Li, H.; Wang, H.; Wang, Y.; Li, Q. Co-oxidative removal of As(III) and tetracycline (TC) from aqueous solutions based on a heterogeneous Fenton's oxidation using Fe nanoparticles (Fe NP)-impregnated solid digestate. *Environ. Pollut.* **2021**, *290*, 118062. [CrossRef]
67. Inglezakis, V.J.; Diamandis, N.A.; Loizidou, M.D.; Grigoropoulou, H.P. Effects of pore clogging on kinetics of lead uptake by clinoptilolite. *J. Coll. Int. Sci.* **1999**, *215*, 54–57. [CrossRef] [PubMed]
68. Inglezakis, V.J.; Papadeas, C.D.; Loizidou, M.D.; Grigoropoulou, H.P. Effects of pretreatment on physical and ion exchange properties of natural clinoptilollite. *Environ. Technol.* **2001**, *22*, 75–82. [CrossRef]

69. Malliou, E.; Loizidou, M.; Spyrellis, N. Uptake of lead and cadmium by clinoptilolite. *Sci. Total Environ.* **1994**, *149*, 139–144. [CrossRef]
70. Ouki, S.K.; Cheeseman, C.R.; Perry, R. Effects of conditioning and treatment of chabazite and clinoptilolite prior to lead and cadmium removal. *Environ. Sci. Technol.* **1993**, *27*, 1108–1116. [CrossRef]
71. Semmens, M.J.; Martin, W.P. The influence of pretreatment on the capacity and selectivity of clinoptilolite for metal removal. *Water Res.* **1988**, *22*, 537–542. [CrossRef]
72. Vaca-Mier, M.; Callejas, R.L.; Gehr, R.; Cisneros, B.E.J.; Alvarez, P.J.J. Heavy metal removal with Mexican clinoptilolite: Multi-component ionic exchange. *Water Res.* **2001**, *35*, 373–378. [CrossRef]
73. Ibrahim, K.M.; Nasser, E.T.; Khoury, H. Use of natural chabazite-philipsite tuff in wastewater treatment from electroplating factories in Jordan. *Environ. Geol.* **2002**, *41*, 547–551. [CrossRef]
74. Dun, F.; Kurniawan, T.A.; Avtar, R.; Xu, P.; Othman, M.H.D. Recovering heavy metals from electroplating wastewater and their conversion into Zn_2Cr-layered double hydroxide (LDH) for pyrophosphate removal from industrial wastewater. *Chemosphere* **2021**, *271*, 129861. [CrossRef]
75. Dun, F.; Kurniawan, T.A.; Lan, L.; Yaqiong, L.; Avtar, R.; Othman, M.H.D. Arsenic removal from aqueous solution by FeS_2. *J. Environ. Manag.* **2021**, *286*, 112246. [CrossRef]
76. Yang, B.; Sun, G.; Quan, B.; Tang, J.; Zhang, C.; Jia, C.; Tang, Y.; Wang, X.; Zhao, M.; Wang, W.; et al. An experimental study of fluoride removal from wastewater by Mn-Ti modified zeolite. *Water* **2021**, *13*, 3343. [CrossRef]
77. Hamd, A.; Dryaz, A.R.; Shaban, M.; AlMohamadi, H.; Abu Al-Ola, K.A.; Soliman, N.K.; Ahmed, S.A. Fabrication and application of zeolite/*Acanthophora Spicifera* nanoporous composite for adsorption of Congo red dye from wastewater. *Nanomaterials* **2021**, *11*, 2441. [CrossRef]
78. Madhuranthakam, C.M.R.; Thomas, A.; Akhter, Z.; Fernandes, S.Q.; Elkamel, A. Removal of chromium(VI) from contaminated water using untreated moringa leaves as biosorbent. *Pollutants* **2021**, *1*, 51–64. [CrossRef]
79. Chen, M.; Nong, S.; Zhao, Y.; Riaz, M.S.; Xiao, Y.; Molokeev, M.S.; Huang, F. Renewable P-type zeolite for superior absorption of heavy metals: Isotherms, kinetics, and mechanism. *Sci. Total Environ.* **2020**, *726*, 138535. [CrossRef]
80. Hong, M.; Yu, L.; Wang, Y.; Zhang, J.; Chen, Z.; Dong, L.; Zan, Q.; Li, R. Heavy metal adsorption with zeolites: The role of hierarchical pore architecture. *Chem. Eng. J.* **2019**, *359*, 363–372. [CrossRef]
81. Li, Y.; Bai, P.; Yan, Y.; Yan, W.; Shi, W.; Xu, R. Removal of Zn^{2+}, Pb^{2+}, Cd^{2+}, and Cu^{2+} from aqueous solution by synthetic clinoptilolite. *Microporous Mesoporous Mater.* **2019**, *273*, 203–211. [CrossRef]
82. Wahono, S.K.; Stalin, J.; Addai-Mensah, J.; Skinner, W.; Vinu, A.; Vasilev, K. Physico-chemical modification of natural mordenite-clinoptilolite zeolites and their enhanced CO_2 adsorption capacity. *Microporous Mesoporous Mater.* **2020**, *294*, 109871. [CrossRef]
83. Figueiredo, H.; Quintelas, C. Tailored zeolites for the removal of metal oxyanions: Overcoming intrinsic limitations of zeolites. *J. Hazard. Mater.* **2014**, *274*, 287–299. [CrossRef] [PubMed]
84. Kurniawan, T.A.; Lo, W.; Othman, M.H.D.; Goh, H.H.; Chong, K.K. Influence of Fe_2O_3 and bacterial biofilms on Cu(II) distribution in a simulated aqueous solution: A feasibility study to sediments in the Pearl River Estuary (PR China). *J. Environ. Manag.* **2023**, *329*, 117047. [CrossRef]
85. Kurniawan, T.A.; Lo, W.; Othman, M.H.D.; Goh, H.H.; Chong, K.K. Biosorption of heavy metals from aqueous solutions using activated sludge, *Aeromasss hydrophyla*, and *Branhamella* spp based on modeling with *GEOCHEM*. *Environ. Res.* **2022**, *214*, 114070. [CrossRef] [PubMed]
86. Kurniawan, T.A.; Othman, M.H.D.; Goh, H.H.; Gikas, P.; Kusworo, T.D.; Anouzla, A.; Chew, K.W. Decarbonization in waste recycling industry using digital technologies to contribute to carbon neutrality and its implications on sustainability. *J. Environ. Manag.* **2023**, *338*, 117765. [CrossRef] [PubMed]
87. Velarde, L.; Nabavi, M.S.; Escalera, E.; Antti, M.L.; Akhtar, F. Adsorption of heavy metals on natural zeolites: A review. *Chemosphere* **2023**, *328*, 138508. [CrossRef]
88. Saleh, T.S.; Badawi, A.K.; Salama, R.S.; Mostafa, M.M.M. Design and development of novel composites containing nickel ferrites supported on activated carbon derived from agricultural wastes and its application in water remediation. *Materials* **2023**, *16*, 2170. [CrossRef]
89. Alasri, T.M.; Ali, S.L.; Salama, R.S.; Alshorifi, F.T. Band-structure engineering of TiO_2 photocatalyst by AuSe quantum dots for efficient degradation of malachite green and phenol. *J. Inorg. Organomet. Polym. Mater.* **2023**. [CrossRef]
90. Baaloudj, O.; Nasrallah, N.; Kenfoud, H.; Bourkeb, K.W.; Badawi, A.K. Polyaniline/$Bi_{12}TiO_{20}$ hybrid system for cefixime removal by combining adsorption and photocatalytic degradation. *ChemEngineering* **2023**, *7*, 4. [CrossRef]
91. Kane, A.; Assadi, A.A.; El Jery, A.; Badawi, A.K.; Kenfoud, H.; Baaloudj, O.; Assadi, A.A. Advanced photocatalytic treatment of wastewater using immobilized titanium dioxide as a photocatalyst in a pilot-scale reactor: Process intensification. *Materials* **2022**, *15*, 4547. [CrossRef]
92. Shahzad, W.; Badawi, A.K.; Rehan, Z.A.; Khan, A.M.; Khan, R.A.; Shah, F.; Ali, S.; Ismail, B. Enhanced visible light photocatalytic performance of Sr0.3 (Ba, Mn) 0.7 ZrO_3 perovskites anchored on graphene oxide. *Ceram. Int.* **2022**, *48*, 24979–24988. [CrossRef]

Disclaimer/Publisher's Note: The statements, opinions and data contained in all publications are solely those of the individual author(s) and contributor(s) and not of MDPI and/or the editor(s). MDPI and/or the editor(s) disclaim responsibility for any injury to people or property resulting from any ideas, methods, instructions or products referred to in the content.

Article

Selection of Wastewater Treatment Technology: AHP Method in Multi-Criteria Decision Making

Jasmina Ćetković [1,*], Miloš Knežević [2], Radoje Vujadinović [3], Esad Tombarević [3] and Marija Grujić [4]

1. Faculty of Economics Podgorica, University of Montenegro, 81000 Podgorica, Montenegro
2. Faculty of Civil Engineering, University of Montenegro, 81000 Podgorica, Montenegro
3. Faculty of Mechanical Engineering, University of Montenegro, 81000 Podgorica, Montenegro
4. Faculty of Civil Engineering, University of Belgrade, 11000 Belgrade, Serbia
* Correspondence: jasmina@ucg.ac.me; Tel.: +382-67-652-016

Citation: Ćetković, J.; Knežević, M.; Vujadinović, R.; Tombarević, E.; Grujić, M. Selection of Wastewater Treatment Technology: AHP Method in Multi-Criteria Decision Making. *Water* **2023**, *15*, 1645. https://doi.org/10.3390/w15091645

Academic Editors: Alexandre T. Paulino, Yung-Tse Hung, Hamidi Abdul Aziz, Issam A. Al-Khatib, Rehab O. Abdel Rahman and Tsuyoshi Imai

Received: 27 February 2023
Revised: 11 April 2023
Accepted: 19 April 2023
Published: 23 April 2023

Copyright: © 2023 by the authors. Licensee MDPI, Basel, Switzerland. This article is an open access article distributed under the terms and conditions of the Creative Commons Attribution (CC BY) license (https://creativecommons.org/licenses/by/4.0/).

Abstract: Wastewater treatment is a process that reduces pollution to those quantities and concentrations at which purified wastewater is no longer a threat to human and animal health and safety and does not cause unwanted changes in the environment. Municipal wastewater is classified as biodegradable water. Special importance should be given to wastewater with a high content of organic matter (COD), phosphorus (P) and nitrogen (N). MBBR technology, developed on the basis of the conventional activated sludge process and the bio filter process, does not take up much space and does not have problems with activated sludge, as in the case of conventional biological reactors, and has shown good results for the removal of organic matter, phosphorus and nitrogen. The aim of this paper is to optimize the wastewater treatment process in the municipality of Dojran, North Macedonia. Three alternative solutions for improving the capacity for wastewater treatment in the municipality of Dojran were analyzed. The shortlist of variants was made on the basis of several criteria, including: analysis of the system in the tourist season and beyond, assessment of the condition and efficiency of the existing wastewater treatment plant (WWTP) in combination with a new treatment plant, treatment efficiency when using different wastewater treatment technologies, the size of the site needed to accommodate the capacity, as well as the financial parameters for the proposed system. The selection of the most favorable solution for the improvement of the wastewater treatment system was made using the AHP (analytic hierarchy process) method. In order to select the optimal solution, a detailed analysis was conducted, considering several decision-making criteria, namely the initial investment, operating costs and management complexity. Based on the obtained results, Variant 3 was recommended, that is, the construction of a completely new station with MBBR technology, with a capacity for 6000 equivalent inhabitants.

Keywords: AHP method; MBBR technology; process optimization; wastewater treatment plant

1. Introduction

Choosing the optimal solution for wastewater treatment is a key stage in the optimization of the wastewater treatment process. This is because, according to some research, the world's population is expected to face the problem of water shortage if consumption remains at the current level [1]. Therefore, water reclamation and its reuse are the only possible solutions to this problem.

Wastewater treatment using moving bed biofilm reactor (MBBR) technology is used to filter wastewater in the industrial and municipal sectors. MBBR is the state-of-the-art wastewater treatment process that uses specialized biological technologies. This treatment can be used in the municipal and industrial sectors for nitrification, BOD removal and water purification, but it can also be integrated with other systems to achieve better results in pollutant removal. MBBR includes a simplified operating system that increases water purity beyond the conventional wastewater treatment limits. This biological treatment technology is preferred over conventional approaches due to several comparative advantages, such as health and ecological advantages, high efficiency, convenience, small space requirements, cost effectiveness, flexibility and ease of operation, etc.

Before the advent of these reactors, many other types of conventional wastewater treatment systems were in use [2], which had significant disadvantages compared to MBBR. In response to those shortcomings, the moving bed biofilm reactor appeared in Norway in the late 1980s and early 1990s [3,4], as along with the first pioneering work on this technology [5]. The new MBBR has long been suggested to be suitable for dairy wastewater treatment [6], whereby the preference for the MBBR system was originally linked to the absence of sludge recycling and its ease of operation [7]. Meanwhile, in different parts of the world, different prominent variants of MBBR technology have been developed with the same basic principle [8,9]. The development of this technology was supposed to solve the problems of small communities regarding the need for small, easy-to-install wastewater treatment plants. Not long after, research reported the commercial success of MBBR technology in a significant number of countries around the world [10–13], which was followed by their even more extensive use in the wastewater treatment process [14]. In addition, some earlier studies examined the efficiency of upgrading an existing plant with MBBR technology in different climatic conditions and seasonal temperature fluctuations, and confirmed the justification of its use even at lower temperatures [15]. In the meantime, there was a need to increase the capacity of existing plants, which necessitated the further development of MBBR technology. Although MBBR technology has some drawbacks [16,17], it has numerous advantages compared to conventional biological treatments, such as space savings, improvement of performance and capacity with minimal additional costs, less clogging, the sludge does not require recirculation, the footprint is consistently reduced, increased biofilm resistance to temperatures, shock loads, toxic compounds, pH, etc. [18–20]. Thanks to its simplicity, flexibility, robustness and compactness, MBBR technology has seen a growth of its application [21] and has become a widely recognized technology for wastewater treatment [22], where it has shown enormous potential in reducing the load of contamination and pollution of municipal and industrial waters [23–27]. The advantages of MBBR technology have been demonstrated in agriculture, the denitrification of drinking water [11,28], and oil refinery wastewater [29], as well as in hospital wastewater treatment [30]. In addition, with small modifications, it is possible to adapt the existing infrastructure to host MBBR [31]. Such high-performance capability in carbon and nitrogen removal and a compact footprint enable MBBR technology to be a good solution for either small decentralized facilities or for upgrading existing centralized facilities [32].

Appreciating the fact that finding suitable technologies for efficient wastewater treatment and its reuse is important for the sustainability of the industry [33–35], a number of studies examined the use of MBBR in specific industries [36–40], but also compared the economic and environmental advantages and disadvantages of using different biological methods in the treatment of industrial wastewater, suggesting that the application of MBBR in some industries has advantages over other methods and technologies [41]. In one of the most recent studies, the outstanding results of the application of MBBR technology for municipal and industrial wastewater were highlighted and it was confirmed that the maximum removal efficiencies are BOD of 97%, COD of 96%, phosphorus of 99% and oxygen of 99%, at an HRT of 2–6 h [42]. At the same time, it is a technology that is applicable for a wide range of wastewater flows, from 10 to 150 thousand $m^3 day^{-1}$ [43], therefore, various mathematical methods were developed for calculating the reactor volume, organic effluent concentration and substrate removal rate [44]. Kawan et al. [45] pointed out the advantages of MBBR technology as a highly modular system that can be used for very low or very high concentrations, as well as for polishing, therefore techno-economic analyses are necessary. Khudhair et al. [46] indicated in their recent research the problem of excess sludge production, which represents one of the limitations of the biological activated sludge process, which can be overcome by upgrading the MBBR process. Namely, their analysis showed that the variant of upgrading the MBBR process to the integrated fixed-film activated sludge (IFAS) completely eliminates sludge and that the system achieved low effluent pollutants concentrations.

One of the key advantages of MBBR technology is its high level of treatment efficiency. Parivallal et al. [47] pointed out in their recent study the high efficiency of MBBR technology in wastewater treatment processes. Namely, a treatment plant with a capacity of one million liters per day meets all standards in terms of basic water quality parameters, such as BOD, COD, TKN and TSS. Similar results, regarding high treatment efficiency, were confirmed by Masłoń and Tomaszek [48], which involved a 15 L-laboratory scale MBSBBR (moving bed sequencing batch biofilm reactor) model. The results of this study indicate a high level of average efficiency in removing COD, total nitrogen (TN) and total phosphorus (TP), ranging from 97.7 ± 0.5%, 87.8 ± 2.6% and 94.3 ± 1.3%, respectively, while the nitrification efficiency reached a level in the range of 96.5–99.7%. Zhou et al. [49] confirmed the feasibility of the two-stage anoxic/oxic moving bed biofilm reactor (TS-A/O-MBBR) in a full-scale municipal wastewater treatment plant (WWTP), as well as its high efficiency and ability to meet high standards in the biological nitrogen removal process. The maximum removal efficiency of total nitrogen and the minimum concentration of total nitrogen in the effluent reached 91.76% and 4.12 mg/L, respectively. Czarnota and Masłoń [50] showed that not all MBBR systems are equally efficient. Their study aimed to assess the effectiveness of MBBR reactors with EvU-Perl carriers. The results of this study, which was aimed at improving efficiency, indicate the need for better sludge management, increasing the volume of nitrification chambers or replacing the biofilm carrier, as annual analyses showed a decrease in biogenic compounds below the prescribed level.

The treatment efficiency and cost-effectiveness of MBBR technology primarily depend on the type of the biofilm carriers that can be modified according to the process, which is the main advantage of this technology. For example, Chu and Wang [51] compared the efficiency between two different biofilm carriers (polymer polycaprolactone—PCL and inert poly-urethane foam—PUF) and gave preference to PCL carriers with a low C/N ratio in terms of TN removal. In a recent study, Ashkanani et al. [52], using MBBR with three AnoxKaldnes media, determined the influence of the shape and surface of biocarriers on efficiency, favoring a biocarrier that has a smaller specific surface due to less clogging. Additionally, Maziotti et al. [53] preferred AnoxKaldnes K3 over Mutag BioChip in their study due to higher COD removal efficiency. Shitu et al. [54] concluded that novel sponge biocarriers (SB) in MBBR increase the diversity of the functional microbial communities and achieve the highest nitrification performance.

Nevertheless, regardless of the wide practical use of MBBR technology, a review of the literature reveals that the research of this technology is limited compared to the literature focusing on conventional systems [18].

Sequencing batch reactors (SBR) are a variation of activated sludge and have been widely applied in wastewater treatment for almost a century due to their unique advantages [55]. As environmental standards became stricter and the number of new pollutants grew exponentially, SBR technology, as a modification of the popular activated sludge process (ASP), gained in importance and application. From small community use to the treatment of high-hardness industrial wastes, the use of SBRs has expanded to the biological treatment of industrial waters containing organic chemicals that are difficult to remove. As one of the integrated systems for anaerobic–aerobic bioreactors, SBR processes are often used in industrial wastewater treatment due to their compactness and high efficiency [56–58], and are used less often for domestic wastewater treatment [59]. The advantages of this technology are not only the high performance in low or varying flow patterns, but also the lower costs over a longer period of time. Applying technology SBR in municipal wastewater treatment with biological process nitrification/denitrification is a major opportunity and a good chance for developing countries to reach sustainable development and ecological balance in their urban areas. Therefore, Quan and Gogina [60] indicated more stable efficiency in the removal of pollutants and decreased environmental damage by 8–11 times while achieving optimal operation by bio-film in their study on the technical economic efficiency of SBR. Dutta and Sarcar [61] emphasized that SBR technologies save more than 60% of the operating expenses for a conventional ASP, and high effluent quality is achieved in a very short aeration time. Ćetković et al. [62] considered the financial and socio-economic feasibility of SBR as one of the variant solutions in the CBA implemented. Due to their excellent process control capabilities and operational flexibility, SBRs are widely used to treat wastewater, but future research on SBR control strategies and the development of intelligent control systems can make SBRs more adaptable to changing environmental conditions and changing wastewater quality in order to maintained optimal and reliable effluent quality. Alagha et al. [63] investigated the performance of a pilot-scale SBR process for the treatment of municipal wastewater quality parameters in terms of two scenarios, namely, pre-anoxic denitrification and a post-anoxic denitrification scenario. Their results confirm that the post-anoxic denitrification scenario was more efficient for higher qualify effluent, which is why the suitability of using this technology in remote areas in arid regions with a high reusability potential is suggested. Fernandes et al. [64] analyzed the microbial diversity and performance in the SBR of a decentralized full-scale system for urban wastewater treatment under limited aeration and confirmed the viability and efficiency of the reactor to treat domestic wastewater. Numerous papers have shown that SBR use results in a more efficient process that requires less energy consumption than conventional systems [65–67].

The aim of this paper is to optimize the wastewater treatment process in the municipality of Dojran, North Macedonia, i.e., to choose between the three offered variant solutions. It was created as a result of the author's involvement in the preparation of a feasibility study related to the improvement of the wastewater treatment system in this municipality, within a wider project financed by SIDA, entitled "Building Municipal Capacity for Project Implementation". The optimization of the wastewater treatment process in the municipality of Dojran should contribute to the implementation process of European standards for environmental protection [68,69] to ensure that the maximum allowed concentrations of pollutants in the wastewater discharged into the recipient are not exceeded. This study analyzes several proven wastewater treatment technologies that are applicable to the location in question.

The article is organized into several sections. In the Introduction, certain aspects of the application of MBBR technology in the wastewater treatment process have been articulated through a review of the relevant literature. The second section of the paper presents the current situation and problems regarding the wastewater treatment process in the municipality of Dojran, as well as the methodology we used to select the optimal variant of wastewater treatment technology in Nov Dojran. In addition, in this section, we determine the equivalent inhabitants and the amount of wastewater that will be generated in the municipality of Dojran as key input for the analysis in the paper. In the third section of the paper, the analysis of three technical variant solutions related to the problem of improving the capacity for wastewater treatment for the municipality of Dojran is presented. In the fourth section, the selection of the optimal variant of wastewater treatment in the municipality of Dojran is made using the AHP (analytic hierarchy process) method. The last section provides a concluding summary of the research, points out the limitations of the approach and suggests ideas for future research on this topic.

2. Data and Methodology

The first point of this section is to provide an overview of the situation and challenges in the wastewater treatment process in the municipality of Dojran, North Macedonia. The second part briefly presents the basics of the relevant methodology that we used in order to select the optimal variant of wastewater treatment process in Nov Dojran. In the third part, we determine the equivalent inhabitants and the amount of wastewater that will be generated in the municipality of Dojran as basic input that is necessary for the selection of the optimal variant of wastewater treatment technology in this municipality.

2.1. Overview of the Situation and Challenges in the Wastewater Treatment Process: Municipality of Dojran, North Macedonia

The hydrography of the municipality of Dojran mainly consists of the Dojran Lake, smaller springs and streams, as well as a few artificial reservoirs. Dojran Lake is located at a height of 140 m above sea level. The surface area of the lake is 42.5 km^2, of which 26.58 km^2, or 62.54% belongs to the Republic of North Macedonia, and 15.92 km^2, or 37.46% to Greece. The water volume of the lake is 289.61 million m^3. The length of the lake is 8.9 km, and the greatest width is 7.1 km. The average depth is 6.7 m, and the greatest is about 10 m. In the period from 1988 to 2000, the level of the lake water constantly decreased, reaching the lowest point of −3.88 m below the zero point. The hydrography of Dojran Lake has been significantly enhanced in recent years by the construction of the Gavoto–Dojran canal, which brings additional water into Dojran Lake, increasing the lake level by about 1.80 m from the absolute minimum.

According to the data from the last official population census from 2002 [70], the municipality of Dojran has 3426 inhabitants, of which Nov Dojran has 1100 inhabitants. The company JPKD Komunalec—Polin Star Dojran was established to provide utility services in the area of the municipality of Dojran. In the context of all the services it offers and performs, of special interest are the services of water collection, treatment and supply, wastewater disposal, construction of water supply systems, construction of sewerage systems and construction of a storm water drainage system.

Thus far, the settlements of Star Dojran, Nov Dojran and Sretenovo have access to the sewerage system. The network is divided into main (primary) and secondary systems. The secondary sewerage network in the municipality of Dojran has a total length of 7650 m, with a diameter ranging from Ø 150 mm to Ø 250 mm. It covers the aforementioned settlements and serves to collect wastewater from households and transport it to the main collector. The main collector system was built in 1989 and stretches along the entire length of Dojran Lake on the Macedonian side. The collector is 8340 m long, made of PVC pipe, with a diameter ranging from Ø 250 mm to Ø 500 mm. Submersible fecal pumps installed in ten pumping stations arranged along the length of the collector pump fecal water to WWTP Toplec in Nov Dojran.

The existing WWTP Toplec is located in the suburb of Nov Dojran and represents the completion of the sewerage system in the municipality of Dojran. The process of wastewater disposal ends with a treatment plant from which the treated water is discharged into Dorjan Lake. It was built in 1988 as the last point of the eastern and western collectors. The plant was designed for 8000 equivalent inhabitants and consists of two blocks, the first of which is technically outdated and out of use, while the second block is in operation. A project for the reconstruction of the second block was prepared in order to increase the efficiency of WWTP Toplec by replacing and supplementing the treatment technology. However, sludge treatment would be a problem in the functioning of the plant even after reconstruction. Sludge dewatering is not foreseen, and the sludge is often left to dry in fields. A new technological solution should overcome this problem. In addition, the storm sewer system is not fully developed and covers only a small part of the municipality. Reconstruction of the existing system or construction of a new WWTP should improve the quality of surface and ground water and soil in the wider region. However, there are also possible negative impacts of reconstruction or construction of a new WWTP, as well as from the purification of wastewater during the so-called operational phase. It is expected that reconstruction or new construction will mainly result in waste that is not classified as hazardous (in accordance with the waste management regulation), such as stones, mixed municipal waste, etc. All waste should be disposed of in landfills. The possible amount of hazardous waste will be small.

The composition of wastewater, which should be treated in the planned WWTP in the municipality of Dojran, corresponds to the typical composition of wastewater. It is necessary to ensure the quality of the effluent that is discharged into Dojran Lake in accordance with the standards of the EU Directive for urban wastewater. According to the local regulation [71], Dojran Lake is classified in the II (second) category.

During the exploitation process, that is, the operational phase of the WWTP, several types of waste will be generated, which can be classified into two main types: waste resulting from the wastewater treatment process and waste resulting from the maintenance of the WWTP itself. Other phenomena that could disrupt the comfort of citizens are noise during the period of construction activities and unpleasant odors during the exploitation process. The realization of the project itself will have a positive impact on the environment because the long-standing problem of loading Dojran Lake with organic matter originating from municipal wastewaters will be solved.

2.2. AHP Method

To select the optimal technology for wastewater treatment in Nov Dojran, the analytic hierarchy process (AHP) method was chosen as one of the most frequently used multi-criteria decision-making methods [72,73], which is used when making decisions in complex problems. It is used with a multi-layered hierarchical structure of goals (which we want to achieve), criteria, sub-criteria and alternatives that we consider. The input data were derived through several comparisons. These comparisons were used to define the degree of importance of the criteria used in decision-making, as well as to determine the relative measures for evaluating alternatives according to each separate decision criterion. The method, which is based on a mathematical, but also human approach, deconstructs the problem by hierarchy and enables evaluation according to different criteria. The AHP method includes four main steps:

- Development of a hierarchy of interconnected decision elements that describe the problem;
- Comparing pairs of decision elements, usually using a 1–9 comparison scale, to obtain input data;
- Calculation of relative weightings of decision-making elements, most often using the method of characteristic values;
- Aggregation of relative weightings of decision elements in order to calculate the rating of alternative decision possibilities.

The relative importance of criteria i and j is evaluated with values from 1 to 9 [74–76]. The significance of those values is presented in Table 1.

Table 1. Table of coefficients of importance of criteria according to Saaty [77].

Intensity of Importance	Definition	Explanation
1	Equal importance	Criteria i and j are equally important
3	A moderate advantage of one criteria over another	Criterion i is moderately more important than j
5	Essential, strong importance	Criterion i is significantly more important than j
7	Very strong importance	Criterion i is very significantly more important than j
9	Extreme importance	Criterion i is extremely more important than j
2, 4, 6, 8	Mean values between two adjacent estimates	To define the rating, a comparison of two estimates is needed (a compromise is needed)
Reciprocities	If one activity has one of the numbers above (e.g., 3) compared to a second activity, then the second activity has a reciprocal value (i.e., 1/3) when compared to the first.	
Rationality	Coefficients resulting from forcing consistency of estimation	

2.3. Determining the Equivalent Inhabitants and the Amount of Wastewater

The wastewater collection and disposal system in the municipality of Dojran covers settlements along Dojran Riviera, namely Nov Dojran, Star Dojran and Sretenovo. The system consists of a secondary and primary sewage network, which ends with the WWTP. For other settlements in the municipality, the construction of a fecal sewage network with a small WWTP is planned, which is why these settlements are not included in the calculation for determining the equivalent inhabitants, that is, the amount of wastewater.

Considering that Dojran is a tourist center with the peak season in the summer months (June, July and August), two analyses of the equivalent inhabitants were carried out: out-of-season (with only the permanent population included) and in-season with permanent population and tourists). Determining the equivalent inhabitants out-of-season is presented in Table 2.

Table 2. Determining the equivalent inhabitants for the municipality of Dojran—out-of-season.

1.	Population according to 2002 census [70] Star Dojran Nov Dojran Sretenovo Total population in the area of interest	$N_1 = 363.00$ inhabitants $N_2 = 1100.00$ inhabitants $N_3 = 315.00$ inhabitants $N_{(2002)} = 1778.00$ inhabitants
2	Population at the end of exploitation period [78]	$N_k = 1908.37$ inhabitants
	$N_k = N_0(1 + p/100)^n$	$N_k = 2000.00$ inhabitants
	N_0—current population	$N_0 = 1833.60$ inhabitants
	p—population growth	$p = 0.16\%$
	n—exploitation period	$n = 25.00$ years
3.	Standard for water supply	$Q_0 = 150.00$ l/day/person
4.	Standard for sewerage	$Q_k = 150.00$ l/day/person
5.	Average wastewater emission per day $Q_{av/day} = \frac{Q_k \cdot N_k}{1000}$	$Q_{av/day} = 300.00$ m^3/day
6.	Maximum wastewater emission per day $Q_{max/day} = a_1 Q_{av/day}$ a_1—maximum daily uneven distribution coefficient	$Q_{max/day} = 450.00$ m^3/day $a_1 = 1.50$
7.	Average wastewater emission per hour $Q_{av/h} = \frac{Q_{max/day}}{24}$ $q_{av/sec} = \frac{Q_{av/h}}{3.6}$	$Q_{av/h} = 18.75$ m^3/h $q_{av/sec} = 5.21$ l/sec
8.	Maximum wastewater emission per hour $Q_{max/h} = a_2 Q_{av/h}$ a_2—maximum daily uneven distribution coefficient $q_{max/sec} = \frac{Q_{max/h}}{3.6}$	$Q_{max/h} = 30.00$ m^3/h $a_2 = 1.60$ $q_{max/sec} = 8.33$ l/sec

The wastewater plant is sized for the average wastewater emission per hour, with the possibility of maximum wastewater emission per hour. We should emphasize that the infiltration of water of another origin (e.g., storm water, lake water) is not included in the calculation because it significantly increases the amount of water for purification, which increases the cost of the plant while reducing efficiency.

In order to determine the equivalent inhabitants during the tourist season, that is, the amount of wastewater, data on the hospitality industry is needed, primarily on its nature and capacities [79]. These data are shown in Table 3. The defined growth of seasonal visitors and the hospitality industry is 3% until 2029 and 0.5% in the remaining period [78].

The number of equivalent inhabitants during the tourist season is determined based on the character and capacity of the hospitality industry for the municipality of Dojran (Table 4).

Table 3. Data on the hospitality industry in the municipality of Dojran.

Year	2016	2029	2046	Year	2016	2029	2046
Total catering facilities:	14	20	22	Capacity/number of places:	433	710	772
- Restaurants	6	8	9	- Restaurants	193	284	309
- Fast food	3	4	4	- Fast food	97	142	154
- Dairy restaurants	1	2	2	- Dairy restaurants	48	71	77
- Coffee bars	4	6	7	- Coffee bars	145	213	232
Total hospitality facility with the possibility of an overnight stay:	47	68	73	Capacity/number of beds:	2079	3053	3323
- Hotels	23	34	37	- Hotels	1040	1527	1662
- Resorts	12	17	18	- Resorts	520	763	831
- Other type	12	17	18	- Other type	520	763	831

Table 4. Determining the equivalent inhabitants for the municipality of Dojran—in-season.

1. Determining the equivalent inhabitants			
Description	Number of visitors	Standard	Total amount
/	/	l/day/person	l/day
Restaurants	309	100	30,882.62
Fast food	154	10	1544.131
Dairy restaurants	77	10	772.0654
Coffee bars	232	10	2316.196
Hotels	1662	200	332,323.8
Resorts	831	120	99,697.14
Other types of accomodation	831	120	99,697.14
Total amount of wastewater			567,233.1
Average drainage rate			150
Equivalent inhabitants—seasonal visitors			3781.554
Equivalent inhabitants—everyday visitors			1908.37
Total equivalent inhabitants in season			5689.92
Determined equivalent number of inhabitants in the season			6000
2. Average wastewater emission per day $Q_{av/day} = \frac{Q_k \cdot N_k}{1000}$			$Q_{av/day} = 900$ m³/day
3. Maximum wastewater emission per day $Q_{max/day} = a_1 Q_{av/day}$ a_1—maximum daily uneven distribution coefficient			$Q_{max/day} = 1350$ m³/day $a_1 = 1.50$
4. Average wastewater emission per hour $Q_{av/h} = \frac{Q_{max/day}}{24}$ $q_{av/sec} = \frac{Q_{av/h}}{3.6}$			$Q_{av/h} = 56.25$ m³/h $q_{av/sec} = 15.63$ l/sec
5. Maximum wastewater emission per hour $Q_{max/h} = a_2 Q_{av/h}$ a_2—maximum daily uneven distribution coefficient $q_{max/sec} = \frac{Q_{max/h}}{3.6}$			$Q_{max/h} = 90.00$ m³/h $a_2 = 1.60$ $q_{max/sec} = 25$ l/sec

3. Analysis: Variant Solutions for Improving the Capacity for Wastewater Treatment

Three technical alternative solutions related to the problem of improving wastewater treatment capacity for the municipality of Dojran were analyzed. The short list of variants was made on the basis of several criteria, including system analysis (in and out of the tourist season), assessment of the condition and efficiency of the existing WWTP in combination with a new treatment plant, treatment efficiency when using different wastewater treatment technologies, size of the site required to accommodate the treatment capacity and financial parameters for the proposed system, i.e., the initial investment and the necessary maintenance budget.

As stated in Section 2, the main reconstruction project of the second block of the existing WWTP was carried out, which provided for an increase in efficiency and capacity up to 6000 equivalent inhabitants. In addition, there is a marked increase in the number of equivalent inhabitants during the tourist season (6000 equivalent inhabitants, compared to 2000 out of season). Because of this, the investor insisted that two options should be considered, which are based on the planned reconstruction of the existing WWTP, which would be used during the season for 6000 equivalent inhabitants. In both variants, the construction of new systems is foreseen—MBBR and SBR, which would operate out of season with a capacity adjusted for 2000 equivalent inhabitants. In the third variant, the existing WWTP is provided as a reserve capacity that can be reconstructed if there is an increase in need.

The solutions that are applicable for the given conditions are as follows:

1. Exploitation of the existing WWTP in accordance with the main project for the reconstruction and construction of the new MBBR wastewater treatment system for the calculated equivalent inhabitants of Dorjan, which will be used outside the tourist season—Variant 1;
2. Exploitation of the existing WWTP in accordance with the main project for reconstruction during the tourist season and construction of a new SBR (sequencing batch reactor) for the calculated equivalent inhabitants of Dojran, which will be used outside the tourist season—Variant 2;
3. Construction of a new MBBR wastewater treatment system for 6000 equivalent inhabitants with two modules, of which, module two will be active in the tourist season, while module one will be active only outside the tourist season—Variant 3.

3.1. Variant 1—Combination of the Existing WWTP and the New MBBR Wastewater Treatment System for 2000 Equivalent Inhabitants

As the calculations of the equivalent inhabitants confirm, due to the fact that Dojran is a tourist center, there are large variations in wastewater emissions during the year. Therefore, in order to achieve greater efficiency and economy, it is planned to build a new plant that would operate throughout the year and serve the permanent residents of the municipality of Dojran, while during the tourist season, in accordance with the reconstruction project, the existing WWTP would also be activated [58]. Variant 1 is presented in Figure 1.

The new treatment plant would be located next to the existing one. It is envisaged that the water will be directed to the distribution shaft equipped with valves via the last pumping station of the main collector system, from where one line would lead to the existing WWTP and the other line would lead to the new plant. The area (approx. 660.0 m^2) where the new treatment plant is located is part of the private plot KP 295 and should be subject to expropriation.

The constituent elements of the planned WWTP are separated into common facilities and equipment and, as such, are most often found in this type of plant. Thus, the following elements are planned for the MBBR treatment station for 2000 equivalent inhabitants:

- Distribution shaft with a sluice gate for directing water to the wastewater treatment plant;
- Inlet pump station—reinforced concrete facility with an automatic coarse screen for bulky waste and a panel house mounted on a steel structure above the pump station;
- Flow measuring shaft for measuring the flowrate of wastewater—reinforced concrete facility with a built-in electromagnetic flow meter and the necessary equipment;
- Equalization pool—reinforced concrete facility with a compressor station and built-in air distribution system, diffusers and mixers;
- Modular (assembly–disassembly) container plant, two stage MBBR–BNB bioreactor with a moving bed, in which there is an automatic mixer, a fine screen and a sludge pump;

○ A modular container that houses a tank (assembly–disassembly) with a built-in cartridge for the microfiltration of treated water, a pre-pumping station and another part of the control equipment (PLC system) with a voltage regulator for the entire system;
○ Sludge storage tank (assembly–disassembly) with a compressor unit;
○ Emergency shaft/reinforced concrete facility made from ready-made prefabricated elements;
○ Collecting shaft with a channel for purified water discharge to the recipient/reinforced concrete facility—from ready-made prefabricated elements;
○ Press and dryer—for sludge dewatering and drying.

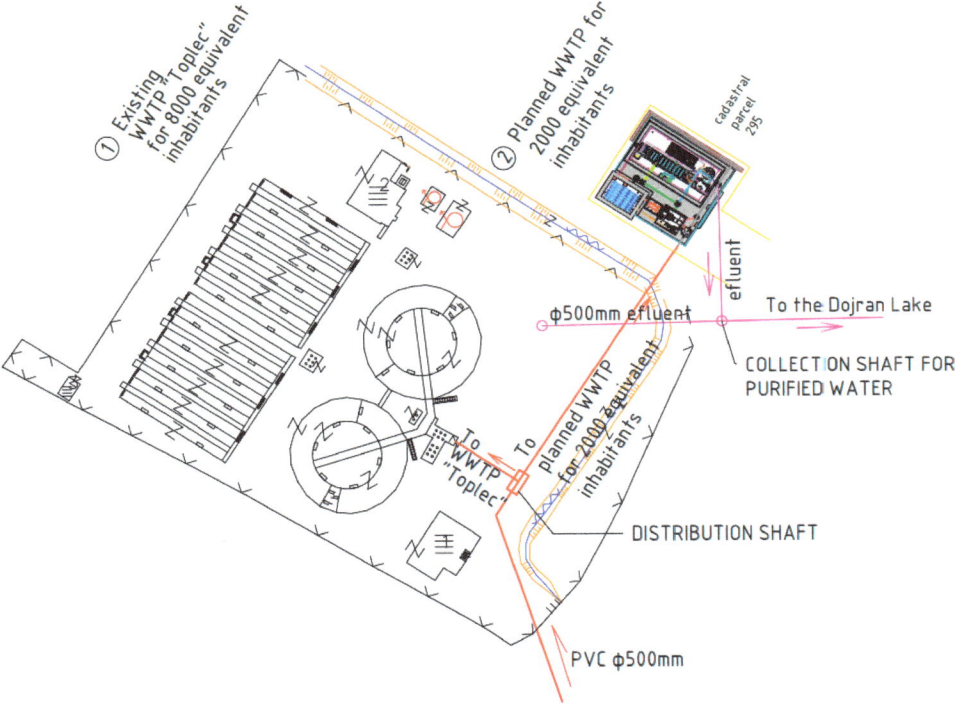

Figure 1. Combination of existing WWTP and new MBBR.

Considering a high level of removal (Table 5), effluent from plants can be discharged into natural streams, as it corresponds to all quality standards.

Table 5. Efficiency of MBBR as a function of filing.

Treatment Efficiency	gBOD5/m²day
75–80%	20.0
80–85%	10.0
85–90%	6.0
90–95%	4.5
95%–100%	2.5

MBBR technology for wastewater treatment has significant advantages. It enables a long retention time for activated sludge, which is good for nitrification. The process can be carried out without a secondary precipitator. Sediment production is reduced. MBBR requires a small area, while the capacity/space ratio of the plant is maximized. It achieves high flexibility in operation in the range of carrier fill from 30 to 70%. A two-stage biological process (high and medium load) increases efficiency and adapts to variable raw water inflow. The carrier material cannot be damaged (there are plants that are up to 20 years old and still use the same carriers). The thickness of the biofilm is controlled and maintained by the continuous separation that occurs under the action of aeration and mixing.

This technology is characterized by a smaller number of disadvantages. Since it is a biological process, the operation requires professional personnel to operate the plant. In addition, it also requires the engagement of qualified operators to ensure that there are no losses.

For the analyzed wastewater treatment technology, the cost calculation of the wastewater treatment plant was made, which included construction and craft work, procurement and installation of the equipment, connection of the plant to the distribution network, as well as its commissioning. The investment costs for this type of wastewater treatment technology were determined on the basis of several previously designed and constructed plants of this type (e.g., WWTP for S. Jurumleri, municipality Gazi Baba for 3500 equivalent inhabitants; WWTP for Novo Konjarevo, municipality Novo Selo for 1000 equivalent inhabitants; WWTP Ilovica and Štuku, municipality Bosilovo; WWTP Stračinci, municipality Gazi Baba). The analysis includes the calculation of plant maintenance and management costs. In accordance with the recommendations of the equipment manufacturer, as well as previous positive engineering practices, the annual cost calculation includes electricity costs (under normal operating conditions of the plant), employee costs (salaries and other related expenses), ongoing equipment maintenance and the servicing and cleaning of plant parts.

According to the current market prices, the investment cost for the construction of a wastewater treatment plant for 2000 equivalent inhabitants with MBBR and all the necessary stages of wastewater treatment is estimated at EUR 1,450,500. In addition, EUR 64,750 should be provided for operating costs on an annual basis. Given that, in this variant, the operation of the existing WWTP is planned during the tourist season, it is necessary to include the costs of its reconstruction, as well as its operating costs in the analysis. They amount to EUR 2,000,000 for reconstruction and EUR 101,500 per year for operating costs. Therefore, for Variant 1, the total investment costs amount to EUR 3,450,500, and the total operating costs are EUR 166,250.

According to the construction dynamics plan (Table 6), the construction would take place in four phases: construction of the MBBR for 2000 equivalent inhabitants with all the elements necessary for pretreatment and biological treatment (phase 1), reconstruction of the existing WWTP (phase 2), construction of the press (phase 3) and construction of a dryer (phase 4).

3.2. Variant 2—Combination of Existing WWTP and New SBR Wastewater Treatment Systems for 2000 Equivalent Inhabitants

The second variant includes the utilization of the existing WWTP (in accordance with the main reconstruction project) during the tourist season, and the construction of a new SBR wastewater treatment system for the calculated equivalent inhabitants of Dojran, which will be used outside the tourist season.

The new treatment plant would be located next to the existing one, as presented in Figure 2. It is envisaged that the water will be directed to the distribution shaft equipped with valves via the last pumping station of the main collector system, from which one line would lead to the existing WWTP and the other line would lead to the new station. The area (about 900 m^2) where the new treatment station is installed is part of the private plot KP295, and it should be subject to expropriation.

Table 6. Construction of new MBBR for 2000 equivalent inhabitants and reconstruction of the existing WWTP.

Dynamics of construction by phases	Costs	Amount (excluding VAT)
	MBBR technology for 2000 equivalent inhabitants	
/	Investment costs	[EUR]
Phase 1	Construction of a WWTP with all the elements required for pre-treatment and biological treatment	800,500
Phase 3	Construction of a sludge dewatering press	300,000
Phase 4	Construction of a dryer for drying sludge	350,000
	Σ	1,450,500
	Operating costs	[EUR/year]
	Electricity consumption - On average, 0.83 kWh/m^3 of purified water and for regular equipment service	10,400
	Ongoing service staff	6000
	Maintenance of the dryer—on average, 23.77 kWh/ton of sludge produced	47,110
	Maintenance of the press—on average, 12.5 kWh (the press would operate 2–3 h per day)	1240
	Σ	64,750
	Existing WWTP "Toplec"	
Dynamics of construction by phases	Costs	Amount (excluding VAT)
	Investment costs	[EUR]
Phase 2	Reconstruction of the station	2,000,000
	Σ	2,000,000
	Operating costs	[EUR/year]
	Electricity consumption and regular equipment servicing	71,500
	Ongoing service staff	30,000
	Σ	101,500

The elements of the planned WWTP are separated into common facilities and equipment that mainly occur for this type of technology. The SBR wastewater treatment system for 2000 equivalent inhabitants includes the following elements: inlet shaft, pumping station with fine screen, flow meter, retention basin, grease and oil trap, biological reactor (aeration and phosphorus elimination), outlet flow meter, clarifier, sludge dewatering press, sludge dryer and service facility.

In an SBR wastewater treatment system, the technological process includes pretreatment (removal of coarse and fine particles, grease and sand), a secondary process (elimination of carbon compounds COD, BOD5, elimination of ammonium) and a tertiary process (dephosphatization, chemical filter) and sludge line (thickening, dewatering/dehydration and drying).

The purification process is divided into two lines: primary (wastewater line) and secondary (sludge line). Wastewater from the municipality of Dojran (to be purified in the new WTTP) flows from the separation shaft directly into the mechanical purification plant, where wastewater is purified from solids. The water is pumped into the reservoir where grease and oils are separated. Wastewater from the grease and oil separator is pumped into the SBR reactors. Air is directly added to the SBR reactor via a blower. The chemical destruction of phosphorus carried out in the SBR reactors is followed by the sludge disposal stage. The excess sludge is pumped into the sludge tank, after which the purified wastewater is discharged by gravity into the well and then into the receiver. Excess sludge from the wastewater treatment process is stored in a tank that is gravity-connected to the detention basin, so that sludge from the top of the tank will overflow into the detention basin. The concentrated activated sludge from the tank will be transported to the press and then to the dryer.

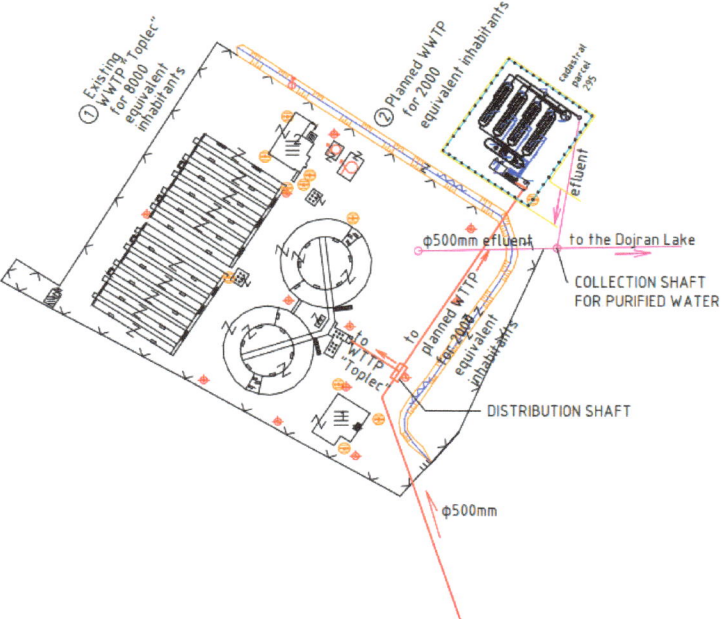

Figure 2. Combination of the existing WWTP and a new SBR.

The application of the SBR system is characterized by certain advantages, but also some disadvantages. One of the advantages is that the levelling of the basin and the primary clarifier (in most cases) can be achieved in one reactor. Biological treatment and secondary sedimentation can be achieved in one reactor. Flexibility and process control is ensured. The SBR system occupies a small area, and capital investment savings are achieved by eliminating the sedimentation tank and other equipment. However, the SBR system has its drawbacks. It requires a higher level of sophistication (compared to conventional systems), especially for larger systems. In addition, it requires a higher level of maintenance (compared to conventional systems), associated with more sophisticated controls, automatic switches and automatic valves. In the decanting phase, it is necessary to avoid capturing floating substances from the water. Depending on the aeration system used by the manufacturer, it may be necessary to include aerobic devices during the selected management cycle. Additionally, a sedimentation basin may be required after the SBR process, depending on the downstream processes.

For the analyzed wastewater treatment technology, the cost for the complete construction of the station with all the construction and craft work, procurement and installation of the equipment, and connection of the station to the electrical distribution network, as well as its commissioning was calculated. The investment costs for this type of wastewater treatment technology were determined on the basis of several previously designed and constructed stations of this type (WWTP for the village of Mavrovi Anovi, municipality of Mavrovo and Rostuša; WWTP for Millennium Cross for 1000 equivalent inhabitants with SBR technology; WWTP for the village of Stenje for 900 equivalent inhabitants with SBR technology). The analysis also includes the calculation of station maintenance and management costs. In accordance with the recommendations of the equipment manufacturer, as well as previous positive engineering practices, the following annual costs are included in the calculation: electricity costs under normal operating conditions of the station, employee costs (salaries and other related expenses), ongoing equipment maintenance and servicing and cleaning of station parts.

According to the current market prices, the investment cost for the construction of a wastewater treatment plant for 2000 equivalent inhabitants with an SBR system and all the necessary stages for wastewater treatment is estimated at EUR 1,456,000 (VAT excluded). Additionally, EUR 69,150 should be provided for operating costs on an annual basis. Given that, in this variant, the operation of the existing WWTP is planned during the tourist season, it is necessary to include the cost of its reconstruction, as well as the operating costs in the analysis. They amount to EUR 2,000,000 for reconstruction and EUR 101,500 per year for operating costs. Therefore, for this variant, the total investment costs are EUR 3,456,000, and the total operating costs are EUR 170,650.

The costs for the construction of a new SBR for 2000 equivalent inhabitants and the reconstruction of the existing WWTP are shown in Table 7. According to the construction dynamics plan, the construction would take place in four phases: construction of the SBR system for 2000 equivalent inhabitants with all the elements necessary for pretreatment and biological treatment (phase 1), reconstruction of the existing WWTP (phase 2), construction of a press (phase 3) and construction of a dryer (phase 4).

Table 7. Construction of new SBR for 2000 equivalent inhabitants and reconstruction of the existing WWTP.

Dynamics of construction by phases	SBR system for 2000 equivalent inhabitants	
	Costs	Amount
/	Investment costs	[EUR]
Phase 1	Construction of a WWTP with all the elements necessary for pretreatment and biological treatment	806,000
Phase 3	Construction of a sludge dewatering press	300,000
Phase 4	Construction of a dryer for drying sludge	350,000
	Σ	1,450,500
	Operating costs	[EUR/year]
	Electricity consumption - On average, 0.90 kWh/m^3 of purified water and for regular equipment service	13,500
	Ongoing service staff	7300
	Maintenance of the dryer—on average, 23.77 kWh/ton of sludge produced	47,110
	Maintenance of the press—on average, 12.5 kWh (press would operate 2–3 h per day)	1240
	Σ	69,150
	Existing WWTP "Toplec"	
Dynamics of construction by phases	Costs	Amount (excluding VAT)
	Investment costs	[EUR]
Phase 2	Reconstruction of the station	2,000,000
	Σ	2,000,000
	Operating costs	[EUR/year]
	Electricity consumption and regular equipment servicing	71,500
	Ongoing service staff	30,000
	Σ	101,500

3.3. Variant 3—Construction of a New MBBR Wastewater Treatment System for 6000 Equivalent Inhabitants

The third variant is essentially the construction of a new MBBR wastewater treatment system for 6000 equivalent inhabitants with two modules, where both modules will be active in the tourist season and only one outside the tourist season. The location of the new station is planned next to the existing one, on part of the parcel KP 295. The area it would occupy is approximately 2400 m^2, and it should be subject to expropriation. Although this variant does not assume the operation of the existing station, it is planned to place a distribution shaft with a gate valve on the collector of the last pumping station, leaving a possibility to activate the existing station, if necessary. The wastewater will be directed from the separation shaft to the newly planned treatment station, as presented in Figure 3.

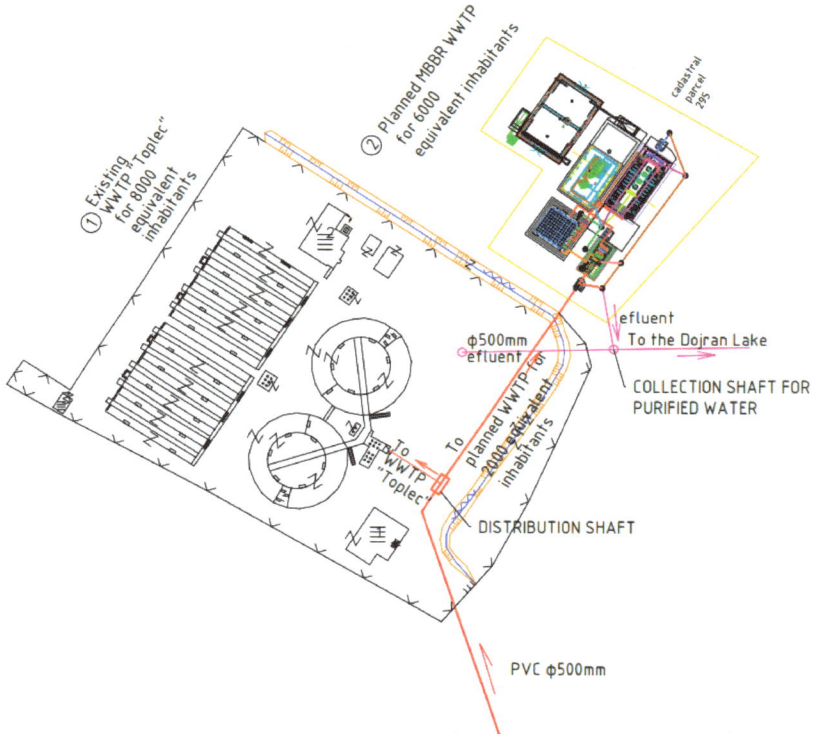

Figure 3. Construction of a new MBBR.

The elements, functioning and efficiency, as well as the advantages and disadvantages of the planned MBBR wastewater treatment system have already been presented in the description of Variant 1.

As for the previous two variants, the cost for the complete construction of the new MBBR wastewater treatment system for 6000 equivalent inhabitants was calculated. In addition, the costs of station maintenance and management were also calculated. According to the current market prices, the investment cost for the construction of a wastewater treatment station for 6000 equivalent inhabitants with an MBBR reactor and all the necessary stages of wastewater treatment is estimated at EUR 3,050,000 (excluding VAT). Additionally, EUR 88,150 should be provided for operating costs on an annual basis. The costs for the construction of a new MBBR system for 6000 equivalent inhabitants and the planned construction dynamics plan are given in Table 8.

Table 8. Construction of a new MBBR wastewater treatment system for 6000 equivalent inhabitants.

Dynamics of construction by phases	Costs	Amount
/	Investment costs	[EUR]
Phase 1	Construction of a WWTP with all the elements necessary for pretreatment and biological treatment	2,400,000
Phase 3	Construction of a sludge dewatering press	300,000
Phase 4	Construction of a dryer for drying sludge	350,000
	Σ	3,050,000
	Operating costs	[EUR/year]
	Electricity consumption - On average, 0.83 kWh/m^3 of purified water and for regular equipment service	29,800
	Ongoing service staff	10,000
	Maintenance of the dryer—on average, 23.77 kWh/ton of sludge produced	47,110
	Maintenance of the press—on average, 12.5 kWh (press would operate 2–3 h per day)	1240
	Σ	88,150

4. Results: Selection of the Optimal Variant Using the AHP Method

In order to select the optimal technology for wastewater treatment, a complex analysis was carried out, taking into account several factors, namely, the initial investment, operating costs and complexity of the facility and equipment, as well as the need for qualified staff to operate the station. As mentioned in Section 2, we have chosen the analytic hierarchy process (AHP) method as the optimal variant of wastewater treatment process. In this specific case (Figure 4), the AHP method was implemented in the following stages:

- Setting the target function: selection of a variant solution for wastewater treatment;
- Defining decision-making criteria: initial investment, operating costs management complexity;
- Selection of alternatives that achieve the target function: the existing WWTP and a new MBBR wastewater treatment system for 2000 equivalent inhabitants (V1), the existing WWTP and a new SBR system for 2000 equivalent inhabitants (V2), a new MBBR wastewater treatment system for 6000 equivalent inhabitants (V3).

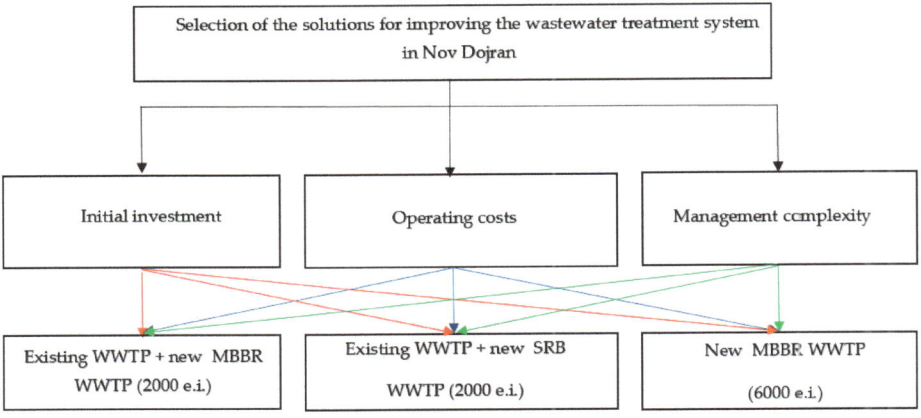

Figure 4. AHP scheme for the specific case.

The weights obtained after normalizing the pairwise comparison of the criteria and the weights obtained after normalizing the pairwise comparison of the alternatives/variants are shown below.

1. Criterion—(lowest) initial investment. When the variants are compared on the basis of this criterion, the sum of the investment costs of each variant appears as the value of the variant. For the first two variants, in addition to the investment cost, the amount required for the reconstruction of the existing WWTP was added.

V1—EUR 3,450,500
V2—EUR 3,456,000
V3—EUR 3,050,000
→ the value of the criteria (in EUR)
▶ Generation of a comparison matrix

Variant	V1	V2	V3	Sum	Weighting Coefficient Average Value
V1	1.00	2.00	0.25	3.25	0.23
V2	0.50	1.00	0.25	1.75	0.13
V3	4.00	4.00	1.00	9.00	0.64
Sum	5.50	7.00	1.50	14.00	1.00

▶ Generation of an induced matrix (normalisation)

Variant	V1	V2	V3	Sum	Weighting Coefficient Average Value
V1	0.18	0.29	0.17	0.63	0.21
V2	0.09	0.14	0.17	0.40	0.13
V3	0.73	0.57	0.67	1.97	0.66
Sum	1.00	1.00	1.00	3.00	1.00

The weighting coefficient of the initial investment by variants is shown in Figure 5.

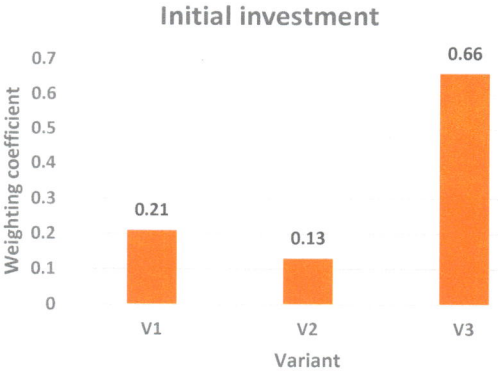

Figure 5. Weighting coefficient of initial investment by variants.

2. Criterion—(lowest) operating costs. In this case, the variants are compared on the basis of the resources required for the maintenance and management of the wastewater treatment system.

V1—166,250 EUR/year
V2—170,650 EUR/year

V3—88,150 EUR/year
→ the value of the criteria (in EUR/year)
▶ Generation of a comparison matrix

Variant	V1	V2	V3	Sum	Weighting Coefficient Average Value
V1	1.00	2.00	0.20	3.20	0.20
V2	0.50	1.00	0.20	1.70	0.11
V3	5.00	5.00	1.00	11.00	0.69
Sum	6.50	8.00	1.40	15.90	1.00

▶ Generation of an induced matrix (normalisation)

Variant	V1	V2	V3	Sum	Weighting Coefficient Average Value
V1	0.15	0.25	0.14	0.55	0.18
V2	0.08	0.13	0.14	0.34	0.11
V3	0.77	0.63	0.71	2.11	0.70
Sum	1.00	1.00	1.00	3.00	1.00

The weighting coefficient of the operating cost by variants is shown in Figure 6.

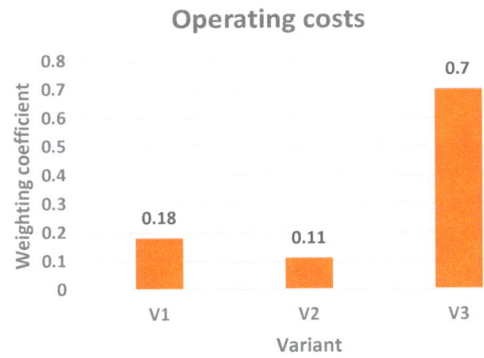

Figure 6. Weighting coefficient of operating costs by variants.

3. Criterion—(lowest) management complexity. In this case, the variants are compared based on the complexity of the system management, for example, starting the system operation in the tourist season and shutting it down at the end of the season, adding the medium that performs biological purification etc.

▶ Generation of a comparison matrix

Variant	V1	V2	V3	Suma	Weighting Coefficient Average Value
V1	1.00	0.50	0.14	1.64	0.08
V2	2.00	1.00	0.14	3.14	0.16
V3	7.00	7.00	1.00	15.00	0.76
Sum	10.00	8.50	1.29	19.79	1.00

▶ Generation of an induced matrix (normalization)

Variant	V1	V2	V3	Sum	Weighting Coefficient Average Value
V1	0.10	0.06	0.11	0.27	0.09
V2	0.20	0.12	0.11	0.43	0.14
V3	0.70	0.82	0.78	2.30	0.77
Sum	1.00	1.00	1.00	3.00	1.00

The weighting coefficient of management complexity by variants is shown in Figure 7.

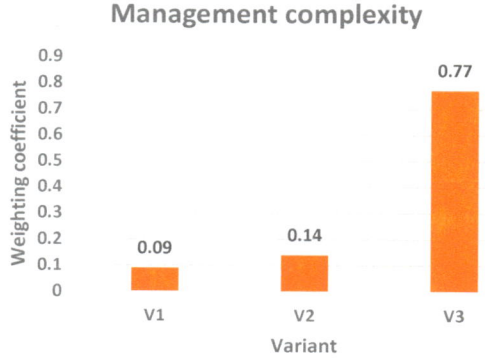

Figure 7. Weighting coefficient of management complexity by variants.

4. Defining the importance of the criteria

▶ Generation of a comparison matrix

Criterium	C1—Initial Investment	C2—Operating Costs	C3—Management Complexity	Sum	Weighting Coefficient Average Value
C1	1.00	0.50	2.00	3.50	0.31
C2	2.00	1.00	3.00	6.00	0.53
C3	0.50	0.33	1.00	1.83	0.16
Sum	3.50	1.83	6.00	11.33	1.00

▶ Generation of an induced matrix (normalisation)

Criterium	C1—Initial Investment	C2—Operating Costs	C3—Management Complexity	Sum	Weighting Coefficient Average Value
C1	0.29	0.27	0.33	0.89	0.30
C2	0.57	0.55	0.50	1.62	0.54
C3	0.14	0.18	0.17	0.49	0.16
Sum	1.00	1.00	1.00	3.00	1.00

For the purpose of greater visibility, the weighting coefficients of the criteria are shown in Figure 8.

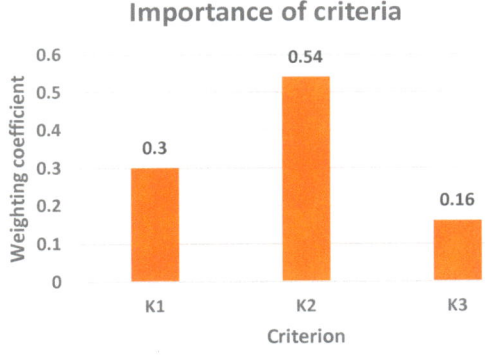

Figure 8. Criterion importance.

The value presented in Table 9 shows that of the three considered criteria, criterion K2—operating costs has the highest weighting coefficient, and therefore, with 54% intensity, has the greatest impact on the ranking decision. Criterion K1—initial investment with 30% intensity is in second place, while Criterion K3—management complexity, with only 16% intensity, is in the third place.

Table 9. Calculation with weighting coefficient.

Criterium	C1—Initial Investment	C2—Operating Costs	C3—Management Complexity
Weighting coefficient	0.30	0.54	0.16

The total priorities of the individual variants were determined so that the priorities of the variants according to each criterion (Table 10) were multiplied by the weights of the criteria (Table 9). These are shown in Table 11 and Figure 9.

Table 10. Priorities of variants according to each criterion.

Variants	Total Priorities of Individual Variants		
V1	0.21	0.18	0.09
V2	0.13	0.11	0.14
V3	0.66	0.70	0.77

Table 11. Total priorities of variants.

Criterium/Variant	C1—Initial Investment	C2—Operating Costs	C3—Management Complexity	Sum	Sum %	Ranking
V1	0.06	0.10	0.01	0.18	17.58	2.00
V2	0.04	0.06	0.02	0.13	12.50	3.00
V3	0.19	0.38	0.13	0.70	69.92	1.00
			Σ	1.00	100.00	

Figure 9. Total priorities of variants—ranking of variants.

The total priorities of the individual variants are shown in Figure 9.

According to the presented analysis, using the AHP methodology, the ranking of the variant solutions is as follows:

- Variant 1 (existing WWTP and new MBBR for 2000 equivalent inhabitants)—17.58%
- Variant 2 (existing WWTP and new SBR for 2000 equivalent inhabitants)—12.50%
- Variant 3 (new MBBR for 6000 equivalent inhabitants)—69.92%.

It can be concluded, based on the obtained results, that the AHP method suggests variant 3 as the best in all considered cases or ranking. This is a consequence of the "crossing" of the ranking of variants according to the physical characteristics of the considered parameters, independent of the ranking of the parameters according to the decision-maker's preferences. Based on the results presented above, it is obvious that Variant 3 is ahead of Variant 1 and Variant 2 according to all three selected criteria.

Through comparative analysis, it can be determined that the obtained results are consistent with the results of other research studies. Namely, the results of both older and more recent research indicate the numerous advantages of using MBBR technology in the wastewater treatment process. Yang et al. [41] concluded in their study that MBBR technology is an economically more attractive option than others, pointing to the significant savings of the capital expenditures (CAPEX). In addition to other advantages, cost savings with MBBR technology are also suggested by the results of other numerous studies [18–20], as well as its simplicity, flexibility, robustness and compactness [21].

5. Conclusions

In order to optimize the wastewater treatment process in the area of the municipality of Dojran, and based on the obtained results, it is recommended to design Variant 3, that is, the construction of a completely new system with MBBR technology for 6000 equivalent inhabitants. According to the AHP method, the best option is Variant 3—the construction of a new MBBR for 6000—which achieved the highest rank with 69.92%. The second-best option is Variant 1—a combination of the existing WWTP and a new MBBR for 2000—with 17.58%. Variant 2—a combination of the existing WWTP and a new SBR for 2000—achieved the lowest rank with 12.50%.

Certain limitations in the study may be partly related to the narrower selection of criteria used to select a variant solution for wastewater treatment. The limitations of the study can be partly attributed to the limitations of the application of the AHP method itself, which have already been discussed in the literature [80–82]. Therefore, the results obtained using the AHP method in this paper can be compared with the results of other methods, both in the assessment of the weight coefficients and in their use in terms of selection, ranking and preference results. Nevertheless, we believe that the mentioned limitations cannot call into question the results of the research, but rather serve as a catalyst for future research.

Author Contributions: Conceptualization, J.Ć., M.K. and R.V.; methodology, J.Ć., M.K. and R.V.; validation, J.Ć, E.T. and M.G.; formal analysis, J.Ć. and E.T.; data curation, E.T. and M.G.; writing—original draft preparation, J.Ć., M.K. and R.V.; writing—review and editing, J.Ć., E.T. and M.G.; visualization, E.T. and M.G.; supervision, J.Ć. and M.K. All authors have read and agreed to the published version of the manuscript.

Funding: This research received no external funding.

Data Availability Statement: Data may be made available on request from the corresponding author.

Conflicts of Interest: The authors declare no conflict of interest.

References

1. Judd, S. *The MBR Book: Principles and Applications Membrane Bioreactors in Water and Wastewater Treatment*, 1st ed.; Elsevier Ltd.: Oxford, UK, 2006; Available online: https://www.elsevier.com/books/the-mbr-book/judd/978-1-85617-481-7 (accessed on 16 February 2023).
2. Rittmann, B.E. Comparative performance of biofilm reactor types. *Biotechnol. Bioeng.* **1982**, *24*, 1341–1370. [CrossRef]
3. Ødegaard, H. *The Moving Bed Biofilm Reactor*; Water Environmental Engineering and Reuse of Water; Hokkaido Press: Hokkaido, Japan, 1999; Volume 575314, pp. 205–305. Available online: https://netedu.xauat.edu.cn/jpkc/netedu/jpkc2009/szylyybh/content/wlzy/7/3/The%20Moving%20Bed%20Biofilm%20Reactor.pdf (accessed on 20 February 2023).
4. Weiss, J.S.; Alvarez, M.; Tang, C.C.; Horvath, R.W. Evaluation of moving bed biofilm reactor technology for enhancing nitrogen removal in a stabilization pond treatment plant. *Proc. Water Environ. Fed.* WEFTEC **2005**, *14*, 2085–2102. [CrossRef]
5. Ødegaard, H. A new MBBR: Applications and results. *Water Sci. Technol.* **1994**, *29*, 157–165. [CrossRef]
6. Rusten, B.; Ødegaard, H.; Lundar, A. Treatment of dairy wastewater in a novel moving bed bioflm reactor. *Water Sci. Technol.* **1992**, *26*, 703–711. [CrossRef]
7. Rusten, B.; Kolkinn, O.; Ødegaard, H. Moving bed biofilm reactors and chemical precipitation for high efficiency treatment of wastewater from small communities. *Wat. Sci. Technol.* **1997**, *35*, 71–79. [CrossRef]
8. Gilligan, T.P.; Morper, M.A. Unique Process for Upgrading Conventional Activated Sludge Systems for Nitrogen Removal [online]. Paper presented at NE WEA. October 1999. Available online: http://www.dep.state.pa.us/dep/DEPUTATE/Watermgt/wsm/WSM_TAO/InnovTechForum/InnovTechForum-IIA-Gilligan_2.pdf (accessed on 25 February 2023).
9. Benakova, A.; Johanidesova, I.; Kelbich, P.; Pospisil, V.; Wanner, J. The increase of process stability in removing ammonia nitrogen from wastewater. *Water Sci. Technol.* **2018**, *77*, 2213–2219. [CrossRef]
10. Rusten, B.; Eikebrokk, B.; Ulgenes, Y.; Lygren, E. Design and operations of the Kaldnes moving bed biofilm reactors. *Aquacult. Eng.* **2006**, *34*, 322–331. [CrossRef]
11. Kermani, M.; Bina, B.; Movahedian, H.; Amin, M.M.; Nikaein, M. Application of moving bed biofilm process for biological organics and nutrients removal from municipal wastewater. *Am. J. Environ. Sci.* **2008**, *4*, 675. [CrossRef]
12. Zafarzadeh, A.; Bina, B.; Nikaeen, M.; Attar, H.M.; Nejad, M.H. Performance of moving bed bioflm reactors for biological nitrogen compounds removal from wastewater by partial nitrifcationdenitrifcation process. *Iran J. Environ. Health Sci. Eng.* **2010**, *7*, 353.
13. Koupaie, E.; Alavimoghadam, M. Comparison of overall performance between "Moving-bed" and "Conventional" sequencing batch reactor. *J. Environ. Health Sci. Eng.* **2011**, *8*, 235–244.
14. Biswas, K.; Taylor, M.W.; Turner, S.J. Successional development of biofilms in moving bed biofilm reactor (MBBR) systems treating municipal wastewater. *Appl. Microbiol. Biotechnol.* **2014**, *98*, 1429–1440. [CrossRef]
15. Andreottola, G.; Foladori, P.; Ragazzi, M. Upgrading of a small wastewater treatment plant in a cold climate region using a moving bed biofilm reactor (MBBR) system. *Wat. Sci. Technol.* **2000**, *41*, 177–185. [CrossRef]
16. Ødegaard, H. A road-map for energy-neutral wastewater treatment plants of the future based on compact technologies (including MBBR). *Front. Environ. Sci. Eng.* **2016**, *10*, 2. [CrossRef]
17. Dias, J.; Bellingham, M.; Hassan, J.; Barrett, M.; Stephenson, T.; Soares, A. Impact of carrier media on oxygen transfer and wastewater hydrodynamics on a moving attached growth system. *Chem. Eng. J.* **2018**, *335*, 399–408. [CrossRef]
18. Dezotti, M.; Geraldo, L.; Bassin, J.P. *Advanced Biological Processes for Wastewater Treatment*; Springer: Cham, Switzerland, 2017. [CrossRef]
19. Castro, F.D.; Bassin, J.P.; Dezotti, M. Treatment of a simulated textile wastewater containing the Reactive Orange 16 azo dye by a combination of ozonation and moving-bed biofilm reactor: Evaluating the performance, toxicity, and oxidation by-products. *Environ. Sci. Pollut. Res.* **2017**, *24*, 6307–6316. [CrossRef]
20. Yang, X.; Crespi, M.; López-Grimau, V. A review on the present situation of wastewater treatment in textile industry with membrane bioreactor and moving bed biofilm reactor. *Desalin. Water Treat.* **2018**, *103*, 315–322. [CrossRef]
21. Rodgers, M.; Zhan, X.M. Moving-medium biofilm reactors. *Rev. Environ. Sci. Bio Technol.* **2003**, *2*, 213–224. [CrossRef]

22. Jenkins, A.M.; Sanders, D. *Introduction to Fixed-Film Bio-Reactors for Decentralized Wastewater Treatment*; Contech Engineered Solutions: West Chester, OH, USA, 2012.
23. Das, A.; Naga, R.N. Activated Sludge Process with MBBR Technology at ETP. *IPPTA J.* **2011**, *23*, 135–137. Available online: https://ippta.co/wp-content/uploads/2021/01/2011_Issue_2_IPPTA_Articel_08.pdf (accessed on 4 February 2023).
24. Brinkley, J.; Johnson, C.H.; Souza, R. Moving bed biofilm reactor technology—A full scale installation for treatment of pharmaceutical wastewater, North Carolina American water works association-water environment federation. In *Annual Conference Technical Program*; NC AWWA-WEA: Raleigh, NC, USA, 2007.
25. Vaidhegi, K. Treatment of Bagasse Based Pulp and Paper Industry Efluent Using Moving Bed Biofilm Reactor. *Int. J. ChemTechnol. Res.* **2011**, *5*, 1313–1319. Available online: https://sphinxsai.com/2013/vol_5_3/pdf/CT=29(1313-1319)IPACT.pdf (accessed on 1 February 2023).
26. Singh, A.; Kamble, S.J.; Sawant, M.; Chakravarthy, Y.; Kazmi, A.; Aymerich, E.; Starkl, M.; Ghangrekar, M.; Philip, L. Technical, hygiene, economic, and life cycle assessment of full-scale moving bed biofilm reactors for wastewater treatment in India. *Environ. Sci. Pollut. Res.* **2013**, *25*, 2552–2569. [CrossRef]
27. Santos, A.D.; Martins, R.C.; Quinta-Ferreira, R.M.; Castro, L.M. Moving bed biofilm reactor (MBBR) for dairy wastewater treatment. *Energy Rep.* **2020**, *6*, 340–344. [CrossRef]
28. McQuarrie, J.P.; Boltz, J.P. Moving bed biofilm reactor technology: Process applications, design, and performance. *Water Environ. Res.* **2011**, *83*, 560–575. [CrossRef] [PubMed]
29. Schneider, E.E.; Cerqueira, A.C.F.P.; Dezotti, M. MBBR evaluation for oil refinery wastewater treatment, with post-ozonation and BAC, for wastewater reuse. *Water Sci. Technol.* **2011**, *63*, 143–148. [CrossRef]
30. Ooi, G.T.H.; Tang, K.; Chhetri, R.K.; Kaarsholm, K.M.S.; Sundmark, K.; Kragelund, C.; Litty, K.; Christensen, A.; Lindholst, L.; Sund, C.; et al. Biological removal of pharmaceuticals from hospital wastewater in a pilot-scale staged moving bed biofilm reactor (MBBR) utilising nitrifying and denitrifying processes. *Bioresour. Technol.* **2018**, *267*, 677–687. [CrossRef]
31. Salvetti, R.; Azzellino, A.; Canziani, R.; Bonomo, L. Effects of temperature on tertiary nitrification in moving-bed biofilm reactors. *Water Res.* **2006**, *40*, 2981–2993. [CrossRef]
32. Biasea, D.A.; Kowalskia, S.M.; Devlina, R.T.; Oleszkiewicza, J.A. Moving bed biofilm reactor technology in municipal wastewater treatment: A review. *J. Environ. Man.* **2019**, *247*, 849–866. [CrossRef]
33. Tarantini, M.; Scalbi, S.; Misceo, M.; Verità, S. Life Cycle Assessment as a Tool for Water Management Optimization in Textile Finishing Industry. In Proceedings of the SPIE—The International Society for Optical Engineering, Bellingham, WA, USA, 8 December 2004; Volume 5583, pp. 163–170. [CrossRef]
34. Nakhate, P.H.; Moradiya, K.K.; Patil, H.G.; Marathe, K.V.; Yadav, G.D. Case study on sustainability of textile wastewater treatment plant based on lifecycle assessment approach. *J. Clean. Prod.* **2020**, *245*, 118929. [CrossRef]
35. Cetinkaya, A.Y.; Bilgili, L. Life Cycle Comparison of Membrane Capacitive Deionization and Reverse Osmosis Membrane for Textile Wastewater Treatment. *Water. Air. Soil Pollut.* **2019**, *230*, 149. [CrossRef]
36. Park, H.O.; Oh, S.; Bade, R.; Shin, W.S. Application of A2O moving-bed biofilm reactors for textile dyeing wastewater treatment. *Korean J. Chem. Eng.* **2010**, *27*, 893–899. [CrossRef]
37. Mozia, S.; Janus, M.; Brozek, P.; Bering, S.; Tarnowski, K.; Mazur, J.; Morawski, A.W. A system coupling hybrid biological method with UV/O3 oxidation and membrane separation for treatment and reuse of industrial laundry wastewater. *Environ. Sci. Pollut.* **2016**, *23*, 19145–19155. [CrossRef]
38. Bering, S.; Mazur, J.; Tarnowski, K.; Janus, M.; Mozia, S.; Morawski, A.W. The application of moving bed bio-reactor (MBBR) in commercial laundry wastewater treatment. *Sc. Total Environ.* **2018**, *627*, 1638–1643. [CrossRef] [PubMed]
39. Altenbaher, B.; Levstek, M.; Neral, B.; Šostar-Turk, S. Laundry Wastewater Treatment in a Moving Bed Biofilm Reactor. *Tekst.-Zagreb* **2010**, *59*, 333–339. Available online: https://www.researchgate.net/publication/287930362_Laundry_wastewater_treatment_in_a_moving_bed_biofilm_reactor (accessed on 31 January 2023).
40. Erkan, H.S.; Çağlak, A.; Soysaloglu, A.; Takatas, B.; Engin, G.O. Performance evaluation of conventional membrane bioreactor and moving bed membrane bioreactor for synthetic textile wastewater treatment. *J. Water Process Eng.* **2020**, *38*, 101631. [CrossRef]
41. Yang, X.; López-Grimau, V.; Vilaseca, M.; Cresp, M. Treatment of Textile Wastewater by CAS, MBR, and MBBR: A Comparative Study from Technical, Economic, and Environmental Perspectives. *Water* **2020**, *12*, 1306. [CrossRef]
42. Madan, S.; Madan, R.; Hussain, A. Advancement in biological wastewater treatment using hybrid moving bed biofilm reactor (MBBR): A review. *App. Water Sci.* **2022**, *12*, 141. [CrossRef]
43. Barkman, E. *Moving Bed Bioreactor Technology for Secondary Treatment of Wastewater*; Moltoni Infra Tech Pty Ltd.: Perth, Australia, 2010.
44. Barwal, A.; Chaudhary, R. To study the performance of biocarriers in moving bed biofilm reactor (MBBR) technology and kinetics of biofilm for retrofitting the existing aerobic treatment systems: A review. *Rev. Environ. Sci. Bio/Technol.* **2014**, *13*, 285–299. [CrossRef]
45. Kawan, J.A.; Suja, F.; Pramanik, S.K.; Yusof, A.; Rahman, R.A.; Hasan, H.A. Effect of Hydraulic Retention Time on the Performance of a Compact Moving Bed Biofilm Reactor for Effluent Polishing of Treated Sewage. *Water* **2022**, *14*, 81. [CrossRef]
46. Khudhair, D.N.; Hosseinzadeh, M.; Zwain, H.M.; Siadatmousavi, S.M.; Majdi, A.; Mojiri, A. Upgrading the MBBR Process to Reduce Excess Sludge Production in Activated Sludge System Treating Sewage. *Water* **2023**, *15*, 408. [CrossRef]

47. Parivallal, G.; Govindaraju, R.A.; Nagalingam, A.; Devarajan, S. Performance evaluation of 1 MLD M3BR type sewage treatment plant. *Indian J. Microbiol. Res.* **2022**, *9*, 149–152.
48. Masłoń, A.; Tomaszek, J.A. A study on the use of the BioBall® as a biofilm carrier in a sequencing batch reactor. *Bioresour. Technol.* **2015**, *196*, 577–585. [CrossRef]
49. Zhou, X.; Zhang, Y.; Li, Z.; Sun, P.; Hui, X.; Fan, X.; Bi, X.; Yang, T.; Huang, S.; Cheng, L.; et al. A novel two-stage anoxic/oxic-moving bed biofilm reactor process for biological nitrogen removal in a full-scale municipal WWTP: Performance and bacterial community analysis. *J. Water Process. Eng.* **2022**, *50*, 103224. [CrossRef]
50. Czarnota, J.; Masłoń, A. Evaluation of the Effectiveness of a Wastewater Treatment Plant with MBBR Technology. *Middle Pomeranian Sci. Soc. Environ. Prot.* **2019**, *21*, 906–925.
51. Chu, L.; Wang, J. Comparison of polyurethane foam and biodegradable polymer as carriers in moving bed biofilm reactor for treating wastewater with a low C/N ratio. *Chemosphere* **2011**, *83*, 63–68. [CrossRef] [PubMed]
52. Ashkanani, A.; Almomani, F.; Khraisheh, M.; Bhosale, R.; Tawalbeh, M.; AlJaml, K. Bio-carrier and operating temperature effect on ammonia removal from second-ary wastewater effluents using moving bed biofilm reactor (MBBR). *Sci. Total. Environ.* **2019**, *693*, 133425. [CrossRef] [PubMed]
53. Mazioti, A.A.; Koutsokeras, L.E.; Constantinides, G.; Vyrides, I. Untapped potential of moving bed biofilm reactors with different biocarrier types for bilge water treatment: A laboratory-scale study. *Water* **2021**, *13*, 1810. [CrossRef]
54. Shitu, A.; Zhu, S.; Qi, W.; Tadda, M.A.; Liu, D.; Ye, Z. Performance of novel sponge bio-carrier in MBBR treating recirculating aquaculture systems wastewater: Microbial community and kinetic study. *J. Environ. Manag.* **2020**, *275*, 111264. [CrossRef] [PubMed]
55. Blackburne, R.; Yuan, Z.; Keller, J. Demonstration of nitrogen removal via nitrite in a sequencing batch reactor treating domestic wastewater. *Water Res.* **2008**, *42*, 2166–2176. [CrossRef]
56. Khursheed, A.; Gaur, R.Z.; Sharma, M.K.; Tyagi, V.K.; Khan, A.A.; Kazmi, A.A. Dependence of enhanced biological nitrogen removal on carbon to nitrogen and rbCOD to sbCOD ratios during sewage treatment in sequencing batch reactor. *J. Clean Prod.* **2018**, *171*, 1244–1254. [CrossRef]
57. Mees, J.B.; Gomes, S.D.; Hasan, S.D.; Gomes, B.M.; Boas, M.A. Nitrogen removal in a SBR operated with and without pre-denitrification: Effect of the carbon:nitrogen ratio and the cycle time. *Environ. Technol.* **2014**, *35*, 115–123. [CrossRef]
58. Mees, J.B.R.; Gomes, S.D.; Vilas Boas, M.A.; Gomes, B.M.; Passig, F.H. Kinetic behavior of nitrification in the post-treatment of poultry wastewater in a sequential batch reactor. *Eng. Agric.* **2011**, *31*, 954–964. [CrossRef]
59. Mace, S.; Mata-Alvarez, J. Utilization of SBR Technology for Wastewater Treatment: An Overview. *Ind. Eng. Chem. Res.* **2002**, *41*, 5539–5553. [CrossRef]
60. Quan, T.H.; Gogina, E. The assessment of technical—Economic efficiency of Sequencing Batch Reactors in municipal wastewater treatment in developing countries, *IOP Conf. Ser. Mater. Sci. Eng.* **2021**, *1030*, 012060. [CrossRef]
61. Dutta, A.; Sarkar, S. Sequencing Batch Reactor for Wastewater Treatment: Recent Advances. *Curr. Pollut. Rep.* **2015**, *1*, 177–190. [CrossRef]
62. Ćetković, J.; Knežević, M.; Lakić, S.; Žarković, M.; Vujadinović, R.; Živković, A.; Cvijović, J. Financial and Economic Investment Evaluation of Wastewater Treatment Plant. *Water* **2022**, *14*, 122. [CrossRef]
63. Alagha, O.; Allazem, A.; Bukhari, A.A.; Anil, I.; Muazu, N.D. Suitability of SBR for Wastewater Treatment and Reuse: Pilot-Scale Reactor Operated in Different Anoxic Conditions. *Int. J. Environ. Res. Public Health* **2020**, *17*, 1617. [CrossRef] [PubMed]
64. Fernandes, H.; Jungles, M.K.; Hoffmann, H.; Antonio, R.V.; Costa, R.H.R. Full-scale sequencing batch reactor (SBR) for domestic wastewater: Performance and diversity of microbial communities. *Bioresour. Technol.* **2013**, *132*, 262–258. [CrossRef]
65. Liping Fan, F.; Xie, Y. Optimization Control of SBR Wastewater Treatment Process Based on Pattern Recognition. *Procedia Environ. Sci.* **2011**, *10*, 20–25. [CrossRef]
66. Showkat, U.; Najar, I.A. Study on the efficiency of sequential batch reactor (SBR)-based sewage treatment plant. *Appl. Water Sci.* **2019**, *9*, 2. [CrossRef]
67. Eslami, H.; Hematabadi, P.T.; Ghelmani, S.V.; Salehi, V.A.; Derakhshan, Z. The Performance of Advanced Sequencing Batch Reactor in Wastewater Treatment Plant to Remove Organic Materials and Linear Alkyl Benzene Sulfonates. *Jundishapur J. Health Sci.* **2015**, *7*, e29620. [CrossRef]
68. European Council. Council Directive 91/271/EEC of 21 May 1991 Concerning Urban Waste Water Treatment. Available online: https://eur-lex.europa.eu/legal-content/EN/TXT/PDF/?uri=CELEX:31991L0271&from=EN (accessed on 2 February 2023).
69. European Commission. Commission Directive 98/15/EC of 27 February 1998 Amending Directive 91/271/EEC. Available online: http://extwprlegs1.fao.org/docs/pdf/eur18544.pdf (accessed on 2 February 2023).
70. State Statistical Office. Census of the Population, Households and Apartments in the Republic of Macedonia, Book H: Total Population, Households and Apartments. 2002. Available online: https://www.stat.gov.mk/PrikaziPublikacija.aspx?id=54&rbr=222 (accessed on 12 February 2023). (In Macedonian)
71. Government of North Macedonia. Decree for the Categorization of Watercourses, Lakes, Reservoirs and Underground Water. *Off. Gaz. Maced.* **1999**, *18*. Available online: https://www.moepp.gov.mk/wp-content/uploads/2014/09/Uredba%20za%20klasifikacija%20na%20vodite%20Sl.vesnik%2018-1999.pdf (accessed on 1 February 2023).
72. Vaidya, O.S.; Kumar, S. Analytic hierarchy process: An overview of applications. *Eur. J. Oper. Res.* **2006**, *169*, 1–29. [CrossRef]

73. Ho, W.; Ma, X. The state-of-the-art integration and applications of the analytic hierarchy process. *Eur. J. Oper. Res.* **2018**, *267*, 399–414. [CrossRef]
74. Špendl, R.; Bohanec, M.; Rajkovič, V. Comparative Analysis of AHP and DEX Decision Making Methods. In *Program and Abstracts of the International Conference on Methodology and Statistics*; Mrvar, A., Ferligoj, A., Eds.; Center of Methodology and Informatics, Institute of Social Sciences at Faculty of Social Science: Ljubljana, Slovenia, 2003; pp. 65–66. Available online: https://kt.ijs.si/MarkoBohanec/pub/DEXAHP2003.pdf (accessed on 18 February 2023).
75. Nolberto, M. *A Strategy for Using Multicriteria Analysis in Decision-Making*; Springer Science Business Media B.V.: Berlin/Heidelberg, Germany, 2011. [CrossRef]
76. Alessio, I.; Nemery, P. *Multi-Criteria Decision Analysis: Methods and Software*; John Wiley & Sons Ltd.: Hoboken, NJ, USA, 2013; Available online: https://www.wiley.com/en-sg/Multi+criteria+Decision+Analysis:+Methods+and+Software-p-9781119974079 (accessed on 18 February 2023).
77. Saaty, T.; Kearns, K.P. *Analytical Planning: The Organization of Systems*; Chapter 3—The Analytic Hierarchy Process Series; Elsevier: Amsterdam, The Netherlands, 1985. [CrossRef]
78. Municipality Dojran. Feasibility Study for the Establishment of a Public-Private Partnership Agreement for Financing, Designing, Projecting, Reconstruction, Renovation, Construction, Use, Management and Maintenance of Public Water Supply and Wastewater Disposal and Treatment Systems in the Municipality of Dojran. 2019; p. 38. Available online: https://dojran.gov.mk/wp-content/uploads/2019/12/A1.FEASIBILITY-STUDY-MK.docx (accessed on 18 February 2023). (In Macedonian)
79. Statistical Office of North Macedonia. List of Capacities in Hospitality in the Republic of Macedonia. 2016. Available online: https://www.stat.gov.mk/Publikacii/8.4.17.05.pdf (accessed on 18 February 2023).
80. Zahedi, F. The Analytic Hierarchy Process—A Survey of the Method and its Applications. *Interfaces* **1986**, *16*, 96–108. [CrossRef]
81. Dyer, J.S. Remarks on the analytic hierarchy process. *Manag. Sci.* **1990**, *36*, 249–258. Available online: https://www.jstor.org/stable/2631946 (accessed on 26 February 2023). [CrossRef]
82. Belton, V. A comparison of the analytic hierarchy process and simple multi-attribute value function. *Eur. J. Oper. Res.* **1986**, *26*, 7–21. [CrossRef]

Disclaimer/Publisher's Note: The statements, opinions and data contained in all publications are solely those of the individual author(s) and contributor(s) and not of MDPI and/or the editor(s). MDPI and/or the editor(s) disclaim responsibility for any injury to people or property resulting from any ideas, methods, instructions or products referred to in the content.

Review

Emerging Contaminants and Their Removal from Aqueous Media Using Conventional/Non-Conventional Adsorbents: A Glance at the Relationship between Materials, Processes, and Technologies

Cristina E. Almeida-Naranjo [1,2], Víctor H. Guerrero [3,*] and Cristina Alejandra Villamar-Ayala [4,5,*]

Citation: Almeida-Naranjo, C.E.; Guerrero, V.H.; Villamar-Ayala, C.A. Emerging Contaminants and Their Removal from Aqueous Media Using Conventional/Non-Conventional Adsorbents: A Glance at the Relationship between Materials, Processes, and Technologies. *Water* 2023, *15*, 1626. https://doi.org/10.3390/w15081626

Academic Editors: Hamidi Abdul Aziz, Issam A. Al-Khatib, Rehab O. Abdel Rahman, Tsuyoshi Imai and Yung-Tse Hung

Received: 3 March 2023
Revised: 22 March 2023
Accepted: 25 March 2023
Published: 21 April 2023

Copyright: © 2023 by the authors. Licensee MDPI, Basel, Switzerland. This article is an open access article distributed under the terms and conditions of the Creative Commons Attribution (CC BY) license (https://creativecommons.org/licenses/by/4.0/).

[1] Departament of Mechanical Engineering, Escuela Politécnica Nacional, Ladrón de Guevara E11-253, Quito 170525, Ecuador
[2] Faculty of Engineering and Sciences—Biotechnology Engineering, Universidad de las Américas, Redondel del Ciclista Antigua Vía a Nayón, P.C., Quito 170124, Ecuador
[3] Department of Materials, Escuela Politécnica Nacional, Ladrón de Guevara E11-253, Quito 170525, Ecuador
[4] Departamento de Ingeniería en Obras Civiles, Facultad de Ingeniería, Universidad Santiago de Chile (USACH), Av. Victor Jara 3659, Estación Central, Santiago 9170022, Chile
[5] Programa para el Desarrollo de Sistemas Productivos Sostenibles, Facultad de Ingeniería, Universidad Santiago de Chile (USACH), Av. Victor Jara 3769, Estación Central, Santiago 9170022, Chile
* Correspondence: victor.guerrero@epn.edu.ec (V.H.G.); cristina.villamar@usach.cl (C.A.V.-A.); Tel.: +56-22-71-82-810 (C.A.V.-A.)

Abstract: Emerging contaminants (ECs) are causing negative effects on the environment and even on people, so their removal has become a priority worldwide. Adsorption and the associated technologies where this process occurs (filtration/biofiltration) have gained great interest, due to its low cost, easy operation, and effectiveness mainly in the removal (up to 100%) of lipophilic ECs (log K_{ow} > 4). Activated carbon continues to be the most efficient material in the removal of ECs (>850 mg/g). However, other conventional materials (activated carbon, clays, zeolites) and non-conventional materials (agro-industrial/forestry/industrial residues, nanomaterials, among others) have shown efficiencies greater than 90%. Adsorption depends on the physicochemical properties of the materials and ECs. Thus, physical/chemical/thermal modifications and nanomaterial synthesis are the most used procedures to improve adsorption capacity. A material with good adsorptive properties could be used efficiently in filtration/biofiltration technologies. Agro-industrial residues are promising alternatives to be used in these technologies, due to their high availability, low toxicity, and adsorption capacities (up to 350 mg/g). In filtration/biofiltration technologies, the material, in addition to acting as adsorbent, plays a fundamental role in operation and hydraulics. Therefore, selecting the appropriate material improves the efficiency/useful life of the filter/biofilter.

Keywords: conventional/non-conventional adsorbents; nanocomposites; lipophilic contaminants; filtration

1. Introduction

ECs are organic, pseudo-persistent, and unregulated "new" contaminants detected in water/wastewater in trace concentrations (ng/L–µg/L) [1]. Pharmaceutical and personal care products, hormones, pesticides, and microplastics, among other chemical substances, are some examples of ECs. They reach the environment through effluents from municipal wastewater treatment plants (WWTPs), septic tanks, hospital effluents, livestock activities, and subsurface storage of household and industrial wastes [1,2]. In fact, several antibiotics, such as azithromycin, amoxicillin, and ciprofloxacin, have been found in influents wastewater from Asian, European, and North American countries at concentrations (ng/L) between 3 and 303,500, 0.4 and 13,625, and 6.1 and 246,100, respectively [3,4].

After reaching the environment, or even inside the WWTPs, some ECs should be degraded/biodegraded to metabolites or gases. An environmental degradation route of ECs is photochemical transformation, which can occur directly by solar UV radiation absorption [5] and indirectly by photosensitized species reaction [6]. For example, diclofenac has demonstrated being photochemically degraded to 1-(8-chlorocarbazolyl) acetic acid and carbazole. ECs' biodegradability depends on their bioavailability, being a more viable process for ECs with octanol-water partitioning coefficients (log K_{ow}) between < 1 and > 4 and high polarity (pKa > 0.5) [5,6]. Ibuprofen, natural estrogens, bisphenol A (BPA), and triclosan are some examples of biodegradable ECs, which generate transformation products or metabolites (e.g., triclosan to 2,8-dichlorodibenzo-p-dioxin under UV light), achieving degradation percentages greater than 50% [6]. However, since ECs with Henry's constant values from 10^{-2} to 10^{-3} mol/m³Pa do not degrade but rather volatilize, ECs such as dibutyl phthalate, di(2-ethylhexyl) phthalate, and nonylphenol have been found in air samples (0.031–0.055 ng/L) [5–7]. Another important factor controlling the fate/behavior of ECs (removal/bioavailability/degradation/volatilization/transport) is their ease of being sorbed–desorbed. Sorption–desorption processes are related to the partitioning soils/sediments coefficient measured by K_d [8]. Some antibiotics, hormones, biocides, and artificial sweeteners with low K_d (300–500 L/kg MLSS) have showed insignificant sorption into sludge [1].

Differences between properties/characteristics/behavior/mobility of different ECs have turned them into substances with potential environmental and human risk, despite their low concentrations. Figure 1 shows the main routes of entry of ECs into the environment, their behavior in it, and the possible effects they could cause in different living beings. Conventional WWTPs do not perform effectively for the removal of ECs. Moreover, conventional wastewater treatment processes/systems/technologies are generally not available or efficient for developing countries, and thus it is necessary to search for options that fit each place [9].

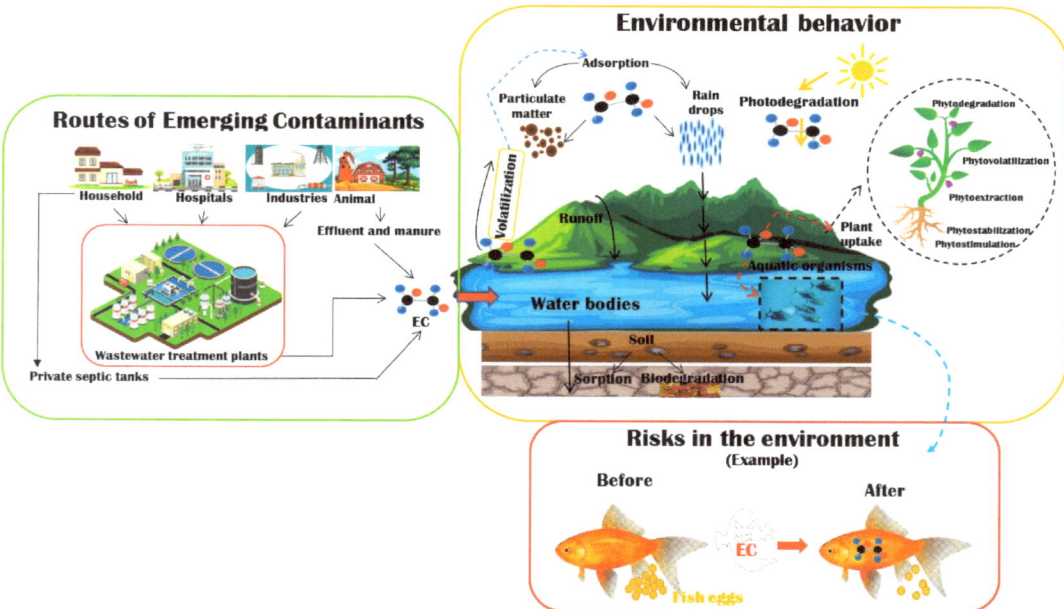

Figure 1. Routes of entry of ECs into the environment, its behavior, and effects.

Adsorption has proven to be the most effective, sustainable, renewable, and selective method with low cost and energy consumption [10]. It allows the removal of contaminants such as heavy metals (metal/adsorbent: nickel, lead/nano polypyrrole-polyethyleneimine [11,12]; zinc/modified fruit peels, graphene oxide + magnetite [13,14]) and dyes (dye/adsorbent: methylene blue, acid orange 10/activated carbon from oil palm waste; methylene blue and crystal violet/Bauhinia forficata residual fruit) [10,15]. Activated carbon is the most used adsorbent, but the high costs of this adsorbent, added to the loss of efficiency in the regeneration process, have made it necessary to search for other alternatives [16].

Although some reviews/research studies have dealt with the removal of ECs through adsorption processes, no bibliographic reviews exist that address the relationship between adsorbent materials and the technologies in which they could be used. This is a limitation because it is necessary to propose real-scale alternatives to remove contaminants from wastewater. Therefore, the objectives of this review are (i) to identify the different adsorbents used in the removal of ECs, (ii) to find the mechanisms involved in the removal of ECs, and (iii) to analyze the relationship between the adsorbent material and the different technologies in which it can be used. To achieve these objectives, the issues presented in Figure 2 are addressed. The different types of ECs, their characteristics, concentration in water resources, and toxicity to some species are first presented. Subsequently, different conventional/non-conventional adsorbents used in the removal of ECs from aqueous media in batch adsorption processes are described. The adsorbent characteristics (e.g., surface area, porosity, functional groups) and their efficiencies/adsorption capacities in the removal of different ECs are shown. The limitation of batch adsorption processes is the volume of treated wastewater/water. Therefore, a relationship between the material and filtration/biofiltration technologies (where adsorption processes occur) used in the removal of ECs is presented. Finally, the conclusions of the work and the outlooks are discussed. To this end, reviews/research papers investigating conventional and non-conventional materials published in the last 10 years were analyzed. However, very little information was found in some materials reviewed, and therefore the search criteria were extended to articles published in the last 20 years.

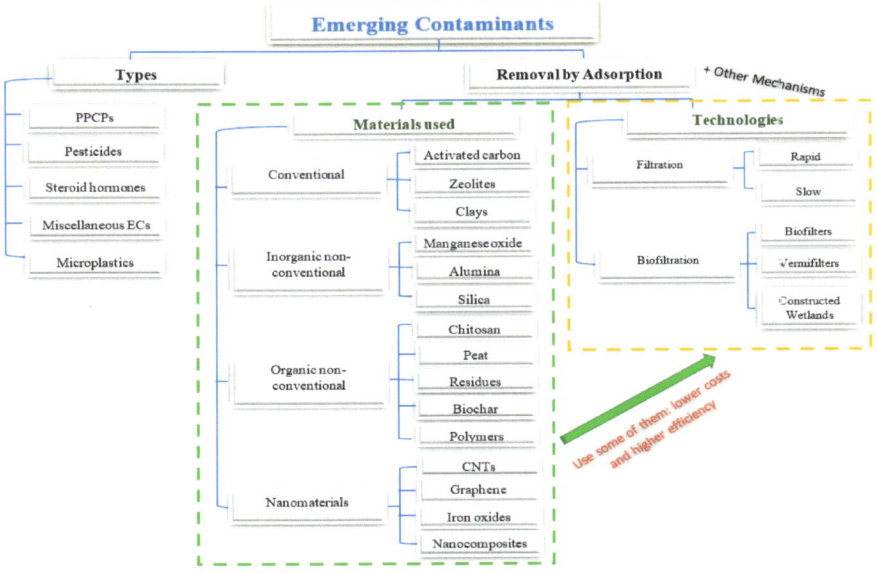

Figure 2. Topics featured in this review.

2. Characteristics, Concentration, and Toxicity of Some ECs in Water Bodies

2.1. Pharmaceuticals and Personal Care Products (PPCPs)

There are more than 3000 substances (50–150 g/inhab) used as analgesics/anti-inflammatories, antibiotics, contraceptives, antidepressants, and pressure regulators, among others [17]. Meanwhile, around 2000 chemical compounds are used as personal care products, which include fragrances, preservatives, disinfectants, and sun protection agents, among others [18]. Indeed, the PPCPs production has reached about 20 million tons/year [19]. PPCPs have been found in surface water (up to 10,000 mg/L), groundwater (up to 100 mg/L), influents (75–73,730 ng/L), and effluents (24–4800 ng/L) of WWTP, sludge, sediments, and even in living beings (e.g., triclosan in fish at 2100 ng/g) [20,21]. In general, PPCPs can be made of several complex molecules with different structures and shapes, which vary widely according to their molecular weight (88.5–>900.0 g/mol) [4,14]. Moreover, they are polar molecules with ionizable groups, i.e., in the solid phase, they have different adsorption mechanisms (e.g., ion exchange). Most of them are lipophilic [4]. However, PPCPs have log K_{ow} values in a wide range (−2.4–13.1), from acid to basic substances under environmental conditions (pKa = 0.6–13.9) [4,22]. PPCPs reach K_d values between 1.9 and 39,000, are partially/completely soluble in water (0.02–3.12 × 10^5 mg/L, T = 20 °C) [4,19], and have dissipation times between <3 and up to 300 days [18].

Recurrent discharges of PPCPs could cause endocrine problems, genotoxicity, aquatic toxicity, and resistance in pathogenic microorganisms [23]. For instance, diclofenac at concentrations between 5 and 50 mg/L can increase the plasma vitellogenin in fish. It has even caused effects on steppe eagles and vultures [1,18]. Meanwhile, ciprofloxacin, tetracycline, ampicillin, trimethoprim, erythromycin, and sulfamethoxazole can increase the antibiotic resistance of *E. coli* and *Xanthomonas maltophilia* [18,19]. Personal care products, such as benzophenone-3 threaten coral reefs and 4-methylbenzylidene camphor, have been demonstrated to generate embryonic malformations in frogs. Additionally, triclosan produces adverse effects on the first stages of the frog's life [21,24,25]. However, PPCPs also affect humans, as their presence has been detected in breast milk, blood, and urine of children [21]. Furthermore, benzyl paraben and benzophenone-4 were found in the placenta, which could indicate a transfer from mother to fetus [26].

2.2. Pesticides

Until 2020, a use of around 3.5 million tons/year of pesticides was estimated, but only less than 0.1% was used for plants [27]. Some pesticides, such as glyphosate or its main metabolite (amino methyl-phosphonic acid), have been found in surface waters (0.02–6.0 g/L), soil (15.9–1025.5 kg/kg), deep waters (0.1 µg/L), and sediments (0.1–100 mg/L). Moreover, pesticides with high vapor pressure (1.51 × 10^{-7}–1.29 × 10^{-1} Pa)/high volatility (e.g., pentachlorophenol) are released to the air during their application (between 5 and 90% of them), moving long distances and even reaching pristine areas [28–30]. They have been found in rainfall, e.g., methyl parathion, and also near agricultural sites where they were not applied (~23 µg/L) [28]. According to their molecular structure, pesticides have different chemical properties, reaching weak to strong acid character (pKa = 0.7–9.1), medium to high solubility in water (2.0–1.2 × 10^6 mg/L, T = 20–25 °C), very low to very high bioaccumulation (log K_{ow} = −4.6–8.0), and high persistence (7 days–>5 years) [28,30].

Pesticides mainly affect non-target organisms, e.g., atrazine (concentrations in water >200 ng/L) [29] causes sex change in male frogs and affected/altered the reproductive system and fertility of mice, fish, and humans [28,29,31]. Another widely used herbicide is glyphosate, which affects the entire food chain, delaying periphytic colonization and reducing the abundance of aquatic organisms such as *Pseudokirchneriella subcapitata* and *Lemma minor* ($EC_{50\text{-}7d}$ = 11.2–46.9 and $EC_{50\text{-}4d}$ = 64.7–270.0 mg a.i./L, respectively) [30,31]. Paraquat produced neurotoxicity and systemic and pulmonary inflammation (inhalation for 16 days) in rats [32]. Furthermore, its use in edible crops is related to antibiotic resistance in humans, even it was classified as a carcinogenic by the World Health Organization [33]. The byproducts/metabolites of pesticides also cause negative effects [28]. For example, aminomethylphosphonic acid produced acute toxicity in *Vibrio fischeri* at concentrations between 50 and 167 mg/L [30].

2.3. Steroid Hormones

The human population generates about 30,000 and 700 kg/year of natural and synthetic estrogens, respectively. Natural estrogens are excreted by the adrenal cortex, testes, ovaries, and placenta of humans and animals (e.g., estrone/E1, 17β-estradiol/E2, estriol/E3) [34]. Meanwhile, synthetic hormones are synthesized from cholesterol (e.g., 17α-ethinylestradiol/EE2) [34,35]. In specific, synthetic estrogens come from oral contraceptives. However, the estrogen generated by cattle is much higher, amounting to 83,000 kg/year in the US and the European Union alone [36].

WWTPs or feedlot effluents are the main pathways for hormones entering soil, surface water, sediment, and groundwater. They have been found within WWTP influents (>802 ng/L), WWTP effluents (>275 ng/L), surface waters (0.04–667 ng/L), groundwater (5–> 1000 ng/L), drinking water (up to 3500 ng/L), and livestock waste (14–533 ng/g) [3,4]. Steroid hormones are characterized by having a molecular weight between 242 and >296 g/mol, being poorly soluble in water (8.8–441.0 mg/L, at 20 °C), with low to moderate hydrophobicity (log K_{ow} = 1.6–4.7), weak acid character (pKa = 10.3–18.9), and non-volatile (vapor pressure= 9×10^{-13}–3×10^{-8} Pa) [34,36]. Despite their low persistence ($t_{half-life}$ = 0.2–9.0 days in water/sediments) compared with other ECs, they cause negative effects on ecosystems and humans (cancer and infertility) by their endocrine-disrupting character [4,36]. Chronic exposure of fathead minnow to EE2 (5–6 ng/L) led to feminization of male fish by the development of the ovary cavity, and impaired oogenesis in females [4,34]. E2 at 1 ng/L induced changes in vitellogenin production in males, and produced the feminization of some species of male fish (1–10 ng/L) [34,35]. Levonorgestrel at 6.5 ng/L inhibited spermatogenesis, reduced fish egg production and reproduction, increased female weight and length, and promoted female masculinization [34]. Likewise, EE2 at 10 ng/L affected cardiac function in bullfrog tadpoles [34], reducing the fish biomass and interrupting the aquatic food chain [36].

2.4. Micellaneous ECs

The use/consumption of different products such as fire retardants, food additives, plasticizers, solvent stabilizers, surfactants/detergents, industrial additives (fluorinated organic compounds), and corrosion inhibitors is considerable worldwide [3,37]. In fact, the worldwide production of surfactants was projected to 24.2 million tons for 2022 [38], and plasticizers such as BPA had a production of 2.0 million metric tons/year [5]. Moreover, around 159,000 metric tons/year of synthetic non-nutritional sweeteners are consumed [39,40].

The physicochemical characteristics of these substances are very varied due to the differences between them, which also makes their behavior in the environment different. For instance, artificial sweeteners present between medium and very high solubility (4–1000 g/L, 20 °C), from low to high acid character (pKa = 1.9–11.8), and very low bioaccumulation (log K_{ow} = −1.61–0.91) [6,39]. However, other substances such as BPA (plasticizer), nonylphenol (detergent), and Tris(2-chloroethyl) phosphate (TCEP, fire retardant) have from low to high bioaccumulation with log K_{ow} values of 3.2, 4.4, and 1.8, respectively [5,41]. Moreover, BPA and phthalates are semi-volatiles, so they easily move into the environment. They have a short half-life (5–18 days) in air because they could be photodegraded [5]. Likewise, fluorinated organic compounds have longer chains exhibiting persistence, from moderately solubility (e.g., perfluoro octane sulfonate = 1.08 g/L) to high solubility (e.g., perfluorooctanoic acid > 20 mg/L) in water, long-distance mobility, high to very high bioaccumulation (e.g., log K_{ow} of perfluoro octane sulfonate = 4.5–6.9), and toxic effects [42,43].

In turn, plasticizers are commonly found in surface water (<1–12,000 ng/L), runoff (50–2410 ng/L), and other water sources [5,7]. They include bisphenol type -A/-S/-F and phthalates [5,7,41]. Fluorinated organic compounds have been detected in surface water

(0.09–578,970 ng/L), groundwater (0.17–8.54 ng/L), WWTP influents (65–112 µg/L) and WWTPs effluents (43–78 µg/L), runoff (~2 ng/L), and drinking water (~54 ng/L) [5,41]. Furthermore, traces of fluorides (0.023–>1600 ng/g) have been found in some species of fish, amphibians, crustaceans, seals, whales, and polar bears [41]. Artificial sweeteners were found in drinking water (36–2400 ng/L), surface water (0.03–9600 ng/L), groundwater (non-detected–33,600 µg/L), seawater (200–393 ng/L), and lakes (non-detected–780 ng/L) [6,39].

Ethoxylated alcohol (surfactant) has been reported to affect fish and invertebrates. In fathead minnows, it affects egg production/larval survival, with a non-observed effect concentration (NOEC) ~0.73 mg/L. In species such as bluegills, it affects survival and growth at concentrations ~5.7 mg/L [44]. Non-ionic surfactants and nonylphenol ethoxylates have exhibited acute toxic effects (EC_{50} = 1.1–25.4 mg/L) on tadpoles of four Australian and two exotic frogs [39,44]. In turn, artificial sweeteners such as aspartame (2000–50,000 mg/L) have been reported to cause cancer in rats and headaches, nausea, and vomiting in humans. Meanwhile, 5% sucralose produced thymus shrinkage and migraine in rats and humans, respectively [39].

2.5. Microplastics

Microplastics (size = 1–5000 nm) are classified as primary (microbeads from personal care products and cosmetics) and secondary (degradation/fragmentation/leaching of plastics) [45]. Approximately 4130 tons/year of microbeads are used in different personal care products in EU countries plus Norway and Switzerland [45]. In total, plastic waste has reached values around 6300 million tons (oceans = 1.6–4.1%) between 1950 and 2015 [45,46]. More than 400,000 tons microplastics/year (95% from WWTPs) could enter the environment [45]. They were found in lakes and rivers (0.05–320 particles/L) and sediments from shore, water, and benthic at concentrations of 75–1300 particles/m^2, 2.5–25,800.0 particles/m^3, and 6.2–980 particles/kg, respectively [45]. A problem associated with microplastics is that those with low density (910–2200 kg/m^3) could move hydrophobic contaminants long distances [47]. Dichlorodiphenyltrichloroethane (DDT), polycyclic aromatic hydrocarbons, chlorinated benzenes, polychlorinated biphenyls (PCBs), organohalogenated pesticides, and endocrine disruptors have been found in microplastics such as polyethylene, polypropylene, PVC, and polystyrene [47]. BPA, polybrominated diphenyl ethers and phthalates, and microplastic additives also have been found in plastic debris at concentrations from 0.1 to 700 ng/g, up to 990 ng/g, and up to 3940 ng/g, respectively [48].

Microplastics produce digestive and locomotion problems, and changes in metabolic profiles in organisms, such as planktonic, crustaceans (e.g., *Daphnia magna*), fish (e.g., zebrafish), turtles, and whales [47,49]. These ECs could even compromise human health since they have been found in consumer species such as brown shrimp [47,48]. The surface area/shape/size/texture of microplastics is also related to toxicity. Particularly, the smallest microplastics in the form of fibers and a greater surface area have been reported to generate greater toxicity [48]. High-density polyethylene (0–80 µm) accumulated in the lysosomal system of blue mussels after 1.0 h of exposure. In addition, earthworms of the species *Lumbricus terrestris* exposed to 450 and 600 g/kg of polyethylene increased their mortality rate between 8 and 25%, respectively. There was also a decrease in their growth and negative effects on the construction of burrows [42]. Polypropylene fibers were more toxic than polyethylene spheres for *Hyalella Azteca* [45]. Among the most used methods to eliminate microplastics from aqueous medium are physical (sedimentation, membrane filtration), chemical (photocatalytic oxidation, coagulation, ozonation), and biological (conventional activated sludge systems, activated sludge systems + membrane bioreactor, microorganism biodegradation) [44,47–50]. The use of materials is also essential in the elimination of this type of EC, mainly through magnetic separation. Nanomaterials such as magnetic carbon nanotubes and magnetite have been used efficiently [51]. Shia et al. (2022) removed polyethylene, polypropylene, polystyrene, and polyethylene terephthalate (size: 200–900 µm), reaching efficiencies between 62.8 and 86.9% [52].

Table 1 shows the features of some ECs and their concentrations for different water sources. It is observed that there is a wide range of ECs found in different water bodies in their common concentrations (0.01 ng/L–>6010 mg/L). The EC characteristics are also very varied (log K_{ow} = −3.4–13.9, water solubility = 0.49–21,600 mg/L), which could be a complication when it comes to removing them from water. Lipophilic substances (high log K_{ow}) and low water solubility such as carbamazepine (anticonvulsant), triclosan (disinfectant), EE2 (hormones), and musk xylene (fragrance) suggest higher removal efficiency through adsorption processes. Meanwhile, hydrophilic substances (low log K_{ow}) and high-water solubility, such as caffeine (stimulant), clofibric acid (lipid regulator), atenolol (beta blocker), and glyphosate (herbicide), are more difficult to remove through adsorption processes. This was verified in the research carried out by León et al. [53], in which triclosan (48.6–76.4%) was more easily removed than caffeine (40.1–67.4%) using agro-industrial residues, under the same operating conditions. Therefore, lipophilic ECs can be removed more easily than hydrophilic by adsorption process.

Table 1. Physical and chemical characteristics of some ECs and their concentrations in water bodies.

EC Type	EC Subtype	Contaminant	Water Solubility [mg/L]	Log K_{ow}	Concentration Found [ng/L]	Type of Water	Reference
PPCPs	Stimulant	Caffeine	21,600	−0.07	753,500–1.0 × 10^6 20–23,970 500–5000	Surface water Ground water Drinking water	[22,54,55]
	Anti-inflammatory	Ibuprofen	21	0.35	13.5–89,500	Effluent WWTP	[19,22,54,56]
	Anticonvulsant	Carbamazepine	112–236	13.90	589–3.5 × 10^8	Influent WWTP	[19,57,58]
	Lipid regulator	Clofibric acid	214,650	2.88	1200–6.6 × 10^7 nd–420	Effluent WWTP Influent WWTP	[19,59]
	Antibiotic	Ciprofloxacin	650	0.28	2200–14,000 1100–44,000	Influent WWTP Hospital wastewater	[19,60,61]
	Diagnostic Contrast Media	Iopromide	23.8	−2.1	780–11,4000 1170–4030	River water Urban effluents	[19,62]
	Antidepressant	Diazepan	50	3.08	nd–100	Surface water	[19,63]
	Beta blocker	Atenolol	300	0.16	nd–100	Influent WWTP	[19,64]
	Fragrance	Musk xylene	0.49–1.0	4.40	90–255	Influent WWTP	[19,65]
	Disinfectant	Triclosan	10.00	4.76	200,000–400,000 200–1854	Drinking water Influent WWTP	[24,25,54]
Pesticides	Herbicide	Glyphosate	15.70	−3.40	up to 6.01 × 10^9	Surface water	[30,66]
	Insecticide	Acetamiprid	2950	0.80	0.08–249	Surface water	[27,29]
		Clothianidin	304	0.70	1.0–740	Surface water	[27,29,67,68]
		Thiamethoxam	4100	−0.13	1.0–914	Surface water	
Steroid hormones	Natural	E2	13	2.45	3.8–188.0	Influent WWTP	[19,58,60,69]
		E1	13	3.43	12–196		
		17α-estradiol	13.3	4.01	6.4–12.6		
	Synthetic	EE2	4.8	3.67	0.59–5.6	Effluent WWTP	[23]
Industrial and chemical substances	Anticorrosive	Methylbenzotriazole	366	2.72	nd–2900	Effluent WWTP	[29]
	Plasticizer	BPA	120	2.2–3.4	0.01–8800 140–12,000	Bottled water Surface waster	[5,70]
Microplastics	-	-	-	-	8.3–3.1 × 10^5 * 0.27–30 * 0.05–320 *	Effluent WWTP Lake water River water	[45,49]

Notes: Units for data with * = million pieces/m^3.

3. Adsorbents Used to Remove ECs from Water

The most used adsorbent in the EC removal is activated carbon, due to its high efficiency (up to 100%), large specific surface area (300–2500 m^2/g), and hydrophobic interactions [71,72]. The use of alumina, soil, zeolites, clays, composites, metal-organic frameworks, and nanoparticles (carbon nanotubes, magnetic nanoparticles, graphene, etc.) has also been reported [1,35]. However, the widespread use of activated carbon and other adsorbents is limited by their high costs (2000–4000 to 20,000–22,000 USD/ton) [71,72]. This has promoted the search for alternative materials that are inexpensive, widely available, require little processing, and are environmentally friendly [17]. This section describes the most common conventional/non-conventional materials used for EC removal.

3.1. Conventional Materials

3.1.1. Activated Carbon

Activated carbon is the most used adsorbent material within WWTPs, with between 5.5 and 8.1% per year between 2008 and 2018, and 3311 thousand metric tons in 2021 [73]. Activated carbon has been recognized as the best adsorbent (adsorption capacity = 12–up to 7800 mg/g, efficiencies = 20–100%) for EC removal [22,71,72]. It has been used for the removal of diazinon, carbamazepine, diclofenac, ibuprofen, naproxen, ketoprofen, triclosan, p-Chloro-m-xylenol [71,72], acetaminophen, androstenedione, E1, E2, E3, EE2, progesterone, testosterone [71], metronidazole [74], paracetamol [75], and nimesulide [76]. Nevertheless, activated carbon is a costly adsorbent whose recovery is difficult and costly, and its efficiency decreases after regeneration (<40%) [72].

3.1.2. Zeolites

Zeolites are natural (Si/Al ratio = 1–5) and low-cost adsorbents (30–120 USD/ton) that are available worldwide [77]. Furthermore, they are synthesized at the laboratory scale, even from residues such as fly ash to obtain higher Si/Al ratios [77]. Zeolites with Si/Al >5 are more efficient in EC removal because they show a high level of hydrophobicity [58]. High hydrophobicity (lower water absorption) allows more zeolite pores to be available for the diffusion/adsorption of ECs. However, zeolites with high silica content are used only at the laboratory scale [78]. Zeolites such as Faujasite, Mordenite, Beta, and ZSM-5 were efficient in the removal of phenol, dichlorophenol [77], tetracycline, oxytetracycline [79], nitrobenzene, carbamazepine, nicotine, erythromycin, nitrosamines [78], and 2,4,6-trichlorophenol [58]. They reached adsorption capacities between 2.4 and 833 mg/g [58,78,80,81] and removal efficiencies from 45% to 90% [77–79,82,83].

3.1.3. Clays

Clays are widely used in contaminant removal due to their worldwide abundance and low cost (40–460 USD/ton), being at least 20 times cheaper than activated carbon. Clays were used efficiently (0.1–1900 mg/g) in the removal of ECs [71,84–87], such as amitriptyline, atenolol, metformin, atenolol, buspirone, ciprofloxacin, ranitidine, timolol [87,88], amoxicillin [71,87], diazepam, carbamazepine, clofibric acid, naproxen, salicylic acid, carbamazepine, diclofenac, ibuprofen, naproxen, phenol, atrazine, trimethoprim [86], methomyl [89], atenolol, sulfamethoxazole, and diclofenac [64]. Furthermore, clays have efficiencies like those of activated carbon (>98%) but achieved in longer times. This is because clays have smaller areas (~11 times) than activated carbon [72,86,87,90].

Some results of previous research about EC adsorption using conventional materials previously described are shown in Table 2. Conventional materials are characterized by a high surface area (10–2500 m^2/g), porosity (0.1–16.9 cm^3/g) [72,91,92], and the presence of several functional groups (OH, COOH, CO, NH$_2$, SiO$_2$, cations, etc.) [86,88], which gives them high efficiency (up to 100%). In some studies, efficiencies of conventional materials used in EC removal have been established to range from 20 to 100% [71,77–79]. However, the most efficient conventional adsorbent is still activated carbon [59]. Therefore, the adsorption parameters experience high variation, such as in EC type, concentration, material dose, pH, and contact time, with the exception of temperature, which is maintained between 20 and 30 °C [64,71,80,87,93]. The diversity in the parameters used in the adsorption processes can be associated with the researcher's aim, which is generally to optimize the process. The differences that exist between the adsorbents and the ECs used in the different studies influence optimization. Temperature is similar in most studies, except those that analyze the thermodynamics of the adsorption process. Because of increasing, it would not be efficient on a larger scale, due to the high costs that it could imply [52,59,67,81]. In turn, some authors have made physical/chemical modifications (e.g., thermal/acid-basic treatments) to the material in order to increase the EC removal efficiency or decrease the adsorbent dose/contact time [84,92].

Table 2. Conventional materials used for ECs removal from synthetic water in batch processes.

Adsorbent Characteristics				Adsorption Behavior				
Adsorbent	Composition/ Functional Groups/Ions	$SA [m^2/g]$ $P [cm^3/g]$	Adsorbent	EC Removed	Adsorption Conditions	Adsorption Mechanism	Removal [%]	Adsorption Capacity [mg/g]	Reference
Activated carbon (AC): Carbonized carbonaceous materials (S.O.) [71]	C, H, N, S and O [78] COOH, C=O, OH−, −NH$_2$, CHO, etc. [91]	SA = 300–2500 [78] P = 0.1–16.9 [91,93-95]	AC from argan waste	Paracetamol Amoxicillin	CT = 0.8–12, T = 25, Ph = 3–11, AD = 0.1, [EC] = 100	van der Waals forces, H-bonding, dipole–dipole interactions, ion exchange, covalent bonding, cation bridging and water bridging [16,63,79,93]	~95	512 319	[96]
			Commercial AC	Caffeine	CT = 4, T = 30 ± 2, pH = 3–8, AD = 0.1, [EC] = 1.125–1.252		44.1	71.7	[22]
				Ibuprofen			52.7	72.3	
				Triclosan			60.8	70.0	
			Commercial GAC	Tetracycline	CT = 180, T = 30 AD = 2.4, [EC] = 100		-	845.9	[91]
				Ibuprofen			-	239.8	
			AC from olive pomace	Niesulide	AD = 0.1–0.5, CT = 6, [EC] = 10–30, pH = 8–11		-	353.3	[76]
			Commercial GAC	Metronidazole	CT = 2, T = 20 ± 1 pH = 3.9–10.2, AD = 5 [EC] = 99.27		69.0–80.0	-	[74]
			Modified commercial AC	Triclosan	CT = 4, T = 30 ± 2 pH = 6, AD = 0.07, [EC] = 1.0		98.3	395.2	[24]
			AC from biomass	Paracetamol	CT = 240, T = 15–35, AD = 0.1, [EC] = 1–20		-	100	[75]
			AC from fruit of Butiacapitate	Paracetamol Ketoprofenon	T = 25–55, AD = 0.9, [EC] = 50–300 pH = 7		73.0–98.2	101.2–134.5	[97]
			GAC from coconut shells	Mix: Caffeine, hydro-chlorothiazide, saccharin, sucralose, sulfamethoxazole	T = 25, AD = 0.07, [EC] = 0.5, pH = 2–12, CT = up to 5.5		-	1.21–4.33	[37]

Table 2. Cont.

Adsorbent Characteristics			Adsorption Behavior					
Adsorbent	Composition/ Functional Groups/Ions	SA [m^2/g] P [cm^3/g]	Adsorbent	EC Removed	Adsorption Conditions	Adsorption Mechanism	Removal [%]	Adsorption Capacity [mg/g]	Reference
Zeolites: Crystalline microporous aluminosilicate (S.O. and N.O.) [72]	Si, Al, O and cations [72] K$^+$, Na$^+$, Ca^{2+}, and Mg^{2+} [92]	SA = 300–2300 [72] P = 0.10–0.35 [92]	Zeolite + Hexadecyltrimethylammonium bromide or chloride	Diclofenac	CT = 24, Ta = 32, Tb = 42, Tc = 52, pH = 2–11, AD = 25, [EC] = 636.28		-	Up to 47.4	[81]
			Natural zeolite	Tetracycline Oxytetracycli-ne	CT = 2, pH = 6, AD = 6.0, [EC] = 44.4–46.0		90 75	-	[79]
			Modified zeolite	Mesosulfuron-methyl	CT = 72, T = 25, pH = 6, AD = 2.0, [EC] = 8	van der Waals forces and acid-base forces, donor-acceptor interaction [60,66]	-	2.4–3.4	[80]
			Cu^{2+} zeolite 4A	Glyphosate	CT = 2, T = room, pH = 6–8, AD = 2.0, [EC] = 100		-	112.7	[33]
			Zeolite Y (ZY)	2,4-dichlorphenoxyacetic acid, paraquat	CT = 24, T = 28 ± 2, pH = 3, AD = 1.0, [EC] = 20–250		-	82.6–175.4 - 71.4–92.6	[82]
			Powdered zeolites: FAU 1/FAU 2	Azithromycin, ofloxacin, sulfamethoxazole	CT = 6, pH = 6.5, [EC] = 0.1		≥80	7.0–8.5 25.3–31.2 -	[83]

Table 2. *Cont.*

Adsorbent	Adsorbent Characteristics			Adsorption Behavior				
	Composition/ Functional Groups/Ions	SA [m²/g] P [cm³/g]	Adsorbent	EC Removed	Adsorption Conditions	Adsorption Mechanism	Removal [%]	Adsorption Capacity [mg/g]	Reference
Clays: Hydrous aluminosilicates (N.O.) [86,87,98]	SiO_2, CO_3^{2-}, MO_x [86,98]/ Ca^{2+}, Mg^{2+}, H^+, K^+, NH^{4+}, Na^+, SO_4^{2-}, Cl^-, PO_4^{3-}, NO_3^-, and OH^- [73,84]	SA: 10–426 [73,84] P: 0.1–0.5 [99]	Natural clay	Methomyl	CT = 3, T = 25 ± 1 pH = 6.6, AD = 10 [EC] = 19.99–43.71		27.6–32.9	0.3–0.5	[85]
			Bentonite	Atenolol, sulfamethoxazole, diclofenac	CT = 24, T = 20 pH = 6.5–8.0, AD = 1.6 [EC] = 1–50	Ion exchange, coordination, dipole–dipole interactions, van der Waals forces, H-bonding, acid-base forces, donor-acceptor interaction, acid-base forces [87]	-	5.3–24.5 1.1–3.5 1.1–4.0	[64]
			Sodium smectite	Tramadol, doxepin	CT = 16, T = 20 pH = 6–7, AD = - [EC] = 10		-	210.7 279.4	[88]
			Bentonite + Surfactant	Diuron and their byproducts	CT = 24, T = 18–20 pH = 3, AD = 0.5 [EC] = 0.25		-	0.1–0.6	[84]
			Organo-montmorillonites, organo-pillared-montmorillon, organo-acid-activated montmorillonite	2,4,5-trichlorophenol	CT = 2, T = 20 ± 1 pH = 4, AD = - [EC] = 10–200		-	374.9	[90]
			Natural/modified bentonite	Methomyl	CT = 2, T = 20, 30, 40 pH = 4, AD = 10 [EC] = 0.1		66–76	5	[89]

Notes: S.O. = synthetic origin, N.O. = natural origin, CT = contact time [h], T = temperature [°C], AD = adsorbent dose [g/L], [EC] = Emerging contaminant concentration [mg/L], SA = surface area, P = porosity.

3.2. Inorganic Non-Conventional Materials

The use of inorganic non-conventional materials is most common in the removal of dyes and heavy metals. For EC removal, it is preferred to use inorganic materials with acid/basic modifications or impregnated (metal oxides or other functional groups) since this increases their ability to remove them. The modifications can also reduce the contact time/dosage of the adsorbent, just like conventional materials [63,100]. The most widely used non-conventional inorganic materials are alumina, manganese oxides, and silica.

Alumina is an amphoteric oxide (pH < 7 = charge (+) and pH > 7 = charge (−)) [101] not commonly used in EC removal. However, there are some studies in which tetracycline [102], atrazine, norfloxacin, ciprofloxacin [103], propazine, prometryne, propachlor, propanil, molinate and phenols/chlorinated phenols [104,105], acrylic acid [106] and surfactants (sodium octyl sulfate, sodium decyl sulfate, sodium dodecyl sulfate, sodium tetradecyl sulfate) [107] were removed. The efficiency/adsorption capacities have reached values from 15% to 95% and 0.07 to 312.02 mg/g, respectively [104–109]. Meanwhile, manganese oxides (MnO_x) are considered efficient adsorbents due to their polymorphism, natural availability, easy synthesis at the industrial scale, low cost (50 USD/ton), and non-toxicity [100]. Some ECs, such as mercaptobenzothiazole, EE2, triclosan, tetracycline, endocrine disruptors, steroid estrogens, BPA, glyphosate, chlorophene, oxytetracycline [100], carbamazepine, niclosamide [110], clarithromycin and roxithromycin [111,112], diclofenac [111,113], and resorcinol [114], have been removed using MnO_x, reaching removal efficiencies between 40% and >90% [100,111,115]. Likewise, silica-based adsorbents are inexpensive and highly available material [102] used in the removal of ECs such as carbamazepine, clofibric acid, diclofenac, ibuprofen, ketoprofen, cloprop, norfloxacin, ciprofloxacin, delta-9-tetrahydrocannabinol [63,108], naproxen, phenols, cloprop(2-(3-chlorophenoxy) propionic acid, and dihydrocarbamazepine [108]. Silica achieves removal efficiencies from 0% to >90% with adsorption capacities from 2.1 to 429.0 mg/g [63,108,116].

Table 3 summarizes previous studies using inorganic materials for the removal of ECs. Inorganic materials are less used in EC removal, where fewer previous studies were found probably due to their lower adsorption capacity (0.16–52.00 mg/g). This could be associated with the lower surface area (even 12 m^2/g), porosity (0.53–1.03), and low variety of functional groups. Like adsorption with conventional materials, there are also very varied parameters, but the temperature has been kept in a range between 19 °C and 30 °C.

Table 3. Inorganic non-conventional materials in ECs removal using synthetic water in batch processes.

Adsorbent Characteristics				Adsorption Behavior					
Adsorbent	Composition/ Functional Groups/Ions	SA[m²/g] P [cm³/g]	Adsorbent	EC Removed	Adsorption Conditions	Adsorption Mechanism	Removal [%]	Adsorption Capacity [mg/g]	Reference
Manganese oxide: N.O. and S.O. [111]	Mn(II), Mn(III), Mn(IV) or Mn(VII) [100,111]/ Mn–O, OH⁻ [115]	SA: 12–236 [111]	Manganese oxide	Diclofenac	CT = 33, T = 30, pH = 7.0, AD = 0.61 [EC] = 49,946.41	Electrostatic, cation-exchange interactions, and van der Waals forces [113]	Up to 90 ± 0.7	-	[113]
			Manganese oxide	Phenol	CT = 1, T = 19, pH = 6.7 ± 0.5, AD = 10.0 [EC] = 0.5		~40	-	[114]
			Manganese oxide	Clarithromycin Roxithromycin	CT = 24, T = 20, pH = 5.0, AD = 1.0 [EC] = 4.94–83.70		85–90	-	[112]
			Manganese oxide + Fe	Resorcinol	CT = 0.33–2.0, pH = 5.0, AD = 0.0157 [EC] = 35.0		~100	-	[114]
			Manganese oxide birnessite	Niclosamide	CT = 24, T = 22 ± 2, pH = 5.0, AD = 0.015 [EC] = 0.042		-	-	[110]

Table 3. Cont.

Adsorbent	Adsorbent Characteristics			EC Removed	Adsorption Behavior				
	Composition/ Functional Groups/Ions	SA [m^2/g] P [cm^3/g]	Adsorbent		Adsorption Conditions	Adsorption Mechanism	Removal [%]	Adsorption Capacity [mg/g]	Reference
Alumina: N.O. and S.O. [105,117]	MO$_x$, FeO, SO$_4^{2-}$, Al$_2$O$_3$ (α, β, γ) [101]/ O-Al-O, OH$^-$ [115]	AS: 50–300 [101,118]	Alumina and surfactant modified alumina	Ortho-Nitro-Phenol	CT = 1, pH = 6.0, AD = 5.0 [EC] = 55.64		-	4.4–7.3	[104]
			α-Alumina nanoparticles and Modified α-Alumina nanoparticles	Ciprofloxacin	CT = 1.5, T = 25 ± 2, pH = 6, AD = 5, [EC] = 10	Electrostatic interactions and hydrogen bonding [103,119]	33.6–97.8	34.5	[103]
			MCM-41	Norfloxacin	CT = 1, T = 15, pH = 3.0–7.0, AD = 0.5 [EC] = 60.0		>80	52	[106]
			Raw alumina and Raw alumina modified with HCl	Acrylic acid	CT = 0.5–120, T = 20–50, pH = 4.5, AD = 66.7, [EC] = 0.08		32.1–36.2	0.29–0.31	[109]
			Alumina and HDTMA modified alumina	Metha-nitrophenol	CT = 1, T = 25–45, pH = 6.0, AD = 5.0 [EC] = 0.4		-	3.0–8.1	[104]
Silica: N.O. and S.O. [63,108]	SiO$_2^-$ [63]/ Si-O-H, OH$^-$ [108]	SA: 7.5–up to 1000 [108,116] P: 0.53–1.03 [109]	Silica	Tetracycline	T = 23, pH = 6.0, AD = - [EC] = -	H bonds or cationic exchange [108,109]	-	-	[102]
			Mesoporous silica SBA-15	Carbamazepine Diclofenac Ibuprofen Ketoprofen Clofibric acid	CT = 2, T = 25, pHa-d = 5, pHe = 3, AD = 2.0 [EC] = 0.1		35.6–94.3	0.16 0.34 0.41 0.28 0.07	[108]

Notes: S.O. = synthetic origin, N.O. = natural origin, CT = contact time [h], T = temperature [°C], AD = adsorbent dose [g/L], [EC] = emerging contaminant concentration [mg/L], SA = surface area, P = porosity.

3.3. Organic Non-Conventional Materials

3.3.1. Chitosan

Chitosan is a highly available (1818 tons/year) and non-porous material from a deacetylated derivative of chitin (1362×10^6 tons/year), which is the second most abundant natural organic resource [119]. Chitosan is a relatively low-cost (2200–4400 USD/ton) adsorbent that is non-toxic, biocompatible, biodegradable, and reactive, and that can be easily modified through physical/chemical methods [120–122]. Due to its low adsorption capacity, which is associated with its crystalline and swellable nature, low porosity, hydrophilicity, surface area, and stability in acid media, chitosan is used in composites within wastewater treatment [119]. Chitosan has been used in the removal of phenols [122,123], pramipexole [124,125], alkylbenzene sulfonate, caffeine, sulpiride, bezafibrate [122], sulfamethoxazole, BPA [126], 2,4-dichlorophenol and 2,4,6-trichlorophenol, clofibric acid, tannic acid, and alkylbenzene sulfonate [127]. This material has reached adsorption capacities from 27 to >1500 mg/g and adsorption efficiencies between 11% and 96% [126–128].

3.3.2. Peat

Peat is a complex product of soil and organic matter decomposition, being available throughout the world from a few meters to tens of meters. Thus, it is a low-cost material (40 USD/ton) [120]. Moreover, peat could compete industrially with adsorbents, such as activated carbon and zeolite, due to their high cation exchange capacity. Peat has been used efficiently (>70%, and 1.71–31.40 mg/g) [8,120,128,129] in the removal of sulfamethoxazole, sulfapyridine [130], metolachlor, phenol, p-chlorophenol [120], BPA [129], p-nitrophenol, tri (n-butyl) phosphate, tris (2-butoxyethyl) phosphate, and tris (2-chloroethyl) phosphate [8].

3.3.3. Agricultural/Agro-Industrial Residues

Agricultural/agro-industrial residues, such as fruit peels and seeds, and husk and shells of legume and cereal, are favorable for EC removal. These materials show a high availability, and chemical stability with the presence of lignin (20–30%), cellulose (35–50%), and hemicellulose (15–30%) [55,131]. They have a renewable nature, and low or no cost (e.g., rice husk between 1.6 and 2.7 USD/ton) [131,132]. Moreover, they are environmentally friendly, require little processing (washing, drying, grinding, and sieving), and add value to material that is generally not used to remove contaminants [16]. Agro-industrial residues have been reported to remove acetaminophen, atenolol, caffeine, carbamazepine, diclofenac, ibuprofen, sulfamethoxazole, tetracycline, levofloxacin, ciprofloxacin, atrazine, clofibric acid [17], oxytetracycline, florfenicol [133], norfloxacin [16], and phenolic compounds [95,134–136]. Their removal efficiencies and adsorption capacities have achieved values between 60% and >90% and from 0.37 to 689 mg/g, respectively [16,17,134,137].

3.3.4. Biochar

Biochar exhibits some advantages over common activated carbon, such as high availability, low cost of renewable raw material (350–1200 USD/ton), high surface density of polar functional groups, and condensed structure [138]. Additionally, raw/modified biochar has more environmental applications than activated carbon, e.g., it behaves as a soil amendment and carbon sequestration agent in fuel cells and supercapacitors [138,139]. Biochar is widely used in EC removal at the lab level, achieving similar efficiencies (~95%) and adsorption capacities (>640 mg/g) to those of activated carbon [139–141]. Sulfamethoxazole, sulfathiazole, sulfamethazine, ibuprofen, triclosan, diclofenac, imidacloprid, atrazine, dibutyl phthalate, dimethyl phthalate [141], tetracyclinbe [142,143], benzophene, benzotriazole [144], BPA, E2 [141,144], carbamazepine, metolachlor, EE2, propranolol, phenols, and somepesticides are among the ECs removed [140].

3.3.5. Industrial Waste

Fly ash (combustion residues), red mud (aluminum industry residues), slag (steel industry residues), slurry, and sewage are industrial waste [77]. The management of these residues can be costly. For instance, sludge treatment represents between 25% and 65% of the water treatment costs [77], making it a problem for municipalities/industries due to the contaminant loading and huge production of them (e.g., municipal sewage sludge from 27 countries of the European Union ~13 million tons/year for 2020) [145]. A possible use for these residues is EC removal, since they are considered good adsorbents due to their characteristics (Table 4) and low cost (20–100 USD/ton) [77,146]. Industrial waste has been used to remove tetracycline [147], sulfamethoxazole, trimethoprim [148,149], dichlorophenoxyacetic acid, phenols [77], carbamazepine [149], BPA, 17-beta-estradiole, 17-alpha-ethinylestradiol, sulfamethyldiazine, sulfamethazine, sulfathiazole, fluoxetine, ibuprofen, carbofuran [145,150], and nimesulide [151]. Removal efficiencies have oscillated from 2% to >90%, while adsorption capacities are between 0.6 and 212 mg/g [77,145,149,152].

3.3.6. Polymeric Adsorbents

Polymeric adsorbents are easily regenerated (soft washing) and their surface could be modified (polar/non-polar) to remove specific contaminants [63]. Furthermore, polymeric adsorbents have lower costs than activated carbon (up to four times). They are produced in a sustainable way and have higher adsorption (15–200 times faster) than activated carbon [63,153]. ECs, such as cephalosporin C, penicillin V, delta-9-tetrahydrocannabinol, nalidixic acid [63], ibuprofen, cephalexin, caffeine, phenols, cefadroxil, erythromycin, BPA, alachlor, trifluralin, prometryn, amitrole [154], diclofenac, and BPA [153,155], have been removed using polymeric adsorbents. In their removal, polymers, such as β-cyclodextrin polymer [153], post-cross-linked polystyrene/poly (methyl acryloyl diethylenetriamine) [156], polystyrene, polyacrylic ester, polyacrylamide, resins (Amberlite XAD-16, XAD-4, XAD-2, XAD-7), and polymer-based inorganic hybrids (polymeric matrix + inorganic nanoparticles), have been used. They have achieved adsorption capacities between 22.2 and 1401 mg/g and removals higher than 90% [63,153,154].

Table 4 summarizes the performance of non-conventional organic materials used as adsorbents for EC removal. Non-conventional organic materials have been widely used in the ECs removal because some of them have a comparable efficiency with conventional materials (up to 100%). This is because they have comparable physical and chemical characteristics, except for their smaller surface area compared to conventional materials. Moreover, non-conventional organic materials exhibit other advantages, such as high availability and low/null cost, and even their use can avoid final disposal problems since some of these materials are residues [55]. Therefore, the use of materials such as these can be an alternative for the treatment of wastewater in developing countries [157].

Table 4. Organic non-conventional materials used for EC removal from synthetic water in batch processes.

Adsorbent	Adsorbent Characteristics		Adsorbent	EC Removed	Adsorption Conditions	Adsorption Behavior			
	Composition/ Functional Groups/Ions	SA[m²/g] P [cm³/g]				Adsorption Mechanism	Removal [%]	Adsorption Capacity [mg/g]	Reference
Biochar: S.O. [139]	C, H, O, N [139]/ COOH, C=O, OH−, RCOOR, C₆H₅OH, (RCO)²O etc. [158]	AS: 2.46–1500 [139] P: 0.21–0.95 [144,158]	Methanol-modified biochar from rice husk	Tetracycline	CT = 0.33, T = 30, pH = 2.0, AD = 1.0 [EC] = 100	Ion exchange, electrostatic, hydrophobic, pore-filling, and bridging interactions [144]	-	~80	[143]
			Modified biochar + *Spiruline sp.*	Tetracycline	CT = 120, T = 20–40, pH = 3–9, AD = 0.1 [EC] = 100		-	132	[142]
Chitosan: S.O. [131]	α-D-glucosamine [159,160]/ OH⁻, -NH₂ [131]	AS: 1.1–23.4 [122,127] P: 0.002–0.7 [122,127]	Chitosan and Modified chitosan with 2-hydroxy-1-naphthaldehyde	Phenol 2-chlorophenol 4-chlorophenol 2,4-dichlorophenol 2,4,6-trichlorophenol	CT = 3, T = 20, pH = 7.0, AD = 1.0 [EC] = 150		-	59.7 70.5 96.4 315.5 375.9	[122]
			Chitosan grafted with sulfonic acid and cross-linked with glutaraldehyde	Pramipexole	CT = 24, T = 25, pH = 10, [EC] = 0–500	Electrostatic and hydrophobic interactions and hydrogen bonding [131]	-	181	[125]
			Sulfonate-grafted chitosan	Pramipexole	CT = 0–24, T = 25, 45, 65, pH = 2–12, AD = 1.0 [EC] = 0–500		11–82	181–367	[124]
			Magnetic modified chitosan	Phenol BPA	CT = 0.67, T = 45, pH = 4.5, AD = 0.6 [EC] = 376–913		96 85.5	-	[123]
Peat: N.O. [71]	Cellulose, lignin, humic acid, fulvic acid [127,131]/ alcohols, aldehydes, ketones, carboxylic acids, and phenolic hydroxides [71,131]	AS: 0.9–>200 [71] P: 70–95% [71]	Fibric peat modified.	BPA	CT = 4, T = 25, pH = 6.9, AD = 0.05, [EC] = 45.0	Hydrophobic interactions, hydrogen bonding [128,129]	-	31.4	[128]
			Raw/modified peat	BPA	CT = 4, T = 25, pH = 7.0, AD = 1.0 [EC] = 2.0		80	1.7	[129]
			Commercial peat soil	Sulfamethoxazole, sulfapyridine	CT = 168, T = 25 ± 3, AD = 20.0, pH = 4.4–9.5 [EC] = 0.15–13.46		-	Up to 4.05 Up to 0.40	[130]

Table 4. Cont.

Adsorbent	Adsorbent Characteristics			Adsorption Behavior				Reference
	Composition/ Functional Groups/Ions	SA [m^2/g] P [cm^3/g]	Adsorbent	EC Removed	Adsorption Conditions	Adsorption Mechanism	Removal [%]	Adsorption Capacity [mg/g]	
Agro-industrial residues: N.O. [131]	Cellulose (35–50%), lignin (20–30%), hemicellulose (15–30%), and pectin, among others [131]/ OH$^-$, COH, COOH, ROR, RCO-H, RCOR, C$_6$H$_5$OH, etc. [16]	SA: 0.034–120 [131,134] P: 0.03–4.8 [17,134]	Rice husk	Norfloxacin	CT = 1, T = 25, pH = 6.2, AD= 3.0 [EC] = 5		96.95	20.1	[16]
			Coffee husk				99.66	33.6	
			Rice husk	E1 + E2 + E3	CT = 4, T = 25, AD = 12, [EC] = 3.5–7		45–90	1.0–2.7	[69]
			Pinecone + Pig manure (BCP)	Sulfamethazine, ciprofloxacin, oxytetracycline, florfenicol	CT = 48, T = 20, pH = 7.6–8.5, AD = 1.0 w/w %, [EC] = 1.2		-	-	[133]
			Mansonia wood sawdust	4-nitrophenol	CT = 2.5, T = 26 ± 4, pH = 4.0, AD = 1.5 [EC] = 120	Π-π interactions, the formation of H-, -COOH, and C=O bonds, and electrostatic interactions [29,160,161]	22.5–55.5	7.4–18.0	[136]
			Granulated cork	Phenols Phenol, 2-chlorophenol, 2-nitrophenol, 2,4-dichlorophenol, Pentachlorophenol PPCPs Carbamazepine, naproxen, ketoprofen, diclofenac, triclosan, methyl paraben	CT = 0.5, T = 20 ± 2, pH = 6.0, AD = 5–20 [EC](phenols) = 30.0, [EC](PPCPs) = 1.0		Phenols 20–100 PPCPs 50–100	Phenols 0.6–1.6 PPCPs 1.8–3.6	[135]
			Charred sawdust of sheesham	Phenol	CT = 1.5, T = 25 ± 2, pH = 2–6, AD = 0.1–10 [EC] = 10–1000		>95.0	300.6–337.5	[95]
			Sawdust from Finland wood	Phenol	CT = 3, T = 22, pH = 5.79, AD = 10 [EC] = 20		Up to 70.4	Up to 5.5	[137]

Table 4. Cont.

Adsorbent	Adsorbent Characteristics			Adsorption Behavior					
	Composition/ Functional Groups/Ions	SA [m²/g] P [cm³/g]	Adsorbent	EC Removed	Adsorption Conditions	Adsorption Mechanism	Removal [%]	Adsorption Capacity [mg/g]	Reference
Industrial residues: N.O. [145,148]	SiO_2, Fe_2O_3, Al_2O_3, CaO, MgO, organic compounds [148] OH^-, $C=O$, COOH, lactone, quinone, etc. [148,162]	Raw: 2–34 Treated or activated: up to 1800. [145,149,161] Treated or activated: 0.098–0.145 [149,161]	Fertilizer and steel industry wastes	2,4-dichlorophenoxyacetic acid, carbofuran	CT = 1.67, T = 25, pH = 6.5–7.5, AD = 1.0 [EC] = 88.01–132.76		-	212 208	[150]
			Sewage Sludge and fish waste	Carbamazepine+ sulfamethoxazole+ trimethoprim	CT = 5, T = 30, pH = 9.39–11.82, AD = 5.0 [EC] = 100 (Of each one)		-	Up to 41.3 Up to 3.8 Up to 13.6	[149]
			Sewage Sludge and fish waste	Trimethoprim, sulfamethoxazole	CT = 180, T = 30, pH = 4.53–7.64, AD = 5.0, [EC] = 100		-	90.0 5.3	[148]
			Paper pulp-based adsorbents + H_3PO_4	Carbamazepine, sulfamethoxazole	T = 25, pH = 7.5, [EC] = 5	Polar interaction, hydrogen bonding, π–π electron-donor-acceptor, acid–base interactions [133,136,137]	-	92 ± 19 13 ± 0.6	[163]
			Phosphoric acid activated corn porous straw carbon	Tetracycline	CT = 12, T = 30, pH = 4, AD = 0.2–1.0 [EC] = 50		8.03–97.0	227.3	[147]
			Activated carbon. from gas mask	Triclosan	CT = 10, T = 24 ± 1, pH = -, AD = 1 × 10⁻³ [EC] = 10–400		2–100	85	[152]
			Carbon slurry waste	2-bromophenol, 4-bromophenol, 2,4-dibromophenol	CT = 11, T = 25 ± 2, pH = 5.8–6.8, AD = 1.0 [EC] = 69.2–100.76		-	40.7 170.4 190.2	[162]
			Activated carbon from sewage sludge	Nimesulide	CT = 2, pH = 8, AD = 1.0		98.6	275.0	[151]
Polymeric adsorbents: S.O. [63,152]	Depends on the polymer [63,154] Tertiary amino, Carboxyl, sulfonic acid, dicyandiamide, amidocyanogen, polyethylene glycol, and 2-Carboxy-3/4-nitrobenzoyl, among others [63]	SA: 15–1000 [63,152] P: 0.16–1.22 [63,152]	Post-cross-linked polystyrene/poly (methyl acryloyl diethylenetriamine)	Phenol, benzoic acid, P-hydroxybenzoic acid	CT = 2.0, T = 25, 35, 45, AD = 2.0 [EC] = 507.3	Hydrophobic interaction, π–π interaction, hydrogen attraction, ionic bonding; complex formation [63,154]	>90	50 190 242.1	[155]
			Hyper-crosslinked β-cyclodextrin	BPA	CT = 0.5–12, T = 25, AD = 0.25, [EC] = 40		94.45	278	[156]
			Hyper-crosslinked β-cyclodextrin porous polymer	BPA	CT = 0.17, T = room, AD = 1 [EC] = 22.8		80–95	22.2	
			Molecularly Imprinted Polymer	Diclofenac	CT = 1, T = room temperature, pH = 7, AD = 5 [EC] = 1–25		100	160	[164]

Notes: S.O. = synthetic origin, N.O. = natural origin, CT = contact time [h], T = temperature [°C], AD = adsorbent dose [g/L], [EC] = emerging contaminant concentration [mg/L], SA = surface area, P = porosity.

3.4. Nanomaterials

Nanomaterials have at least one size dimension between 1 and 100 nm. In recent years, nanomaterials (nanoparticles, nanotubes, nanofilms, and nanowires) have attracted a lot of interest as wastewater treatment applications, mainly in adsorption and photocatalysis at the lab scale [165,166]. They are called new-generation adsorbents due to their high performance, large surface area (up to 3200 m^2/g), appropriate dispersibility, catalytic potential, large surface energy, abundant reactive sites, rapid dissolution, high reactivity, and free surface energy (>surface reactivity) [166]. Nevertheless, the recovery of non-magnetic nanoparticles after the EC adsorption is still complicated and some of them have been found to be toxic [167]. Therefore, adequate techniques must be used to separate the nanoparticles from the aqueous solution to take advantage of the benefits that these materials offer.

EC removal efficiencies close to 100% have been achieved using adsorbent doses smaller than micro-adsorbents [168]. Chemical modifications in nanoparticles have been conducted to improve their adsorption capacities [169]. Nevertheless, no study has used modified materials to explain whether or not modifications are sustainable processes to improve EC removal. Thus, there is not enough information about the residues produced after the modification or to verify if the higher efficiency obtained is comparable with the raw material and is costly representative [55,168]. Furthermore, all studies have been carried out at the lab scale, and none of them has analyzed the feasibility of full-scale synthesis of nanomaterials [168]. Carbon-based nanoparticles, metal oxides, metal nanoparticles, and nanocomposites are among the most used nanomaterials in the EC removal. Some of these nanomaterials are described below.

3.4.1. Carbon-Based Nanomaterials

Carbon-based nanomaterials are the most used to remove contaminants from wastewater, due to their good adsorptive characteristics for the removal of organic/inorganic contaminants and microorganisms [62].

a. Carbon Nanotubes (CNTs)

CNTs are the most anisotropic materials available. They are formed by hollow and layered structures with a length from nanometers to millimeters [170], and can be single wall (SWCNT, cylindrical) or multiple wall (MWCNT, concentric cylinders), with diameters from 0.4–2.0 to 2.0–25.0 nm, respectively [170,171]. CNTs have chemically inert surfaces that promote physical adsorption. However, their surface could be modified by incorporating heteroatoms, thereby increasing their affinity with different ECs, the selectivity of adsorption through families of compounds, and improving its performance in the desorption process [62]. CNTs have been used to remove ECs such as pharmaceuticals (tetracycline, oxytetracycline, sulfamethoxazole, sulfapyridine, sulfamethazine, ciprofloxacin, norfloxacin, ofloxacin, lincomycin, caffeine, etc.) [62,93,166], and personal care products (triclosan) [166], hormones (EE2), phenols, pesticides (atrazine, diuron, dichlobenil, isoproturon) [72,172], and roxarsone [173]. These materials have achieved removals between 67.5% and 99.8% and adsorption capacities between 8.6 and 554.0 mg/g [72,93,166,173,174].

b. Graphene

Graphene is a new material which is considered the thinnest that exists (~0.03 nm). As a nano adsorbent, it includes three forms: pristine graphene, graphene oxide (GO), and reduced graphene oxide (rGO) [175]. Moreover, graphene has faster diffusion or surface reactions of antibiotics, which allows rapid and effective adsorption, and is cheaper at full-scale production than other adsorbents such as CNTs [176,177]. Nevertheless, the hydrophobicity and limited dispersibility of graphene decrease its potential for adsorbent purposes [178]. Graphene is considered a good adsorbent (large delocalized π-electron system and tunable chemical properties) [93,176] for the ECs removal (100% and 19–3710 mg/g) [166,176,179,180]. Graphene has been reported for its use for removal of tetrabromobisphenol A, bisphenol A, phenol [9], EE2, E2 [181], diclofenac, levofloxacin, metformin, nimesulide, sulfamethoxazole, efalexin, ofloxacin, amoxilin, tetracycline, ciprofloxacin [176], atenolol, clofibric acid, aminoglycoside, β-lactams, glycopeptide, macrolide, quinolone, sulphonamide [177], acetaminophen [181], metformin [182], and nicotine [183].

3.4.2. Metal-Based Nano Adsorbents

Metal-based nano adsorbents are characterized by their high efficiency for contaminant removal, relatively low cost, and short distance of intraparticle diffusion. Moreover, they can be compressed without changes in their surface area, and are resistant to abrasion, magnetic, and photocatalytic [167]. Additionally, metal oxide nanomaterials are environmentally friendly. Thus, these nanomaterials (iron oxides, aluminum, manganese, titanium, magnesium, and cerium oxides) could be used combined with living beings in filter media, slurry reactors, powders, and pellets [167,184].

Iron oxide nanoparticles are characterized by their relatively easy synthesis, magnetism, recycling capability, relative low cost (e.g., zero-valent iron 1.66 USD/m^3 of treated water), fast kinetics, and biocompatibility [185–187]. The iron nanoparticles most used in the adsorption processes for EC removal is zero-valent iron. However, magnetite (Fe_3O_4), maghemite (γ-Fe_2O_3), hematite (α-Fe_2O_3), and goethite (α-FeOOH) are also used, but mainly as composites and in degradation processes [167,186].

Zero-valent iron (ZVI) is characterized by having an iron oxide envelope and a Fe^0 core, which exhibits a core shell structure. ZVI shows a high reactivity, higher than conventional granular iron, dual properties of adsorption and reduction (-0.44 V), and easy dispersion [187]. The removal of amoxicillin, norfloxacin, ampicillin, ciprofloxacin, chloramphenicol, dichloroacetamide, metronidazole, diazepam, tetracycline, oxytetracycline, and cytostatic drugs using ZVI has reported values between 50% and 100% [50]. To improve the capacity of zero-valent iron to remove ECs such as florfenicol, tetracycline, metronidazole, and enrofloxacin, chemical modifications were made, achieving removal efficiencies between more than 90% and 100% [50]. ZVI has also been used in continuous adsorption processes for the removal of pharmaceuticals (carbamazepine, caffeine, sulfamethoxazole, 3,4-methylenedioxyamphetamine, 3,4-methylenedioxymethamphetamine, ibuprofen, gemfibrozil, and naproxen) and sweeteners (acesulfame-K and sucralose), reaching efficiencies greater than 97% and 76%, respectively [188]. However, ZVI presents some disadvantages associated with its very short lifetime since it needs stabilization/surface modification and forms clumps due to the van der Waals and magnetic forces [50].

Magnetite is commonly used in the adsorption of contaminants (mainly heavy metals) due to its low cost, environmentally friendly nature, and the possibility of treating large volumes of wastewater [167]. Magnetite was used in the removal of ECs such as levofloxacin [189], caffeine [167], nalidixic acid, salicylic acid, flumequine, benzotriazole [190], paracetamol, ciprofloxacin, diclofenac, and oxytetracycline [191], reporting removal rates between 36.7% and 100.0%, and adsorption capacities from 6.1 to 100.0 mg/g [189–191]. Maghemite is a low-cost material that is highly available. It has been used in the removal of diclofenac, achieving adsorption capacities between 120 and 261 mg/g [192]. Magnetite and maghemite are characterized by their magnetism, which facilitates their separation after the adsorption process [184]. However, these nanoparticles are very unstable because they can change to other phases, and therefore they need to be stabilized before their use. To avoid this problem, these nanoparticles are employed mainly in composites form [191,192].

The other iron oxides used in EC removal are goethite and hematite. Goethite is not magnetic, so its separation from treated wastewater is more complicated than that of magnetite and maghemite [193]. Goethite is an abundant hydrated iron oxide from soils that is used as a model adsorbent due to its thermodynamic stability at room temperature; PPCPs adsorption has also been studied in iron-rich soils [193,194]. Thus, ECs such as levofloxacin [194], diclofenac, ibuprofen [166,195], tetracycline, and flumequine [166] were removed using goethite, showing removal efficiencies between 25% and 90% and adsorption capacities from 0.025 to 0.72 mg/g [166,194,195]. Hematite is the most stable phase of magnetic iron oxide [185], being used in the removal of cephalexin [196] and carbamazepine [197], and achieving removal efficiencies higher than 90% with adsorption capacities between 2.8 and 70.0 mg/g [196,197].

There are other metal oxides, such as aluminum oxides, zinc oxides, magnesium oxides, and cerium oxides, that are used in contaminant removal. These oxides are characterized by their low cost, thermal stability, easy synthesis and regeneration, surface reactivity, and versatility, among others [118,198,199]. Nano-alumina (dichlorodiphenyltrichloroethane, polychlorinated biphenyls, ciprofloxacin) [103,199], nano-silica (ciprofloxacin) [168], zinc oxide (naphthalene) [31], and magnesium oxide (linezolid) [199] are some nano-oxides used in EC removal that have reached removal efficiencies between 55% and 100% [31,168,198,199].

3.4.3. Nanocomposites

A nanocomposite is a multiphase material where one or more materials are deposited on a support material. The most used support materials are polymers, graphene, zeolites, biochar, clay, CNT, activated carbon, silica, biopolymers (chitosan/cellulose/alginate), membranes, and magnetic substrates [200–202]. Nanocomposites are used in some environmental applications, including the removal of contaminants from wastewater (e.g., heavy metals, ECs, dyes) [202], since they exhibit better characteristics than many adsorbents, including activated carbon. Some properties that are improved when forming nanocomposites and make them better adsorbents are selectivity, stability (mechanical/chemical), porosity, separation of the aqueous medium (magnetism), reduction of adsorption time/adsorbent dose, and cost, among others [200,201].

Another advantage of nanocomposites is their photocatalytic properties. In the EC adsorption processes, these contaminants are only transferred from the aqueous medium to the adsorbent, but they are not degraded [167]. Removing the ECs from the adsorbent to reuse it could be a difficult and expensive process. Therefore, combining adsorption with photocatalysis could be efficient because the ECs will be degraded into less toxic substances [8]. However, nanocomposites/nanoparticles can also have disadvantages, such as a complex synthesis, probably using toxic substances [203]. Non-magnetic nanomaterials are difficult to separate from the aqueous medium. Even previous studies include nanomaterials within ECs [204]. Furthermore, the use of nanomaterials on a large scale is not possible yet.

Table 5 summarizes some studies about the use of nanomaterials for EC removal. Iron oxide nanoparticles, their composites (magnetic character), graphene, and their composites are the nanomaterials commonly used in EC removal. The former is useful for separating the adsorbent from the aqueous medium (magnetic separation), while the latter and its composites can be associated with the high efficiency of the materials; in fact, they have even been used in the adsorption of microplastics. ECs such as tetracycline, sulfachloropyradazine [205], ametryn, prometryn, simazine, simeton and atrazine [203], metolachlor, BPA, tonalide, triclosan, ketoprofen, estriol [206], linezolid [198], carbamazepine, ibuprofen, clofibric acid [57,207], ciprofloxacin, erythromycin, amoxicillin [200,208], diclofenac [209], and sulfamethoxazole [210] and microplastics such as polystyrene [211,212] have been removed using different nanomaterials. They have achieved removal efficiencies between 25% and 100% with adsorption capacities between 0.9 and 3070.0 mg/g [61,121,177,207,213–217]. The application of nanomaterials in adsorption processes has been suggested as a promising alternative, due to adsorption capacities comparable, and even higher, than conventional materials. Indeed, in some studies, the adsorbent doses are lower than other doses. However, the costs and commercial production of nanomaterials are not discussed in these studies [218].

Table 5. Nanomaterials used in ECs removal using synthetic water in batch processes.

Adsorbent	Adsorbent Characteristics			Adsorption Behavior					
	Composition/ Functional Groups/ions	SA [m²/g] P [cm³/g]	Adsorbent	EC Removed	Adsorption Conditions	Adsorption Mechanism	Removal [%]	Adsorption Capacity [mg/g]	Reference
CNT: S.O. [170,171]	Graphene or graphite sheet with π conjugative structure and highly hydrophobic surface [170]/ –OH, –C=O, –COOH [72]	SWCNT SA: 400–1020 [62,93] MWCNTs SA: 38.7–>500 [93] P = 0.59 [93]	*SWCNT	17α-ethinyl estradiol BPA	CT = 4, pH = 3.5–11, T = room, AD = 0.05 [EC] = 2.28–2.96	Hydrophobic effect, π-π interactions, hydrogen bonding, covalent bonding, and electrostatic interactions [165,166]	95–98 75–80	35.5–35.7 13.4–16.1	[171]
			SWCNT MWCNT	Lincomycine, Sulfamethoxazole, iopromide	CT = 72, T = 20 ± 1, pH = 6.0 ± 0.2, AD = – [EC] = 12,000		-	-	[62]
			MWCNT	Roxarsone	CT = –, T = 10, pH = 2–11.7, AD = 2 [EC] = 10		-	Up to 13.5	[172]
			MWCNT MWCNT-COOH MWCNT-NH₂ N-CNT	1,8dichlorooctane, nalidixic acid, 2-(4-ethylphenoxy) ethanol	CT = 72, T = 25, pH = 7.0 ± 0.5, AD = 0.02–0.2 [EC] = 20–80		-	248–380 79–111–	[173]
			MWCNT modified with HNO₃	Diclofenac	CT = 1, T = 25, pH = 7.0, AD = 5.4, [EC] = 50		Up to 95	Up to 8.6	[174]
Graphene: S.O. [180]	2D single layer of sp² hybridized carbon atom [180]/ Epoxide, carbonyl, carboxyl, and hydroxyl groups [180]	SA: 46.4–2630 (theoretical) [179,180] P: 0.065 [219,220]	Graphene oxide	β-estradiol 17α-ethynyl estradiol	CT = 0.83, T = 25, pH = 3.0, AD = 0.40 [EC] = 8.0	Hydrophobic effect, π-π interactions, hydrogen bonding, covalent bonding, and electrostatic interactions [153,180]	97.2 98.5	- -	[180]
			Graphene oxide	Diclofenac	CT = 0.25, T = 60, pH = 6.0, AD = 0.16 [EC] = 400		96.2	653.9	[175]
			Graphene oxide	TetrabromoBPA BPA	CT = 24, T = 15–35, pH = 6.0, AD = 1.25 [EC] = 20		-	19.1 17.5	[9]
			Graphene oxide	Metformin	CT = 1–3, T = 1545, pH = 4.5–8.5, AD = 0.05–0.15 [EC] = 300–700		59–97.6	122.6	[182]
			Graphene oxide	Nicotine	CT = 0.5, T = 25–55 pH = 3–10.5 AD = 0.1 [EC] = 5–150		-	96.5	[183]
			Double-oxidized graphene oxide	Acetaminophen	CT = 0.17, T = 25, pH = 8.0, AD = 0.02 [EC] = 10		83.7	704	[181]
			Graphene oxide nanoflakes	BPA, 4-nonylphenol, tetrabromineBPA	CT = 0.08–24, T = 25, pH = 4–9, AD = 1.25, [EC] = 20		-	19–30	[220]
			3D Graphene oxide	Polystyrene microplastics	CT = 2, T = 76, pH = ~7, AD = 1.25, [EC] = 600		-	617.3	[212]

Table 5. Cont.

Adsorbent	Adsorbent Characteristics			Adsorption Behavior				
	Composition/ Functional Groups/ions	SA [m²/g] P [cm³/g]	EC Removed	Adsorbent	Adsorption Conditions	Adsorption Mechanism	Removal [%]	Adsorption Capacity [mg/g]	Reference
Iron oxides: N.O. and S.O. [192,193]	FeO, [193,221]/ FeO, OH-, COOH-, C=O [193]	SA: ZVI > 10 Hematite ~30 Goethite 18–83.5 Magnetite 40–300 Maghemite 31–178 [40,167,192,193] P: Magnetite ~0.22 [216]	Levofloxacin	Magnetite	CT = 0–4, T = 15–45, pH = 6.5, AD = 1.0, [EC] = 2.5–20		36.7–80.1	6.1–6.8	[189]
			Diclofenac	Maghemite	CT = 6.25, T = 25 ± 1, pH = 7, AD = 1.0, [EC] = 100–500		~90	261	[192]
			Ibuprofen	Goethite	CT = 2, T = 25, pH = 7, AD = 0.25, [EC] = 1.65–2.06	Ion-dipole, van der Waals forces, or ion exchange, hydroxyethyl cleavage and chelation [40,167,192,193,196]	89–91	0.72–3.47	[195]
			Diclofenac	Goethite	CT = 0.08–48, T = 25, pH = 5.3–10, AD = 10, [EC] = 1		75	-	[193]
			Cephalexin	Hematite	CT = 0.08–6.67, T = 25, pH = 2–10, AD = 2, [EC] = 25–250		~99.7	70	[196]
			4-n-nonylphenol	Hematite and goethite	CT = 24, T = Room T, pH = 7.0–9.0, AD = 100, [EC] = 1.5		-	-	[222]
			Oxytetracycline	Magnetite	T = 5–35, pH = 5.6, AD = 2.5–20, [EC] = 868.86		60–100	-	[191]
Nano-alumina: S.O. [101]	MO_x, FeO, SO_4^{2-}, Al_2O_3 (α, β, γ) [101]/ O-Al-O, OH-, C=O [101]	SA: 143.7 [223] P: 3 [223]	Dichlorodiphenyltrichloroethane, Polychlorinated biphenyls	Nano-alumina	CT = 0.33, T = 28–34, [EC] = 10–60	Hydrophobic, hydrogen bonding, and van der Waals, Electrostatic interactions [63,101,102,223] Π-π interaction, electron donor acceptors [200]	54–68	Up to 0.18 Up to 0.16	[224]
Nano-silica: S.O. [63,102]	SiO_2 [102]/ Si-O-H, OH⁻ [63]	Information not available	Ciprofloxacin	Nano-silica/modified nano-silica	CT = 3.33, T = 25 ± 2, pH = 3–7, AD = 1–50, [EC] = 20		56.8–89.9	85	[168]
ZnO: S.O.	Zn_2O/ Zn-O, O-H, C-H	SA: ~14.8	Naphthalene	ZnO	CT = 1, T = 25 ± 2, pH = 2–12, AD = 600 [EC] = 25		100	148.3	[26]
MgO: S.O. [200]	MgO [202] Mg-O, OH- [200]	SA: 48–108 [200] P: 0.3–24.8 [200]	Linezolid	MgO	CT = 1, T = 35, pH = 3–8, AD = 0.2–0.8 [EC] = 10–100		~62	123.4	[198]

Table 5. Cont.

Adsorbent	Adsorbent Characteristics			Adsorption Behavior				
	Composition/ Functional Groups/ions	SA [m²/g] P [cm³/g]	Adsorbent	EC Removed	Adsorption Conditions	Adsorption Mechanism	Removal [%]	Adsorption Capacity [mg/g]	Reference
Nano-composites: S.O. [177,179,202,203]	Depends on the components of the composite, e.g., carbon atom, FeO$_x$, graphene, and graphite sheet, among others/ Depends on the components of the composite, e.g, OH-, COOH-, C=O, epoxy, and amino, among others [178,179,223,225]	SA: 3.18- around 1260 [201] P: 0.15-0.72 [178,223,225]	MOFs UiO-66	Sulfachloropyra-dazine	CT = 2, T = 25, pH = 5.5, AD = 0.1, [EC] = 10-100	Depends on the components of the composite, are an example hydrogen bonding, π–π interaction, cation–π bonding, and amidation reaction, electrostatic interaction, hydrophobic interaction, ligand exchange, cation–π bonding, chemisorption, etc. [178,225,226]	80	417	[205]
			Magnetic activated carbon	Triclosan Bisphenol-A Tonalide Metolachlor Ketoprofen E2	CT = 1, T = 25-45, pH = 7, AD = -, [EC] = 0.025-0.25		96-98	21.32 31.05 29.41 22.37 28.49 20.20	[206]
			Magnetic activated carbon	Carbamazepine	CT = 0.5, T = 25, pH = 6.65, AD = 0.05, [EC] = 20.0		93	189.5	[214]
			Triethoxyphenylsilane (TEGs)-functionalized magnetic palm-based powdered AC	BPA, carbamazepine, ibuprofen, clofibric acid	CT = 6, T = Room T, pH = 7, AD = 0.1, [EC] = 10		-	58.1-166.7	[207]
			Magnetic activated carbon/chitosan	Ciprofloxacin, erythromycin, amoxicillin	CT = 2, T = 25, pH = -, AD = 1.5, [EC] = -		54-82	90.1 178.6 526.3	[200]
			Magnetic cellulose ionomer/layered double hydroxide	Diclofenac	CT = 0.5, T = -, pH = 9, AD = 1.0, [EC] = 0.5		~100	268	[209]
			Magnetic chitosan grafted graphene oxide	Ciprofloxa-cin	CT = 8, T = -, pH = 5, AD = 0.33, [EC] = 20		-	282.9	[208]
			Fe$_3$O$_4$/graphene oxide reduced	Ametryn, atrazine prometryn, simazine, simeton	CT = 1.17, T = 25, pH = 5.0, AD = 0.5, [EC] = 10		93.6	54.8-63.7	[203]
			Magnetically modified graphene nanoplatelets	Amoxicillin	CT = 1.5, T = 20, pH = 5, AD = 2.0, [EC] = 10.0		84	14.1	[225]

Table 5. Cont.

Adsorbent Characteristics				Adsorption Behavior				
Adsorbent	Composition/ Functional Groups/ions	SA [m²/g] P [cm³/g]	EC Removed	Adsorption Conditions	Adsorption Mechanism	Removal [%]	Adsorption Capacity [mg/g]	Reference
			Tetracycline	CT = 8, T = 25, pH = 6, AD = 0.4, [EC] = 44.44		>90	-	[176]
Fe₃O₄@SiO₂-Chitosan/GO			Tetracycline	CT = 0.25, T = 22, pH = 6.5, AD = 0.25, [EC] = 20–100		~100%	-	
Fe/Cu-GO			Sulfamethazine	CT = 12, T = 25, pH = 6, AD = 1, [EC] = 2031.8–20,332		-	-	
GO-BC			Tetracycline	CT = 2.0, T = 22 ± 2, pH = 7.8, AD = 4.0, [EC] = 20		-	~5.4	[213]
Sawdust+ FeCl₃			Ciprofloxacin	CT = 48, T = 20, pH = 3.0, AD = 5.0, [EC] = 160		-	>76	[217]
Biochar + Chitosan			Ciprofloxacin, norfloxacin	CT = 24, T = 25, pH = 6, AD = 2.0, [EC] = 300		-	245.6 293.2	[215]
Magnetic bamboo-based activated carbon			Sulfamethoxazole	CT = 24, pH = 7 ± 0.10, [EC] = 0.17–78		-	51.8–87.9	[216]
Magnetic pine sawdust biochar			Tylosin	CT = 0.25–24, T = 25, pH = 3–11, AD = 2, [EC] = 20		-	3070	[226]
Maize straw + manganese/iron oxides			BPA	CT = 24, T = 24 ± 1, pH = 4.5, AD = 0.1, [EC] = 10–100		-	168.4	[227]
Palm-shell waste AC + magnesium silicate			Carboxylate-modified PE, amine-modified PE, PE (~1 μm)	CT = 48, T = 25, pH = 7, [EC] = 1.0		71.6–92.1	5.4–9.7	[213]
Chitin + GO+ O-C₃N₄								

Notes: S.O. = synthetic origin, N.O. = natural origin, CT = contact time [h], T = temperature [°C], AD = adsorbent dose [g/L], [EC] = emerging contaminant concentration [mg/L], SA = surface area, P = porosity, PE = polystyrene.

After reviewing a series of adsorbents used in the removal of ECs, it is observed that some parameters that influence the adsorption process are the nature of the adsorbent (morphology, SA, P, functional groups), the adsorbent dose (a higher dose favors adsorption), the contact time (longer contact times do not necessarily favor adsorption), particle size (smaller size favor the adsorption), and pH [52,54,98,168]. pH depends on the isoelectric point of the adsorbent; however, there is no mention of whether the pH modification would be profitable and friendly at real scale [222]. Furthermore, characteristics such as the solubility and lipophilicity of EC influence its removal through adsorption processes [4]. Likewise, it is observed that the removal mechanisms are related to the characteristics/properties/structure of the adsorbent and contaminant [32]. The dominant mechanisms in the removal of ECs are physical interactions such as electrostatic interactions, hydrophobic effect, π–π interactions, hydrogen bonds, etc. [32,35,99,222–224]. However, graphene oxide, silica, and some nanocomposites also show chemical interactions such as ion exchange and covalent bonding. Iron oxides also appeared to chelation forces by removing some ECs [167,179,180,182,183].

On the other hand, the regeneration of the adsorbents occurs only in some studies. Among the commonly regenerated materials are activated carbon, zeolite, modified adsorbents, and some nanomaterials [71,158,168,213]. Adsorbent regeneration is associated with the adsorbent cost. In fact, there are materials such as agro-industrial residues that are not regenerated. Probably, it is related to their high availability and low cost [55,161]. In addition, although the efficiency obtained by the adsorbent after regeneration is analyzed, the environmental impact that this process can generate is not indicated. Likewise, the reuse of the saturated material (after the adsorption process) is not shown in the bibliography used in this study [158,169,213].

The results in batch adsorption processes allow orientations to work that involve continuous adsorption processes [55]. However, the optimal conditions of the batch adsorption process cannot be transferred to fixed-bed columns. In fact, other evaluations are carried out, such as the saturation time of the filter material, determination of the rupture curve, etc. [75]. Furthermore, few studies present an adsorption study in batch process and fixed-bed columns [75,98,152].

4. Technologies Using Adsorbent Materials to Remove ECs from Wastewater

Adsorption processes alone or in combination with other mechanisms are used and studied in different filtration and biofiltration technologies at different levels (lab/pilot/full). In biofiltration technologies, the materials used, in addition to adsorbing the contaminants, fulfill other fundamental functions. They act as a support medium for microorganisms (bacteria) and plants and retain nutrients, organic matter, and solids, among others [55,157]. Therefore, the adsorbent materials must meet several requirements to achieve high efficiencies in the contaminant removal and for the proper operation of the technologies. Material characteristics, cost, availability, hydraulic performance (clogging), feasibility, adsorption capacity, toxicity with living beings, chemical/mechanical stability, recoverability, and disposal ease are factors considered [43]. Adequate adsorbent materials improve the operation and the efficiency of filtration/biofiltration technologies. Some of the alternative materials and mechanisms for EC removal are shown in Figure 3. However, very little research has been conducted on the adsorbent material–technology relationship. Therefore, in this section, a review of the subject is carried out.

4.1. Filtration Technologies

Fixed-bed columns or filtration technologies are low-cost and easy to operate technologies (lab/full-scale) that have low energy consumption and are easily scaled. Filtration technologies are used to treat secondary/tertiary effluents, achieving good efficiencies (up to 100%) for organic matter and specific contaminants [157]. The mechanisms for the contaminant removal are produced on the materials used as filter beds. These mechanisms depend on the contaminant nature, bed depth, quantity, packaging, size, and the feature of the material.

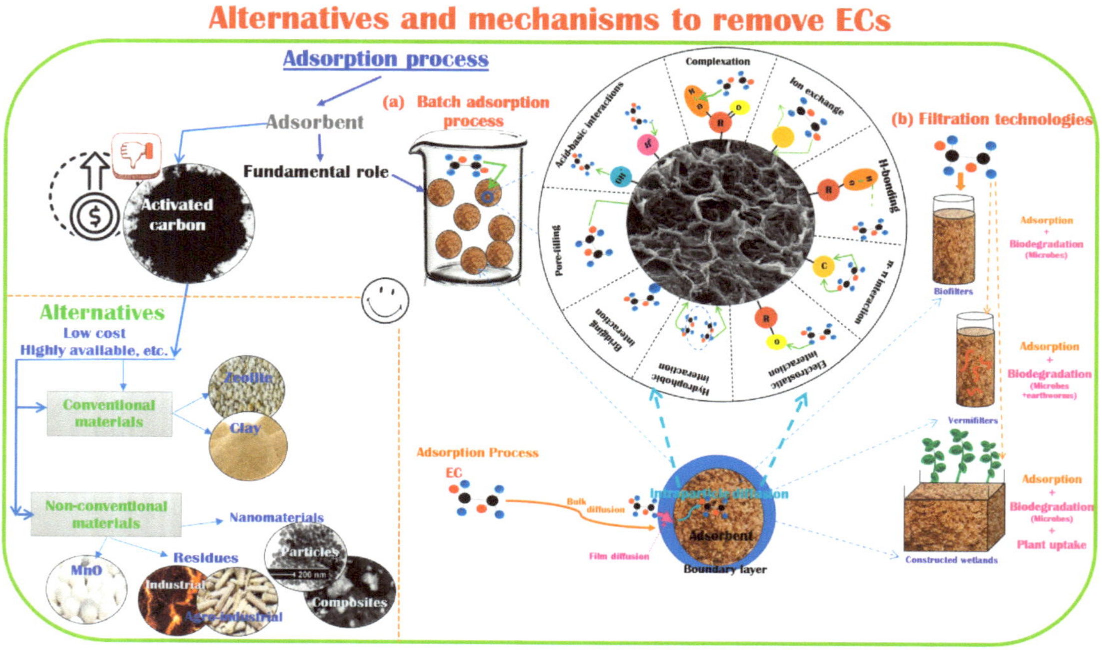

Figure 3. Alternatives and mechanisms to remove ECs in batch and filtration technologies.

To improve contaminant removal in fixed-bed columns, they are conditioned with microorganisms, giving origin to biofilters. Biofilters are more used than filters in EC removal. Moreover, depending on the retention time (RT) or hydraulic load rate (HLR) used in filters/biofilters, there could be rapid or slow filtration [228]. Considering the feeding type, filters/biofilters could be continuous or intermittent. In the next paragraphs, the main characteristics of these processes are summarized.

4.1.1. Rapid Filtration

Rapid filtration is widely used worldwide in water purification processes (low consumption of energy and chemical products) [228]; they operate with a TOC loading ~3.1 mg/L and organic loading = 3.7–36.7 g/m^{3}d [229]. Rapid filters are already used in existing drinking water treatment plants; despite not being designed to eliminate Ecs, rapid filters have partially degraded (even >50%) several of them [230]. Materials such as activated carbon, anthracite, garnet, and pumice have been used. However, sand (0.4–1.5 mm) is the most used material in them [228]. The mechanisms observed in EC removal are adsorption on sand, oxidation, and adsorption by metal oxides (FeO$_x$/MnO$_x$/bio-MnO$_x$), and biodegradation by autotrophic and heterotrophic bacteria. However, biodegradation has been determined to be more significant than adsorption (only 10–15%) [228].

The hydraulic retention time (3.3–33.3 h) in rapid filters is essential in EC removal. Reducing the time by half decreases the removal efficiency (>10%) of triclosan, galaxolide, tonalide, and celestolide [231]. Other ECs such as bentazone [231,232], carbofuran, triclosan, gemfibrozil, ketoprofen, caffeine, erythromycin, naproxen, carbamil, benzenesulfonamide, microsistin-LR [228], dichlorprop [232], atrazine, bentazon, and carbamazepine [231], present in surface and groundwater, were removed. The removal efficiencies in fast filters (microcosm, columns, field) have reported to be variable (0–99%) [228,231–234].

4.1.2. Slow Filtration

Slow filtration has a lower filtration rate (1/20 or less) than rapid filtration and does not require prior chemical coagulation [235]. Moreover, slow filtration is a technology with low operational costs (low energy consumption) and simplicity in operation and maintenance. The most common filter material is also sand (0.1–0.4 mm); however, coarse sand and other materials (e.g., GAC, quartz/silica) have been used [236]. Slow filtration is combined with microbiological action, so biosorption/biodegradation occurs, phenomena that predominate over adsorption. Adsorption, mechanical filtration, and degradation processes could occur after biodegradation. In this case, the material not only absorbs ECs, but also fulfills other functions such as being a support for microorganisms and retaining their food until they consume it [237]. Slow filtration has been used in the removal of ECs such as paracetamol, diclofenac, naproxen, ibuprofen, methylparaben, benzophenone-3, E1, E3, EE2 [235], propranolol, iopromide, diclofenac, tebuconazole, and propiconazole [238], among others, achieving removal rates between <15 and >98% [235].

4.2. Biofiltration

Biofiltration is a biological filtration. Initially, the adsorption in the filter bed material is an exclusive process. However, over time, the active sites of the material become saturated, and this mechanism diminishes, so that other mechanisms begin to predominate. In the second stage, biological adsorption–degradation occurs due to the presence of aerobic, anaerobic, facultative microorganisms, bacteria, fungi, algae, and protozoa [55]. At this stage, the function of the adsorbent is also to offer a specific area for the growth of bacteria/plants/earthworm. Adsorption decreases until only biological degradation occurs, in the third stage. Therefore, the parameters that determine the biofilter efficiency are the surface characteristics of the material (pore size, specific surface area, functional groups), the degree of compaction, the hydrophobicity of the bacteria, and the adsorbent [157]. Biofiltration technologies for EC removal are generally used as secondary treatment when there is a high load of organic matter or tertiary treatment. Constructed wetlands (CWs), vermifiltres, and biofilters are the types of biofiltration technology [239–241].

4.2.1. Biofilters

In biofilters such as water percolates through filter bed material, microorganisms attach to the material surface (diffusion, convection, sedimentation, and active mobility of microbial cells), colonize it, and form stable biofilms (able to resist/degrade even toxic contaminants) [241]. Materials such as clay, anthracite, activated carbon, and sand are used conventionally as adsorbents to remove ECs in biofilters. Activated carbon has been shown to be efficient in removing pesticides, but not for personal care products and endocrine disruptors. Other materials such as biochar, rice husk (raw/biochar), peanut shells, fruit peels, sawdust, wood chips, or mixtures thereof are used [157,158,240]. These materials have proven to be efficient for removing some ECs (0–100%) [242–245] such as 17β-estradiol-17α-acetate, pentachlorophenol, 4-tert-octylphenol, caffeine, gemfibrozil, BPA, benzophenone, atrazine, dicamba, triclosan [246], acetaminophen, erythromycin, sulfamethoxazole, cotinine, aminotrizaole, ibuprofen, atrazine, and naproxen, among others [244]. The removal mechanisms of ECs in biofilters are adsorption and biodegradation, produced by the filter bed material/biofilm and microorganisms, respectively [241]. Nevertheless, oxidation (in the material) could occur if other filter materials are used (e.g., manganese oxide) [244].

4.2.2. Constructed Wetlands

Constructed wetlands (generally used as secondary treatment) are systems formed by plants (macrophytes/ornamental), substrates (support materials), native microorganisms, and water interacting with each other [246]. The substrate is fundamental in the efficiency of CWs since it fulfills physical, chemical, and biological functions to remove contaminants. Furthermore, the materials are the support medium (allow growth) of plants and microorganisms [2]. Other support material functions include physical sedimentation, filtration, and gas diffusion between the material particle gap [247].The conventional materials used are soil, sand, and gravel (8–16 mm), even in the removal of ECs [245,248]. Moreover, red soil, volcanic rock, stone, vesuvianite, zeolite, and brick were used. At the lab scale, materials such as rice husk, pine bark, and granulated cork were used for EC removal [247].

Different types of CWs (surface free water, horizontal groundwater flow, vertical groundwater flow, and hybrid CWs) at full/mesocosm/microcosm/pilot/lab scale were used for the removal of several ECs [245,247,248]. The removal mechanisms of the ECs are produced by their sorption in the material support (hydrophobic partitioning, van der Waals interaction, electrostatic interaction, ion exchange, and surface complexation), plant uptake (phytostabilization, phytoaccumulation, phytodegradation), and/or biodegradation aerobic/anaerobic process. However, the main mechanisms in the EC removal are biodegradation and sorption [53,246]. Among the ECs removed (0 and >99%) were phenols, diclofenac, naproxen, atrazine, endosulfan, erythromycin, clarithromycin, azithromycin, E1, E2, carbamazepine, gemfibrozil, sulfamethoxazole, sulfapyridine, ibuprofen, acetaminophen, triclosan, and BPA, among others [245,247,248].

4.2.3. Vermifilters

Vermifilters are engineering systems made up of earthworms, microorganisms (biotic component), and a filter material (abiotic component) that maintain symbiotic relationships. The main function of earthworms is to regulate microbial activity and biomass while microorganisms biodegrade waste materials/contaminants [249]. Earthworm species (*Eisenia foetida, Lumbricus rubellus, Eudrilus eugeniae,* and *Eisenia Andrei*) are suspended on a filtration bed (active zone of earthworms) that can be soil, compost, and cow manure, where the degradation of contaminants occurs [250]. However, alternative materials (in toxicity tests and mesocosm scale) such as coconut fiber, corn cob, peanut shells, and rice husks are also used [55,157]. In turn, sand, gravel, cobblestone, and quartz sand are used as filter beds [250]. Vermifilters were used in the removal of ECs such as ciprofloxacin, ofloxacin, sulfamethoxazole, trimethoprim, tetracycline, metronidazole [251], amoxicillin, ampicillin, ticarcillin, ceftazidime, cefotaxime, ceftriaxone, streptomycin, gentamicin, erythromycin, tetracycline, chloramphenicol, and ciprofloxacin [249]. Removal efficiencies between 40% and 98% were reached [251]. The mechanisms involved in the removal of ECs were absorption/degradation by earthworms, adsorption/degradation in the biofilm, biodegradation under the load of the microorganisms, and sorption in the bed filter material [249,251].

Table 6 shows the operating conditions of different types of filters/biofilters that were used in the removal of some ECs. In addition, the material(s) used in the technologies and the mechanism by which the ECs were removed are indicated. It is observed that biofilters are more used than filters for the removal of ECs. Biofilters (fast and slow) that use sand as a filter bed are used to remove ECs present mainly in surface and groundwater. Additionally, the presence of living organisms (microorganisms, plants, and earthworms) improves the efficiency of EC removal. However, vermifilters are the least used biofilters (there is little previous research) in the removal of ECs compared to CWs and biofilters.

Table 6. Technologies using materials to remove ECs from wastewater.

Type of Filtration (Scale)	Bed Filter Material (Size, mm)	Influent Type	EC Removed	Operational Conditions	Removal Mechanism	Removal [%]	Reference
Aerated rapid filtration. (microcosms scale)	Sand (3–5)	Anaerobic groundwater	Mecoprop, bentazone, glyphosate, p-nitrophenol	[EC] = 3×10^{-5}–2.4×10^{-3}, RT = 0.17	Biodegradation	1–85	[232]
			Dichlorprop	[EC] = 2×10^{-4}, RT = 0.93		>50	[232]
Rapid filtration (microcosms scale)	Sand (3–5)	Anaerobic groundwater	Bentazone	[EC] = 5, RT = 312	Biodegradation	92	[231]
Rapid filtration (microcosms scale)	Filtralite clay (0.8–1.6)	Groundwater enriched with Ecs	2,6-dichlorobenzamide, bromoxynil, chlorotoluron, diuron, ioxynil, isoproturon, linuron, 4-chloro-2-methylphenoxy acetic acid	[EC] = 2.1×10^{-3}–6.6×10^{-3}, RT = 0.023, FR = 21	Biodegradation	13–98	[234]
Rapid filtration (lab scale)	Sand	Influent water from the RSF filters (WRK, Nieuwegein, The Netherlands) enriched with Ecs	Atrazine, bentazon, metolachlor and clofibric acid, carbamazepine	[EC] = 0.01, RT = 8 and 96, FR = 80	Biodegradation/Sorption	-	[233]
Rapid filtration: downflow, upflow, dual media down flow (field scale)	Sand (0.7–2.5) Sand + hydroanthracite (1.4–2.5) Sand + anthracite (1.6–2.5)	Surface water from The Netherlands and Belgium	Caffeine, acesulfame-K, sucralose, metformin, phenazone, chloridazon, valsartan, sulfadiazine, sotalol, etc	[EC] ≤ 1×10^{-5}–5.7×10^{-4}, RT = 15–240	Biodegradation/Sorption (probably)	0–93	[230]
Slow filtration with rapid pulses of a carbon source (lab scale)	Quartz sand (0.210–0.297)	WWTP effluent	Atenolol, metoprolol, iopromide, tomeprol, carbamazepine, diclofenac, sulfadiazine, sulfamethoxazole, etc.	RT = 150, FR = 0.15	Biodegradation	-	[229]
GAC sandwich slow filtration (lab scale)	Coarse sand (0.6) GAC (0.4–1.7) Coarse sand + GAC	Synthetic wastewater	Mix of DEET, paracetamol, caffeine and triclosan	[EC] = 0.025, HLR = 5, 10, 20	Adsorption (GAC) + Biodegradation	18.8–100	[236]
Slow filtration (pilot scale)	Silica sand (0.15–0.30) Support: pea gravel	Stream water/Stream water + 1% of primary effluent added, both enriched with Ecs	Caffeine, carbamazepine, 17-β estradiol, EI, gemfibrozil, phenazone	[EC] = 0.05, HLR = 5	Sorption and/or biodegradation	<10–100	[252]
Household slow filtration with intermittent and continuous flows (pilot scale)	Sand (0.09–0.5) Support: coarse sand (1–3) + fine gravel (3–6) + coarse gravel (10–12) Top: non-woven polyester	Synthetic wastewater	BPA	[EC] = 2.35 Continuous flow HLR = 1.58 Intermittent flow HLR = 0–875	Biodegradation	Continuous flow 14 ± 6 Intermittent flow 3 ± 8	[253]

Table 6. Cont.

Type of Filtration (Scale)	Bed Filter Material (Size, mm)	Influent Type	EC Removed	Operational Conditions	Removal Mechanism	Removal [%]	Reference
Biofilters (bench scale)	GAC (1.0–1.2) and anthracite (0.8–2.0)/sand (0.55–0.65) dual media	Municipal waste streams	Acetaminophen, ibuprofen, erythromycin, sulfamethoxazole, trimethoprim, carbamazepine, atenolol, gemfibrozil, tri(2-chloroethyl) phosphate, DEET, cotinine, aminotriazole, atrazine, caffeine, E2, iopromide	[EC] = 2.27×10^{-4}–6.44×10^{-3}, RT = 0.17 and 0.30	Biodegradation	>75	[242]
Biofilters (pilot scale)	Anthracite/sand (1.07/0.52)	Superficial water of Grand River enriched with Ecs	DEET, atrazine, naproxen, ibuprofen, nonylphenol, carbamazepine	[EC] = 5×10^{-4}–5×10^{-3}, RT = 0.08 and 0.23, HLR = 500	Adsorption (non-biodegradable Ecs) and biodegradation (biodegradable Ecs)	<20–100	[243]
Biofilters (pilot scale)	GAC/sand and anthracite/sand	Water from the full-scale recarbonation chambers enriched with Ecs	Atenolol, atrazine, carbamazepine, fluoxetine, gemfibrozil, metolachlor, sulfamethoxazole, tris(2-chloroethyl) phosphate	[EC] = 1×10^{-4}–2×10^{-4}, 1×10^{-3}–3×10^{-3}, RT = 8.4 and 4.2, HLR = 488 and 976	Adsorption and biodegradation	GAC/sand: 49.1–94.4 anthracite/sand: 0–66.1	[254]
Biofilters (pilot scale)	Anthracite-sand and previously used biological activated carbon (BAC)-sand dual media, BAC = (0.9)	Raw surface water (Colorado River) enriched with Ecs	Sulfamethoxazole, caffeine, gemfibrozil, naproxen, DEET, trimethoprim, acetaminophen, ibuprofen, sucralose, meprobamate	[EC]= 1×10^{-4}–$\times 10^{-3}$, RT = 0.17, HLR = 904.56	Biodegradation and BAC sorption	<50–>99	[242]
Biofilters (pilot scale)	Natural manganese oxides (3–5)	Secondary effluent of WWTP	1-hydroxybenzotriazol, 4′-hydroxydiclofenac, 10,11-dihydro-10,11-dihydroxycarbamazepine, acyclovir, benzotriazole, diclofenac, carbamazepine, carboxy-acyclovir, diatrizoic acid, erythromycin, gabapentin, iomeprol, tolyltriazole, sulfamethoxazole, tramadol,	RT = 5 and 10, FR = 8000, HLR = 400,	Adsorption, biodegradation, oxidation	70–98	[244]
Horizontal/vertical subsurface flow and hybrid CWs, aerated/unaerated (mesocosm scale)	Zeolite (20–30)	Domestic sewage enriched with Ecs	Sulfamonomethoxine, sulfamethazine, sulfameter, trimethoprim, norfloxacin, ofloxacin, enrofloxacin, erythromycin-H_2O, roxithromycin, oxytetracycline, lincomycin	[EC] = 5×10^{-3}, HLR = 1.67, PT = *Iris tectorum*	Sorption and biological processes	87.4–99.1	[255]

Table 6. Cont.

Type of Filtration (Scale)	Bed Filter Material (Size, mm)	Influent Type	EC Removed	Operational Conditions	Removal Mechanism	Removal [%]	Reference
Four CWs of subsurface horizontal flow (pilot scale)	Gravel (12.7–19.05)	Synthetic wastewater	Carbamazepine, sildenafil, methylparaben	[EC] = 0.2, FR = 15, RT = 72, PT = *Heliconia Zingiberales* and *Cyperus Haspan*	Biodegradation, adsorption, plant absorption	<10–97	[2]
Combination of partially saturated and unsaturated vertical subsurface flow CWs (experimental scale)	Top: sand layer (1–2) Underneath: gravel (3–8)	Urban wastewater (surrounding residential area) from primary treatment	Ciprofloxacin, ofloxacin, pipemidic acid, azithromycin	[EC] = 5 × 10^{-4}, HLR = 0.55, PT = *Phragmites australis*	Sorption and biodegradation	<−200–>90	[246]
Line 1: Partially vertical flow Line 2: unsaturated vertical flow + horizontal subsurface flow + free water surface CWs (experimental scale)	-	Urban wastewater from primary treatment	Caffeine, trimethoprim; sulfamethoxazole, DEET, sucralose	HLR = 0.55, FR = 138.89	Biodegradation	<10–100	[256]
Vertical flow CW	Top: gravel (4.8–9.5) Filter media: sand (0.27) Bottom: medium gravel (4.8–9.5) + coarse gravel (25–32)	Wastewater	Ibuprofen and caffeine	[EC] ≤ 0.1, HLR = 16, RT = 168, PT = *Heliconia rostrata*	Biodegradation, adsorption, plant absorption	90–97	[257]
Vermifiltration (pilot scale)	Soil Sand (0.1–0.8) Detritus (3–10) Support: cobblestone (10–50)	Hospital effluent from sedimentation basin	Ciprofloxacin, ofloxacin, sulfamethoxazole, trimethoprim, tetracycline, metronidazole	HLR = 4.17, ET = *Eisenia foetida*, ED = 10,000	Adsorption, earthworm absorption (mineralization/transformation), biodegradation	40–98	[251]
Integrated CW	Chaff and soil	Domestic + livestock wastewater	Androsta-1,4-diene-3,17-dione, 17α-trenbolone, 17α-boldenone, 17β-boldenone, testosterone, stanozolol, progesterone, ethynyl testosterone, 19-norethindrone, norgestrel, medroxyprogesterone, cortisol, cortisone, prednisone, miconazole, fluconazole, itraconazole, etc.	[EC] = 6.3 × 10^{-7}–1.05 × 10^{-4}, RT = 36, PT = *Myriophyllum verticillatum* L. and *Pontederia cordata*	Biodegradation, adsorption, plant absorption	<10–97.6	[249]
Vermifiltration (pilot scale)	Top: vermigrating,s (0.110) and cow-dung (0.05–5), Small gravel (2–4), Medium gravel (6–8) Support: coarse gravel (12–14)	Clinical laboratory wastewater	Amoxicillin, ampicillin, ticarcillin, ceftazidime, cefotaxime, ceftriaxone, streptomycin, gentamicin, erythromycin, tetracycline, chloramphenicol, ciprofloxacin	HLR = 4.17, RT = 7–8, ET = *Eisenia foetida*, ED = 10,000	Earthworms/microorganisms degradation, biofilm adsorption, filter media sorption	-	[250]

Notes: [EC] = emerging contaminant concentration [mg/L], FR = flow rate [mL/min], HLR = hydraulic loading rate [cm/h], GAC = granular activated carbon, PT = plant type, ET = earthworm type, ED = earthworm density (worms/m^3).

Likewise, the most used materials in the filter bed of filtration/biofiltration technologies are mostly gravel and sand (classified as conventional materials). However, in a few studies, other materials such as granular activated carbon, silica, and anthracite are used, showing to be very efficient. This makes it necessary to study other adsorbents such as agro-industrial residues, which have proved to be efficient in the removal of organic matter and nutrients.

4.3. Outlook and Future Perspectives

Based on the findings of this bibliographical review, it is essential to continue researching/developing efficient adsorbents to remove Ecs. These adsorbents should be friendly to the environment, low cost, and available in the local market to reduce transportation costs and take advantage of resources/value waste. Moreover, the adsorbents will be biocompatible to be used in biofiltration technologies. This would imply, for example: i) Deepening the modification/regeneration of materials and the synthesis of nanomaterials using environmentally friendly substances. Ii) Analyzing the technical and economic feasibility of its production/synthesis on an industrial scale to be tested/used in technologies such as filtration/biofiltration. This is to remove contaminants from real wastewater. Iii) Investigating the efficiency, chemical/mechanical stability, and behavior (toxicity) of highly available materials (agro-industrial residues, industrial waste) when used in filtration/biofiltration technologies. In addition, it is necessary to deepen into how the presence of several ECs or their coexistence with other contaminants (e.g., metals, dyes, organic matter, nutrients) influences the behavior/efficiency of the adsorbents. Finally, the possible applications that the filter bed/adsorbent material may have once its useful life is over should be researched, since there is little to no evidence on the topic.

5. Concluding Remarks

The removal of ECs is an emerging concern since adsorption and the technologies in which this process occurs are efficient and low-cost alternatives. However, finding an adsorbent with good adsorption characteristics for different ECs is challenging. Lipophilic (log K_{ow} > 4) and poorly soluble in water ECs are the most easily removed (efficiency up to 100%), while hydrophilic ECs are more difficult to remove (greater amount of adsorbent/contact time). Therefore, it can be suggested that there is not yet an ideal adsorbent for the removal of all ECs. However, as observed in previous studies, when the optimal adsorption conditions are determined (adsorbent type, particle size/adsorbent dose, contact time, pH), the adsorption capacity of the material increases.

Undoubtedly, activated carbon has demonstrated to be the best adsorbent (up to 100%, >850 mg/g) for Ecs. Nevertheless, the challenge of this material is associated with the reduction of its costs, the use/exploitation of other materials for its production, the use of more environmentally friendly substances for its activation/regeneration, and maintaining its adsorption efficiency after regeneration. Nanomaterials also suggest being a promising alternative for the removal of Ecs, but it is necessary to produce them on a larger scale and improve their separation from the aqueous medium. In turn, industrial waste and agro-industrial residues (rice husks, coconut fibers, corn cobs, peanut shells, sugarcane bagasse, and fruit shells/seeds) are promising alternatives to replace activated carbon. This is due to its low/zero cost, high availability, and relatively high adsorption capacities (up to 300 mg/g). In addition, the reuse of agro-industrial waste would also solve its management problem and is aligned with the circular economy and the objectives of sustainable development. Furthermore, due to the low toxicity of agro-industrial residues (e.g., 14d-LC_{50} = 82–97%), they could be used (alone/mixed) in biofiltration technologies.

Biofiltration technologies are characterized by their efficiency (up to 100%), low cost, and easy operation/maintenance, which are reasons why they are widely used in decentralized wastewater treatment systems in developing countries. However, research about this type of technology is limited to the use of gravel and sand, classified as conventional materials and representing between 50 and 60% of technological costs. Although adsorption is not the fundamental mechanism in biofiltration technologies, it does become representative (up to 20%). Furthermore, the role of the material is not only limited to the removal of contaminants but also has other functions that are essential for the performance/efficiency of biofilters. Thus, it is important to continue testing alternative materials that are capable of meeting these requirements.

Author Contributions: Conceptualization, C.A.V.-A.; writing—original draft preparation, C.E.A.-N.; writing—review and editing, C.A.V.-A. and V.H.G. All authors have read and agreed to the published version of the manuscript.

Funding: This word is supported from Escuela Politécnica Nacional through the projects PIS-18-01. This work was also supported by FONDECYT (Grant project 11190352) from Chile.

Data Availability Statement: Not applicable.

Conflicts of Interest: The authors declare no conflict of interest.

References

1. Rout, P.R.; Zhang, T.C.; Bhunia, P.; Surampalli, R.Y. Treatment technologies for emerging contaminants in wastewater treatment plants: A review. *Sci. Total Environ.* **2021**, *753*, 141990. [CrossRef]
2. Delgado, N.; Bermeo, L.; Hoyos, D.A.; Peñuela, G.A.; Capparelli, A.; Marino, D.; Navarro, A.; Casas-Zapata, J.C. Occurrence and removal of pharmaceutical and personal care products using subsurface horizontal flow constructed wetlands. *Water Res.* **2020**, *187*, 116448. [CrossRef]
3. Stuart, M.; Lapworth, D. Emerging Organic Contaminants in Groundwater. In *Smart Sensors, Measurement and Instrumentation*; Mukhopadhyay, S., Mason, A., Eds.; Springer: Berlin/Heidelberg, Germany, 2013; pp. 259–284. ISBN 9783642370069.
4. Tran, N.H.; Reinhard, M.; Gin, K.Y.H. Occurrence and fate of emerging contaminants in municipal wastewater treatment plants from different geographical regions-a review. *Water Res.* **2018**, *133*, 182–207. [CrossRef]
5. Wilkinson, J.; Hooda, P.S.; Barker, J.; Barton, S.; Swinden, J. Occurrence, fate and transformation of emerging contaminants in water: An overarching review of the field. *Environ. Pollut.* **2017**, *231*, 954–970. [CrossRef]
6. Luo, Y.; Guo, W.; Ngo, H.H.; Nghiem, L.D.; Hai, F.I.; Zhang, J.; Liang, S.; Wang, X.C. A review on the occurrence of micropollutants in the aquatic environment and their fate and removal during wastewater treatment. *Sci. Total Environ.* **2014**, *473–474*, 619–641. [CrossRef]
7. Wilkinson, J.L.; Hooda, P.S.; Swinden, J.; Barker, J.; Barton, S. Spatial (bio) accumulation of pharmaceuticals, illicit drugs, plasticisers, per fl uorinated compounds and metabolites in river sediment, aquatic plants and berthic organisms *. *Environ. Pollut.* **2018**, *234*, 864–875. [CrossRef]
8. Zheng, C.; Feng, S.; Liu, P.; Fries, E.; Wang, Q.; Shen, Z.; Liu, H.; Zhang, T. Sorption of Organophosphate Flame Retardants on Pahokee Peat Soil. *Clean-Soil Air Water* **2016**, *44*, 1163–1173. [CrossRef]
9. Catherine, H.N.; Tan, K.-H.; Shih, Y.; Doong, R.; Manu, B.; Ding, J. Surface interaction of tetrabromobisphenol A, bisphenol A and phenol with graphene-based materials in water: Adsorption mechanism and thermodynamic effects. *J. Hazard. Mater. Adv.* **2023**, *9*, 100227. [CrossRef]
10. Baloo, L.; Isa, M.H.; Sapari, N.B.; Jagaba, A.H.; Wei, L.J.; Yavari, S.; Razali, R.; Vasu, R. Adsorptive removal of methylene blue and acid orange 10 dyes from aqueous solutions using oil palm wastes-derived activated carbons. *Alex. Eng. J.* **2021**, *60*, 5611–5629. [CrossRef]
11. Birniwa, A.H.; Abubakar, A.S.; Huq, A.K.O.; Mahmud, H.N.M.E. Polypyrrole-polyethyleneimine (Ppy-PEI) nanocomposite: An effective adsorbent for nickel ion adsorption from aqueous solution. *J. Macromol. Sci. Part A Pure Appl. Chem.* **2021**, *58*, 206–217. [CrossRef]
12. Birniwa, A.H.; Kehili, S.; Ali, M.; Musa, H.; Ali, U.; Rahman, S.; Kutty, M.; Jagaba, A.H.; Sa, S.; Tag-eldin, E.M.; et al. Polymer-Based Nano-Adsorbent for the Removal of Lead Ions: Kinetics Studies and Optimization by Response Surface Methodology. *Separations* **2022**, *9*, 356. [CrossRef]
13. Castro, D.; Rosas-Laverde, N.M.; Aldás, M.B.; Almeida-Naranjo, C.; Guerrero, V.H.; Pruna, A.I. Chemical modification of agro-industrial waste-based bioadsorbents for enhanced removal of Zn(II) ions from aqueous solutions. *Materials* **2021**, *14*, 2134.
14. Almeida-Naranjo, C.E.; Morillo, B.; Aldás, M.B.; Garcés, N.; Debut, A.; Guerrero, V.H. Zinc removal from synthetic waters using magnetite/graphene oxide composites. *Remediation* **2023**, *33*, 135–150. [CrossRef]
15. Sellaoui, L.; Bouzidi, M.; Franco, D.S.P.; Alshammari, A.S.; Gandouzi, M.; Georgin, J.; Mohamed, N.B.H.; Erto, A.; Badawi, M. Exploitation of Bauhinia angmuir residual fruit powder for the adsorption of cationic dyes. *Chem. Eng. J.* **2023**, *456*, 141033. [CrossRef]
16. Paredes-Laverde, M.; Silva-agredo, J.; Torres-palma, R.A. Removal of norfloxacin in deionized, municipal water and urine using rice (Oryza sativa) and coffee (Coffee angmu) husk wastes as natural adsorbents. *J. Environ. Manag.* **2018**, *213*, 98–108. [CrossRef]

17. Quesada, H.B.; Alves Baptista, A.T.; Cusioli, L.F.; Seibert, D.; de Oliveira Bezerra, C.; Bergamasco, R. Surface water pollution by pharmaceuticals and an alternative of Removal by low-cost adsorbents: A review. *Chemosphere* **2019**, *222*, 766–780. [CrossRef] [PubMed]
18. Ebele, A.J.; Abou-Elwafa Abdallah, M.; Harrad, S. Pharmaceuticals and personal care products (PPCPs) in the freshwater aquatic environment. *Emerg. Contam.* **2017**, *3*, 1–16. [CrossRef]
19. Wang, J.; Wang, S. Removal of pharmaceuticals and personal care products (PPCPs) from wastewater: A review. *J. Environ. Manag.* **2016**, *182*, 620–640. [CrossRef] [PubMed]
20. Brausch, J.M.; Rand, G.M. A review of personal care products in the aquatic environment: Environmental concentrations and toxicity. *Chemosphere* **2011**, *82*, 1518–1532. [CrossRef]
21. Dhillon, G.S.; Kaur, S.; Pulicharla, R.; Brar, S.K.; Cledón, M.; Verma, M.; Surampalli, R.Y. Triclosan: Current status, occurrence, environmental risks and bioaccumulation potential. *Int. J. Environ. Res. Public Health* **2015**, *12*, 5657–5684. [CrossRef]
22. Kaur, H.; Bansiwal, A.; Hippargi, G.; Pophali, G.R. Effect of hydrophobicity of pharmaceuticals and personal care products for adsorption on activated carbon: Adsorption isotherms, kinetics and mechanism. *Environ. Sci. Pollut. Res.* **2017**, *25*, 20473–20485. [CrossRef]
23. Adeel, M.; Song, X.; Wang, Y.; Francis, D.; Yang, Y. Environmental impact of estrogens on human, animal and plant life: A critical review. *Environ. Int.* **2017**, *99*, 107–119. [CrossRef]
24. Kaur, H.; Hippargi, G.; Pophali, G.R.; Bansiwal, A. Biomimetic lipophilic activated carbon for enhanced removal of triclosan from water. *J. Colloid Interface Sci.* **2019**, *535*, 111–121. [CrossRef]
25. Zepon Tarpani, R.R.; Azapagic, A. Life cycle environmental impacts of advanced wastewater treatment techniques for removal of pharmaceuticals and personal care products (PPCPs). *J. Environ. Manag.* **2018**, *215*, 258–272. [CrossRef] [PubMed]
26. Valle-Sistac, J.; Molins-delgado, D.; Díaz, M.; Ibáñez, L.; Barceló, D.; Díaz-cruz, M.S. Determination of parabens and benzophenone-type UV filters in human placenta. First description of the existence of benzyl paraben and benzophenone-4. *Environ. Int.* **2016**, *88*, 243–249. [CrossRef] [PubMed]
27. Pietrzak, D.; Kmiecik, E.; Malina, G.; Wa, K. Fate of selected neonicotinoid insecticides in soil e water systems: Current state of the art and knowledge gaps. *Chemosphere* **2020**, *255*, 126981. [CrossRef]
28. Glinski, D.A.; Purucker, S.T.; Van Meter, R.J.; Black, M.C.; Henderson, W.M. Analysis of pesticides in surface water, stemflow, and throughfall in an agricultural area in South Georgia, USA. *Chemosphere* **2018**, *209*, 496–507. [CrossRef]
29. de Souza, R.M.; Seibert, D.; Quesada, H.B.; de Jesus Bassetti, F.; Fagundes-Klen, M.R.; Bergamasco, R. Occurrence, impacts and general aspects of pesticides in surface water: A review. *Process Saf. Environ. Prot.* **2020**, *135*, 22–37. [CrossRef]
30. Villamar-Ayala, C.A.; Carrera-Cevallos, J.V.; Espinoza-Montero, P.J.; Alejandra, C.; Carrera-Cevallos, J.V. Technology Fate, eco-toxicological characteristics, and treatment processes applied to water polluted with glyphosate: A critical review. *Crit. Rev. Environ. Sci. Technol.* **2019**, *49*, 1476–1514. [CrossRef]
31. Kaur, Y.; Bhatia, Y.; Chaudhary, S.; Chaudhary, G.R. Comparative performance of bare and functionalize ZnO nanoadsorbents for pesticide removal from aqueous solution. *J. Mol. Liq.* **2017**, *234*, 94–103. [CrossRef]
32. Franco, D.S.P.; Georgin, J.; Lima, E.C.; Silva, L.F.O. Advances made in removing paraquat herbicide by adsorption technology: A review. *J. Water Process Eng.* **2022**, *49*, 102988. [CrossRef]
33. Zavareh, S.; Farrokhzad, Z.; Darvishi, F. Modification of zeolite 4A for use as an adsorbent for glyphosate and as an antibacterial agent for water. *Ecotoxicol. Environ. Saf.* **2018**, *155*, 1–8. [CrossRef] [PubMed]
34. Ilyas, H.; van Hullebusch, E.D. Performance comparison of different constructed wetlands designs for the removal of personal care products. *Int. J. Environ. Res. Public Health* **2020**, *17*, 3091. [CrossRef] [PubMed]
35. Ahmad, J.; Naeem, S.; Ahmad, M.; Usman, A.R.A.; Al-wabel, M.I. A critical review on organic micropollutants contamination in wastewater and removal through carbon nanotubes. *J. Environ. Manag.* **2019**, *246*, 214–228. [CrossRef] [PubMed]
36. Adeel, M.; Yang, Y.S.; Wang, Y.Y.; Song, X.M.; Ahmad, M.A.; Rogers, H. Uptake and transformation of steroid estrogens as emerging contaminants influence plant development. *Environ. Pollut.* **2018**, *243*, 1487–1497. [CrossRef]
37. Diniz, V.; Gasparini Fernandes Cunha, D.; Rath, S. Adsorption of recalcitrant contaminants of emerging concern onto activated carbon: A laboratory and pilot-scale study. *J. Environ. Manag.* **2023**, *325*, 116489. [CrossRef]
38. Palmer, M.; Hatley, H. The role of surfactants in wastewater treatment: Impact, removal and future techniques: A critical review. *Water Res.* **2018**, *147*, 60–72. [CrossRef]
39. Praveena, S.M.; Cheema, M.S.; Guo, H.-R. Non-nutritive artificial sweeteners as an emerging contaminant in environment: A global review and risks perspectives. *Ecotoxicol. Environ. Saf.* **2019**, *170*, 699–707. [CrossRef]
40. Luo, J.; Zhang, Q.; Cao, M.; Wu, L.; Cao, J.; Fang, F.; Li, C. Ecotoxicity and environmental fates of newly recognized contaminants-artificial sweeteners: A review. *Sci. Total Environ.* **2019**, *653*, 1149–1160. [CrossRef]
41. Lapworth, D.J.; Baran, N.; Stuart, M.E.; Ward, R.S. Emerging organic contaminants in groundwater: A review of sources, fate and occurrence. *Environ. Pollut.* **2012**, *163*, 287–303. [CrossRef]
42. Kumar, K.S. Fluorinated Organic Chemicals: A Review. *Res. J. Chem. Environ.* **2005**, *9*, 50–79.
43. Wang, Y.; Chang, W.; Wang, L.; Zhang, Y.; Zhang, Y.; Wang, M.; Wang, Y.; Li, P. A review of sources, multimedia distribution and health risks of novel fluorinated alternatives. *Ecotoxicol. Environ. Saf.* **2019**, *182*, 109402. [CrossRef]
44. Ivanković, T.; Hrenović, J. Surfactants in the environment. *J. Surfactants Environ.* **2010**, *61*, 95–110. [CrossRef] [PubMed]

45. Horton, A.A.; Walton, A.; Spurgeon, D.J.; Lahive, E.; Svendsen, C. Microplastics in freshwater and terrestrial environments: Evaluating the current understanding to identify the knowledge gaps and future research priorities. *Sci. Total Environ.* **2017**, *586*, 127–141. [CrossRef] [PubMed]
46. Silva, A.B.; Bastos, A.S.; Justino, C.I.L.; Duarte, A.C.; Rocha-santos, T.A.P. Microplastics in the environment: Challenges in analytical chemistry—A review. *Anal. Chim. Acta* **2018**, *1017*, 1–19. [CrossRef]
47. Avio, C.G.; Gorbi, S.; Regoli, F. Plastics and microplastics in the oceans: From emerging pollutants to emerged threat. *Mar. Environ. Res.* **2017**, *128*, 2–11. [CrossRef]
48. Padervand, M.; Lichtfouse, E.; Robert, D.; Wang, C. Removal of microplastics from the environment. A review. *Environ. Chem. Lett.* **2020**, *18*, 807–828. [CrossRef]
49. Li, J.; Liu, H.; Paul Chen, J. Microplastics in freshwater systems: A review on occurrence, environmental effects, and methods for microplastics detection. *Water Res.* **2018**, *137*, 362–374. [CrossRef]
50. Zhou, Y.; Wang, T.; Zhi, D.; Guo, B.; Zhou, Y. Applications of nanoscale zero-valent iron and its composites to the removal of antibiotics: A review. *J. Mater. Sci.* **2019**, *54*, 12171–12188. [CrossRef]
51. Abuwatfa, W.H.; Al-Muqbel, D.; Al-Othman, A.; Halalsheh, N.; Tawalbeh, M. Insights into the removal of microplastics from water using biochar in the era of COVID-19: A mini review. *Case Stud. Chem. Environ. Eng.* **2021**, *4*, 100151. [CrossRef]
52. Shi, X.; Zhang, X.; Gao, W.; Zhang, Y.; He, D. Removal of microplastics from water by magnetic nano-Fe3O4. *Sci. Total Environ.* **2022**, *802*, 149838. [CrossRef] [PubMed]
53. León, G.R.; Aldás, M.B.; Guerrero, V.H.; Landázuri, A.C.; Almeida-Naranjo, C.E. Caffeine and irgasan removal from water using bamboo, laurel and moringa residues impregnated with commercial TiO_2 nanoparticles. *MRS Adv.* **2019**, *4*, 3553–3567. [CrossRef]
54. Gavrilescu, M.; Demnerová, K.; Aamand, J.; Agathos, S.; Fava, F. Emerging pollutants in the environment: Present and future challenges in biomonitoring, ecological risks and bioremediation. *New Biotechnol.* **2015**, *32*, 147–156. [CrossRef] [PubMed]
55. Almeida-Naranjo, C.E.; Frutos, M.; Tejedor, J.; Cuestas, J.; Valenzuela, F.; Rivadeneira, M.I.; Villamar, C.A.; Guerrero, V.H. Caffeine adsorptive performance and compatibility characteristics (Eisenia foetida Savigny) of agro-industrial residues potentially suitable for vermifilter beds. *Sci. Total Environ.* **2021**, *801*, 149666. [CrossRef]
56. Ghauch, A.; Tuqan, A.M.; Kibbi, N. Ibuprofen removal by heated persulfate in aqueous solution: A kinetics study. *Chem. Eng. J.* **2012**, *197*, 483–492. [CrossRef]
57. Deng, Y.; Ok, Y.S.; Mohan, D.; Pittman, C.U.; Dou, X. Carbamazepine removal from water by carbon dot-modified magnetic carbon nanotubes. *Environ. Res.* **2018**, *169*, 434–444. [CrossRef]
58. Jiang, N.; Erdős, M.; Moultos, O.A.; Shang, R.; Vlugt, T.J.H.; Heijman, S.G.J.; Rietveld, L.C. The adsorption mechanisms of organic micropollutants on high-silica zeolites causing S-shaped adsorption isotherms: An experimental and Monte Carlo simulation study. *Chem. Eng. J.* **2020**, *389*, 123968. [CrossRef]
59. Boudrahem, N.; Delpeux-ouldriane, S.; Khenniche, L. Single and mixture adsorption of clofibric acid, tetracycline and paracetamol onto Activated carbon developed from cotton cloth residue. *Process Saf. Environ. Prot.* **2017**, *111*, 544–559. [CrossRef]
60. Duan, W.; Wang, N.; Xiao, W.; Zhao, Y.; Zheng, Y. Ciprofloxacin adsorption onto different micro-structured tourmaline, halloysite and biotite. *J. Mol. Liq.* **2018**, *269*, 874–881. [CrossRef]
61. Peng, J.; Wang, X.; Yin, F.; Xu, G. Characterizing the removal routes of seven pharmaceuticals in the activated sludge process. *Sci. Total Environ.* **2019**, *650*, 2437–2445. [CrossRef]
62. Kim, H.; Hwang, Y.S.; Sharma, V.K. Adsorption of antibiotics and iopromide onto single-walled and multi-walled carbon nanotubes. *Chem. Eng. J.* **2014**, *255*, 23–27. [CrossRef]
63. Akhtar, J.; Aishah, N.; Amin, S.; Shahzad, K. A review on removal of pharmaceuticals from water by adsorption. *Desalin. Water Treat.* **2015**, *57*, 12842–12860. [CrossRef]
64. Lozano-Morales, V.; Gardi, I.; Nir, S.; Undabeytia, T. Removal of pharmaceuticals from water by clay-cationic starch sorbents. *J. Clean. Prod.* **2018**, *190*, 703–711. [CrossRef]
65. Käfferlein, H.U.; Göen, T.; Angerer, J. Musk xylene: Analysis, occurrence, kinetics, and toxicology. *Crit. Rev. Toxicol.* **1998**, *28*, 431–476. [CrossRef] [PubMed]
66. Meffe, R.; Bustamante, I. De Science of the Total Environment Emerging organic contaminants in surface water and groundwater: A first overview of the situation in Italy. *Sci. Total Environ.* **2014**, *481*, 280–295. [CrossRef]
67. Shoiful, A.; Ueda, Y.; Nugroho, R.; Honda, K. Degradation of organochlorine pesticides (OCPs) in water by iron (Fe)-based materials. *J. Water Process Eng.* **2016**, *11*, 110–117. [CrossRef]
68. Man, Y.B.; Chow, K.L.; Cheng, Z.; Kang, Y.; Wong, M.H. Profiles and removal efficiency of organochlorine pesticides with emphasis on DDTs and HCHs by two different sewage treatment works. *Environ. Technol. Innov.* **2018**, *9*, 220–231. [CrossRef]
69. Honorio, J.F.; Veit, M.T.; Regina, C.; Tavares, G. Alternative adsorbents applied to the removal of natural hormones from pig farming effluents and characterization of the biofertilizer. *Environ. Sci. Pollut. Res.* **2018**, *26*, 28429–28435. [CrossRef]
70. Akhbarizadeh, R.; Dobaradaran, S.; Schmidt, T.C.; Nabipour, I.; Spitz, J. Worldwide bottled water angmuire of emerging contaminants: A review of the recent scientific literature. *J. Hazard. Mater.* **2020**, *392*, 122271. [CrossRef]
71. Grassi, M.; Kaykioglu, G.; Belgiorno, V.; Lofrano, G. Removal of Emerging Contaminants from Water and Wastewater by Adsorption Process. In *Emerging Compounds Removal from Wastewater: Natural and Solar Based Treatments*; Springer: Berlin/Heidelberg, Germany, 2012; pp. 15–37.

72. Sophia A., C.; Lima, E.C. Removal of emerging contaminants from the environment by adsorption. *Ecotoxicol. Environ. Saf.* **2018**, *150*, 1–17. [CrossRef] [PubMed]
73. Pallarés, J.; González-Cencerrado, A.; Arauzo, I. Production and characterization of activated carbon from barley straw by physical activation with carbon dioxide and steam. *Biomass Bioenergy* **2018**, *115*, 64–73. [CrossRef]
74. Forouzesh, M.; Ebadi, A.; Aghaeinejad-meybodi, A. Separation and Purification Technology Degradation of metronidazole antibiotic in aqueous medium using activated carbon as a persulfate activator. *Sep. Purif. Technol.* **2019**, *210*, 145–151. [CrossRef]
75. García-Mateos, F.J.; Ruiz-Rosas, R.; Marqués, M.D.; Cotoruelo, L.M.; Rodríguez-Mirasol, J.; Cordero, T. Removal of paracetamol on biomass-derived activated carbon: Modeling the fixed bed breakthrough curves using batch adsorption experiments. *Chem. Eng. J.* **2015**, *279*, 18–30. [CrossRef]
76. Raupp, Í.N.; Filho, A.V.; Arim, A.L.; Muniz, A.R.C.; da Rosa, G.S. Development and characterization of activated carbon from olive pomace: Experimental design, kinetic and equilibrium studies in nimesulide adsorption. *Materials* **2021**, *14*, 6820. [CrossRef]
77. Lin, S.H.; Juang, R.S. Adsorption of phenol and its derivatives from water using synthetic resins and low-cost natural adsorbents: A review. *J. Environ. Manag.* **2009**, *90*, 1336–1349. [CrossRef]
78. Jiang, N.; Shang, R.; Heijman, S.G.J.; Rietveld, L.C. High-silica zeolites for adsorption of organic micro-pollutants in water treatment: A review. *Water Res.* **2018**, *144*, 145–161. [CrossRef]
79. Lye, J.W.P.; Saman, N.; Sharuddin, S.S.N.; Othman, N.S.; Mohtar, S.S.; Noor, A.M.M.; Buhari, J.; Cheu, S.C.; Kong, H.; Mat, H. Removal performance of tetracycline and oxytetracycline from aqueous solution via natural zeolites: An equilibrium and kinetic study. *Clean-Soil Air Water* **2017**, *45*, 1600260. [CrossRef]
80. Rasamimanana, S.; Mignard, S.; Batonneau-Gener, I. Hierarchical zeolites as adsorbents for mesosulfuron-methyl removal in aqueous phase. *Microporous Mesoporous Mater.* **2016**, *226*, 153–161. [CrossRef]
81. Sun, K.; Shi, Y.; Wang, X.; Li, Z. Sorption and retention of diclofenac on zeolite in the presence of cationic surfactant. *J. Hazard. Mater.* **2017**, *323*, 584–592. [CrossRef]
82. Pukcothanung, Y.; Siritanon, T.; Rangsriwatananon, K. The efficiency of zeolite Y and surfactant-modified zeolite Y for removal of 2,4-dichlorophenoxyacetic acid and 1,1″-dimethyl-4,4″-bipyridinium ion. *Microporous Mesoporous Mater.* **2018**, *258*, 131–140. [CrossRef]
83. de Sousa, D.N.R.; Insa, S.; Mozeto, A.A.; Petrovic, M.; Chaves, T.F.; Fadini, P.S. Equilibrium and kinetic studies of the adsorption of antibiotics from aqueous solutions onto powdered zeolites. *Chemosphere* **2018**, *205*, 137–146. [CrossRef]
84. Bouras, O.; Bollinger, J.C.; Baudu, M.; Khalaf, H. Adsorption of diuron and its degradation products from aqueous solution by surfactant-modified pillared clays. *Appl. Clay Sci.* **2007**, *37*, 240–250. [CrossRef]
85. Nassar, M.M.; Farrag, T.E.; Ahmed, M.H. Removal of an insecticide (methomyl) from aqueous solutions using natural clay. *Alexandria Eng. J.* **2012**, *51*, 11–18. [CrossRef]
86. Srinivasan, R. Advances in application of natural clay and its composites in removal of biological, organic, and inorganic contaminants from drinking water. *Adv. Mater. Sci. Eng.* **2011**, *2011*, 872531. [CrossRef]
87. Thiebault, T. Raw and modified clays and clay minerals for the removal of pharmaceutical products from aqueous solutions: State of the art and future perspectives. *Crit. Rev. Environ. Sci. Technol.* **2019**, *50*, 1451–1514. [CrossRef]
88. Thiebault, T.; Guégan, R.; Boussafir, M. Adsorption mechanisms of emerging micro-pollutants with a clay mineral: Case of tramadol and doxepine pharmaceutical products. *J. Colloid Interface Sci.* **2015**, *453*, 1–8. [CrossRef] [PubMed]
89. León, H.; Almeida-Naranjo, C.; Aldás, M.B.; Guerrero, V.H. Methomyl removal from synthetic water using natural and modified bentonite clays. *IOP Conf. Ser. Earth Environ. Sci.* **2021**, *776*, 012002. [CrossRef]
90. Zaghouane-Boudiaf, H.; Boutahala, M. Preparation and characterization of organo-montmorillonites. Application in adsorption of the 2,4,5-trichlorophenol from aqueous solution. *Adv. Powder Technol.* **2011**, *22*, 735–740. [CrossRef]
91. Torrellas, S.Á.; Lovera, R.G.; Escalona, N.; Sepúlveda, C.; Sotelo, J.L.; García, J. Chemical-activated carbons from peach stones for the adsorption of emerging contaminants in aqueous solutions. *Chem. Eng. J.* **2015**, *279*, 788–798. [CrossRef]
92. Reeve, P.J.; Fallow, H.J. Natural and surfactant modified zeolites: A review of their applications for water remediation with a focus on surfactant desorption and toxicity towards microorganisms. *J. Environ. Manag.* **2018**, *205*, 253–261. [CrossRef]
93. Yu, F.; Li, Y.; Han, S.; Ma, J. Adsorptive removal of antibiotics from aqueous solution using carbon materials. *Chemosphere* **2016**, *153*, 365–385. [CrossRef]
94. Álvarez-Torrellas, S.; Rodríguez, A.; Ovejero, G.; Gómez, J.M.; García, J. Removal of caffeine from pharmaceutical wastewater by adsorption: Influence of NOM, textural and chemical properties of the adsorbent. *Env. Tech. J.* **2016**, *37*(13), 1618–1630. [CrossRef]
95. Mubarik, S.; Saeed, A.; Mehmood, Z.; Iqbal, M. Phenol adsorption by charred sawdust of sheesham (Indian rosewood; Dalbergia sissoo) from single, binary and ternary contaminated solutions. *J. Taiwan Inst. Chem. Eng.* **2012**, *43*, 926–933. [CrossRef]
96. Benjedim, S.; Romero-cano, L.A.; Pérez-cadenas, A.F.; Bautista-toledo, M.I.; Lotfi, E.M. Removal of emerging pollutants present in water using an E-coli biofilm supported onto activated carbons prepared from argan wastes: Adsorption studies in batch and fixed bed. *Sci. Total Environ.* **2020**, *720*, 137491. [CrossRef]
97. Yanan, C.; Srour, Z.; Ali, J.; Guo, S.; Taamalli, S.; Fèvre-Nollet, V.; da Boit Martinello, K.; Georgin, J.; Franco, D.S.P.; Silva, L.F.O.; et al. Adsorption of paracetamol and ketoprofen on activated charcoal prepared from the residue of the fruit of Butiacapitate: Experiments and theoretical interpretations. *Chem. Eng. J.* **2023**, *454*, 139943. [CrossRef]
98. Thiebault, T.; Boussafir, M. Adsorption Mechanisms of Psychoactive Drugs onto Montmorillonite. *Colloid Interface Sci. Commun.* **2019**, *30*, 100183. [CrossRef]

99. Kuila, U.; Prasad, M. Specific surface area and pore-size distribution in clays and shales. *Geophys. Prospect.* **2013**, *61*, 341–362. [CrossRef]
100. Islam, M.A.; Morton, D.W.; Johnson, B.B.; Mainali, B.; Angove, M.J. Manganese oxides and their application to metal ion and contaminant removal from wastewater. *J. Water Process Eng.* **2018**, *26*, 264–280. [CrossRef]
101. Kasprzyk-Hordern, B. Chemistry of alumina, reactions in aqueous solution and its application in water treatment. *Adv. Colloid Interface Sci.* **2004**, *110*, 19–48. [CrossRef]
102. Turku, I.; Sainio, T.; Paatero, E. Thermodynamics of tetracycline adsorption on silica. *Environ. Chem. Lett.* **2007**, *5*, 225–228. [CrossRef]
103. Nguyen, N.T.; Dao, T.H.; Truong, T.T.; Minh, T.; Nguyen, T.; Pham, T.D. Adsorption characteristic of ciprofloxacin antibiotic onto synthesized alpha alumina nanoparticles with surface modification by polyanion. *J. Mol. Liq.* **2020**, *309*, 113150. [CrossRef]
104. Aazza, M.; Ahla, H.; Moussout, H.; Maghat, H. Adsorption of metha-nitrophenol onto alumina and HDTMA modified alumina: Kinetic, isotherm and mechanism investigations. *J. Mol. Liq.* **2018**, *268*, 587–597. [CrossRef]
105. Aazza, M.; Ahlafi, H.; Moussout, H.; Maghat, H. Ortho-Nitro-Phenol adsorption onto alumina and surfactant modified alumina: Kinetic, isotherm and mechanism. *J. Environ. Chem. Eng.* **2017**, *5*, 3418–3428. [CrossRef]
106. Khan, H.; Gul, K.; Ara, B.; Khan, A.; Ali, N.; Ali, N.; Bilal, M. Adsorptive removal of acrylic acid from the aqueous environment using raw and chemically modified alumina: Batch adsorption, kinetic, equilibrium and thermodynamic studies. *J. Environ. Chem. Eng.* **2020**, *8*, 103927. [CrossRef]
107. Doan, T.H.Y.; Le, T.T.; Nguyen, T.M.T.; Chu, T.H.; Pham, T.N.M.; Nguyen, T.A.H.; Pham, T.D. Simultaneous adsorption of anionic alkyl sulfate surfactants onto alpha alumina particles: Experimental consideration and modeling. *Env. Tech. & Innovation.* **2021**, *24*, 101920. [CrossRef]
108. Bui, T.X.; Choi, H. Adsorptive removal of selected pharmaceuticals by mesoporous silica SBA-15. *J. Hazard. Mater.* **2009**, *168*, 602–608. [CrossRef]
109. Chen, W.; Li, X.; Pan, Z.; Bao, Y.; Ma, S.; Li, L. Efficient adsorption of Norfloxacin by Fe-MCM-41 molecular sieves: Kinetic, isotherm and thermodynamic studies. *Chem. Eng. J.* **2015**, *281*, 397–403. [CrossRef]
110. Tran, T.H. Adsorption and transformation of the anthelmintic drug niclosamide by manganese oxide. *Chemosphere* **2018**, *201*, 425–431. [CrossRef]
111. Remucal, C.K.; Ginder-Vogel, M. A critical review of the reactivity of manganese oxides with organic contaminants. *Environ. Sci. Process. Impacts* **2014**, *16*, 1247–1266. [CrossRef]
112. Feitosa-Felizzola, J.; Hanna, K.; Chiron, S.; Poincare, H. Adsorption and transformation of selected human-used macrolide antibacterial agents with iron (III) and manganese (IV) oxides. *Environ. Pollut.* **2009**, *157*, 1317–1322. [CrossRef]
113. Liu, W.; Langenhoff, A.A.M.; Sutton, N.B.; Rijnaarts, H.H.M. Application of manganese oxides under anoxic conditions to remove diclofenac from water. *J. Environ. Chem. Eng.* **2018**, *6*, 5061–5068. [CrossRef]
114. Zhao, F.; Li, X.; Graham, N. Treatment of a model HA compound (resorcinol) by potassium manganate. *Sep. Purif. Technol.* **2012**, *91*, 52–58. [CrossRef]
115. Zhang, L.; Ma, J.; Yu, M. The microtopography of manganese dioxide formed in situ and its adsorptive properties for organic micropollutants. *Solid State Sci.* **2008**, *10*, 148–153. [CrossRef]
116. Diagboya, P.N.E.; Dikio, E.D. Silica-based mesoporous materials; emerging designer adsorbents for aqueous pollutants removal and water treatment. *Microporous Mesoporous Mater.* **2018**, *266*, 252–267. [CrossRef]
117. Gupta, V.K. Suhas Application of low-cost adsorbents for dye removal—A review. *J. Environ. Manag.* **2009**, *90*, 2313–2342. [CrossRef] [PubMed]
118. Dao, T.H.; Vu, T.Q.M.; Nguyen, N.T.; Pham, T.T.; Nguyen, T.L.; Yusa, S.I.; Pham, T.D. Adsorption characteristics of synthesized polyelectrolytes onto alumina nanoparticles and their application in antibiotic removal. *Langmuir* **2020**, *36*, 13001–13011. [CrossRef] [PubMed]
119. Wang, J.; Zhuang, S.; Wang, J. Removal of various pollutants from water and wastewater by modified chitosan adsorbents. *Crit. Rev. Environ. Sci. Technol.* **2018**, *47*, 2331–2386. [CrossRef]
120. Ali, I.; Asim, M.; Khan, T.A. Low cost adsorbents for the removal of organic pollutants from wastewater. *J. Environ. Manag.* **2012**, *113*, 170–183. [CrossRef]
121. Lessa, E.F.; Nunes, M.L.; Fajardo, A.R. Chitosan/waste coffee-grounds composite: An efficient and eco-friendly adsorbent for removal of pharmaceutical contaminants from water. *Carbohydr. Polym.* **2018**, *189*, 257–266. [CrossRef]
122. Zhou, L.-C.; Meng, X.; Fu, J.; Yang, Y. Highly efficient adsorption of chlorophenols onto chemically modified chitosan. *Appl. Surf. Sci.* **2014**, *292*, 735–741. [CrossRef]
123. Tarasi, R.; Alipour, M.; Gorgannezhad, L.; Imanparast, S.; Yousefi-Ahmadipour, A.; Ramezani, A.; Ganjali, M.R.; Shafiee, A.; Faramarzi, M.A.; Khoobi, M. Laccase Immobilization onto Magnetic β-Cyclodextrin-Modified Chitosan: Improved Enzyme Stability and Efficient Performance for Phenolic Compounds Elimination. *Macromol. Res.* **2018**, *26*, 755–762. [CrossRef]
124. Kyzas, G.Z.; Kostoglou, M.; Lazaridis, N.K.; Lambropoulou, D.A.; Bikiaris, D.N. Environmental friendly technology for the removal of pharmaceutical contaminants from wastewaters using modified chitosan adsorbents. *Chem. Eng. J.* **2013**, *222*, 248–258. [CrossRef]
125. Kyzas, G.Z.; Bikiaris, D.N.; Lambropoulou, D.A. Effect of humic acid on pharmaceuticals adsorption using sulfonic acid grafted chitosan. *J. Mol. Liq.* **2017**, *230*, 1–5. [CrossRef]

126. Zhou, A.; Chen, W.; Liao, L.; Xie, P.; Zhang, T.C.; Wu, X.; Feng, X. Comparative adsorption of emerging contaminants in water by functional designed magnetic poly (N-isopropylacrylamide)/chitosan hydrogels. *Sci. Total Environ.* **2019**, *671*, 377–387. [CrossRef]
127. Nie, Y.; Deng, S.; Wang, B.; Huang, J.; Yu, G. Removal of clofibric acid from aqueous solution by polyethylenimine-modified chitosan beads. *Front. Environ. Sci. Eng.* **2014**, *8*, 675–682. [CrossRef]
128. Zhou, Y.; Chen, L.; Lu, P.; Tang, X.; Lu, J. Removal of bisphenol A from aqueous solution using modified fibric peat as a novel biosorbent. *Sep. Purif. Technol.* **2011**, *81*, 184–190. [CrossRef]
129. Zhou, Y.; Lu, P.; Lu, J. Application of natural biosorbent and modified peat for bisphenol a removal from aqueous solutions. *Carbohydr. Polym.* **2012**, *88*, 502–508. [CrossRef]
130. Chen, K.; Liu, L.; Chen, W. Adsorption of sulfamethoxazole and sulfapyridine antibiotics in high organic content soils. *Environ. Pollut.* **2017**, *231*, 1163–1171. [CrossRef]
131. Mo, J.; Yang, Q.; Zhang, N.; Zhang, W.; Zheng, Y.; Zhang, Z. A review on agro-industrial waste (AIW) derived adsorbents for water and wastewater treatment. *J. Environ. Manag.* **2018**, *227*, 395–405. [CrossRef]
132. FAO. *Setting up Fuel Supply Strategies for Large-Scale Bio-Energy Projects Using Agricultural and Forest Residues*; Junginger, M., Ed.; FAO: Bangkok, Thailand, 2000; ISBN 907395858X.
133. Ngigi, A.N.; Ok, Y.S.; Thiele-Bruhn, S. Biochar-mediated sorption of antibiotics in pig manure. *J. Hazard. Mater.* **2018**, *364*, 663–670. [CrossRef] [PubMed]
134. Bhatnagar, A.; Sillanpää, M.; Witek-Krowiak, A. Agricultural waste peels as versatile biomass for water purification—A review. *Chem. Eng. J.* **2015**, *270*, 244–271. [CrossRef]
135. Mallek, M.; Chtourou, M.; Portillo, M.; Monclús, H.; Walha, K.; Salvadó, V. Granulated cork as biosorbent for the removal of phenol derivatives and emerging contaminants. *J. Environ. Manag.* **2018**, *223*, 576–585. [CrossRef] [PubMed]
136. Ofomaja, A.E. Kinetics and pseudo-isotherm studies of 4-nitrophenol adsorption onto mansonia wood sawdust. *Ind. Crops Prod.* **2011**, *33*, 418–428. [CrossRef]
137. Larous, S.; Meniai, A.-H. The use of sawdust as by product adsorbent of organic pollutant from wastewater: Adsorption of phenol. *Energy Procedia* **2012**, *18*, 905–914. [CrossRef]
138. Thompson, K.A.; Shimabuku, K.K.; Kearns, J.P.; Knappe, D.R.U.; Summers, R.S.; Cook, S.M.; Cook, S.M. Environmental Comparison of Biochar and Activated Carbon for Tertiary Wastewater Treatment. *Environ. Sci. Technol.* **2016**, *50*, 11253–11262. [CrossRef]
139. Rangabhashiyam, S.; Balasubramanian, P. Industrial Crops & Products The potential of lignocellulosic biomass precursors for biochar production: Performance, mechanism and wastewater application—A review. *Ind. Crops Prod.* **2019**, *128*, 405–423. [CrossRef]
140. Li, L.; Zou, D.; Xiao, Z.; Zeng, X.; Zhang, L.; Jiang, L.; Wang, A.; Ge, D.; Zhang, G.; Liu, F. Biochar as a sorbent for emerging contaminants enables improvements in waste management and sustainable resource use. *J. Clean. Prod.* **2018**, *210*, 1324–1342. [CrossRef]
141. Bedia, J.; Peñas-garz, M.; Almudena, G.; Rodriguez, J.J. A Review on the Synthesis and Characterization of Biomass-Derived Carbons for Adsorption of Emerging Contaminants from Water. *J. Carbon Res.* **2018**, *4*, 63. [CrossRef]
142. Choi, Y.K.; Choi, T.R.; Gurav, R.; Bhatia, S.K.; Park, Y.L.; Kim, H.J.; Kan, E.; Yang, Y.H. Adsorption behavior of tetracycline onto Spirulina sp. (microalgae)-derived biochars produced at different temperatures. *Sci. Total Environ.* **2020**, *710*, 136282. [CrossRef]
143. Jing, X.R.; Wang, Y.Y.; Liu, W.J.; Wang, Y.K.; Jiang, H. Enhanced adsorption performance of tetracycline in aqueous solutions by methanol-modified biochar. *Chem. Eng. J.* **2014**, *248*, 168–174. [CrossRef]
144. Kim, E.; Jung, C.; Han, J.; Her, N.; Park, C.; Jang, M.; Son, A.; Yoon, Y. Sorptive removal of selected emerging contaminants using biochar in aqueous solution. *J. Ind. Eng. Chem.* **2016**, *36*, 364–371. [CrossRef]
145. Devi, P.; Saroha, A.K. Utilization of sludge based adsorbents for the removal of various pollutants: A review. *Sci. Total Environ.* **2016**, *578*, 16–33. [CrossRef]
146. Tran, H.N.; Nguyen, H.C.; Woo, S.H.; Nguyen, T.V.; Vigneswaran, S.; Hosseini-Bandegharaei, A.; Rinklebe, J.; Kumar Sarmah, A.; Ivanets, A.; Dotto, G.L.; et al. Removal of various contaminants from water by renewable lignocellulose-derived biosorbents: A comprehensive and critical review. *Crit. Rev. Environ. Sci. Technol.* **2019**, *49*, 2155–2219. [CrossRef]
147. Yang, Q.; Wu, P.; Liu, J.; Rehman, S.; Ahmed, Z.; Ruan, B. Batch interaction of emerging tetracycline contaminant with novel phosphoric acid activated corn straw porous carbon: Adsorption rate and nature of mechanism. *Environ. Res.* **2019**, *181*, 108899. [CrossRef]
148. Nielsen, L.; Bandosz, T.J. Analysis of sulfamethoxazole and trimethoprim adsorption on sewage sludge and fish waste derived adsorbents. *Microporous Mesoporous Mater.* **2016**, *220*, 58–72. [CrossRef]
149. Nielsen, L.; Bandosz, T.J. Analysis of the competitive adsorption of pharmaceuticals on waste derived materials. *Chem. Eng. J.* **2015**, *287*, 139–147. [CrossRef]
150. Gupta, V.K.; Ali, I.; Saini, V.K. Adsorption of 2,4-D and carbofuran pesticides using fertilizer and steel industry wastes. *J. Colloid Interface Sci.* **2006**, *299*, 556–563. [CrossRef]
151. Filho, A.V.; Tholozan, L.V.; Arim, A.L.; de Almeida, A.R.F.; da Rosa, G.S. High-performance removal of anti-inflammatory using activated carbon from water treatment plant sludge: Fixed-bed and batch studies. *Int. J. Environ. Sci. Technol.* **2023**, *20*, 36333644. [CrossRef]

152. Sharipova, A.A.; Aidarova, S.B.; Bekturganova, N.E.; Tleuova, A.; Schenderlein, M.; Lygina, O.; Lyubchik, S.; Miller, R. Triclosan as model system for the adsorption on recycled adsorbent materials. *Colloids Surfaces A Physicochem. Eng. Asp.* **2016**, *505*, 193–196. [CrossRef]
153. Alsbaiee, A.; Smith, B.J.; Xiao, L.; Ling, Y.; Helbling, D.E.; Dichtel, W.R. Rapid removal of organic micropollutants from water by a porous β-cyclodextrin polymer. *Nature* **2016**, *529*, 190–194. [CrossRef]
154. Pan, B.; Pan, B.; Zhang, W.; Lv, L.; Zhang, Q.; Zheng, S. Development of polymeric and polymer-based hybrid adsorbents for pollutants removal from waters. *Chem. Eng. J.* **2009**, *151*, 19–29. [CrossRef]
155. Li, X.; Zhou, M.; Jia, J.; Ma, J.; Jia, Q. Design of a hyper-crosslinked β-cyclodextrin porous polymer for highly efficient removal toward bisphenol a from water. *Sep. Purif. Technol.* **2018**, *195*, 130–137. [CrossRef]
156. Li, H.; Fu, Z.; Yan, C.; Huang, J.; Liu, Y.N.; Kirin, S.I. Hydrophobic–hydrophilic post-cross-linked polystyrene/poly (methyl acryloyl diethylenetriamine) interpenetrating polymer networks and its adsorption properties. *J. Colloid Interface Sci.* **2016**, *463*, 61–68. [CrossRef] [PubMed]
157. Tejedor, J.; Cóndor, V.; Almeida-Naranjo, C.E.; Guerrero, V.H.; Villamar, C.A. Performance of wood chips/peanut shells biofilters used to remove organic matter from domestic wastewater. *Sci. Total Environ.* **2020**, *738*, 139589. [CrossRef]
158. Ahmed, M.B.; Zhou, J.L.; Ngo, H.H.; Guo, W.; Chen, M. Progress in the preparation and application of modified biochar for improved contaminant removal from water and wastewater. *Bioresour. Technol.* **2016**, *214*, 836–851. [CrossRef] [PubMed]
159. Zhai, X.; Ren, Y.; Wang, N.; Guan, F.; Agievich, M.; Duan, J.; Hou, B. Microbial Corrosion Resistance and Antibacterial Property of Electrodeposited Zn–Ni–Chitosan Coatings. *Molecules* **2019**, *24*, 1974. [CrossRef]
160. Anastopoulos, I.; Pashalidis, I.; Orfanos, A.G.; Manariotis, I.D.; Tatarchuk, T.; Sellaoui, L.; Bonilla-Petriciolet, A.; Mittal, A.; Núñez-Delgado, A. Removal of caffeine, nicotine and amoxicillin from (waste) waters by various adsorbents. A review. *J. Environ. Manag.* **2020**, *261*, 110236. [CrossRef]
161. Silva, C.P.; Jaria, G.; Otero, M.; Esteves, V.I.; Calisto, V. Waste-based alternative adsorbents for the remediation of pharmaceutical contaminated waters: Has a step forward already been taken? *Bioresour. Technol.* **2017**, *250*, 888–901. [CrossRef]
162. Bhatnagar, A. Removal of bromophenols from water using industrial wastes as low cost adsorbents. *J. Hazard. Mater.* **2007**, *139*, 93–102. [CrossRef]
163. Oliveira, G.; Calisto, V.; Santos, S.M.; Otero, M.; Esteves, V.I. Paper pulp-based adsorbents for the removal of pharmaceuticals from wastewater: A novel approach towards diversification. *Sci. Total Environ.* **2018**, *631–632*, 1018–1028. [CrossRef]
164. Samah, N.A.; Sánchez-Martín, M.J.; Sebastián, R.M.; Valiente, M.; López-Mesas, M. Molecularly imprinted polymer for the removal of diclofenac from water: Synthesis and characterization. *Sci. Total Environ.* **2018**, *631–632*, 1534–1543. [CrossRef] [PubMed]
165. Zhao, L.; Deng, J.; Sun, P.; Liu, J.; Ji, Y.; Nakada, N.; Qiao, Z.; Tanaka, H.; Yang, Y. Nanomaterials for treating emerging contaminants in water by adsorption and photocatalysis: Systematic review and bibliometric analysis. *Sci. Total Environ.* **2018**, *627*, 1253–1263. [CrossRef] [PubMed]
166. Zhang, Y.; Wu, B.; Xu, H.; Liu, H.; Wang, M.; He, Y.; Pan, B. Nanomaterials-enabled water and wastewater treatment. *NanoImpact* **2016**, *3–4*, 22–39. [CrossRef]
167. Almeida-Naranjo, C.E.; Aldás, M.B.; Cabrera, G.; Guerrero, V.H. Caffeine removal from synthetic wastewater using peel composites: Material characterization, isotherm and kinetic studies. *Environ. Challenges* **2021**, *5*, 100343. [CrossRef]
168. Pham, T.D.; Vu, T.N.; Nguyen, H.L.; Le, P.H.P.; Hoang, T.S. Adsorptive removal of antibiotic ciprofloxacin from aqueous solution using protein-modified nanosilica. *Polymers* **2020**, *12*, 57. [CrossRef]
169. Yu, F.; Ma, J.; Han, S. Adsorption of tetracycline from aqueous solutions onto multi-walled carbon nanotubes with different oxygen contents. *Sci. Rep.* **2014**, *4*, 5326. [CrossRef]
170. Gupta, V.K.; Kumar, R.; Nayak, A.; Saleh, T.A.; Barakart, M.A. Adsorptive removal of dyes from aqueous solution onto carbon nanotubes: A review. *Adv. Colloid Interface Sci.* **2013**, *193–194*, 24–34. [CrossRef] [PubMed]
171. Joseph, L.; Heo, J.; Park, Y.; Flora, J.R.V.; Yoon, Y. Adsorption of bisphenol A and 17 α-ethinyl estradiol on single walled carbon nanotubes from seawater and brackish water. *Desalination* **2011**, *281*, 68–74. [CrossRef]
172. Hu, J.; Tong, Z.; Hu, Z.; Chen, G.; Chen, T. Adsorption of roxarsone from aqueous solution by multi-walled carbon nanotubes. *J. Colloid Interface Sci.* **2012**, *377*, 355–361. [CrossRef]
173. Patiño, Y.; Díaz, E.; Ordóñez, S.; Gallegos-Suarez, E.; Guerrero-Ruiz, A.; Rodríguez-Ramos, I. Adsorption of emerging pollutants on functionalized multiwall carbon nanotubes. *Chemosphere* **2015**, *136*, 174–180. [CrossRef]
174. Hu, X.; Cheng, Z. Removal of diclofenac from aqueous solution with multi-walled carbon nanotubes modified by nitric acid. *Chinese J. Chem. Eng.* **2015**, *23*, 1551–1556. [CrossRef]
175. Hiew, B.Y.Z.; Lee, L.Y.; Lee, X.J.; Gan, S.; Thangalazhy-Gopakumar, S.; Lim, S.S.; Pan, G.T.; Yang, T.C.K. Adsorptive removal of diclofenac by graphene oxide: Optimization, equilibrium, kinetic and thermodynamic studies. *J. Taiwan Inst. Chem. Eng.* **2018**, *98*, 150–162. [CrossRef]
176. Li, M.; Liu, Y.; Zeng, G.; Liu, N.; Liu, S. Graphene and graphene-based nanocomposites used for antibiotics removal in water treatment: A review. *Chemosphere* **2019**, *226*, 360–380. [CrossRef]
177. Naseem, T.; Waseem, M. A comprehensive review on the role of some important nanocomposites for antimicrobial and wastewater applications. *Int. J. Environ. Sci. Technol.* **2021**, *19*, 2221–2246. [CrossRef]

178. Baig, N.; Sajid, M.; Saleh, A. Graphene-based adsorbents for the removal of toxic organic pollutants: A review. *J. Environ. Manag.* **2019**, *244*, 370–382. [CrossRef] [PubMed]
179. Wang, X.; Yin, R.; Zeng, L.; Zhu, M. A review of graphene-based nanomaterials for removal of antibiotics from aqueous environments. *Environ. Pollut.* **2019**, *253*, 100–110. [CrossRef]
180. Borthakur, P.; Boruah, P.K.; Das, M.R.; Kulik, N.; Minofar, B. Adsorption of 17α-ethynyl estradiol and β-estradiol on graphene oxide surface: An experimental and computational study. *J. Mol. Liq.* **2018**, *269*, 160–168. [CrossRef]
181. Moussavi, G.; Hossaini, Z.; Pourakbar, M. High-rate adsorption of acetaminophen from the contaminated water onto double-oxidized graphene oxide. *Chem. Eng. J.* **2016**, *287*, 665–673. [CrossRef]
182. Balasubramani, K.; Sivarajasekar, N.; Naushad, M. Effective adsorption of antidiabetic pharmaceutical (metformin) from aqueous medium using graphene oxide nanoparticles: Equilibrium and statistical modelling. *J. Mol. Liq.* **2020**, *301*, 112426. [CrossRef]
183. Liu, S.; Tang, W.; Yang, Y. Adsorption of nicotine in aqueous solution by a defective graphene oxide. *Sci. Total Environ.* **2018**, *643*, 507–515. [CrossRef]
184. Kunduru, K.R.; Nazarkovsky, M.; Farah, S.; Pawar, R.P.; Basu, A.; Domb, A.J. Nanotechnology for water purification: Applications of nanotechnology methods in wastewater treatment. In *Water Purification*; Elsevier Inc.: Amsterdam, The Netherlands, 2017; pp. 33–74. ISBN 9780128043004.
185. Hamdy, A.; Mostafa, M.K.; Nasr, M. Zero-valent iron nanoparticles for methylene blue removal from aqueous solutions and textile wastewater treatment, with cost estimation. *Water Sci. Technol.* **2018**, *78*, 367–378. [CrossRef] [PubMed]
186. Kumar, B.; Smita, K.; Cumbal, L.; Debut, A.; Galeas, S.; Guerrero, V.H. Phytosynthesis and photocatalytic activity of magnetite (Fe_3O_4) nanoparticles using the Andean blackberry leaf. *Mater. Chem. Phys.* **2016**, *179*, 310–315. [CrossRef]
187. Crane, R.A.; Scott, T.B. Nanoscale zero-valent iron: Future prospects for an emerging water treatment technology. *J. Hazard. Mater.* **2012**, *211–212*, 112–125. [CrossRef] [PubMed]
188. Liu, Y.; Blowes, D.W.; Ptacek, C.J.; Groza, L.G. Removal of pharmaceutical compounds, artificial sweeteners, and perfluoroalkyl substances from water using a passive treatment system containing zero-valent iron and biochar. *Sci. Total Environ.* **2019**, *691*, 165–177. [CrossRef] [PubMed]
189. Al-Jabari, M.H.; Sulaiman, S.; Ali, S.; Barakat, R.; Mubarak, A.; Khan, S.A. Adsorption study of levofloxacin on reusable magnetic nanoparticles: Kinetics and antibacterial activity. *J. Mol. Liq.* **2019**, *291*, 111249. [CrossRef]
190. Minh, T.D.; Lee, B.K.; Nguyen-Le, M.T. Methanol-dispersed of ternary Fe_3O_4@γ-APS/graphene oxide-based nanohybrid for novel removal of benzotriazole from aqueous solution. *J. Environ. Manag.* **2018**, *209*, 452–461. [CrossRef]
191. Rakshit, S.; Sarkar, D.; Punamiya, P.; Datta, R. Kinetics of oxytetracycline sorption on magnetite nanoparticles. *Int. J. Environ. Sci. Technol.* **2013**, *11*, 1207–1214. [CrossRef]
192. Leone, V.O.; Pereira, M.C.; Aquino, S.; Oliveira, L.; Correa, S.; Ramalho, T.C.; Gurgel, L.; da Silva, A.C. Adsorption of diclofenac on a magnetic adsorbent based on maghemite: Experimental and theoretical studies. *New J. Chem.* **2018**, *42*, 437–449. [CrossRef]
193. Zhao, Y.; Liu, F.; Qin, X. Adsorption of diclofenac onto goethite: Adsorption kinetics and effects of pH. *Chemosphere* **2017**, *180*, 373–378. [CrossRef]
194. Qin, X.; Liu, F.; Wang, G.; Weng, L.; Li, L. Adsorption of levofloxacin onto goethite: Effects of pH, calcium and phosphate. *Colloids Surfaces B Biointerfaces* **2014**, *116*, 591–596. [CrossRef]
195. Yin, R.; Sun, J.; Xiang, Y.; Shang, C. Recycling and Reuse of Rusted Iron Particles Containing Core-shell Fe-FeOOH for Ibuprofen Removal: Adsorption and Persulfate-Based Advanced Oxidation. *J. Clean. Prod.* **2018**, *178*, 441–448. [CrossRef]
196. Nassar, M.Y.; Ahmed, I.S.; Hendy, H. A facile one-pot hydrothermal synthesis of hematite (α-Fe_2O_3) nanostructures and cephalexin antibiotic sorptive removal from polluted aqueous media. *J. Mol. Liq.* **2018**, *271*, 844–856. [CrossRef]
197. Rajendran, K.; Sen, S. Adsorptive removal of carbamazepine using biosynthesized hematite nanoparticle. *Environ. Nanotechnol. Monit. Manag.* **2018**, *9*, 122–127. [CrossRef]
198. Fakhri, A.; Behrouz, S. Comparison studies of adsorption properties of MgO nanoparticles and ZnO-MgO nanocomposites for linezolid antibiotic removal from aqueous solution using response surface methodology. *Process Saf. Environ. Prot.* **2015**, *94*, 37–43. [CrossRef]
199. Sivaselvam, S.; Premasudha, P.; Viswanathan, C.; Ponpandian, N. *Enhanced Removal of Emerging Pharmaceutical Contaminant Ciprofloxacin and Pathogen Inactivation Using Morphologically Tuned MgO Nanostructures*; Elsevier B.V.: Amsterdam, The Netherlands, 2020; Volume 8, ISBN 9142224284.
200. Danalıoğlua, S.T.; Bayazit, Ş.S.; Kuyumcu, Ö.K.; Salam, M.A. Efficient removal of antibiotics by a novel magnetic adsorbent: Magnetic activated carbon/chitosan (MACC) nanocomposite. *J. Mol. Liq. J.* **2017**, *240*, 589–596. [CrossRef]
201. Lompe, K.M.; Vo Duy, S.; Peldszus, S.; Sauvé, S.; Barbeau, B. Removal of micropollutants by fresh and colonized magnetic powdered activated carbon. *J. Hazard. Mater.* **2018**, *360*, 349–355. [CrossRef] [PubMed]
202. Wang, B.; Wan, Y.; Zheng, Y.; Lee, X.; Liu, T.; Yu, Z.; Huang, J.; Ok, Y.S.; Chen, J.; Gao, B.; et al. Alginate-based composites for environmental applications: A critical review. *Crit. Rev. Environ. Sci. Technol.* **2018**, *49*, 318–356. [CrossRef]
203. Boruah, P.K.; Sharma, B.; Hussain, N.; Das, M.R. Magnetically recoverable Fe_3O_4/graphene nanocomposite towards efficient removal of triazine pesticides from aqueous solution: Investigation of the adsorption phenomenon and specific ion effect. *Chemosphere* **2016**, *168*, 1058–1067. [CrossRef] [PubMed]
204. Sauvé, S.; Desrosiers, M. A review of what is an emerging contaminant. *Chem. Cent. J.* **2014**, *8*, 15. [CrossRef]

205. Azhar, M.R.; Abid, H.R.; Periasamy, V.; Sun, H.; Tade, M.O.; Wang, S. Adsorptive removal of antibiotic sulfonamide by UiO-66 and ZIF-67 for wastewater treatment. *J. Colloid Interface Sci.* 2017, *500*, 88–95. [CrossRef] [PubMed]
206. Alizadeh Fard, M.; Barkdoll, B. Magnetic activated carbon as a sustainable solution for removal of micropollutants from water. *Int. J. Environ. Sci. Technol.* 2018, *16*, 1625–1636. [CrossRef]
207. Wong, K.T.; Yoon, Y.; Snyder, S.A.; Jang, M. Phenyl-functionalized magnetic palm-based powered activated carbon for the effective removal of selected pharmaceutical and endocrine-disruptive compounds. *Chemosphere* 2016, *152*, 71–80. [CrossRef] [PubMed]
208. Wang, F.; Yang, B.; Wang, H.; Song, Q.; Tan, F.; Cao, Y. Removal of ciprofloxacin from aqueous solution by a magnetic chitosan grafted graphene oxide composite. *J. Mol. Liq.* 2016, *222*, 188–194. [CrossRef]
209. Hossein, M.; Mohammadirad, M.; Shemirani, F.; Akbar, A. Magnetic cellulose ionomer/layered double hydroxide: An efficient anion exchange platform with enhanced diclofenac adsorption property. *Carbohydr. Polym.* 2017, *157*, 438–446. [CrossRef]
210. Ma, Y.; Yang, L.; Wu, L.; Li, P.; Qi, X.; He, L.; Cui, S.; Ding, Y.; Zhang, Z. Carbon nanotube supported sludge biochar as an efficient adsorbent for low concentrations of sulfamethoxazole removal. *Sci. Total Environ.* 2020, *718*, 137299. [CrossRef]
211. Yuan, F.; Yue, L.; Zhao, H.; Wu, H. Study on the adsorption of polystyrene microplastics by three-dimensional reduced graphene oxide. *Water Sci. Technol.* 2020, *81*, 2163–2175. [CrossRef]
212. Sun, C.; Wang, Z.; Zheng, H.; Chen, L.; Li, F. Biodegradable and re-usable sponge materials made from chitin for efficient removal of microplastics. *J. Hazard. Mater.* 2021, *420*, 126599. [CrossRef]
213. Alidadi, H.; Dolatabadi, M.; Davoudi, M.; Askari, F.B.; Farideh, J.-B.; Hosseinzadeh, A. Enhanced Removal of Tetracycline Using Modified Sawdust: Optimization, Isotherm, Kinetics, and Regeneration Studies. *Process Saf. Environ. Prot.* 2018, *117*, 51–60. [CrossRef]
214. Baghdadi, M.; Ghaffari, E.; Aminzadeh, B. Removal of carbamazepine from municipal wastewater effluent using optimally synthesized magnetic activated carbon: Adsorption and sedimentation kinetic studies. *J. Environ. Chem. Eng.* 2016, *4*, 3309–3321. [CrossRef]
215. Peng, X.; Hu, F.; Zhang, T.; Qiu, F.; Dai, H. Amine-functionalized magnetic bamboo-based activated carbon adsorptive removal of ciprofloxacin and norfloxacin: A batch and fixed-bed column study. *Bioresour. Technol.* 2018, *249*, 924–934. [CrossRef]
216. Reguyal, F.; Sarmah, A.K. Site energy distribution analysis and influence of Fe3O4 nanoparticles on sulfamethoxazole sorption in aqueous solution by magnetic pine sawdust biochar. *Environ. Pollut.* 2018, *233*, 510–519. [CrossRef]
217. Zaheer, M.; Sun, X.; Liu, J.; Song, C.; Wang, S.; Javed, A. Enhancement of ciprofloxacin sorption on chitosan/biochar hydrogel beads. *Sci. Total Environ.* 2018, *639*, 560–569. [CrossRef]
218. Liu, Y.; Tourbin, M.; Lachaize, S.; Guiraud, P. Nanoparticles in wastewaters: Hazards, fate and remediation. *Powder Technol.* 2014, *255*, 149–156. [CrossRef]
219. Wanjeri, V.W.O.; Sheppard, C.J.; Prinsloo, A.R.E.; Ngila, J.C.; Ndungu, P.G. Isotherm and kinetic investigations on the adsorption of organophosphorus pesticides on graphene oxide based silica coated magnetic nanoparticles functionalized with 2-phenylethylamine. *J. Environ. Chem. Eng.* 2018, *6*, 1333–1346. [CrossRef]
220. Catherine, H.N.; Ou, M.-H.; Manu, B.; Shih, Y. Adsorption mechanism of emerging and conventional phenolic compounds on graphene oxide nano flakes in water. *Sci. Total Environ.* 2018, *635*, 629–638. [CrossRef]
221. Cao, Z.; Liu, X.; Xu, J.; Zhang, J.; Yang, Y.; Zhou, J.; Xu, X.; Lowry, G.V. Removal of Antibiotic Florfenicol by Sulfide-Modified Nanoscale Zero-Valent Iron. *Environ. Sci. Technol.* 2017, *51*, 11269–11277. [CrossRef]
222. Al-Ahmari, S.D.; Watson, K.; Fong, B.N.; Ruyonga, R.M.; Ali, H. Adsorption Kinetics of 4-n-Nonylphenol on Hematite and Goethite. *Environ. Chem. Eng.* 2018, *6*, 4030–4036. [CrossRef]
223. Marcelo, L.R.; de Gois, J.S.; Araujo, A.; Vargas, D. Synthesis of iron-based magnetic nanocomposites and applications in adsorption processes for water treatment: A review. *Environ. Chem. Lett.* 2020, *19*, 1229–1274. [CrossRef]
224. Taha, M.R.; Mobasser, S. Adsorption of DDT and PCB by nanomaterials from residual soil. *PLoS ONE* 2015, *10*, e0144071. [CrossRef]
225. Kerkes-Kuyumcu, Ö.; Bayazit, S.S.; Alam, M.A. Antibiotic amoxicillin removal from aqueous solution using magnetically modified graphene nanoplatelets. *J. Ind. Eng. Chem.* 2016, *36*, 198–205. [CrossRef]
226. Yin, Y.; Guo, X.; Peng, D. Iron and manganese oxides modified maize straw to remove tylosin from aqueous solutions. *Chemosphere* 2018, *205*, 156–165. [CrossRef]
227. Choong, C.E.; Ibrahim, S.; Yoon, Y.; Jang, M. Removal of lead and bisphenol A using magnesium silicate impregnated palm-shell waste powdered activated carbon: Comparative studies on single and binary pollutant adsorption. *Ecotoxicol. Environ. Saf.* 2018, *148*, 142–151. [CrossRef] [PubMed]
228. Wang, J.; de Ridder, D.; van der Wal, A.; Sutton, N.B. Harnessing biodegradation potential of rapid sand filtration for organic micropollutant removal from drinking water: A review. *Crit. Rev. Environ. Sci. Technol.* 2021, *51*, 2086–2118. [CrossRef]
229. Zhang, L.; Carvalho, P.N.; Bollmann, U.E.; Ei-taliawy, H.; Bester, K. Enhanced removal of pharmaceuticals in a biofilter: Effects of manipulating co-degradation by carbon feeding. *Chemosphere* 2019, *236*, 124303. [CrossRef]
230. Di Marcantonio, C.; Bertelkamp, C.; van Bel, N.; Pronk, T.E.; Timmers, P.H.A.; van der Wielen, P.; Brunner, A.M. Organic micropollutant removal in full-scale rapid sand filters used for drinking water treatment in The Netherlands and Belgium. *Chemosphere* 2020, *260*, 127630. [CrossRef]

231. Hedegaard, M.J.; Prasse, C.; Albrechtsen, H.-J. Microbial degradation pathways of the herbicide bentazone in filter sand used for drinking water treatment. *Environ. Sci. Water Res. Technol.* **2019**, *5*, 521–532. [CrossRef]
232. Hedegaard, M.J.; Albrechtsen, H. Microbial pesticide removal in rapid sand filters for drinking water treatment e Potential and kinetics. *Water Res.* **2014**, *48*, 71–81, Erratum in *Water Res.* **2017**, *122*, 708–713. [CrossRef]
233. Brunner, A.M.; Vughs, D.; Siegers, W.; Bertelkamp, C.; Kolkman, A.; Laak, T. Monitoring transformation product formation in the drinking water treatments rapid sand filtration and ozonation. *Chemosphere* **2018**, *214*, 801–811. [CrossRef]
234. Papadopoulou, A.; Hedegaard, M.J.; Musovic, S.; Smets, B.F. Methanotrophic contribution to biodegradation of phenoxy acids in cultures enriched from a groundwater-fed rapid sand filter. *Appl. Microbiol. Biotechnol.* **2018**, *103*, 1007–1019. [CrossRef]
235. Pompei, C.M.E.; Ciric, L.; Canales, M.; Karu, K.; Vieira, E.M.; Campos, L.C. Influence of PPCPs on the performance of intermittently operated slow sand filters for household water purification. *Sci. Total Environ.* **2017**, *581–582*, 174–185. [CrossRef] [PubMed]
236. Li, J.; Zhou, Q.; Campos, L.C. The application of GAC sandwich slow sand filtration to remove pharmaceutical and personal care products. *Sci. Total Environ.* **2018**, *635*, 1182–1190. [CrossRef]
237. Verma, S.; Daverey, A.; Sharma, A. Slow sand filtration for water and wastewater treatment—A review. *Environ. Technol. Rev.* **2017**, *6*, 47–58. [CrossRef]
238. Georgin, J.; Franco, D.S.P.; Da Boit Martinello, K.; Lima, E.C.; Silva, L.F.O. A review of the toxicology presence and removal of ketoprofen through adsorption technology. *J. Environ. Chem. Eng.* **2022**, *10*, 107798. [CrossRef]
239. Georgin, J.; Franco, D.S.P.; Netto, M.S.; Gama, B.M.V.; Fernandes, D.P.; Sepúlveda, P.; Silva, L.F.O.; Meili, L. Effective adsorption of harmful herbicide diuron onto novel activated carbon from Hovenia dulcis. *Colloids Surfaces A Physicochem. Eng. Asp.* **2022**, *654*, 129900. [CrossRef]
240. Zhou, Y.; Zhang, L.; Cheng, Z. Removal of organic pollutants from aqueous solution using agricultural wastes: A review. *J. Mol. Liq.* **2015**, *212*, 739–762. [CrossRef]
241. Thuptimdang, P.; Siripattanakul-Ratpukdi, S.T.R.; Youngwilai, A.; Khan, E. Biofiltration for Treatment of Recent Emerging Contaminants in Water: Current and Future Perspectives. *Water Environ. Res.* **2020**, *93*, 972–992. [CrossRef]
242. Greenstein, K.E.; Lew, J.; Dickenson, E.R.V.; Wert, E.C. Investigation of biotransformation, sorption, and desorption of multiple chemical contaminants in pilot-scale drinking water biofilters. *Chemosphere* **2018**, *200*, 248–256. [CrossRef]
243. Zhang, S.; Courtois, S.; Gitungo, S.; Raczko, R.F.; Dyksen, J.E.; Li, M.; Axe, L. Microbial community analysis in biologically active filters exhibiting efficient removal of emerging contaminants and impact of operational conditions. *Sci. Total Environ.* **2018**, *640–641*, 1455–1464. [CrossRef]
244. Zhang, Y.; Zhu, H.; Szewzyk, U.; Lübbecke, S.; Geissen, S.U. Removal of emerging organic contaminants with a pilot-scale biofilter packed with natural manganese oxides. *Chem. Eng. J.* **2017**, *317*, 454–460. [CrossRef]
245. Ahmed, M.B.; Zhou, J.L.; Ngo, H.H.; Guo, W.; Thomaidis, N.S.; Xu, J. Progress in the biological and chemical treatment technologies for emerging contaminant removal from wastewater: A critical review. *J. Hazard. Mater.* **2017**, *323*, 274–298. [CrossRef] [PubMed]
246. Ávila, C.; García-Galán, M.J.; Borrego, C.M.; Rodríguez-Mozaz, S.; García, J.; Barceló, D. New insights on the combined removal of antibiotics and ARGs in urban wastewater through the use of two configurations of vertical subsurface flow constructed wetlands. *Sci. Total Environ.* **2021**, *755*, 142554. [CrossRef]
247. Gorito, A.M.; Ribeiro, A.R.; Almeida, C.M.R. A review on the application of constructed wetlands for the removal of priority substances and contaminants of emerging concern listed in recently launched EU legislation *. *Environ. Pollut.* **2017**, *227*, 428–443. [CrossRef] [PubMed]
248. Chen, J.; Liu, Y.; Deng, W.; Ying, G. Removal of steroid hormones and biocides from rural wastewater by an integrated constructed wetland. *Sci. Total Environ.* **2019**, *660*, 358–365. [CrossRef]
249. Arora, S.; Saraswat, S.; Rajpal, A.; Shringi, H.; Mishra, R.; Sethi, J.; Rajvanshi, J.; Nag, A.; Saxena, S.; Kazmi, A.A. Effect of earthworms in reduction and fate of antibiotic resistant bacteria (ARB) and antibiotic resistant genes (ARGs) during clinical laboratory wastewater treatment by vermifiltration. *Sci. Total Environ.* **2021**, *773*, 145152. [CrossRef]
250. Bhat, S.A.; Vig, A.P.; Li, F.; Ravindran, B. (Eds.) *Earthworm Assisted Remediation of Effluents and Wastes*; Springer: Singapore, 2020; ISBN 9789811545214.
251. Shokouhi, R.; Ghobadi, N.; Godini, K.; Hadi, M.; Atashzaban, Z. Antibiotic detection in a hospital wastewater and comparison of their removal rate by activated sludge and earthworm-based vermifilteration: Environmental risk assessment. *Process Saf. Environ. Prot.* **2019**, *134*, 169–177. [CrossRef]
252. Alessio, M.D.; Yoneyama, B.; Kirs, M.; Kisand, V.; Ray, C. Pharmaceutically active compounds: Their removal during slow sand filtration and their impact on slow sand filtration bacterial removal. *Sci. Total Environ.* **2015**, *524–525*, 124–135. [CrossRef] [PubMed]
253. Sabogal-Paz, L.P.; Cintra, L.; Bogush, A.; Canales, M. Household slow sand filters in intermittent and continuous flows to treat water containing low mineral ion concentrations and Bisphenol A. *Sci. Total Environ.* **2020**, *702*, 135078. [CrossRef] [PubMed]
254. Ma, B.; Arnold, W.A.; Hozalski, R.M. The relative roles of sorption and biodegradation in the removal of contaminants of emerging concern (CECs) in GAC-sand biofilters. *Water Res.* **2018**, *146*, 67–76. [CrossRef]
255. Chen, J.; Deng, W.; Liu, Y.; Hu, L.; He, L.; Zhao, J.; Wang, T.; Ying, G. Fate and removal of antibiotics and antibiotic resistance genes in hybrid constructed wetlands. *Environ. Pollut.* **2019**, *249*, 894–903. [CrossRef] [PubMed]

256. Sgroi, M.; Pelissari, C.; Roccaro, P.; Sezerino, P.H.; García, J.; Vagliasindi, F.G.A.; Ávila, C. Removal of organic carbon, nitrogen, emerging contaminants and fluorescing organic matter in different constructed wetland configurations. *Chem. Eng. J.* **2018**, *332*, 619–627. [CrossRef]
257. Oliveira, M.D.; Atalla, A.A.; Emanuel, B.; Frihling, F.; Cavalheri, P.S.; Migliolo, L.; Filho, F.J.C.M. Ibuprofen and caffeine removal in vertical flow and free-floating macrophyte constructed wetlands with Heliconia rostrata and Eichornia crassipes. *Chem. Eng. J.* **2019**, *373*, 458–467. [CrossRef]

Disclaimer/Publisher's Note: The statements, opinions and data contained in all publications are solely those of the individual author(s) and contributor(s) and not of MDPI and/or the editor(s). MDPI and/or the editor(s) disclaim responsibility for any injury to people or property resulting from any ideas, methods, instructions or products referred to in the content.

Review

Physicochemical Technique in Municipal Solid Waste (MSW) Landfill Leachate Remediation: A Review

Hamidi Abdul Aziz [1,2,*], Siti Fatihah Ramli [2] and Yung-Tse Hung [3]

1. School of Civil Engineering, Engineering Campus, Universiti Sains Malaysia, Nibong Tebal 14300, Pulau Pinang, Malaysia
2. Solid Waste Management Cluster, Science & Engineering Research Centre, Engineering Campus, Universiti Sains Malaysia, Nibong Tebal 14300, Pulau Pinang, Malaysia
3. Department of Civil and Environmental Engineering, Cleveland State University, Cleveland, OH 44115, USA
* Correspondence: cehamidi@usm.my; Tel.: +60-4599-6215; Fax: +60-4599-6906

Abstract: Leachate generation is among the main challenging issues that landfill operators must handle. Leachate is created when decomposed materials and rainwater pass through the waste. Leachate carries many harmful pollutants, with high concentrations of BOD, COD, colour, heavy metals, ammoniacal nitrogen (NH_3-N), and other organic and inorganic pollutants. Among them, COD, colour, and NH_3-N are difficult to be completely eliminated, especially with a single treatment. They should be handled by appropriate treatment facilities before being safely released into the environment. Leachate remediation varies based on its properties, the costs of operation and capital expenditures, as well as the rules and regulations. Up until now, much scientific and engineering attention was given to the development of comprehensive solutions to leachate-related issues. The solutions normally demand a multi-stage treatment, commonly in the form of biological, chemical, and physical sequences. This review paper discussed the use of contemporary techniques to remediate landfill leachate with an emphasis on concentrated COD, colour, and NH_3-N levels with low biodegradability that is normally present in old landfill or dumping grounds in developing countries. A semi-aerobic type of landfill design was also discussed, as this concept is potentially sustainable compared to others. Some of the challenges and future prospects were also recommended, especially for the case of Malaysia. This may represent landfills or dumpsites in other developing countries with the same characteristics.

Keywords: physicochemical; solid waste; COD; colour; landfill; leachate; semi-aerobic

Citation: Aziz, H.A.; Ramli, S.F.; Hung, Y.-T. Physicochemical Technique in Municipal Solid Waste (MSW) Landfill Leachate Remediation: A Review. *Water* **2023**, *15*, 1249. https://doi.org/10.3390/w15061249

Academic Editor: Christos S. Akratos

Received: 23 February 2023
Revised: 16 March 2023
Accepted: 20 March 2023
Published: 22 March 2023

Copyright: © 2023 by the authors. Licensee MDPI, Basel, Switzerland. This article is an open access article distributed under the terms and conditions of the Creative Commons Attribution (CC BY) license (https://creativecommons.org/licenses/by/4.0/).

1. Introduction

The number of residents in Malaysia in the third quarter of 2022 was reported to be 32.9 million [1]. This results in a massive quantity of municipal solid waste (MSW) production of around 38,427 metric tonnes per day (1.17 kg/capita/day). A total of 82.5% of waste ends up in landfills. In the same year, the annual MSW collection reached 14 million metric tonnes [2]. Organic waste stands between 40% and 60% of waste in most developing countries. Although there are many advantages to landfilling for waste disposal, it raised serious issues because of the highly polluted leachate it creates. Leachate contaminants include more than 200 different chemicals, most of which are hazardous to the environment. Rainwater infiltrating the deposited waste at the landfill or dumpsite results in the formation of leachate. Additionally, leachate may be produced from a number of different sources, such as transpiration, groundwater intake, storing wet materials, evaporation losses, surface flow, and the hydrolysis and biodegradation of organic molecules [3]. Leachate may drain as runoff or move to the bottom of the waste body. These may pollute groundwater and surface water, endangering both aquatic life and human health. The amount and composition of leachate are influenced by a variety of variables, including seasonal weather fluctuations, landfilling methods, nature of waste type and composition, and landfill design. Due to excessive rainfall during the wet season and high evaporation rates in tropical nations, such as Malaysia, landfill leachate is quickly produced.

Knowledge in improving the treatment for this landfill leachate management control is still an ongoing process and attracted significant attention from scientists, engineers, and technologists throughout the world. Various leachate treatment techniques were developed in many different ways, including those that incorporate biological, physical, and chemical processes and their combinations. Further improvements, especially on the optimisation for cost saving, are part of the continuing efforts.

The main problem with landfill leachate is the amount and level of variability that it exhibits [4]. In order to choose the best leachate treatment method, it is necessary to characterise the leachate and estimate its risk. In addition, treating the high concentration of NH_3-N in landfill leachate is a challenging process. The physicochemical method successfully eliminates heavy metals, inorganic macro-components, and refractory organic molecules from leachate. On the other hand, biological processes effectively remove dissolved organics and nutrients from leachates [5].

The age and biodegradability of the leachate limit the biological process for leachate treatment [6,7]. Physical-chemical techniques are normally necessary for lowering hazardous and refractory substances [8]. As a result, an integrated strategy combining biological with either pre- or post-physical-chemical processes is an effective choice that offers higher effluent quality [9]. The purpose of this paper is to review and summarise various physicochemical technologies in leachate treatment and compare their performances and limitations. There are many physicochemical treatments in the literature that provide effective techniques to deal with substantial organic content in leachate. The review also helps to better understand the suitable methods for specific types of leachate which may be applied in the field.

2. Landfilling

Landfilling stands among the main elements in a solid waste management strategy even after the implementation of the 3Rs: reduce, recover, recycle. This is due to the fact that, in most cases, not all waste can be recovered and recycled. Open dumping or regulated dumping are the two main methods used by most developing countries to dispose of their waste. These countries normally have financial constraints to apply costly treatment systems such as materials recovery facilities, waste-to-energy technology, etc. For the same reason, Malaysia currently relies primarily on landfilling as its main way of disposing of its MSW, and this is expected to still be the favoured option for the next 10 to 15 years to come. However, due to economic limitations, proper sanitary landfill concepts have not yet been fully implemented in the whole country. There are many old and improperly designed landfill sites that are still in operation to date, and some of them are almost reaching their end of life. The leachate is still being produced and must be properly controlled.

The category of landfill could be generally grouped into anaerobic, semi-aerobic, and aerobic. In developing countries, the anaerobic landfill is the most common, where the waste is commonly discarded and covered and, sometimes, left uncovered. Open dumping is still being practised, but the trend is now towards control tipping, and more countries are moving forward towards the sanitary landfill. The anaerobic landfills produce concentrated leachate, which is difficult to treat by a conventional method to up to the standard discharge limits. This type of landfill is further constrained by fire incidents and greenhouse gas emissions, which primarily contain methane (CH_4) and carbon dioxide (CO_2). Quite commonly, these gases are untapped and just released untreated in developing countries due to many limitations, especially the cost factor.

On the other hand, the Fukuoka technique, or a semi-aerobic method, was established at Fukuoka University over 20 years ago and was employed in numerous sites around Japan, China, Iran, and Malaysia. However, this method was not widely adopted by many other countries. Semi-aerobic landfilling was first used in Malaysia in 1988, and since then, the quality of the leachate has improved noticeably. The Fukuoka approach can be used in developing nations in a variety of situations for various goals, such as creating new landfill sites, improving existing ones, or effectively closing ones that were already constructed. Leachate collection pipes are built beneath the semi-aerobic landfill, as shown in Figure 1.

Leachate is removed by this conduit from a disposal location. Air from an open pit is extracted into this leachate collection pipe, which then moves into the waste body. In this manner, greenhouse gases such as methane and carbon dioxide are produced less, as the process encourages aerobic biodegradation or organic matter and allows early waste stabilisation [10]. Figure 2 illustrates a typical layout of a semi-aerobic landfill.

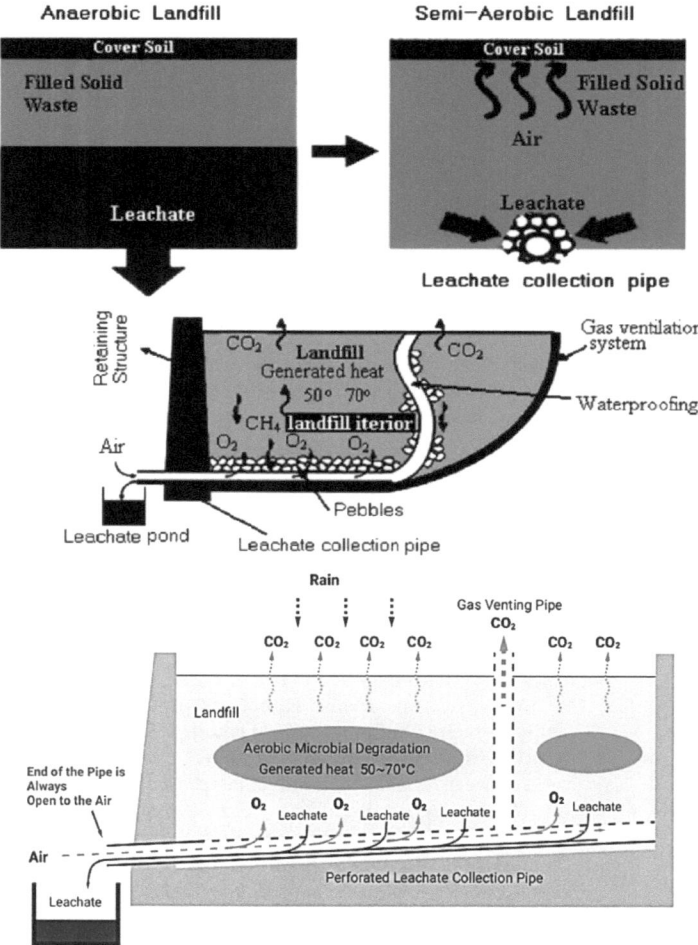

Figure 1. Conceptual landfill design using a semi-aerobic concept [11].

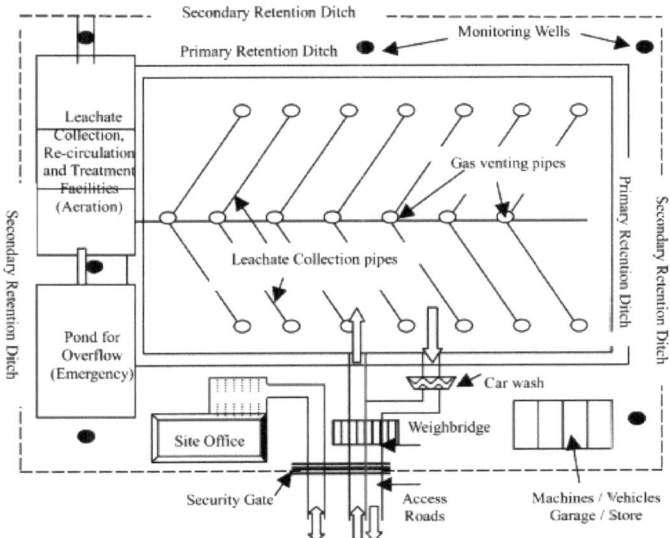

Figure 2. An illustration of a landfill that uses a semi-aerobic process [12].

Gas venting and leachate collecting pipelines are crucial components of a semi-aerobic landfill system. Figure 3 shows semi-aerobic landfill leachate pipes which provide air passage and remove leachates from the waste body through the natural convection process of cold air (outside) and hot air (within the waste body). This process mimics human blood veins [13]. Additionally, these pipelines have a number of benefits. Leachate, for instance, is evacuated more quickly than it would be in landfills without these pipelines. As a result, leachate fouling in waste materials is avoided and landfills are conveniently accessible to fresh air. Aerobic environments promote microbial activity and enhance waste decomposition. Due to their placement within rocks, collection pipelines are shielded both from clogging and operational harm. Leachate seepage is less likely because leachate is quickly drained, which lowers the pressure brought on by water on the ground [14].

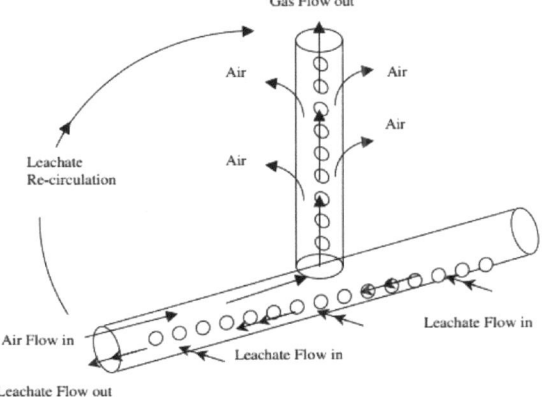

Figure 3. Leachate collection and gas ventilation pipes in a semi-aerobic landfill [12].

Landfills that employ a semi-aerobic concept are potentially sustainable, as they could offer various advantages over alternative solutions. When leachate flows through the pipes and is released from the sites, it lowers water pressure and prevents seepage. Garbage naturally allows fresh air to pass through it. Thus, leachate cleaning and waste stabilisation only take a short while. The amount of methane released decreases, despite an increase in carbon dioxide content. Semi-aerobic landfills also require straightforward technology, simple setup and operation, a small number of engineering protocols, equipment, and machinery, easier maintenance and operation, and inexpensive startup costs. Semi-aerobic landfills also contribute to a reduction in global warming by limiting the release of methane [13].

3. Characteristics of Landfill Leachate

Leachate is abundant in a wide variety of compounds (organic and inorganic), including humic and fulvic substances, heavy metals, fatty acids, and other potentially harmful compounds. Table 1 outlines the typical characteristics of landfill leachate. Numerous papers noted considerable variations in the components of the leachate. However, landfill age was used to identify three categories of leachates: fresh (under 5 years), transitional (5–10 years), and stabilised or old (over 10 years). Leachate quality can generally be controlled by a range of elements, such as the age of the site, rainfall, weather changes (which are seasonal), the nature of waste, and the waste properties. Fresh leachate is rich in organics and is highly biodegradable. NH_3-N is dominant in aged and stabilised landfill; it is normally non-biodegradable. Transitional landfill leachate has an intermediate quality between the young leachate and the mature leachate [15].

Table 1. Typical compositions of landfill leachate.

No.	Parameter	Unit	Type of Landfill Leachate		
			Young (<5 Years)	Intermediate (5–10 Years)	Stabilised (>10 Years)
1	pH		<6.5	6.5–7.5	>7.5
2	COD	mg/L	>10,000	4000–10,000	<4000
3	BOD_5/COD		0.5–1.0	0.1–0.5	<0.1
4	Organic compound		80% VFA [a]	5–30% VFA [a] + HFA [b]	HFA [b]
5	NH_3-N	mg/L	<400	NA [c]	>400
6	TOC/COD		<0.3	0.3–0.5	>0.5
7	Kjeldahl nitrogen	g/L	0.1–0.2	NA [c]	NA [c]
8	Heavy metals	mg/L	Low to medium	Low	Low
9	Biodegradability		Important	Medium	Low

Note: Source: [15]. [a] VFA is volatile fatty acid. [b] HFA is humic and fulvic acid. [c] NA is not available.

High colour intensity indicates that leachate consists significant content of organic substance, as colour is one of the important indicators of organic loading [16]. Concentrated COD, colour, and ammoniacal nitrogen (NH_3-N) were discovered during long-term landfill leachate monitoring in Malaysia. These high concentrations are regularly regarded by landfill operators as urgent problems that require proper attention.

4. Landfill Leachate Treatment

Among the biggest issues in managing landfill is determining how to deal with enormous and considerable amounts of leachate. A variety of techniques, including physicochemical, biological, and chemical processes, are normally required to remediate leachate effectively. These techniques typically involve numerous expensive and labour-intensive operations. Leachate treatments are difficult to perform owing to their hefty loading, complicated compounds, and flow that changes with the seasons [17]. The effectiveness of leachate treatment is generally improved with the right combination of treatment methods. Various combinations were reported [18]. Table 2 listed the options for treating leachate based on landfill age. A few of the typically employed methods to treat landfill leachate are shown in Figure 4. This is followed by Table 3, which provides different leachate treatment results according to landfill age. Cherni et al. [19], Mojiri et al. [7], and Teng et al. [5] reviewed the effectiveness of several leachate treatments from various literature sources. This is followed by Table 4, which presents the removal of leachate parameters by numerous applications by Matsufuji [20]. In Table 5, a summary of various advanced oxidation processes (AOPs) employed in processing leachate is presented as sourced and reviewed by Cherni et al. [19] and Teng et al. [5].

Table 2. Options for treating leachate based on landfill age.

Leachate Treatment	Landfill Age (Years)		
	Young (<5)	Intermediate (5–10)	Mature (>10)
Co-treatment with domestic wastewater	Good	Fair	Poor
Recycling	Good	Fair	Poor
Aerobic process (suspended growth)	Good	Fair	Poor
Aerobic process (fixed film)	Good	Fair	Poor
Anaerobic process (suspended growth)	Good	Fair	Poor
Anaerobic process (fixed film)	Good	Fair	Poor
Natural evaporation	Good	Good	Good
Coagulation/flocculation	Poor	Fair	Fair
Chemical precipitation	Poor	Fair	Poor
Carbon adsorption	Poor	Fair	Good
Oxidation	Poor	Fair	Fair
Air stripping	Poor	Fair	Fair
Ion exchange	Good	Good	Good
Microfiltration	Poor	-	-
Ultrafiltration	Poor	-	-
Nanofiltration	Good	Good	Good
Reverse osmosis	Good	Good	Good

Note: Adapted from reference [21].

Table 3. An overview of various technologies in treating leachate.

	Coagulation Flocculation				
Parameter (Removals)	Turbidity (90%) NH_3–N (46.7%) COD (53.9%)	TP (47%) TOC (15%) NH_3–N (20%) TN (4%)	COD (61.9%) Colour (98.8%) SS (99.5%)	Organic Matter (22.57)	
	Electrocoagulation (EC)				
Parameter (Removals)	(with Al electrodes) COD (70%) TN (24%) Colour (56%) Turbidity (60%)	(with Fe electrodes) COD (68%) TN (15%) Colour (28%) Turbidity (16%)	COD (60%) NH_3–N (37%) Colour (94%) Turbidity (88%) SS (89%)	heavy metals Cr (51%) As (59%) Cd (71%) Zn (72%) Ba ((95%) Pb (>99%)	
	Adsorption				
Parameter (Removals)	COD (77.3%) Colour (82.5%)	COD (93.6%) NH_3–N (84.8%)	Colour (100%) COD (~80%) $NH3^+$-N (100%)	COD (36%) NH_3–N (99%) Cl (18%)	COD (51.0%) NH_3–N (32.8%) Cl (66.0%) Br (81.0%) Cu (97.1%)

Note: Adapted from: [5,7,19].

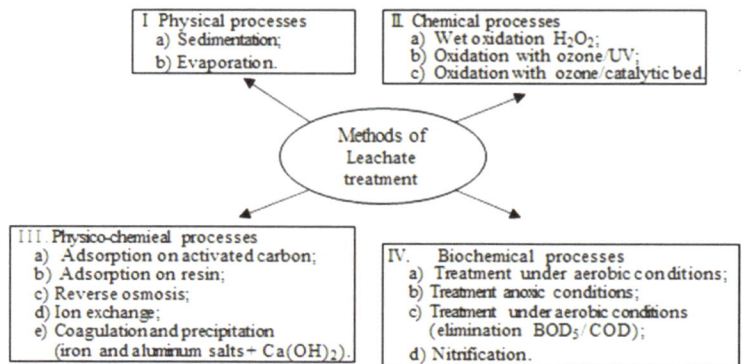

Figure 4. Techniques frequently employed at municipal landfills to treat leachate [3].

Table 4. Removal of leachate parameters by various applications performance [20].

Treatment Method	Leachate Parameters					
	BOD	COD	SS	NH_3-N	Colour	Heavy Metals
Activated Sludge Process	▲	•	∅	∅	∅	∅
Contact Aeration Process	▲	•	∅	∅	∅	∅
Rotary Biodisk Conductor Process	▲	•	∅	∅	∅	∅
Biological Trickling Process	▲	•	▲	∅	∅	∅
Biological Nitrogen	▲	•	∅	▲	∅	∅
Flocculation-Sedimentation	•	▲	▲	∅	▲	•
Sand filtration	∅	∅	▲	×	•	×
Activated Carbon (Adsorption)	▲	▲	•	∅	▲	•
Chemical Oxidation	×	•	×	×	▲	×

Notes: High (▲) Medium (•) Low (∅).

Table 4 demonstrates the effectiveness of biological processes in treating leachates from newly constructed landfills (less than five years old). Biological treatment is normally employed when the leachate is biodegradable (BOD_5/COD > 0.3). As shown, biological treatment is the most appropriate method for fresh leachates containing high concentrations of organic material (>10,000 mg/L). For mature and stabilised leachates which are in the methanogenic phase (>10 years), the biodegradability ratio (BOD_5/COD) is normally less than 0.1, and it is hard to be biodegraded. It is normally rich in humic and fulvic acids [21]. For leachates with elevated NH_3-N levels and limited biodegradability, a physical chemical approach is the best option.

Malaysia uses a variety of leachate treatment techniques. Figure 5 illustrates one of the promising leachate treatment systems in Malaysia for a moderate site which receives 1500 tons of MSW per day.

Table 5. Summary of advanced oxidation processes (AOPs) in treating leachate.

AOP	Removal Efficiency		AOP	Removal Efficiency	
	Parameters	Removal (%)		Parameters	Removal (%)
Fenton	TOC COD	68.9 69.6	Ozone (O_3)	Colour COD Ammonia	100 88 79
	COD	88.6		COD Colour	70 100
	COD	70		COD Colour	16.5 40.5
	COD UV_{254} Colour	58.70 85.69 88.30		Humic Acid Fulvic Acid	88 83.3
	COD	97.83		COD	43
	COD BOD_5	48 30		COD TOC BOD_5	65 62 36
	COD TOC	97.83 74.24		COD UV_{254}	46 51
	total organic carbon, total inorganic carbon total nitrogen, colour	88.7 100 96.5 98.2		Colour UV	~90 ~70
TiO$_2$ Photocatalysis	COD Colour	58 36	Electro-oxidation	COD TOC	68 40.6
	COD TOC	67 82.5		COD	80
	COD	84		TOC Ammonium nitrogen	40 99
Ferrosonication (FS)	COD BOD_5	46 33	W-doped TiO$_2$	COD	46
Heterogeneous catalytic ozonation (O_3/TiO$_2$)	COD NTU BOD_5	24 94 98	Heterogeneous catalytic ozonation (O_3/ZnO)	COD NTU BOD_5	33 95 98

Note: Adapted from: [5,19].

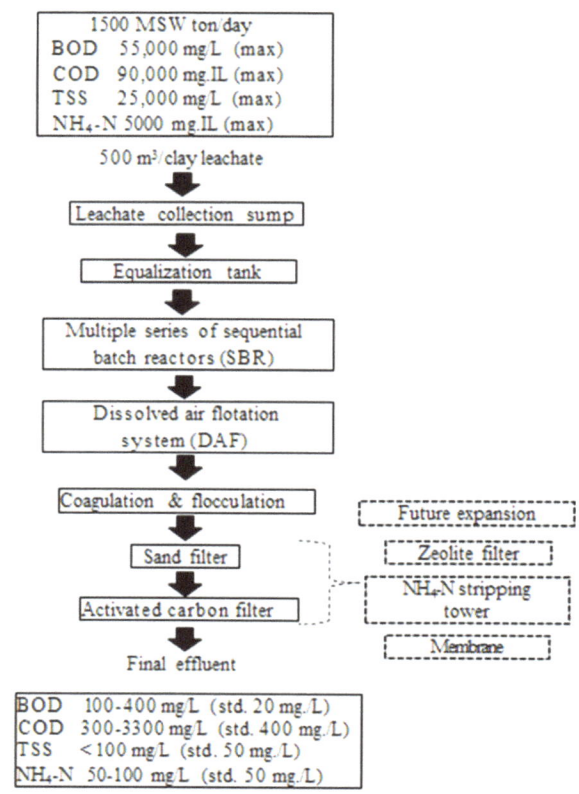

Figure 5. A typical landfill leachate treatment system in Malaysia for a moderate site landfill site (1500 tons/day) [22].

5. Physicochemical Treatment for Landfill Leachate

Biological processes have limitations in treating recalcitrant and non-biodegradable compounds with a BOD_5 to COD ratio of less than 0.5. Leachate treatment using physicochemical methods is often more efficient financially and quicker to complete for this kind of leachate [23]. This physicochemical approach can be used to treat old leachate that has highly elevated levels of COD and NH_3-N, low BOD, and good oxidation-reduction ability. The most popular physicochemical processes are coagulation-flocculation, adsorption, chemical precipitation, reverse osmosis, ammonia stripping, and oxidation [24]. Figure 6 illustrates the criteria that should be considered when selecting a treatment for landfill leachate [21].

The necessity to boost the efficacy of biological systems led to the development of physicochemical procedures. They are, therefore, frequently used following a biological pre-treatment. This technique functions by changing the chemical makeup of specific compounds or the physical components that can trap or remove pollutants. Table 6 lists some of these technologies' benefits and drawbacks [24].

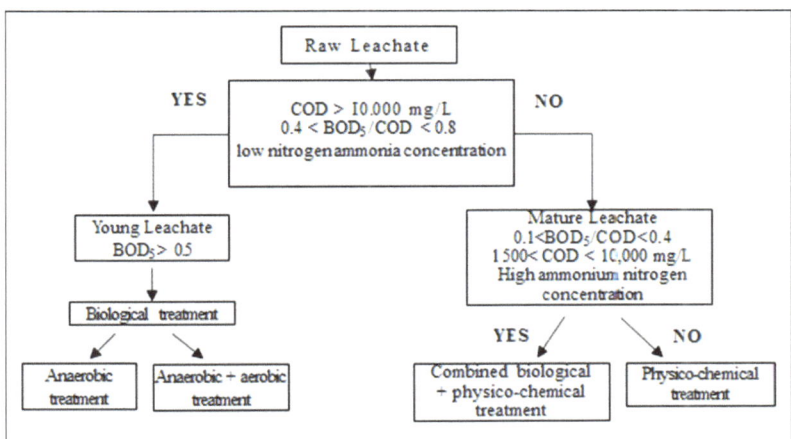

Figure 6. A typical leachate treatment selection protocol. Source [21].

Table 6. Leachate physicochemical treatment system benefits and drawbacks.

Physicochemical	Advantages	Disadvantages	Observations
Coagulation and Flocculation	Effective at removing suspended particles, humic acids, heavy metals, and organic matter.	Owing to the expense of inputs and the handling of the created chemical sludge, the system's functioning requires very high coagulant concentrations, making it economically impracticable to implement this technology on a large scale.	For some membrane systems, this technology serves as a pre-treatment. Some membrane systems appear to use this technique as pre-treatment.
PACT (Powdered Activated Carbon Treatment)	Removes some poisons, chlorine, phenols, ammoniacal nitrogen, colour, odour, and taste. Safeguards the process against BOD and organic toxin shock loads by stabilising it. It is simple to use, operate, and maintain and has inexpensive installation costs. Pre-treatment technology is used with several membrane systems.	High operating expenses with on-site regenerating or coal deployment, as well as outputs with high potential pollutants	Aeration, biological oxidation, and physical adsorption happen at the same time as coal is supplied directly to the reactor.
Advanced Chemical Oxidation	It divides these high molecular weight molecules, which increases their treatability by making them more receptive to microorganisms in biological reactors and partially eliminates recalcitrant organic material and refractory chemicals.	Due to the complexity of the operation and the high cost of operation, such as energy and the value of the inputs necessary in significant doses, a competent technical operator is required.	The most common oxidative technology is ozonisation.
Evaporation	Up to 95% reduction in leachate volume.	Polluting gases are released, and it costs a lot of energy-60 kg of gasoline are required to burn 1 m^3 of leachate. An output of dry sludge equal to around 5% of the entire volume is produced.	The option that is most frequently used is the landfill's own biogas being captured and burned.

Note: Adapted from: [19].

5.1. Ammoniacal Nitrogen (NH$_3$-N) Reduction by Air Stripping

The content level of ammoniacal nitrogen (NH$_3$-N) in landfill leachate level can be decreased or removed effectively using the air stripping technique. Basically, this physical mechanism is effective when the pH is 11. This ammonia loss is caused by desorption through the water surface, and the process is influenced by temperature. The stripping techniques are significantly influenced by the leachate quality and the treatment reactor layout. By raising the flow rate of the incoming air (air: waste flow rate) and employing smaller bubble diffusers, ammonia elimination can be increased [25]. The primary disadvantage of air stripping is the environmental pollution brought on by the release of NH_3 into the atmosphere. Consequently, further exhaust gas treatment is required. However, the NH_3-N becomes gas at this pH, and this gas, combined with the acidic solution, will produce ammonium salts that can be utilised as fertiliser in the mineral form [19]. Another key drawback of this technique is the limited effectiveness of organic matter removal, even though that air stripping is excellent at removing NH_3-N. Despite its ability to remove 90% of the ammonia at 20 °C, air stripping alone normally could not meet effluent discharge restrictions. As a result, the subsequent nitrification and denitrification process via air stripping can probably meet the effluent discharge guidelines. Furthermore, despite the removal of COD and ammonia, air stripping was found to increase toxicity in several laboratory tests. However, air stripping was discovered to be the most economical alternative approach for high ammonia removal when compared to other processes such as membrane filtration [26].

A number of operational conditions, including pH, initial NH_3-N level, hydraulic gas-to-liquid ratio (G/L), loading rate, and recirculation period [27], influence the air stripping system. Their study showed that, regardless of changes in the G/L or hydraulic loading rate (HLR), increasing pH from 9 to 12 resulted in a considerable improvement in ammonia removal efficiency, with the maximum ammonia stripping achieved at pH 12. As the G/L ratio rose, the removal efficacy increased by up to 56% for both HLRs of 57.6 and 172.8 m^3/m^2/day. Under the following conditions: HLR of 172.8 m^3/m^2/day, pH 12, G/L of 728, and liquid recirculation, 99% of its ammonia (2520 mg/L) was stripped within three hours. The ammonia concentration in the final sample was 25.2 mg/L, which is almost as good as the allowable discharge limit.

Leite et al. [28] investigated and characterised the NH_3-N stripping method at open horizontal flow reactors. The study involved the superficial load in three different phases phase 1 (650 kg N-NH_4^+ ha^{-1}.day^{-1}), phase 2 (750 kg N-NH_4^+ ha^{-1}.day^{-1}), and phase 3 (850 kg N-NH_4^+ ha^{-1}.day^{-1}). The procedure demonstrated a removal efficiency of 99.0%, 99.3%, and 99.5% in the first, second, and third phases, respectively. In addition, phases 1 and 2 had removal efficiencies of 69.2%, 40.12%, and 29.23%, respectively, for organic matter reported in terms of total COD. Following a series of tests, the researchers concluded and demonstrated that the effectiveness of ammonia removal was directly connected with the surface load, but the effectiveness of carbonaceous material removal was correlated with the amount of organic matter applied in the influent.

5.2. Coagulation-Flocculation Process

Prior to a biological process, coagulation-flocculation is one of the several physico-chemical treatments that is generally employed in the preparation of stabilised and matured landfill leachates [19]. Generally, the physical-chemical through a coagulation-flocculation process involves the destabilisation of small particles (colloids) in wastewater to create flocs that could be easily precipitated. In an effort to destabilise the colloidal particles, various coagulants react differently with colloidal particles [29]. Coagulants commonly occur in a variety of forms. The most often used coagulants are chemical-based ones, such as alum and ferric salts. When trace metal salts such as ferrous sulphate, aluminium sulphate, and ferric chloride are added during the coagulation-flocculation stage, high valence cations are formed in the solution, which lowers the zeta potential values. Ferric ion salts generally outperform aluminium salts, primarily due to their insoluble nature over a wider pH range [19]. These destabilising occurrences are caused by a variety of mechanisms, such as charge neutralisation, trapping, adsorption, and complexation with the metal ions of the coagulant to create insoluble aggregates [30]. The dosage and types of coagulant/flocculant, as well as the experimental settings such as pH, time, and temperature, have a considerable effect on the coagulation efficacy [31]. In addition, the efficacy of the coagulation method, which removes organic matter and phenolic compounds from wastewater, is also influenced by the mixing conditions and the characteristics of phenolic compounds, such as particle size, charge, and hydrophobicity. The coagulation-flocculation approach was used successfully to eliminate non-biodegradable organics, suspended solids, colloidal particulates, colour, turbidity, and heavy metals, depending on the pollutant and types of coagulant and flocculant [30].

Leachate treatment via the coagulation-flocculation technique was the subject of a great deal of research. Djeffal et al. [32] investigated the efficacy of the coagulation-flocculation technique for the purification of leachate from the Souk-Ahras City Technical Centre landfill in Algeria. Three distinct coagulants; ferric chloride, aluminium sulphate, and alum, as well as two agitation techniques; mechanical and ultrasound, were applied in the study to optimise the operational conditions. With a 15% coagulant dosage, 250 rpm stirring rate, and a response time of roughly 15 min for ferric chloride, a considerable drop in turbidity (99.4%) was made possible. The turbidity was reduced by 98.9% and 98.6%, using aluminium sulphate and alum, respectively, with the other two coagulants having an optimal coagulant-to-leachate volume ratio of one. The results of the bacteriological tests also showed a lack of *Streptococci*, total germs, and faecal coliforms. Furthermore, when a 37 kHz ultrasonic waves frequency of 30 W power was used to treat the leachate, it was discovered that the turbidity of the supernatants greatly decreased.

Mohd-Salleh et al. [33] employed polyaluminium chloride (PAC) in treating leachate in various operating settings (variable dosage and pH). The objectives of the study were to identify the best coagulant dosage in a range of dosages (2250–4500 mg/L), as well as to examine the best pH (pH 3–10). This was carried out through different sets of jar tests to assess the influence of five different leachate parameters, including suspended particles, NH_3-N, COD, and heavy metals, on the removal efficacy. They concluded that the ideal PAC dosage and sample pH was 3750 mg/L and pH 7, respectively. Reductions of 95%, 53%, 97%, and 79% in the suspended particles, COD, Fe, and Cr, respectively, were achieved at this ideal dose and pH.

5.3. Adsorption

Molecules from a normally liquid medium (the adsorbate) are pulled to and maintained on the surface of the other, frequently solid medium (adsorbent) as a surface phenomenon (adsorption). A large surface area of the adsorbent is needed to boost the treatment's efficiency because the process takes place on its surface. The adsorbate can attach to the surface of the adsorbent due to the specific properties of its surface. Because adsorption includes a surface mechanism, the adsorbent surface area is important, and an adsorbent with a greater surface area and increased porosity offers the adsorbate more interaction sites [34]. It is also conceivable for a reversible phenomenon known as desorption to occur when adsorption takes place under specific circumstances. Adsorbates are transported back to the liquid phase during desorption after being liberated from the adsorbent's surface. Figure 7 illustrates the fundamental idea of adsorption [35].

Physical adsorption, also known as physisorption, and chemical adsorption, sometimes referred to as chemisorption, are the two categories into which adsorption can be separated. In essence, the bonding between the two forms of adsorption varies. Van der Waals forces of attraction among the adsorbent and adsorbate produced physical bonding in the process of physisorption; meanwhile, adsorbate and adsorbent in chemisorption were attracted to one another with force comparable to chemical bonding [36].

Due to the massive production of activated carbon (AC) adsorbents, the commonest form of adsorbent, the application of the adsorption technique substantially expanded over the years. Because of its highly porous surface area, thermostability, and exceptional ability, a wide variety of organic and inorganic contaminants dissolved in aqueous media can be removed by AC with remarkable efficiency. The pollutants in AC in columns typically reduce COD levels more efficiently than chemical treatment methods. Because of this, there is now a much higher chance that the high quantities of organic chemicals in leachate may be removed via the adsorption technique [37].

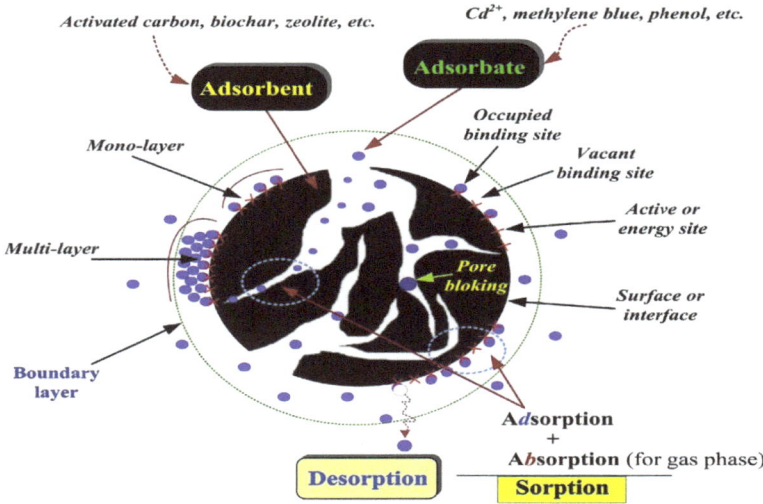

Figure 7. The adsorption fundamentals. Source: [35].

Contemporary porous materials were proposed for improved adsorption capacities and reduced environmental effects [38]. These materials are available in a diverse range of chemical configurations, surface finishes, and geometries. Adsorbents are often constructed from a variety of substances, such as natural substances, enhanced natural substances (such as activated carbon), synthetic substances (such as zeolites and resins), industrial and agricultural wastes and by-products, biological adsorbents, and others [39]. There were numerous different categories and types of adsorbents reported in the literature. Some of them are simplified graphically in Figure 8 [40].

In earlier investigations, many researchers applied and utilised a variety of media, including chitosan, activated carbon, zeolite, clay, and others, to remediate leachate [7,41]. Rohers et al. [42] published their work on the effectiveness of activated carbon in leachate treatment via a column study. Sand filters and activated carbon columns were used as an option for the physical-chemical leachate pre-treatment in their work. The results showed that COD, BOD_5, colour, and NH_3-N decreased by up to 74%, 47%, 93.4%, and 90%, respectively. Their work was based on a wider range of the BOD_5/COD proportion from 0.3 to 0.9. Additionally, the NH_3-N concentration was decreased by 85.37% using an activated carbon post-filtration column. The post-treatment also led to significant heavy metal reductions (60–96%).

Other naturally occurring minerals, including clays and zeolite, are also good adsorbents to complement AC in treating leachate. Natural minerals with excellent adsorption and ion exchange abilities, such as zeolites (clinoptilolite), were proposed for use as adsorbents due to their low cost [33,43]. Many workers researched how natural zeolite affected ammonium ions (NH_4-N) remediation from leachate [44]. The impacts of a number of variables, including reaction time (T), pH, and zeolite concentration (ZC), were examined in a batch process to optimise the process. The first step was to investigate the effects of pH at different pH ranges (pH 5 to 9). The influence of the ZC in the pH-optimal range of 10–200 g/L was next assessed. According to the results, raising the ZC from 10 to 80 g/L improved the elimination of NH_4-N. When the ZC concentration was raised from 80 to 200 g/L, the performance unfortunately decreased due to the overdosage phenomenon. The studies' findings demonstrated that a pH of 7, a ZC of 80 g/L, and a reaction time of 30 min were required for removing NH_4-N (44.49%). The work demonstrated that the clinoptilolite could be used to effectively and economically extract ammonium ions from landfill leachate [44].

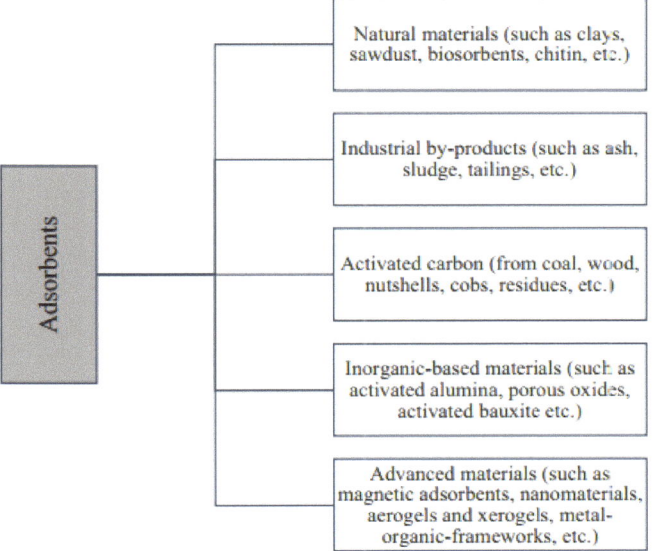

Figure 8. Classification of adsorbents. Source: [40].

The efficacy of raw zeolite and heated activated zeolite to possibly reduce COD, ammoniacal nitrogen, and colour in leachate was investigated by Aziz et al. [34] in 2020. The zeolite applied in the study was heated for three hours at various temperatures for activation. The optimal dosage was determined to be 10 g of raw zeolite, which reduced 53.1% NH_3-N, 22.5% COD, and 46% colour. As much as 24.3% of COD and 73.8% of colour were reduced at pH 4. At the optimal pH of pH 7, roughly 55.8% of the NH_3-N was decreased. The best dosage was 10 g of heated activated zeolite at 150 °C, and this temperature led to reductions in 45.1% NH_3-N, 11.8% COD, and 43.7% colour. The heating temperature of 150 °C exhibited the best performance and was cheaper, which showed potential to be upscaled in the field. Additionally, the zeolite's capacity was improved and increased from 41.30 cmol/kg to 181.90 cmol/kg by heating.

5.4. Integrated Treatment

It is quite common to combine physicochemical and other treatment methods in treating leachate to adhere to the acceptable threshold for effluent release. Combining treatments was shown to be more economical and affordable for treating mature leachate due to their capacity to synergistically enhance the benefits of each of the methods used [45]. The combination of two or more biological, physiochemical, and biological-physical-chemical processes are among the common workable hybrid approach in the treatment. The hybrid approach combines many technologies to produce a product that is better for the environment and could be used at a lower cost at once, as no additional post-treatment is necessary [46]. Many landfill leachate sites already combine a biological treatment with an adsorption pre-treatment [47]. Some of the common hybrid/combined approaches that demonstrated effective treatment of landfill leachate are presented here.

Mohajeri et al. [48] investigated the sequencing batch reactor (SBR) mixed with powder of sawdust-enriched bentonite as an adsorbent in removing organic chemicals from established landfills. Based on their pH values, the sawdust was examined at neutral, alkaline, and acidic conditions. At the ideal aeration speed of 7.5 L/min and reaction time of 22 h, SBR-augmented bentonite treatment eliminated COD and NH_3-N by 99.28% and 95.41%, respectively. It was a notable success that, even with the reaction time decreased to two hours, the removal of both contaminants in the existence of sawdust only decreased to 17%.

de Oliveira et al. [49] reported their findings in remediating landfill leachate employing a combination of a filter-press reactor, a coagulation-flocculation (alum) process, and electrochemical approaches with a boron-doped diamond electrode. pH 6.0 and 20 mL/L $Al_2(SO_4)_3$ (50 g/L) was proven to be the most favourable condition for the coagulation-flocculation process. Three distinct coagulants, ferric chloride, aluminium sulphate, and alum, as well as two agitation techniques, ultrasound and mechanical, were used in the study to optimise the operational conditions. This process used up to 40% less energy to remove the organic load while keeping a similar efficient mineralisation rate (>90% COD reduction). By the end of the electrolysis process, colour, turbidity, and NH_3-N were totally removed.

El Mrabet et al. [50] investigated the application of the Fenton method in conjunction with adsorption onto naturally occurring bentonite clay in a landfill leachate treatment. The optimum Fenton conditions occurred at 2000 mg/L of Fe^{2+}, 2500 mg/L of H_2O_2, and pH 3, which exhibited 92% and 73% reductions in colour and COD, respectively. The pre-treated leachate was then passed through the naturally occurring bentonite clay. The impacts of a number of factors, including pH, reaction time, adsorbent dosage, and temperature, on the adsorption effectiveness were examined. As much as 84% of the total COD and 98% of the colour were removed by the integration of the Fenton and adsorption processes (bentonite dosage: 3 g/L; pH 5; contact time: 5 h; temperature = 35 °C).

Table 7 summarises various combinations in treating landfill leachate as sourced and reviewed by Cherni et al. [19] and Mojiri et al. [7]

Table 7. Combinations of landfill leachate treatments from the literature.

Combination Treatment Category	Removal Efficiency		Combination Treatment Category	Removal Efficiency	
	Parameters	Removal (%)		Parameters	Removal (%)
Advanced oxidation process/coagulation/adsorption	COD	94	Bioreactor/coagulation	Colour	85.8
	As	87		COD	84.8
	Fe	96		Ammonia	94.2
	P	86		TSS	91.8
Advanced oxidation process/adsorption	Ammonia	94.5	Bioreactor/membrane	COD	95
	COD	95.1		Fe	71
	Colour	95.0		Zn	74
	HA (ABS_{254})	97.9			
Advanced oxidation process/adsorption (ion-exchange)	Ammonia	90	Advanced oxidation process/coagulation	COD	68
	Nitrite	100		Colour	97
	Nitrate	98		HA (UV_{254})	83
	Colour	98			
	Turbidity	98			
	COD	74			
Electrodissolution/advanced oxidation process/chemical flocculation	COD	85		COD	90.2
	Colour	96		HA	93.7
	Turbidity	76		COD	91

Note: Adapted from: [7,19].

6. Challenges

1. Treating leachate is tough and demanding owing to its complex compounds, which involve large differences in its volumetric chemical compositions. Selecting an acceptable, cost-effective, and efficient combination procedure is a demanding undertaking.
2. Leachate normally varies in terms of loading due to large fluctuations in water quantity and quality. This is because it is greatly influenced by the amount of waste disposed of daily, season, and weather conditions, which make it difficult to choose and run an effective treatment method and consistent performance.
3. Treatment of leachate depends on its composition. As leachate properties differ, treatment methods for leachate A might not work well for leachate B. Therefore, a treatability study is highly recommended. Experiences and performances of an existing leachate treatment plant will complement this treatability study.
4. It is also not straightforward to determine an appropriate and the best combination of available technologies and how to combine them to achieve a steady operation.
5. Budget restrictions in developing countries make it challenging to establish and maintain an effective treatment system.
6. Treating NH_3-N and total nitrogen is a challenging task. Usually, a nitrification-denitrification system or an ammonia stripping plant is required, although they are a bit costly. Zeolite filters, however, recently became a promising method as an alternative in removing NH_3-N.
7. In addition, some leachate treatment facilities frequently employ post-treatment steps to polish the treated effluent. Nanofiltration and reverse osmosis were employed in some sites to meet discharge limits; however, this is costly and may not be widely affordable in developing countries.
8. There is a limitation of technical knowledge in underdeveloped countries on the management and operation of treatment facilities.

7. Conclusions

Leachate contamination is an increasing threat to the environment and human health, particularly in developing countries. Leachate not only affects underground aquifers and the Earth's ecosystem, but it also releases toxic pollutants and greenhouse gases into the atmosphere. As a result, mitigation of these negative effects is necessary; this requires a cost-effective, sustainable approach and environmentally friendly leachate treatment facilities.

In spite of the presence of different approaches in the treatment, no distinct and single method is normally sufficient, efficient, or economical enough to meet the requirements of effective standards. Further research is still ongoing to meet the demand, especially for developing countries.

In multiple-stage treatment systems, current trends involve a combination of biological, chemical, and physicochemical processes. Chosen techniques depend on many technical factors, which should be properly assessed and examined because one technique may not be adaptable to all situations.

The review assessed numerous leachate treatment technologies, their efficacy, and the benefits and limitations to the environment. Subsequently, it is necessary to research and develop an innovative technology which can optimise the performance of the treatment at an affordable cost, especially in reducing energy and chemical usage.

Author Contributions: Conceptualization, H.A.A. and Y.-T.H., methodology, H.A.A. and Y.-T.H., validation, H.A.A. and Y.-T.H., resources, H.A.A. and Y.-T.H., writing—original draft preparation, H.A.A., S.F.R. and Y.-T.H.; writing—review and editing, H.A.A., S.F.R. and Y.-T.H.; visualization, H.A.A. and Y.-T.H.; supervision, H.A.A. and Y.-T.H.; funding acquisition, H.A.A. All authors have read and agreed to the published version of the manuscript.

Funding: This study was supported and funded by Universiti Sains Malaysia under the RUI-grant (1001/PAWAM/8014081) and RU-bridging grant (304. PAWAM.6316096) for research pertaining to the Solid Waste Management Cluster, Engineering Campus, Universiti Sains Malaysia.

Conflicts of Interest: The authors declare no conflict of interest.

References

1. Department of Statistics, Malaysia (DOSM). Population & Demography. 2022. Available online: https://www.dosm.gov.my/v1/index.php?r=column/ctwoByCat&parent_id=115&menu_id=L0pheU43NWJwRWVSZklWdzQ4TlhUUT09 (accessed on 7 December 2022).
2. Malaysian Investment Development Authority (MIDA). Waste to Energy for a Sustainable Future. 2021. Available online: https://www.mida.gov.my/waste-to-energy-for-a-sustainable-future/ (accessed on 7 December 2022).
3. Jelonek, P.; Neczaj, E. The use of Advanced Oxidation Processes (AOP) for the treatment of landfill leachate. In Proceedings of the 4th International Conference on Advances in Sustainable Sewage Sludge Management, Szczyrk, Poland, 3–5 December 2012.
4. Saxena, V.; Kumar Padhi, S.; Kumar Dikshit, P.; Pattanaik, L. Recent developments in landfill leachate treatment: Aerobic granular reactor and its future prospects. *Environ. Nanotechnol. Monit. Manag.* **2022**, *18*, 100689. [CrossRef]
5. Teng, C.; Zhou, K.; Peng, C.; Chen, W. Characterization and treatment of landfill leachate: A review. *Water Res.* **2021**, *203*, 117525. [CrossRef] [PubMed]
6. Luo, H.; Zeng, Y.; Cheng, Y.; He, D.; Pan, X. Recent advances in municipal landfill leachate: A review focusing on its characteristics, treatment, and toxicity assessment. *Sci. Total Environ.* **2020**, *703*, 135468. [CrossRef] [PubMed]
7. Mojiri, A.; Zhou, J.L.; Ratnaweera, H.; Ohashi, A.; Ozaki, N.; Kindaichi, T.; Asakura, H. Treatment of landfill leachate with different techniques: An overview. *J. Water Reuse Desalin.* **2021**, *11*, 66–96. [CrossRef]
8. Banchon, C.; Cañas, R. Coagulation and oxidation strategies for landfill leachate wastewater. *Res. Sq.* **2022**, 1–16. [CrossRef]
9. Zhang, F.; Peng, Y.; Wang, Z.; Jiang, H.; Ren, S.; Qiu, J. New insights into co-treatment of mature landfill leachate with municipal sewage via integrated partial nitrification, Anammox and denitratation. *J. Hazard. Mater.* **2021**, *415*, 125506. [CrossRef]
10. Rahim, I.R.; Jamaluddin, A. Cost Analysis of The Fukuoka Method Landfill System in North Kolaka Regency, Southeast Sulawesi, Indonesia. In Proceedings of the Indonesia International Conference on Science, Technology and Humanity, Yogyakarta, Jakarta, 7 December 2015; pp. 15–20.
11. Tashiro, T. The "Fukuoka Method": Semi-Aerobic Landfill Technology. In Proceedings of the IRBC Conference, Metro Vancouver, BC, Canada, 20–22 September 2011. Available online: http://www.metrovancouver.org/2011IRBC/Program/IRBCDocs/IRBCFactsheet_FukuokaMethodWasteMgmt_Fukuoka.pdf (accessed on 11 September 2022).
12. Theng, L.C.; Matsufuji, Y.; Mohd, N.H. Implementation of the Semi-Aerobic Landfill System (Fukuoka Method) in Developing Countries: A Malaysia Cost Analysis. *Waste Manag.* **2005**, *25*, 702–711. [CrossRef]
13. Amiri, A.W.; Tsutsumi, J.I.G.; Nakamatsu, R.A. Case Study of Fukuoka Landfill Method and Environmental Impact Assessment of Solid Waste Management in Kabul City. *Int. J. Techn. Res. Appl.* **2016**, *4*, 46–51.
14. Sheppard, D.A. *Practical Guide to Landfill Management in Pacific Island Countries and Territories*; Secretariat of the Pacific Regional Environment Programme: Apia, Samoa, 2010.
15. Anqi, T.; Zhang, Z.; Suhua, H.; Xia, L. Review on landfill leachate treatment methods. In *IOP Conference Series: Earth and Environmental Science*; IOP Publishing: Bristol, UK, 2020; p. 565. [CrossRef]
16. Aziz, H.A.; Alias, S.; Adlan, M.N.; Asaari, A.H.; Zahari, M.S. Colour Removal from Landfill Leachate By Coagulation And flocculation Processes. *Bioresour. Technol.* **2007**, *98*, 218–220. [CrossRef]
17. Bu, L.; Wang, K.; Zhao, Q.L.; Wei, L.L.; Zhang, J.; Yang, J.C. Characterisation of Dissolved Organic Matter During Landfill Leachate Treatment by Sequencing Batch Reactor, Aeration Corrosive Cell-Fenton, And Granular Activated Carbon Inseries. *J. Hazard. Mater.* **2010**, *179*, 1096–1105. [CrossRef]
18. Galeano, L.A.; Vicente, M.A.; Gil, A. Treatment of Municipal Leachate of Landfillby Fenton-Like Heterogeneous Catalytic Wet Peroxide Oxidation Using An Al/Fepillared Montmorillonite As Active Catalyst. *Chem. Eng. J.* **2011**, *178*, 146–153. [CrossRef]
19. Cherni, Y.; Elleuch, L.; Messaoud, M.; Kasmi, M.; Chatti, A.; Trabelsi, I. Recent Technologies for Leachate Treatment: A Review. *Euro-Mediterr. J. Environ. Integr.* **2021**, *6*, 79. [CrossRef]
20. Matsufuji, J. *Technical Guideline on Sanitary Landfill*; Japan International Co.: Tokyo, Japan, 1990.
21. Costa, A.M.; Alfaia, R.G.D.S.M.; Campos, J.C. Landfill Leachate Treatment in Brazil—An Overview. *J. Environ. Manag.* **2019**, *232*, 110–116. [CrossRef]
22. Aziz, H.A.; Ramli, S.F. Recent development in sanitary landfilling and landfill leachate treatment in Malaysia. *Int. J. Environ. Eng.* **2018**, *9*, 201. [CrossRef]
23. Taoufik, M.; Elmoubarki, R.; Moufti, A.; Elhalil, A.; Farnane, M.; Machrouhi, A. Treatment of Landfill Leachate by Coagulation-Flocculation with FeCl$_3$: Process Optimisation Using Box—Behnken Design. *J. Mater. Environ. Sci.* **2018**, *9*, 2458–2467.
24. Lippi, M.; Gaudie Ley, M.B.R.; Mendez, G.P.; Felix Cardoso Junior, R.A. State of Art of Landfill Leachate Treatment: Literature Review and Critical Evaluation. *Ciência Nat.* **2018**, *40*, 78. [CrossRef]
25. Shah, M.; Gami, J. Landfill Leachate Technologies: A Review. *Glob. Res. Dev. J. Eng.* **2019**, 54–58.
26. Rathnayake, W.A.P.P.; Herath, G.B.B. A Review of Leachate Treatment Techniques. In Proceedings of the 9th International Conference on Sustainable Built Environment, Kandy, Sri Lanka, 13–15 December 2018; pp. 97–106. Available online: https://www.researchgate.net/publication/329915923 (accessed on 11 September 2022).
27. Khôi, T.T.; Thúy, T.T.T.; Nga, N.T.; Huy, N.N.; Thúy, N.T. Air Stripping for Ammonia Removal from Landfill Leachate in Vietnam: Effect of Operation Parameters. *TNU J. Sci. Technol.* **2021**, *226*, 73–81. [CrossRef]

28. Leite, V.D.; Paredes, J.M.R.; de Sousa, T.A.T.; Lopes, W.S.; de Sousa, J.T. Ammoniacal Nitrogen Stripping from Landfill Leachate at Open Horizontal Flow Reactors. *Water Environ. Res.* **2018**, *90*, 387–394. [CrossRef]
29. Bahrodin, M.B.; Zaidi, N.S.; Hussein, N.; Sillanpää, M.; Prasetyo, D.D.; Syafiuddin, A. Recent Advances on Coagulation-Based Treatment of Wastewater: Transition from Chemical to Natural Coagulant. *Curr. Pollut. Rep.* **2021**, *7*, 379–391. [CrossRef]
30. Achak, M.; Elayadi, F.; Boumya, W. Chemical Coagulation/Flocculation Processes for Removal of Phenolic Compounds from Olive Mill Wastewater: A Comprehensive Review. *Am. J. Appl. Sci.* **2019**, *16*, 59–91. [CrossRef]
31. Mohd-Salleh, S.N.A.; Mohd-Zin, N.S.; Othman, N. A Review of Wastewater Treatment Using Natural Material and Its Potential As Aid And Composite Coagulant. *Sains Malays.* **2019**, *48*, 155–164. [CrossRef]
32. Djeffal, K.; Bouranene, S.; Fievet, P.; Déon, S.; Gheid, A. Treatment of Controlled Discharge Leachate by Coagulation-Flocculation: Influence Of Operational Conditions. *Sep. Sci. Technol.* **2021**, *56*, 168–183. [CrossRef]
33. Mohd-Salleh, S.N.A.; Mohd-Zin, N.S.; Othman, N.; Mohd-Amdan, N.S.; Mohd-Shahli, F. Dosage and pH Optimisation on Stabilised Landfill Leachate via Coagulation-Flocculation Process. *MATEC Web Conf.* **2018**, *250*, 06007. [CrossRef]
34. Aziz, H.A.; Noor, A.F.M.; Keat, Y.W.; Alazaiza, M.Y.D.; Hamid, A.A. Heat Activated Zeolite for the Reduction of Ammoniacal Nitrogen, Colour, and COD in Landfill Leachate. *Int. J. Environ. Res.* **2020**, *14*, 463–478. [CrossRef]
35. Bello, M.M.; Raman, A.A.A. Synergy of Adsorption and Advanced Oxidation Processes in Recalcitrant Wastewater Treatment. *Environ. Chem. Lett.* **2019**, *17*, 1125–1142. [CrossRef]
36. Abuabdou, S.M.A.; Teng, O.W.; Bashir, M.J.K.; Aun, N.C.; Sethupathi, S. Adsorptive Treatment of Stabilised Landfill Leachate Using Activated Palm Oil Fuel Ash (POFA). In Proceedings of the Conference: International Symposium on Green and Sustainable Technology (ISGST2019) Universiti Tunku Abdul Rahman, Kampar, Malaysia, 23–26 April 2019. [CrossRef]
37. Shehzad, A.; Bashir, M.J.K.; Sethupathi, S.; Lim, J.W. An Overview of Heavily Polluted Landfill Leachate Treatment Using Food Waste as an Alternative and Renewable Source of Activated Carbon. *Process Saf. Environ. Prot.* **2015**, *98*, 309–318. [CrossRef]
38. Torretta, V.; Ferronato, N.; Katsoyiannis, I.A.; Tolkou, A.K.; Airoldi, M. Novel and Conventional Technologies for Landfill Leachates Treatment: A Review. *Sustainability* **2017**, *9*, 9. [CrossRef]
39. Crini, G.; Lichtfouse, E.; Wilson, L.D.; Morin-Crini, N. Conventional and Non-Conventional Adsorbents for Wastewater Treatment. *Environ. Chem. Lett.* **2019**, *17*, 195–213. [CrossRef]
40. Reshadi, M.A.M.; Bazargan, A.; McKay, G. A Review of the Application of Adsorbents for Landfill Leachate Treatment: Focus on Magnetic Adsorption. *Sci. Total Environ.* **2020**, *731*, 138863. [CrossRef]
41. Kasmuri, N.; Sabri, S.N.M.; Wahid, M.A.; Rahman, Z.A.; Abdullah, M.M.; Anur, M.Z.K. Using Zeolite in the Ion Exchange Treatment to Remove Ammonia-Nitrogen, Manganese and Cadmium. *AIP Conf. Proc.* **2018**, *2031*, 020004. [CrossRef]
42. Rohers, F.; Dalsasso, R.L.; Nadaleti, W.C.; Matias, M.S.; de Castilhos, A.B., Jr. Physical–chemical Pre-Treatment of Sanitary Landfill Raw Leachate by Direct Ascending Filtration. *Chemosphere* **2021**, *285*, 131362. [CrossRef] [PubMed]
43. Augusto, P.A.; Castelo-Grande, T.; Merchan, L.; Estevez, A.M.; Quintero, X.; Barbosa, D. Landfill Leachate Treatment by Sorption in Magnetic Particles: Preliminary Study. *Sci. Total Environ.* **2019**, *648*, 636–668. [CrossRef]
44. Mosanefi, S.; Alavi, N.; Eslami, A.; Saadani, M.; Ghavami, A. Ammonium Removal from Landfill Fresh Leachate Using Zeolite as Adsorbent. *J. Mater. Cycles Waste Manag.* **2021**, *23*, 1383–1393. [CrossRef]
45. Ai, J.; Wu, X.; Wang, Y.; Zhang, D.; Zhang, H. Treatment of Landfill Leachate with Combined Biological and Chemical Processes: Changes in the Dissolved Organic Matter and Functional Groups. *Environ. Technol. Innov.* **2019**, *40*, 2225–2231. [CrossRef]
46. Nath, A.; Debnath, A.A. Short Review on Landfill Leachate Treatment Technologies. *Mater. Today Proc.* **2022**, *67*, 1290–1297. [CrossRef]
47. Yenis Septiariva, I.; Padmi, T.; Damanhuri, E.; Helmy, Q. A Study on Municipal Leachate Treatment through a Combination of Biological Processes and Ozonation. *MATEC Web Conf.* **2019**, *276*, 06030. [CrossRef]
48. Mohajeri, P.; Selamat, M.R.; Aziz, H.A.; Smith, C. Removal of COD and Ammonia Nitrogen by a Sawdust/Bentonite-Augmented SBR Process. *J. Clean Energy Technol.* **2019**, *1*, 125–140. [CrossRef]
49. de Oliveira, M.S.; da Silva, L.F.; Barbosa, A.D.; Romualdo, L.L.; Sadoyama, G.; Andrade, L.S. Landfill Leachate Treatment by Combining Coagulation and Advanced Electrochemical Oxidation Techniques. *ChemElectroChem* **2019**, *6*, 1427–1433. [CrossRef]
50. el Mrabet, I.; Benzina, M.; Zaitan, H. Treatment of Landfill Leachate from Fez City by Combined Fenton and Adsorption Processes Using Moroccan Bentonite Clay. *Desalin. Water Treat.* **2021**, *225*, 402–412. [CrossRef]

Disclaimer/Publisher's Note: The statements, opinions and data contained in all publications are solely those of the individual author(s) and contributor(s) and not of MDPI and/or the editor(s). MDPI and/or the editor(s) disclaim responsibility for any injury to people or property resulting from any ideas, methods, instructions or products referred to in the content.

MDPI AG
Grosspeteranlage 5
4052 Basel
Switzerland
Tel.: +41 61 683 77 34

Water Editorial Office
E-mail: water@mdpi.com
www.mdpi.com/journal/water

Disclaimer/Publisher's Note: The title and front matter of this reprint are at the discretion of the Guest Editors. The publisher is not responsible for their content or any associated concerns. The statements, opinions and data contained in all individual articles are solely those of the individual Editors and contributors and not of MDPI. MDPI disclaims responsibility for any injury to people or property resulting from any ideas, methods, instructions or products referred to in the content.

www.ingramcontent.com/pod-product-compliance
Lightning Source LLC
LaVergne TN
LVHW072320090526
838202LV00019B/2319